C000068430

Computational Fluid Dynamics in Fire Engineering

Computational Fluid Dynamics in Fire Engineering
Theory, Modelling and Practice

Edited by

Guan Heng Yeoh and *Kwok Kit Yuen*

AMSTERDAM • BOSTON • HEIDELBERG • LONDON
NEW YORK • OXFORD • PARIS • SAN DIEGO
SAN FRANCISCO • SINGAPORE • SYDNEY • TOKYO

Academic Press is an imprint of Elsevier

ELSEVIER

Butterworth-Heinemann is an imprint of Elsevier
30 Corporate Drive, Suite 400, Burlington, MA 01803, USA
Linacre House, Jordan Hill, Oxford OX2 8DP, UK

Copyright © 2009, Elsevier Inc. All rights reserved.

No part of this publication may be reproduced, stored in a retrieval system, or
transmitted in any form or by any means, electronic, mechanical,
photocopying, recording, or otherwise, without the prior written permission
of the publisher.

Permissions may be sought directly from Elsevier's Science & Technology Rights
Department in Oxford, UK: phone: (+44) 1865 843830, fax: (+44) 1865 853333,
E-mail: permissions@elsevier.com. You may also complete your request online
via the Elsevier homepage (http://elsevier.com), by selecting "Support & Contact"
then "Copyright and Permission" and then "Obtaining Permissions."

Library of Congress Cataloging-in-Publication Data
Application submitted

British Library Cataloguing-in-Publication Data
A catalogue record for this book is available from the British Library.

ISBN: 978-0-7506-8589-4

For information on all Butterworth–Heinemann publications
visit our Web site at www.elsevierdirect.com

Printed in the United States of America
09 10 11 12 13   10 9 8 7 6 5 4 3 2 1

Working together to grow
libraries in developing countries

www.elsevier.com | www.bookaid.org | www.sabre.org

ELSEVIER   BOOK AID
International   Sabre Foundation

Table of Contents

Preface

For many practicing engineers, undergraduate and graduate students, or researchers who are engaged in fire modeling, the evolution of digital computers has further enhanced the reliance on computational models to better understand and predict the fire phenomenon. Modeling of fires is challenging and encompasses a wide spectrum of different length scales, a broad range of engineering disciplines, and a multitude of different computational approaches. Not surprisingly, the number of books dealing with the fundamentals of fire dynamics is staggering; they include the predictive treatment of fires via physical models based on theoretical or analytical approaches. Among the many sweeping changes in the subject of fire dynamics, the use of the field model, a physical model utilizing the computational fluid dynamics (CFD) concepts, techniques, and models along with fire models, is gaining significant momentum and recognition within the fire community. The feasible application of these computational fire models has certainly brought about the modern development of fire safety science and the emergence of such modeling in fire engineering.

Because of the increasing importance on the field modeling of fire dynamics in fire engineering, there is therefore an increasing compelling need to develop a single comprehensive compendium in providing a systematic exposition of the many important aspects of the field modeling approach. The authors can fully understand the predicament and difficulties experienced by fire modelers mustering the knowledge in order to apply the field model with confidence in their investigative studies. For the *uninitiated* fire modeler who is learning about CFD for the first time, the nitty-gritty elements within this particular mathematically sophisticated discipline can be difficult to master. For the *uninitiated* fire modeler who needs to also learn about combustion, radiation, soot production, and solid pyrolysis—the essential elements in fire models—for the first time, the learning of each of these complex disciplines may well be a long laborious process; most will become disheartened toward continuing further. In obtaining the numerical results for a range of fire problems, the modeler may be unwittingly applying the field model without a prior understanding of the basic theory behind the formulation and therefore limits of applicability of the many models within its scope.

The aim of the present text is to try to consolidate the fundamental understanding of the many disciplines in the field modeling approach. There are many textbooks dealing with each individual discipline ranging from introductory to advanced levels. This book attempts, however, to present a unifying approach of the fundamental ideas of fire modeling in one single volume. In a way, it may be argued that some of the materials covered are "old-fashioned" or "common knowledge." Nevertheless, the tried-and-proven ideas actually form a wonderfully intuitive and meaningful learning experience for the *uninitiated* fire modeler. This book as such does not pretend to present a comprehensive review of the background theory and development of the respective models from each discipline. Rather, the material in this book is written with a view to satisfy the *initiated* fire modeler through an intuitive, practically oriented approach to fire modeling. More sophisticated aspects such as the state-of-the-art modeling of soot and solid pyrolysis and the advanced turbulence modeling via the large eddy simulation approach are presented and discussed in Chapters 4 and 5, respectively. In Chapter 6, the authors illustrate other possible modeling approaches of varying degrees of sophistication and scope to describe every aspect of the fire problem. These include the application of artificial neural network to fire predictions, evacuation calculation for the occupant escaping from a building in the event of a fire, and the probabilistic approach to assess the total fire safety system. By the end of the chapter, the authors' primary objective is to expound on the concept of a total fire safety engineering analysis for fire safety assessments and evaluations.

The materials contained in this book have been accumulated through years of research at Commonwealth Science Industry Research Organization (CSIRO), Australia; the Australian Nuclear Science Technology Organization (ANSTO), Australia; and the Department Building and Construction, City University of Hong Kong, Hong Kong; to each of which, we wish to acknowledge our indebtness. This book would not have been possible without the combined efforts of the many colleagues and fellow researchers who have generously assisted us in various ways towards its completion. The authors are particularly indebted to Siu Ming Lo, Eric Lee, Alice Cheung, Sherman Cheung, Mark Ho, Hechao Huang and Chunmei Zhao for their invaluable contributions in giving shape to the final version of the text. Special thanks are given to Jonathan Simpson, senior acquisition editor of Elsevier Science & Technology, who applauded the book's inception and provided the necessary encouragement in bringing to fruition the writing of this book. The authors are also grateful to the publisher for having offered their immense assistance both in academic elucidation and professional skills in the publication process.

This book is dedicated with great affection to our families. Dr Yeoh acknowledges the untiring support and love of his wife, Natalie, and his daughters, Genevieve and Ellana, for their magnanimous understanding and unflinching encouragement during the untold amount of hours spent in preparing and writing this book. Dr Yuen would also like to extend his deepest appreciation to his wife, Irene, and his son, Anthony, for their support in the writing of this text.

To all who have been involved, we express our most heartfelt appreciation.

Guan Heng Yeoh and Kwok Kit Yuen

1 Introduction

Abstract

Modeling of fires has contributed significantly to the modern development of fire safety science and the emergence of the discipline of fire engineering. Suitable numerical simulation tools have effectively taken center stage for practicing fire engineers to exploit the freedoms offered under the performance-based, fire safety engineering approach. The core of all fire modeling activities remains essentially on the proper treatment of the gas phenomena of the fire itself for any subsequent assessment of impact on structures, people, or environment. In this chapter, the deterministic model based on the field model or computational fluid dynamics in fire modeling is introduced by examining its historical development, impact in reviewing major fire disasters, utilization in research, and application in practice. At the end of this chapter, the scope of the book foreshadows the many important aspects of fire modeling that will be covered in later chapters.

1.1 Historical Development of Fire Modeling

Amongst the many incidents of uncontrollable fires, unwanted fires in enclosures are the most frequently encountered. Significant examples of some major fire disasters recorded in history are the Kings Cross Fire in the London Underground, which occurred on 18 November 1987, and the collapse of the World Trade Center Towers in New York on 11 September 2001. The hazard that these fires represent is usually associated with the uncontrolled nature of the exothermic chemical reactions, especially between organic or combustible materials and air and their interaction with the structural components. What follows from the analysis of this fire hazard is that it cannot, in general, be totally eliminated, but it can be reduced to an acceptably low level via appropriate design considerations and procedures.

Fire dynamics embraces numerous complicated physical and chemical interactions, which include fluid dynamics, thermodynamics, combustion, radiation, or even multi-phase effects. During the early investigations of enclosure fire development, a great deal of attention has been focused on better understanding the fire behaviors using experimental techniques and theoretical approaches. Experiments provide useful observations and measurements of the flaming process, while theoretical models employ a mathematical description of the physical phenomena through the input of experimental data. There are, however, limitations in fully applying experimental techniques and theoretical approaches to a range of fire problems. Conducting full-scale experiments can be rather expensive due to the high costs of construction of a fire

Computational Fluid Dynamics in Fire Engineering
Copyright © 2009 by Academic Press. Inc. All rights of reproduction in any form reserved.

facility and the instrumentation and hardware required for data collection. On the other hand, in spite of the low computational costs associated with the use of theoretical approaches, these models are still highly dependent on the experimental data from which they are correlated and the specific geometrical configuration where the fire experiments are carried out.

With the advent of digital computers, the use of numerical methodologies in fire modeling offers fire modelers the flexibility of aptly simulating the fire behaviors in different enclosure configurations, hence overcoming the constraints in experimental techniques and theoretical approaches. There are essentially two major categories of computer models for analyzing enclosure fire development. The first category is the stochastic or probabilistic models, which treat the fire growth as a series of sequential events or states. Here, mathematical rules are established to govern the transition from one event to another—for example, from ignition to established burning—and probabilities are assigned to each transfer point based on the analysis of relevant experimental data, historical fire incident data, and computer model results. The second category, which is the primary focus of this book, is the deterministic models. Through these models, the processes encountered in a compartment fire are represented by interrelated mathematical expressions based on physics and chemistry. Generally speaking, these models—normally known as room fire, computer fire, or mathematical fire models—can provide an accurate estimate of the impact of fire and, more importantly, suggested measures for fire prevention or control.

In fire modeling, the most widely used physically based fire model is the "zone" or "control volume" model. Zone modeling has proven to be a practical methodology in providing estimates to the fire processes in enclosure. Essentially, it solves the conservations equations for distinct and relatively large control volumes. On the basis of the "Two Layers Assumption," the dominant characteristic of this type of model is exemplified in Figure 1.1. The zone model assumes that the burn room is divided into two layers (i.e., the upper layer of hot gases and the bottom layer of cold gases). Within the enclosure, the hot layer contains all the combustion products, which are taken to be well mixed and homogenous in temperature, while the cold layer is filled with the entrained ambient air. The transient layer height and temperature change (i.e., h_L and T_L) are calculated by considering the global conservation of mass and energy. Invoking the mass conservation, the mass accumulated in the hot layer $\dot{m}_L$ is given by

$$\dot{m}_L = \dot{m}_P - \dot{m}_o \qquad\qquad (1.1.1)$$

where $\dot{m}_P$ is the mass flow rate of the combustion products from the plume entering the hot layer and $\dot{m}_o$ is the mass flow rate of the exhausting hot gases. Similarly, the net energy gain in the hot layer $\dot{E}_L$ through applying the energy conservation is calculated according to

$$\dot{E}_L = \dot{E}_P - \dot{E}_o - \dot{E}_c \qquad\qquad (1.1.2)$$

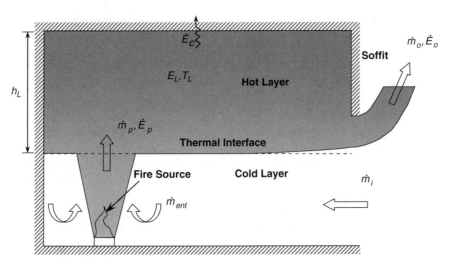

Figure 1.1 Schematic representation of "Two Layers Assumption" taken in zone modeling.

where $\dot{E}_P$ is the energy gain due to the exothermic chemical reaction between the fuel and air, $\dot{E}_o$ is the energy loss to the surroundings, and $\dot{E}_c$ is the energy loss due to the convective heat transfer to the boundaries of enclosure.

The beginnings of zone modeling can be traced back to the mid-1970s with the description of the fundamental equations in Quintiere (1977). Based on these equations, the very first zone model published was RFIRES by Pape et al. (1981). This was followed by the Harvard series of models developed by Emmons, Mitler, and co-workers (Mitler and Emmons, 1981, Mitler and Rockett, 1987), ASET model and ASET-B model in Walton (1985), FPETOOL, a descendant of the FIREFORM model, by Nelson (1986, 1990), CFAST model from the National Institute of Standards and Technology (NIST) as reported in Peacock et al. (1993), and a variety of other different models (Babrauskas, 1979, Davis and Cooper, 1991). The development of these zone models has been facilitated by advancements both in the understanding of the basic physics of fire growth in a compartment and in the computational technology. While most of the zone models are based on the same fundamental principals, significant variation in features exists among these models—single-room or multi-room enclosure, sprinkler/detector activation, smoke filling through openings, and many others. As aforementioned, typical model outputs of the zone models are the prediction of the evolution of the gas temperatures (T_L) and the thickness of the upper smoke layer (h_L). Comprehensive investigations on the use of zone models to specific fire problems can be found in Friedman (1991), Cox (1995), Walton (1995), and Novozhilov (2001).

Although zone models have been widely adopted and have demonstrated considerable success, they still remain a prescriptive approach to fire modeling.

These models generally require the necessity of *a priori* knowledge of the flow pattern and the vanishing of the local effect within the two zones. In spite of their ease of usage, they are very likely to be imprecise in predicting fire scenarios where the empirical correlations are breached—for example, fires that have restricted entrainment areas or irregular geometrical structures. Owing to global averaging that is performed on the variables of interest over the two zones within the computational domain, these models are generally unable to predict the local physical quantities as required. The field model, an alternative to deterministic modeling, improves the spatial resolution of the zone model by further dividing the computational domain into a three-dimensional mesh comprised of many tiny cells. Field modeling of fires calculates changes in each cell by using the fundamental equations of fluid dynamics. They consist generally of a set of three-dimensional, time-dependent equations, non-linear partial differential equations expressing the conservation of mass, momentum, and energy. This process of solving the fundamental dynamics with digital computers is commonly referred as Computational Fluid Dynamics (CFD). Field model calculates the physical conditions in each cell, which results from changes in adjacent cells. In hindsight, the ability to simulate a range of fire scenarios without the limitations associated with empirical correlations and the feasibility of accommodating complex geometries represent some of the many advantages that the field model has over the zone model. Owing to the evolution of computer technology, there have been intensifying activities toward the concerted development of CFD-based fire models. The enormous contribution of CFD in fire modeling is reviewed in the next section.

1.2   Overview of Current Trends in Fire Modeling

Fire modeling, which emphasizes the application of CFD techniques in fire engineering, first appeared in the late 1970s by the development of the computer code UNDSAFE-I (Yang and Chang, 1977) and subsequently in the late 1980s and early 1990s by the application of JASMINE (developed by Fire Research Station, UK) and FLOW-3D (developed by Atomic Energy Authority, Harwell, UK) to unravel the cause of fatalities suffered in the Kings Cross Fire in the London Underground station (Cox et al., 1989, Simcox et al., 1992). Within reasonable limits of computer usage and cost, early field modeling approaches have assumed the fire to be adequately represented by a volumetric heat source, thus removing the need of including combustion in the model. Non-uniform buoyancy forces were, however, allowed to affect both the mean flow and fluctuating motions. This rather simplified approach consisted of only solving the transport equations governing mass, momentum, and energy with the addition of a two-equation k (kinetic energy) and ε (dissipation of kinetic energy) closure model to describe the turbulent flow. In building construction, numerical simulations adopting the volumetric heat source approach provided the feasibility of predicting and analyzing the complicated phenomena of the smoke filling processes for three categories of atria configurations of high, flat, or cubic in Hong Kong, China, with additional considerations of whether the atria are (1) open to adjacent compartments, (2) separated from other parts of the buildings by glazing, and (3) a combination of 1 and 2 (Chow and Wong, 1991).

Treatment of the fire via the volumetric heat source approach, which only affects the temperature distribution by modifying the source term in the energy equation, neglects, however, the important consideration of combustion chemistry between the fuel and air, and is thus unable to predict the necessary concentrations of the products within the smoke layer. For naturally developing fires, radiation heat loss, which is due to the radiation from hot, smoky gases such as carbon dioxide, water vapor, carbon monoxide, as well as finely dispersed soot particles, accounts for a substantial fraction of the total heat release rate. As demonstrated in Markatos et al. (1982), a reduction of 20% of the total net calorific value of the fire was specified in their field model to consider the radiation contribution in simulating the buoyancy-induced flow in the shopping mall fire test facility. Such *ad hoc* prescription is commonplace in field modeling, but the assumption may introduce errors and uncertainties in the predicted results. To overcome the shortcomings of the volumetric heat source approach, combustion models capable of predicting the spatial distribution of the species concentrations that are also required for radiation calculations for non-luminous and luminous flames have been further developed and incorporated in the field model.

Since most naturally developing fires are diffusion flames, the consideration based on the fast chemistry assumption has resulted in a number of practical models to treat the combustion of fires. The conserved scalar approach based on the mixture fraction (Bilger, 1980) and the eddy break-up (Spalding, 1976) or eddy dissipation (Magnussen and Hjertager, 1976) models have been widely adopted and validated rigorously against experimental data. Coupled with useful radiation models such as the discrete transfer radiative method (Lockwood and Shah, 1981) or discrete ordinates method (Jamaluddin and Smith, 1988), the many notable examples of the applications of different combinations of the combustion and radiation models in fire modeling are exemplified by the predictions of wall fires in Wang and Joulain (1996, 2000), Yan and Holmstedt (1996), Jia et al. (1999), and Yuen et al. (2000); tunnel fires in Fletcher et al. (1994) and Woodburn and Britter (1996a, 1996b); and compartment fires in Luo and Beck (1994), Lewis et al. (1997), Wen et al. (2000, 2001), Yeoh et al. (2002a, 2002b, 2003a, 2003b), and Cheung et al. (2004). According to the review performed by Novozhilov (2001), radiation due to luminous diffusion flames contributes rather significantly in lowering the flame temperatures. This thereby constitutes the additional need of incorporating appropriate models to determine the production of soot particles in order to ascertain the net effect of emission due to the concentrations of these minute carbonaceous particles.

Lewis et al. (1997) studied the significance of incorporating radiation in fire modeling for a single-room compartment based on the experimental setup and measurement carried out by Steckler et al. (1984). They discovered that the inclusion of radiation significantly improved the upper-layer temperatures by 12%. The predictions with the eddy-dissipation model and radiative heat exchange provided good agreement with the experimental data. Implementation of the soot radiation model has appeared in Luo and Beck (1996) to investigate the flashover characteristic in a full-scale multi-room building structure. In their simulations, the soot model of Tesner et al. (1971a, 1971b) was employed, and radiation heat transfer was accounted by the use of the discrete transfer radiative method. In general, the field model predictions captured the basic trends and profiles of the measured data. Wen et al. (2001) investigated the effect of microscopic and global radiative heat exchange on the field prediction in a single-room compartment jet fire situation. By accounting the effect of soot contribution via the soot model by Leung et al. (1991), the inclusion of radiation and soot models significantly improved the temperature's prediction. In comparing the predicted results with and without microscopic radiation, temperature differences up to approximately 100 K were attained, and soot concentration was over-predicted by up to 50% when the microscopic radiation was neglected.

Unlike the radiation model adopted in Luo and Beck (1996), the current authors along with other co-workers adopted the discrete ordinates method and included the most established models of combustion and soot for simulating different compartment fire scenarios in their fire model (Cheung et al., 2004, Yeoh et al., 2002b, 2003a, 2003b). With regards to soot, two different models by Moss et al. and Syed et al. have been adopted in the numerical studies. On the basis of the good agreement achieved between the predicted results, especially those with soot consideration and measured data, the presence of soot was seen to greatly augment the radiation in the compartment fire. Without solving the spatial distribution of the soot concentrations, the temperature distributions were inadequately predicted.

Almost all of the entire above-mentioned fire modeling studies applied the standard k–ε model to characterize the turbulence in the fluid flow. With regards to other similar or more complicated turbulence models that could also be feasibly applied in fire modeling, Liu and Wen (2002) have demonstrated the applicability of a modified second order moment closure model originally proposed by Hanjalic and Jakirlic (1993) to aptly simulate a buoyant diffusion flame. Their model predictions revealed a more comparable agreement with the fire measurements of McCaffrey (1979). Another notable contribution is the use of the Algebraic Reynolds Stress Model to predict the fire behavior in an enclosure as exemplified in Than and Savilonis (1993). The Reynolds Stress Model, which is able to capture the anisotropic behavior of the turbulent stresses, represents a more sophisticated turbulent model for fire modeling, but such a model has been known to be rather unstable in character. For example, Woodburn and Drysdale (1998) reported convergence difficulties in achieving reasonable computational results in their numerical study of flame development in trenches. The requirement to solve additional transport equations in the Reynolds Stress Model generally results in higher computational costs; it is therefore not widely considered in many fire-modeling investigations.

The k–ε model of turbulence, the mixture fraction based and eddy dissipation combustion models, the discrete transfer radiative and discrete ordinates methods, and the soot models proposed by Magnussen and Hjertager (1976), Moss et al. (1988), and Syed et al. (1990) can now be regarded as standard models for most field modeling applications. For compartment fires, these models have proven to be appropriate in simulating *pre-flashover* fires such as evidenced by the range of validation studies carried out in Luo and Beck (1996), Lewis et al. (1997), Wen et al. (2000, 2001), and Yeoh et al. (2002a, 2002b, 2003a, 2003b). Particularly in Yeoh et al. (2003b), the current authors have demonstrated the feasibility of extending the application of a complete numerical procedure involving the modeling of the simultaneously occurring flow, convection, combustion, soot generation, and radiation phenomena processes of turbulent buoyant fires to different full-scale compartment configurations: a single-room, two-room, and multi-room compartment. The numerical study clearly showed the presence of soot significantly augmenting the radiation heat exchange in lowering the flame temperatures. Figure 1.2 depicts the predicted distribution of soot mass fraction on the center plane bisecting the burner, doorway, and open end of the two-room compartment structure. As expected, the soot loading increases with increasing fire intensity, reaching a maximum concentration of about 1% for the fire size of 160 kW, which is a typical value commonly registered for weakly sooty flames. Owing to high soot loading in the burn room, the inclusion of soot radiation into the field model is pivotal in attaining accurate temperature predictions. Without the presence of soot contributing to the radiation effect, the absence of such mechanism will inadvertently result in much higher predicted flame temperatures.

Recently, CFD investigations based on the concept of large eddy simulation in turbulence modeling has emerged as the next focus of development in fire modeling. The underlying theory of large eddy simulation includes the use of subgrid-scale models to characterize the small-scale motion, while the large-scale motion is resolved based on as fine a scale as the underlying computational grid will allow. As such, it can provide a more complete description of the transient flow structure instead of the k–ε representation of the turbulent flow, where the flow variables are generally averaged with respect to time. McGrattan et al. (1994) first developed a two-dimensional large eddy simulation fire model using finite difference and vorticity-stream function formulation. This code has been gradually extended to incorporate increasing complexities in handling the fire dynamics such as illustrated in Baum et al. (1994), McGrattan et al. (1996, 1998), and Xin et al. (2002, 2005). Currently, this well-known three-dimensional computer code known as the Fire Dynamic Simulator (FDS) is readily available as a freeware and can be downloaded from the NIST Web page.

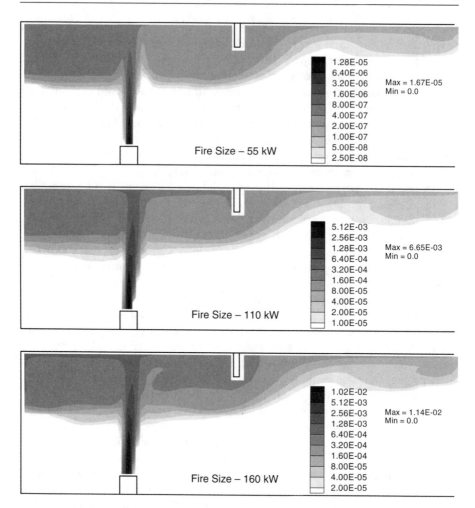

Figure 1.2 Predicted soot distribution through the center of fire source, doorway, and open end for different fire sizes.

As encouraging results are continuously being reported in literatures for the use of the large eddy simulation methodology, fire modelers and researchers are beginning to extensively study a variety of fire-related problems through this particular approach. For example, Wang et al. (2002) examined the fire propagation phenomenon over a vertical wall, Gao et al. (2004) performed large eddy simulation investigations on the smoke movement in a ventilated tunnel fire, Qin et al. (2005) parametrically studied the fire-induced flow through a stairwell, and Chow and Zou (2005) used the large eddy simulation results to derive an empirical equation on the fire-induced air mass flow through the doorway.

On a more fundamental understanding of the fire dynamics, the current authors along with other co-workers have carried out large eddy simulation investigations to capture and identify several key features or physical aspects of a turbulent buoyant fire with a burner diameter of approximately 0.3 m (Cheung et al., 2007a). Generally categorized as a medium-scale fire, the puffing behavior as well as the three distinct regimes of the fire plume—persistent flame, intermittent flame, and buoyant flame—were adequately represented by the large eddy simulation model, which includes the salient features of a two-step predictor-corrector scheme for low Mach number compressible flows, a Smagroinsky subgrid-scale turbulent model, a subgrid-scale combustion model, which explicitly determines the local heat release rate, discrete ordinates method for radiation heat transfer, and the soot model of Moss et al. (1988) and Syed et al. (1990). Figure 1.3 shows the model prediction of the instantaneous temperature contours that clearly indicate the three-zone flame structure as proposed by Cox and Chitty (1980) and McCaffrey (1983). The formation of large vortical structures was seen to be well captured with the predicted puffing frequency agreeing closely with experimentally determined frequencies. Comparisons of instantaneous, mean, and root-mean-square quantities also showed quantitative agreement against other experimental data.

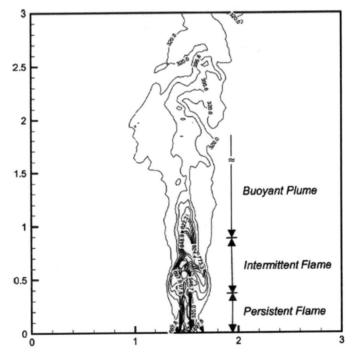

Figure 1.3 Instantaneous temperature contours accompanied by the distinction of the three-zone flame structure predicted by the current authors and co-workers.

Kang and Wen (2004) have applied the large eddy simulation fire model to predict a small-scale buoyant fire (burner diameter of 0.057 m) experimentally tested by Venkatesh et al. (1996). In their fire simulations, they have modified the FDS computer code by incorporating the modified laminar flamelet model based on the Cook and Riley (1998) approach for the subgrid-scale combustion modeling. The numerical predictions have managed to capture unique characteristics specifically belonging to small-scale buoyant fires such as flame anchoring and double flame. The extended five-zone flame structure of Venkatesh et al. (1996) is illustrated in Figure 1.4. According to Venkatesh et al. (1996), the persistent flame region of a small-scale fire can be further divided into three sub-zones: the quenching zone, the primary anchoring zone (PAZ), and the post-PAZ. It is believed that flame anchoring at PAZ is the distinctive characteristic that distinguishes small-scale fires from medium-scale or large-scale fires. Comparing the predicted results with the experimental data, Kang and Wen (2004) have demonstrated that the use of the model improved the predictions of temperature, velocities, and heat release rate by up to 30%. Their study showed that the large eddy simulation approach was capable of capturing the fine details and unique characteristics of a small-scale buoyant fire such as the five-zone structure of the fire plume.

Direct numerical simulation instead of large eddy simulation has been adopted by Luo (2005) to investigate the dynamics of buoyant diffusion flames from rectangular, square, and round fuel sources. Here, the turbulent fluid flow was solved directly by the governing transport equations without undertaking any averaging or approximation other than the consideration of appropriate

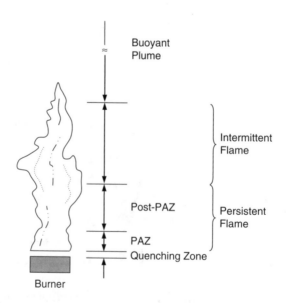

Figure 1.4 Schematic drawing illustrating the five-zone flame structure.

numerical discretisations performed on them. Fully three-dimensional simulations were performed employing numerical methods as high as the sixth order to solve the governing equations for variable-density flow and single-step finite-rate Arrhenius chemistry. Significant differences among the different shape fuel sources as revealed during the numerical simulations were found in the vortex dynamics, entrainment rate of the surrounding air, small-scale mixing and consequently the flame structures. Concentrated regions of fine-scale mixing and intense reactions located around the corners were ascertained for the presence of corners in non-circular flames. Moreover, the rectangular flame exhibited a different dynamic behavior from the square flame by creating different entrainment, mixing, and combustion characteristics between the minor and major axis directions. It was the first time that axis switching was observed by direct numerical simulation in a rectangular flame of an aspect ratio of 3, of which the study by Luo (2005) raised further questions in combustion prediction and control.

CFD in fire modeling has certainly come a long way and is increasingly being employed with greater frequency due to the ever-increasing power of digital computers and the development of quicker numerical algorithms. The feasibility and applicability of the field model is further reviewed in its enormous impact on fire modeling, particularly through the investigations of major fire disasters in the preceding section.

1.3 Review of Major Fire Disasters and Impact on Fire Modeling

1.3.1 Kings Cross Fire

The Kings Cross Fire in the London Underground, which occurred on 18 November 1987, represented one of the most dreadful fires, resulting in 31 fatalities and more than 60 recorded injuries ranging from severe burns to smoke inhalation. The fire incident was most probably caused by a lit match discarded on the wooden escalator, which fell down the side of the escalator. Once the fire started beneath the escalator due to the burning of rubbish and residual grease, the fire spread, and once it had taken hold, the shape and slope of the escalator caused the upwardly moving rising plume to be entrained with the rising escalator itself. This provided a channel for the fire to travel swiftly and eventually flashed over and very rapidly filled the ticket hall with flames and hot, smoky gases. Investigations later revealed that trains that were approaching and departing the actual underground station created the fanning effect that accelerated the combustion process and the spread of flames over the escalator.

One key aspect revealed from the Kings Cross Fire incident was the acute realization of the rapid development of the fire once it had seized the whole escalator, and then propagated like a firestorm engulfing the entire ticket hall.

At that time, fire experts failed to understand and could not explain why the fire was so severe and why it progressed so rapidly in the underground station. The use of computers to investigate the Kings Cross Fire was heralded as a pivotal moment in the development in fire modeling by uncovering the cause of this major disaster. Simcox et al. (1992) carried out computer simulations of the flow of hot gases and drew attention to the indirect consequence of a combustion phenomenon known as the *trench* effect. This mechanism was suggested as the most plausible explanation responsible for the rapid spread of the fire. A CFD fire model was constructed with the emphasis on investigating the flow phenomena within the principal region of interest, which was the Piccadilly line escalator tunnel and the ticket hall. Figure 1.5a shows an isometric view of the surface grid above the ticket hall as well as the location of the tunnel, the exits to the orbital passageway, and the entrance to the Victoria line escalator. A vertical section through the grid in Figure 1.5b illustrates the slope of the tunnel and the ticket hall, while the grid on a perpendicular section through the tunnel in Figure 1.5c describes the treatment of the escalators and the location of the fire source, which was characterized by the time-varying volumetric heat source approach with different heat release rates.

On the basis of the numerical results shown in Figure 1.6, the compelling feature of the fire spread was the way that the hot gases in the buoyant plume laid along the escalator with a velocity through the heat source registering as high as 14.5 m s^{-1}. This clearly illustrated the *trench* effect. According to Simcox et al. (1992), the combination of the Conada and chimney effects was responsible for the trench effect. The former caused the buoyant plume to be attached to the floor of the escalator, since air was unable to be entrained from the side of the wall and it therefore has a tendency to stick to the wall. Once the trench effect was fully established, the hot gases flowing up the trench preheated the wooden escalator and in turn caused the fire to further spread up the escalator much quicker than would be expected. In addition to the strong flow up the trench, secondary flow also existed along the tunnel above the heat source, as shown in Figure 1.7, which exemplified the characteristic of a curling flow in accordance with the eyewitness evidence and experimental results from scale models. More detailed results and in-depth analyses can be referred to in Simcox et al. (1992).

1.3.2 *World Trade Center Fire*

The attack of the twin towers of the World Trade Center (WTC) in New York City due to the impact of two jet airliners on 11 September 2001, as shown in Figure 1.8, is considered the single greatest fire disaster in history, which resulted in a total of 2758 fatalities and another 24 listed as missing and presumed dead. A large proportion of the occupants in the twin towers were killed instantly by the impact, while the rest were trapped and died after the towers collapsed. Following the tragedy, computer simulations of fires were carried out by McGratten et al. (2005) in an attempt to better understand the

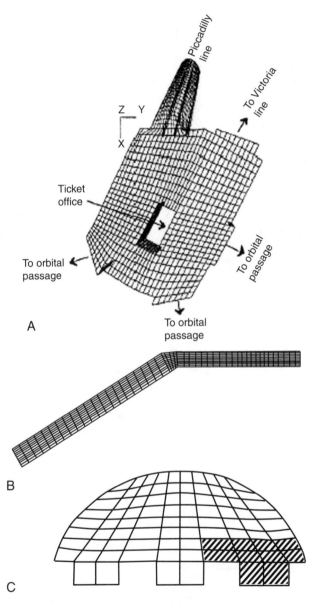

Figure 1.5 Grid layouts of the Kings Cross ticket hall and the inflow and outflow boundaries employed in the computer simulations; (b) grid slice illustrating the tunnel and ticket hall; and (c) cross-sectional grid through the tunnel, where the shaded area indicates the location of the heat source (after Simcox et al., 1992).

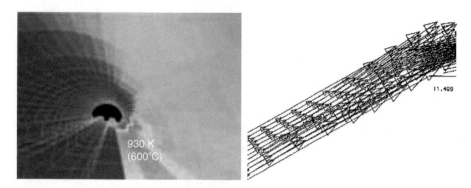

Figure 1.6 (Left) Flooded contours illustrating the temperature distribution within the tunnel, with the lighter-colored regions indicating the hotter temperatures in the escalator trench. (Right) A cross-sectional view of the velocity vectors cutting through the heat source showing the trench effect of the flow of hot gases in the buoyant plume lying along the escalator (after Simcox et al., 1992).

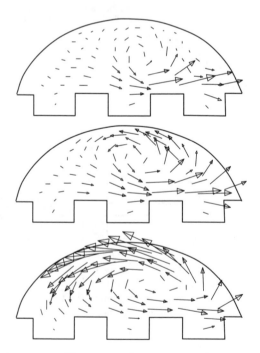

Figure 1.7 Velocity vectors illustrating the presence of secondary flow along the escalator tunnel (after Simcox et al., 1992).

Figure 1.8   Attack on the twin towers of the WTC in New York City.

collapse of the WTC towers. In order to determine the initial and boundary conditions for the establishment of fires in the towers, the collision characteristic of the airplanes damaging the fire structures, opening up vents, rearranging the furnishings, and the distribution of jet fuel (including the jet fuel consumed in the fire ball) were first determined through the structural dynamics software package called LS-DYNA (Kirpatrick et al., 2005). Subject to the availability of these conditions, the large eddy simulation FDS computer code was later employed to model the spreading of fire and determine the temperature and products of combustion from the fire. Figure 1.9 illustrates some sample numerical results of the transient development of the upper-layer temperature of the 94th floor of WTC Tower 1 shortly after impact. The stripes surrounding the images depicted a summary of the visual observations, with the black stripes representing the broken windows and the rest indicating either external flaming or fires that were seen inside the building.

On this floor, a substantial amount of the jet fuel was spread throughout the east side of the floor due to the direct impact by the left wing of the airplane. This particular simulation captured the migration of the fires rather well. For the first 30 minutes, the fires burnt vigorously in the northeast quadrant. After 30 minutes, the fire spread at almost the same rate as the real fires toward the southeast and northwest regions. In McGratten et al. (2005), it was noted that the close match in the spread rate could be attributed by the model of the

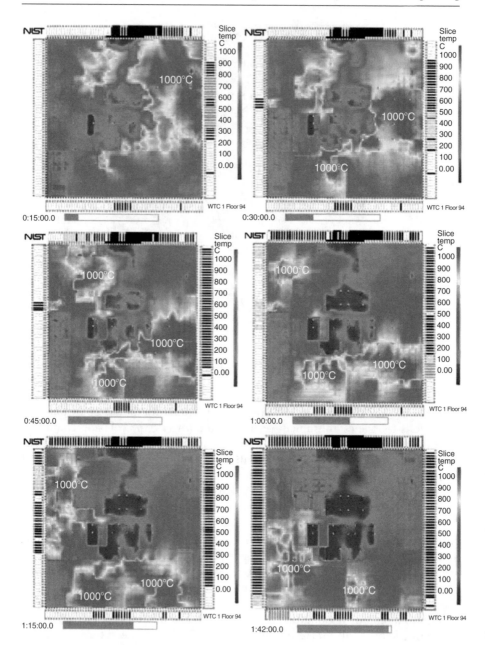

Figure 1.9 Upper-layer temperature contours of WTC Tower 1, 94th Floor (after McGratten et al., 2005).

actual window breakage pattern. Note that the window breaking times were prescribed as model inputs in the fire simulations. In reality, the window breakage phenomenon was caused by the build-up of heat from adjacent fires, which often were seen to flare up due to the increased supply of oxygen from newly broken windows. This was adequately captured in the simulation although it was unlikely that the model would have been capable of replicating the exact window breakage pattern, but rather it served to provide a rough sense of the probable mechanism for the simulated fire spread. The duration of high temperatures exceeding 1000°C, typical of a fully engulfing compartment fire, was largely a function of the presence of combustible load in any particular place. The more furnishings to burn, the longer the fire was sustained. Other analyses were also carried out for other designated floors, which have been detailed in their report. These simulations formed part of a chain of four major modeling efforts that have been run serially: (1) impact analysis, (2) fire simulation, (3) thermal analysis of the structural steel, and (4) mechanical analysis of the collapse sequence. For the third modeling aspect, predicted results from FDS were employed as model inputs such as radiative fluxes and assigned temperatures and temperature gradients to compute the structural response due to the high temperatures and level of deformation weakening the steel structures, which led to the imminent collapse of the WTC towers (Zarghamee et al., 2005a, 2005b).

1.4   Application of Fire Dynamics Tools in Practice

The two major fire disasters, as exemplified in the previous section, describe the increasing usage of the field model as an attractively viable tool in fire investigation. Relatively low computer hardware costs are now ensuring greater accessibility, and inevitable improvements in hardware capacities are also fostering the type of CFD model that can be employed, which is evidenced by the large eddy simulation investigation performed for the WTC fire and the time-averaging approach adopted in the Kings Cross fire. For other fire disasters, lessons are also being learned through the use of fire modeling investigations of the catastrophic tunnel fire in the road tunnel of Mont Blanc between France and Italy on 24 March 1999, which resulted in 39 fatalities, and the nightclub fire in Gothenburg on 30 October 1998 where 63 people were killed. Fire simulations formed part of the recent deliberations of the French courts on the Mont Blanc tunnel fire and on the inquiry into the Gothenburg disco fire.

   In fire safety engineering, a good understanding of a *pre-flashover* fire is vital for providing scientific data for designing workable fire services systems. According to Drysdale (1999), the pre-flashover fire can be taken as the growth stage in which the average compartment temperature is relatively low and the fire is localized in the vicinity of its origin. It is crucial to understand the development and distribution of temperature for a pre-flashover fire, as most of the services systems are supposed to operate during the growth stage. For example, fire detectors are expected to sense smoke or heat and alert the occupants to evacuate via fire alarms, and sprinkler heads should be activated by temperature/smoke sensors to discharge water downward. The key to fire safety design is to adopt practical preventive measures of not allowing the fire to grow to flashover, which can inflict severe damage to the building structure and the contents within. Such a flaming scenario at this flaming stage is known as a *post-flashover* fire. In a more succinct definition according to Drysdale (1999), the fully developed, or post-flashover, fire can be realized as the stage during which all combustible items in the compartment are involved and flames appear to fill the entire volume. The fire behaviors of the Kings Cross fire and WTC fire in the previous section depict the typical prevalent nature of post-flashover fires. Once flashover has occurred, the occupants in the building are threatened directly, and anyone who has not escaped before flashover is unlikely to survive. This explains why much of the field modeling efforts have concentrated in developing suitable models to predict the early stages of a growing or pre-flashover fire.

Knowledge on the distribution of hot air flow from the fire origin during a pre-flashover fire is related to the proper understanding of the smoke spread mechanism. For most applications in practice, the assessment of *smoke control design strategies* with field modeling is increasingly being considered as the method of choice in innovative designs. As building designs become more complex, it is in these kinds of structures where traditional prescriptive-based regulations are often not readily applicable, and an engineered solution obtained from the performance-based, fire safety engineering approach is essential. High-rise buildings, covered shopping malls, airport and railway terminals, tunnels, and atrium hotels are just some of many examples where the utility of field modeling is becoming more prevalent. Often, these structures are unique in their construction, which requires the use of field models for fire safety assessment.

An illustration of the use of CFD fire modeling in practice is provided in Figure 1.10, where the safeness of a refuge floor under a fire situation is investigated for the smoke spilling out of the fire room from the window opening and spreading into the upper refuge floor due to various wind velocities. A refuge floor as stipulated in the building code of a number of countries is considered as part of the exit route in a high-rise building and acts as a safe place for a short rest before continuing escape downward. An in-house computer code FIRE3D considering turbulence with the fire source represented as volumetric heat source producing a heat release rate of 300 kW, which is the typical output from an office fire, as well as a volumetric smoke source, is applied for the numerical simulation. The surface plot corresponding to scalar smoke concentration of 0.01 with a wind velocity of 1 m s^{-1} clearly demonstrates that smoke that emerges from the fire compartment below tends to re-enter the refuge floor at the windward side of the building. Near the lift core, the smoke layer

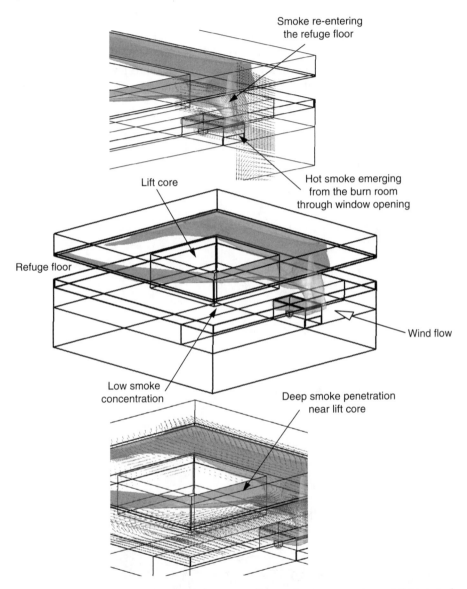

Figure 1.10 Surface plot corresponding to scalar smoke concentration of 0.01 (wind velocity at 1 m s^{-1}) at 5 minutes.

is found to be deep, which could affect the safe evacuation of the occupants. Nevertheless, the smoke layer remains thin in a majority of locations of the refuge floor.

Another practical application of a CFD fire model is demonstrated below for the assessment of the smoke distribution in a covered basketball stadium in Adelaide, South Australia. The configuration for the Adelaide Basketball

Sports Centre has a hexagonal outer perimeter (top view in Figure 1.11) with sloping concourse level seating at the upper half and sloping court level seating at the lower half all the way around the stadium (isometric view in Figure 1.12). Basketball courts are located at the ground level of the building with the fire source centrally located during the fire test, as depicted in

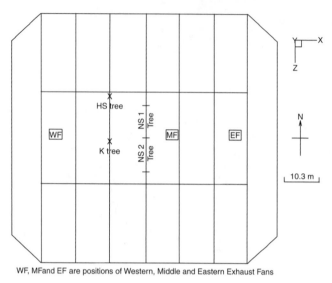

WF, MF and EF are positions of Western, Middle and Eastern Exhaust Fans

Figure 1.11 Stadium configuration and locations of exhaust fans and thermocouple trees.

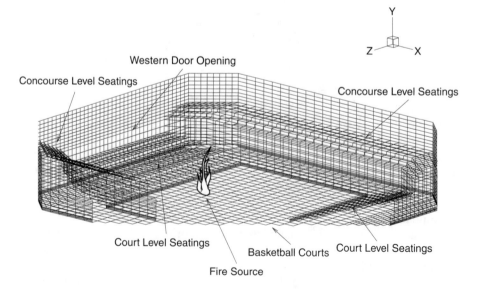

Figure 1.12 Three-dimensional cut-off view of the basketball stadium.

Figure 1.12. Exhaust fans are located above the ceiling, as indicated in Figure 1.11, with an extraction capacity of 40 m³ s⁻¹. All fire escape doors are closed except for the western door opening (see Figure 1.12). Temperature measurements of the hot smoke at various locations in the interior of the stadium are indicated in Figure 1.11. The in-house computer code FIRE3D, considering turbulence, combustion, and radiation models, is applied to provide insights into the transient flow and thermal structures within the stadium, which are represented by the velocity vectors and isotherms after 200 seconds elapsed in Figures 1.13, 1.14, and 1.15, respectively.

For the flow behavior depicted on a plane located 7.7 m above the fire source in Figure 1.13, large entrainment of the cold air enters the stadium through the western door due to the operation of the exhaust fans. Disturbances induced by the operation of the exhaust fans show erratic and irregular flow patterns as exemplified by the large recirculation region. For the constant x-y plane cutting across the fire source in Figure 1.14 (left), the incoming cold air becomes totally mixed with the combustion products at the level where the western door is situated, instead of being completely entrained into the combustion zone. Generated combustion products are seen to be re-circulated back into the region above the western door. In another view, velocity distribution plotted at constant x-z plane cutting across the fire source in Figure 1.15 (left) reveals that the turbulent flow re-circulates at the concourse level seating as well as at the court level seating. The flame above the fire source displays a disorganized plume structure as indicated by the isotherms in Figures 1.14 (right) and 1.15 (right), respectively. As illustrated in Figure 1.16, the computed mean temperatures in the stadium away from the fire of approximately 310 K agree well with the experimental measurements performed during the fire test.

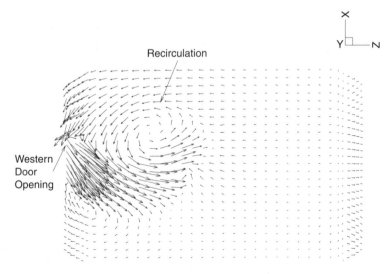

Figure 1.13   Plan view of velocity distribution at 200 seconds.

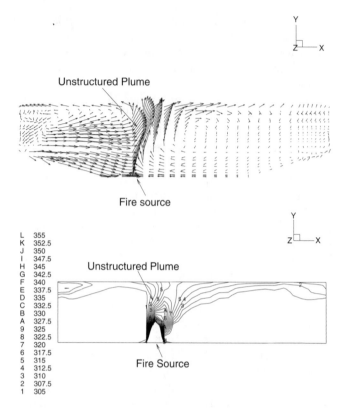

Figure 1.14 Velocity and temperature distributions at a constant *x-y* plane across the fire source at 200 seconds.

In spite of the significant advances that have been achieved, fire modeling remains a developing and evolving technology. The ability to simulate the fundamental fire spread on a condensed solid material is still a challenging prospect, although the current authors have made some considerable strides in combining the gas phase field model with the solid phase pyrolysis model developed for cellulosic fuels, in predicting the flame spread over a vertical combustible wall lining in Yuen et al. (2000) with reasonable success. Suitable models for the prediction of the onset of a flashover and back draft are also other important areas that are currently under development. For structural analysis, time scales that are relevant in affecting the structure behavior imply in most cases fully developed or post-flashover fires. Clear intentions to couple models of the fire and its impact on the structure of a building have been brought into the sharp focus by the requirement to analyze the WTC collapse. Numerous ways of effectively coupling the CFD fire models to Finite Element structural analysis models are underway, and although a seamless coupling has yet to be attained or achieved at this present time, it may possibly be attainable in the not too distant future.

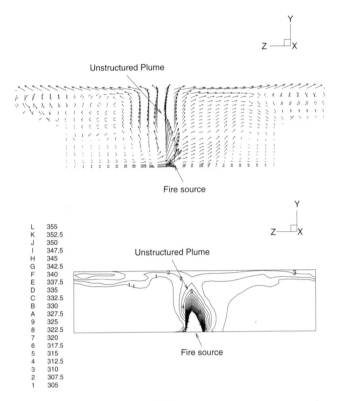

Figure 1.15 Velocity and temperature distributions at a constant x-z plane across the fire source at 200 seconds.

1.5 Validation and Verification of Fire Dynamics Tools

In the context of fire engineering, turbulence treatment achieved either through time averaging or subgrid scaling in large eddy simulation, and models for combustion, radiation, soot production, and solid pyrolysis, need to be validated for their representation of reality. The issue of verification is attested on the model equations, which are then solved to adequate numerical accuracy subject to sources of errors, which could be due to numerical assumptions in the models, numerical solution techniques, computer software and hardware, and application particularly due to human error. According to Beard (1997), these errors are based upon the following limitations:

(i) Numerical assumptions in a model can only ever be an approximation (good or bad) to the real world

(ii) Computational results from the field model are greatly influenced by the numerical techniques adopted, resolution of the grid, and may depend somewhat on the boundary conditions that are assumed

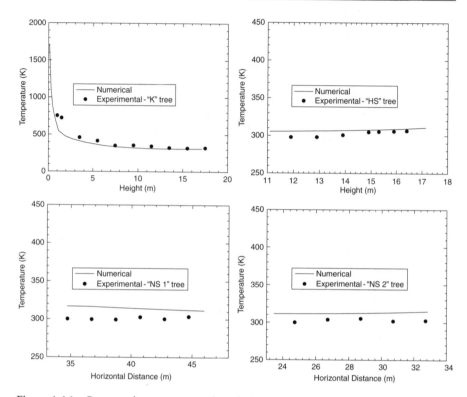

Figure 1.16 Computed temperatures plotted alongside with experimental measurements of "K," "HS," "NS 1," and "NS 2" thermocouple trees at 200 seconds.

(iii) The possibility of computer software error requires assessment on the methods and procedures that have been developed and applied. With regards to computer hardware error, users should be mindful of plausible faults that may exist in the hardware

(iv) Owing to the prevalence of unavoidable human error, probable mistakes in inserting input or in the analysis of output should always be minimized

Verification and validation have very distinct definitions. In spite of the absence of a universal agreement on the details of these definitions, there is a fairly standard and consistent agreement on their usage. Verification can be defined as a process for assessing the numerical approximation and, when conditions permit, estimating the sign and magnitude of the solution error and the uncertainty in that estimated error. On the other hand, validation can be defined as a process for assessing simulation model uncertainty by using benchmark experimental data, and when conditions permit, estimating the sign and magnitude of the simulation modeling error itself.

In the fire literature, there are many comparisons of CFD fire modeling predictions with experimental data. Most of them contain elements of both *validation* and *verification*. A series of experiments have been initially preformed by various researchers (Chitty and Cox, 1979, McCaffrey, 1979, 1995, You and Faeth, 1979) to better understand the flame structure or so-called fire plume. In their experiments, the average flame heights were measured and correlated to the heat release rate of the fire. Based on the measured variation of centerline temperature rise with height, McCaffrey (1979) correlated the temperature variation to the heat release of a free-standing buoyant diffusion flame. Experimental studies preformed by the aforementioned researchers have contributed significantly to the knowledge of fire plumes. Their studies have also led to the development of the fire plume empirical equation. The database is currently utilized in the validation and verification of the large eddy simulation fire model.

For the single-room compartment fire, Heselden (1971) conducted a short series of fire tests in a shopping mall test facility at the Fire Research Station in the early 1970s. As the fire tests were performed just prior to the demolition of the mall, only four experiments were conducted before excessive air leakages occurred through the fabric into the system. Since experimental data of two-dimensional representation of the actual fire tests were only attained, early-fire modeling development has predominantly focused on the validation of two-dimensional numerical models (Markatos et al., 1982, Xue et al., 2001, Yang and Chang, 1977). A more comprehensive three-dimensional experimental study was nonetheless performed by Steckler et al. (1982) of which a total of 55 full-scale experiments were carried out in a single-room compartment configuration. In order to study the behavior of fire-induced flow, a number of geometrical factors were parametrically investigated: fire size and its various locations, door width, and window opening area. Velocity and temperature profiles, mass flow rates, as well as the thermal interface and the height of the neutral plane of room openings were also recorded during the experiments. Their experimental dataset has become a standard benchmark test for validating numerous numerical fire models (Kerrison et al., 1994a, 1994b, Kumar et al., 1991, Lewis et al., 1997, and Yeoh et al., 2002a). Similar experiments can be also found in Nakaya et al. (1985) and Tran and Janssens (1989). All the experimental works as reviewed previously have provided not only a large existing database of reliable results for validation studies, but also contributed to the significant knowledge of the behaviors of enclosure fires.

Following the series of fire tests carried out in single-room compartment situations, a number of fire tests in more complex and sophisticated geometric arrangements have also been investigated. Cooper et al. (1982) have studied the upper hot layer stratification in a multi-room fire experiment, while Luo and Beck (1994, 1996) have obtained useful experimental results from a three-story Experimental Building-Fire Facility (EBFF) by examining the fire spread behavior of two flashover and non-flashover fires in the multi-room compartment configuration. Within the three-story EBFF, measurements of temperatures, radiation heat flux, gas composition, smoke optical density, and soot concentration have been obtained. He and Beck (1997) carried out more experiments and ascertained the smoke spread behavior in the three-story fire facility. These comprehensive experimental datasets have been widely adopted for validating fire models (Luo et al., 1997, Yeoh et al., 2003a, 2003b). Fleischmann and his co-workers have also performed a number of experiments to study various phenomena in enclosure fires (Fleischmann et al., 1990, 1994, Nielsen and Fleischmann, 2000). The work by Nielsen and Fleischmann (2000) studied a pre-flashover fire in a two-compartment structure. By controlling the fire sizes and its location, four different fire scenarios were studied in their experiments. A total of nine thermocouple trees were placed evenly along the centerline of the compartments to measure the temperature development of the hot and cold layer. Variations of the O_2 and CO_2 concentrations were also monitored at the gas sampling points. These comprehensive experimental data have been employed for model comparison and validation (Cheung et al., 2004, Yeoh et al., 2002b, Yeoh et al., 2003a).

1.6 Scope of the Book

Field modeling of the fire dynamics comprises mainly of two components: the CFD methodology and the fire model. The former represents the core of the field model, which provides the basic transport mechanisms for mass, momentum, and energy (including heat transfer due to conduction convection and radiation), while the latter contains the detailed specification of the fire description such as combustion, non-luminous and luminous radiation, soot production, and solid pyrolysis. Acceptable use of the field model therefore requires a good CFD computer code, a good fire model, and ultimately the knowledge as well as proficient use of the CFD-based fire model.

A number of field models currently available fall into one of two distinct categories: general-purpose CFD commercial software packages that can be used for fire modeling applications, and specific field modeling computer codes that are intended only for modeling fires. The general-purpose CFD codes such as PHONEICS, STAR-CD, ANSYS-CFX, and ANSYS-FLUENT are examples of the former. For the latter, examples are JASMINE (developed by Fire Research Station, UK), KAMELON (developed by SINTEF/NTH, Norway), SMARTFIRE (developed by University of Greenwich, UK), SOFIE (developed by Cranfield University/Fire Research Station, UK), FDS (developed by NIST, USA), and our in-house computer code FIRE3D. Regardless of which computer code is employed, it is imperative that intended fire users or modelers must possess the required level of knowledge of the basic "theory," assessing different "modeling" strategies, and how these models of varying degrees of sophistication are employed in "practice" in order to better apply the field model in a suitable manner to deal with a range of fire problems. The scope of this book covering the essential aspects of field modeling pre-flashover fires can thus be envisaged by the road map depicted in Figure 1.17.

On the mathematical level, the fundamental laws of fluid mechanics and heat transfer that relate to the conservation mass, momentum, and energy

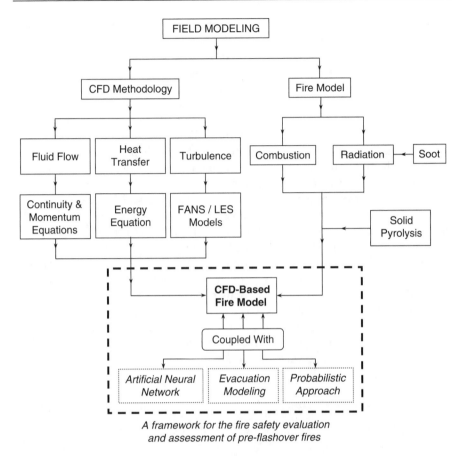

A framework for the fire safety evaluation
and assessment of pre-flashover fires

Figure 1.17 Road map of scope of the book.

equations together with other concepts and equations that supplement the field modeling approach are discussed in Part I of Chapter 2. Since most practical fires are turbulent in nature, the concept of time-averaging based upon Reynolds-averaging and Favre-averaging along with a variety of models ranging from the simple k–ε to the complex Reynolds Stress representations of the turbulence, are described in Part II of Chapter 2. Parts I and II in this chapter represent the essential flow equations of field modeling.

The importance of combustion in fires is dealt with in Part III of Chapter 3. There, the principle knowledge of whether the process is governed by chemical kinetics or turbulent mixing strongly influences the selection of appropriate combustion models for fires. Also, since most practical fires experience significant radiation heat loss during the burning process, radiation heat transfer is treated in Part IV of Chapter 3. The choice of radiation models generally requires considerations on the level of simplification assumed for the radiation properties of absorbing gases corresponding to the level of sophistication adopted for the radiative transfer methods. Parts III and IV in this chapter establish the additional considerations in field modeling.

Part V of Chapter 4 is devoted to the consideration of models determining the concentration of fine carbonaceous particles (soot). Insights into the controlling physical and chemical mechanisms associated with the soot formation and soot oxidation are discussed along with the proposal of suitable models. The ability to simulate and predict the prospect of flame spreading over condensed solids is described in Part VI of Chapter 4. A three-dimensional solid pyrolysis model for charring materials, including moisture migration, is presented. Parts V and VI in this chapter constitute the supplemental considerations in field modeling.

Chapter 5 is dedicated predominantly on the description of the large eddy simulation (LES) technique in fire modeling. Subgrid scale modeling is illustrated with special emphasis on suitable explicit time-marching methods, the effects of subgrid fluctuations on the chemical heat release rate, and appropriate radiation and soot modeling for large eddy simulation of turbulent fires.

The use of CFD fire modeling in conjunction with other important practical aspects in fire engineering such as artificial neural network, evacuation modeling, and probabilistic approach, is discussed in Chapter 6.

In addition to the treatment of basic theory and CFD fire models from the these chapters, worked examples and test cases are included to purposefully demonstrate the utility of the field model in practice, especially tackling practical fires in full-scale configurations.

Review Questions

1.1. In analyzing enclosure fire development using computational models, what are probabilistic methods? Alternatively, what are deterministic methods?

1.2. In deterministic methods, what is the difference between zone models and field models?

1.3. In field modeling, how are fires treated in early computational studies?

1.4. What are the many standard models or features in most current field modeling applications?

1.5. What is large eddy simulation and how is it transforming fire modeling?

1.6. Which fire disaster heralded the use of computer fire models in ascertaining the cause of fatalities and what phenomenon was uncovered in this particular disaster?

1.7. What is a pre-flashover fire?

1.8. What is a post-flashover fire?

1.9. Why is the understanding of a pre-flashover fire important?

1.10. How are field models applied in practice, and from what aspects are they increasingly being considered?

1.11. What are the challenges ahead for fire modeling?

1.12. What are the possible limitations in the application of fire models?

1.13. Provide suitable definitions of verification and validation. Why are they important in fire modeling?

1.14. Field modeling of the fire dynamics consists of two components. What are they?

2 Field Modeling Approach

Abstract

Field modeling is the branch of applied mathematical modeling that fire dynamics is concerned with. The analysis of the behavior of fire begins with the consideration of the fundamental laws of fluid mechanics and heat transfer as encapsulated in the laws of conservation mass, momentum, and energy. This, together with other concepts and equations that supplement the field modeling approach, is the fundamental basis of the field modeling of fire. Since most fires are gaseous by nature, the concern with the static and dynamic behavior of the gases as a continuous fluid with open boundaries is frequently encountered.

Practical fires are generally turbulent. Even at the present level of research computational capacity, it is still not feasible to solve the conservation equations directly to the required accuracy. Turbulent flows are extremely complex time-dependent flows, and to resolve such flows, the use of turbulence models has proven to be reasonably satisfactory. A variety of models are described in this chapter to differentiate their applicability and usefulness to field modeling.

PART I   MATHEMATICAL EQUATIONS

2.1   Computational Fluid Dynamics: Brief Introduction

Computational fluid dynamics is a branch of study that continues to gain in popularity and importance in the modern practice of fluid dynamics. Better known by its acronym CFD, three basic questions are posed following to succinctly describe this terminology.

What is CFD? In essence, CFD is simply the study of fluid systems that could be static or dynamically changing in time and space. The *fluid dynamics* component is performed through numerical methods on high-speed digital computers, which incidentally represents the *computational* description of the terminology. Additionally, the physical characteristics of a fluid in motion can usually be described by the consideration of fundamental mathematical equations, usually in partial differential form, governing a process of interest. In order to solve these mathematical equations, they are converted into discrete forms using high-level computer programming languages into in-house computer programs or commercial CFD software packages. These *algebraic* equations are solved through dedicated techniques, which will be described in

Computational Fluid Dynamics in Fire Engineering
Copyright © 2009 by Academic Press. Inc. All rights of reproduction in any form reserved.

later sections within this chapter. CFD actually covers three major disciplines: fluids engineering, mathematics, and computer science. Acquiring some basic knowledge from each of these disciplines is usually required to better understand CFD.

Where is CFD applied? Through decades of active research and development, this versatile technique has undeniably come of age during this recent era. High-technology industries such as aeronautics and astronautics have heavily integrated CFD techniques into the process of designing and manufacturing aircrafts, spacecrafts, and engines. The long-standing history of applying this powerful methodology is thoroughly demonstrated by the countless aeronautical systems one finds at airports and space-pads. A commercial airliner taking off on a runway or a space shuttle rocketing off into space is an awesome spectacle, symbolic of unprecedented human advancement since the Wright brothers. Another major player in the usage of CFD is none other than the automotive industry. For example, many Formula One vehicles that are shaving off crucial seconds in lap times are due to the systematic application of CFD studies of horizontal drag reduction. As CFD becomes more accessible, many traditional engineering industries are rapidly adopting this analytical tool to solve a variety of complex flow systems. Mechanical, civil, chemical, electrical, electronic, and environmental engineering industries have benefited greatly through the use of CFD, to understand and resolve numerous fluid flow problems that could not have been solved by prescriptive and exhausting trial-and-error approaches. Mechanical ventilation in buildings, molding and extrusion processes, cooling of micro-circuit and computer boards, and distribution of pollutants and effluents in air and water are just some of the typical examples where CFD has played an important role in the design of industrial products and processes. This methodology is also finding its way in a number of non-industrial application areas, not withstanding the engineering industries as aforementioned. Whether from obtaining our daily weather forecast from meteorology, understanding the fluid flows in rivers or oceans in hydrology and oceanography, or even in blood and air flows within our respective vascular and respiratory systems in biomedical research, CFD has increasingly become an economic and robust method of analysis and will only gain in prominence for years to come.

Why CFD? Traditionally, both experimental and analytical methods have been used to study the various aspects of fluid dynamics and assist in the design of equipment and industrial processes involving fluid flow and heat transfer. With the advent of high-speed digital computers, the computational aspect has emerged as another viable approach to resolve complex fluid dynamics issues. With the lowering costs of computer hardware and greater speeds of computer chips, the trend is clearly toward greater reliance on the computational approach for industrial designs, particularly when the fluid flows are very complex. Also, multi-purpose CFD programs have gradually found favor in industrial as well as in academic institutions. With advanced robust models to better encapsulate the flow physics, these commercially available software packages provide incentive to scientifically adopt CFD techniques in the quest to find unique solutions to fluid dynamics and heat transfer problems. It is important to note that analytical methods will continue to be used by many, especially for simple fluid flow problems, and experimental methods will feature significantly for prototype testing as well as for the validation of CFD models. Hence, we still require analytical and experimental methods to complement the CFD analytical tool in some specific investigative studies and analyses. For most flow problems, there are, however, numerous advantages in applying CFD over analytical and experimental methods. First and foremost, the cost-effectiveness of carrying out multiple parametric studies with greater accuracy allows the construction of new and more improved system designs and concerted optimization carried out on existing equipment with substantial reduction of lead times, which results in enhanced efficiency and lower operating costs. Secondly, the primary objective is to gain an increased knowledge of how systems (such as an aircraft) are expected to perform, so evolutionary improvements during the design and optimization process can be made. Based on this second point, CFD therefore continually asks the question "What if ... ?" in all investigative studies and analyses. With such a positive overview, the importance of computational fluid dynamics in field modeling is described in the subsequent section.

2.2 Computational Fluid Dynamics in Field Modeling

In the context of fire engineering, what is the relationship between CFD and field modeling and the role CFD plays in field modeling? Within the fire community, field modeling is a well-recognized terminology. But what does it usually entail? As described in the previous section, CFD encompasses the study of fluids that are in motion through computational means. Conservation equations are solved to describe the transport phenomena usually through numerical techniques. However, a burning fire constitutes more than just a description of the fluid mechanics. Let us further investigate the behavior of a burning fire, as demonstrated by the schematic diagram in Figure 2.1.

Most fires involve combustible gases. According to Drysdale (1999), the term *fuel*, required for combustion, can be defined as the state of matter—be it in the form of gases, liquids, or solids—burning in the atmosphere. Consider the burning of a solid fuel such as a piece of furniture in a room, for the purpose of illustration. For almost all types of solids, chemical decomposition (pyrolysis) is responsible for yielding products of sufficiently low molecular weight that can be volatilized from the surface and enter the flame as fuel. Clearly, the visible flame is a gas phase phenomenon, and flaming combustion of a solid fuel, as described by the pyrolysis process, or a liquid fuel must involve the conversion of fuel to gaseous form. For a burning liquid, the process is generally simpler where volatiles are released due to evaporative boiling at the surface.

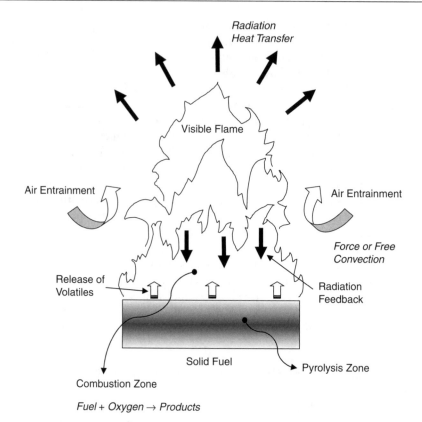

Figure 2.1 Schematic representation of a burning solid fuel in air showing the respective physical processes involved.

As the combustible volatiles fuel the burning fire, as shown in Figure 2.1, they react with the oxygen in the air, which is being entrained into the combustion zone under appropriate conditions in the gas phase. The air entrainment may be driven by force or free convection depending on the environmental conditions surrounding the burning fire. Assuming a carbon-based fuel, combustion products such as carbon dioxide and water vapor are generated for a complete chemical reaction. In reality, such a chemical reaction is an idealization of the actual gas phase combustion, whereby the complexity of the overall process is hidden. The true picture usually consists of a series of elementary reaction steps taking part in the course of combustion. Heat is released following the chemical reaction—an exothermic process. With very few exceptions, particulate smoke is usually produced in all fires. Depending on the nature of different fuels, smoke can contain high concentrations of finely dispersed particles, commonly known as soot, or narcotic gases such as carbon monoxide, in addition to the production of combustion products. Radiation from these hot combustion gases and non-luminous flames also plays an important consideration in fires. For an unabated fire, radiation heat transfer causes the heat supplied by the visible flame onto the solid surface to sustain a continuous supply of volatiles into the combustion zone for flaming combustion.

The example of the burning of a solid fuel in an open space as described in Figure 2.1 can usually be categorized as a free-standing fire. Another important classification, "compartment fire," exhibits some distinguishing features and characteristics that are usually different from a free-standing fire. Here, the confining rigid boundaries surrounding the enclosure greatly influence the fire dynamics.

As shown in Figure 2.2 for a compartment fire, a fire source that could be a gas, liquid, or solid fuel ignites and burns at the center of the enclosure. Combustion takes place resulting in a heat release due to the exothermal reaction; this fire source can now be considered as a heat source discharging energy onto the surrounding fluid or wall boundaries by three modes of heat transfer: conduction, convection, and radiation. For natural fires, the hot fluid heated by the fire source is driven by buoyancy; the development of an upward flow, normally turbulent in nature, transfers the hot fluid into the upper region of the room. At this early stage of the fire growth, the pyrolysis rate and energy release rate are affected only by the burning of the fuel itself, and not by the presence of the boundaries of the compartment. In time, the presence of the wall boundaries and the soffit of the room create a "reservoir" for the accumulation of the hot gases or smoke below the ceiling, which is commonly referred as the "Hot Layer." While hot gases accumulate in the upper region, fresh air from the outside surrounding is entrained into the room through the doorway to supply oxygen to the fire in the lower level. This represents the "Cold Layer" region of the compartment. In fire engineering, the two-layer flow pattern is a prevalent feature identifying compartment fires. The height of this

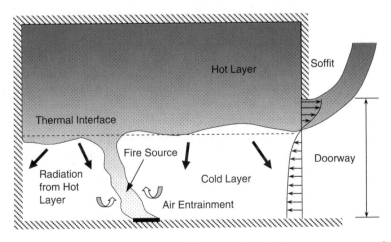

Figure 2.2 Schematic representation of a fire source in a single-compartment enclosure.

interface is often considered as an important datum of measure in fire experiment; the term "Thermal Interface" is usually introduced to characterize the averaged height of the interface. As this interface continues to descend due to increasing concentrations of hot gases, especially down to the level below the depth of the soffit, the amount of hot gases eventually exceeds the volume of the reservoir, and some of these hot gases will spill into the surrounding below the soffit, as shown in Figure 2.2. These two counter flows have a tendency of creating a rapid change of velocity at the doorway. As a result, positive pressure is sustained at the top to exhaust the hot gases, while negative pressure entrains the fresh air at the bottom of the doorway.

Free-standing and *compartment fires* as described are just some of the many representative fires that have been studied extensively to gain invaluable insights into the fire phenomena. From the standpoint of modeling consideration, can all the associated physical processes as described in fires be viably studied using CFD techniques and appropriate numerical models? There have been significant advances achieved in the last two decades, especially through the feasibility of attaining solutions through conservations equations of fluid dynamics and heat transfer coupled with suitable models capturing the complex physical processes associated with combustion, radiation, smoke movement and production, and solid pyrolysis. Increasing usage of these models to resolve many current fire problems—incidentally also recognized as the art of *field modeling*—are indicative of the state-of-the-art development and maturity of CFD techniques and models. With greater speeds expected in digital computers (arriving sooner rather than later), field modeling will undoubtedly play an even greater role and cement its place as the preferred approach in the areas of fire science and fire engineering.

Nevertheless, adopting simplified models to resolve the fire phenomena can at best only provide a general description of the fluid and heat flow distribution. More sophisticated approaches need to take into account the many associated complexities involving combustion, radiation, smoke movement and production, and solid pyrolysis. However, the simplified approach generally gives a basic understanding of the actual dynamics of the real fire, despite its simplistic representation of reality. Therefore, the modeling of fires still remains a very challenging task for fire modelers, because of his or her need to scope the general dynamics of the system before applying analytical techniques to the problem. One of the main difficulties from the modeling perspective lies in the need of acquiring the necessary background knowledge, basic as well as advanced, stemming from different yet integrated disciplines. Consider for the moment the study of physical chemistry of *laminar* or *turbulent* flaming combustion. Within this discipline alone, a large database of literatures exists, and volumes of books have been written primarily to explain and address the fundamental principals and theories of combustion. Such an accumulation of knowledge does not precipitate overnight but is the result of years of dedicated research and to the formulation of techniques devoted to solving a range of combustion problems. Similar inference can also be ascribed to other disciplines such as radiation and solid pyrolysis. So, how can we adequately assimilate and integrate all these important concepts from the varied disciplines into field modeling?

Fortunately, like the advancement in CFD, these disciplines have also reached a level of maturity in establishing stable and robust models for a wide range of applications. The techniques and models from each of these disciplines can now be readily employed to adequately describe the fire phenomena. From this stance, this book therefore aims to consolidate the existing knowledge and technical know-how behind many of these available techniques and models employed in practice. Particularly, the relevance toward modeling and application in fires is emphasized. In addition, some future developments in the context of *field modeling* are provided to further demonstrate the enhancement of the state-of-the-art toward modeling fires. For the rest of this chapter, the formulation of the fundamental transport equations governing the fluid flow and heat transfer as well as turbulence in fires is described. First and foremost, the equation of state is illustrated in the next section.

2.3 Equation of State

In general, thermodynamics is concerned with substances in all three phases: solid, liquid, and gas. Most thermodynamic problems ordinarily involve gases or vapors such as in burning fires, though some of thermodynamic problems encountered may, in a few instances, involve liquids and solid. Such problems generally involve the *intensive* and *extensive* properties of systems. In determining the interrelationship between the various properties of a substance, it shall be assumed in the proceeding discussion that the intensive properties are uniform throughout the system under consideration, meaning that the system is in *thermal* and *mechanical* equilibrium. When the system is in thermal, mechanical, and chemical equilibrium, the system is said to be in a state of thermodynamic equilibrium. In other words, it can be said that although the intensive properties can change rapidly from place to place, the system can thermodynamically adjust to new conditions so quickly that the changes are effectively instantaneous. Hence the system always remains in thermodynamic equilibrium.

According to Kuo (1986), an *intensive property* can be defined as one that is unchanged when the size of the system is increased by adding to it any number of systems that are identical to the original system, while an *extensive property* can be conversely defined as one that increases in proportion to the size of the system in such a process.

Intensive properties commonly found in a fluid in motion are the density (ρ), pressure (p), specific internal energy (e), and temperature (T), while other variables such as volume (V), mass (m), total enthalpy (h), and others are characteristically considered as extensive properties. Invoking the thermodynamic equilibrium provides the means whereby the state of substance can usually be described in terms of only two thermodynamic variables or intensive properties. For an *ideal* or *perfect gas*, the following widely adopted equation of state that obeys the laws of two famous investigators, Boyle and Charles, is

$$pV = nR_u T \qquad (2.3.1)$$

where R_u is the universal gas constant (8.31431 kJ kmol^{-1}K^{-1}). Setting the pressure $p = 101325$ Pa, $T = 273.15$ K (0°C) and $n = 1$ mole, $V = 0.022414$ m^3. For ideal gases, this volume represents the total volume that will be occupied by 28 g N_2 (nitrogen), 32 g O_2 (oxygen), 44 g CO_2 (carbon dioxide), 18 g H_2O (water vapor), or 16 g CH_4 (methane).

The density or concentration of a gas can be determined according to

$$\rho = \frac{nM}{V} = \frac{pM}{R_u T} \qquad (2.3.2)$$

where M is the molecular weight of a gas mixture. It can be shown for air that one mole corresponds to a molecular weight of 0.02895 kg. For a pressure of $p = 101325$ Pa and a temperature of $T = 0$°C (273.15 K), the density will be 1.292 kg m^{-3}.

The combustion process in a fire usually causes substantial changes in the temperature of the surroundings as a result of many combustion products being formed at high temperatures. For a reacting fluid flow in chemical equilibrium, the number of moles for each chemical species can usually be expressed by a known material at a volume V and temperature T:

$$n_i^* = f(V, T) \qquad (2.3.3)$$

where the superscript * denotes the equilibrium values. The equation of state for a system in equilibrium becomes

$$p = f\left(V, T, n_1^*, n_2^*, \ldots, n_N^*\right) \qquad (2.3.4)$$

From Dalton's law of partial pressures p_i, the expression for a mixture of thermally perfect gases in thermodynamic equilibrium can thus be alternatively given similar to equation (2.3.1) as:

$$p = \sum_i p_i = \frac{1}{V} \sum_{i=1}^{N} n_i^* R_u T \qquad (2.3.5)$$

The mixture pressure for a system in chemical non-equilibrium can be simply represented by removing the superscript * in equation (2.3.5).

Gases flowing at high speeds are generally categorized as *compressible* flows. Here, the density varies dramatically across a broad range of pressures as well as temperatures in the flow field. In such flows, the equation of state provides the necessary linkage between the energy conservation equation and the equations for the conservation of mass and momentum. There are also a number of situations where gases flowing at low speeds behave as *incompressible* fluids. The density remains invariant, and the fluid flow can thus be solved by only considering the equations pertaining to the conservation of mass and momentum. For other low-speed flows that are typical of burning fires, these are regarded nonetheless as *weakly compressible*. The term *weakly* refers to the consequence of density change being affected mainly by the substantial temperature variations due to chemical reactions in the flow field but not from the pressure variations, since the pressure remains relatively unperturbed within the surroundings. If a perfect gas is assumed, the equation of state derived in equation (2.3.2) may be simplified under the *weakly compressible* condition by setting the pressure to the fixed ambient pressure P_0—in other words,

$$\rho = \frac{p_0 M}{R_u T} \tag{2.3.6}$$

To establish the appropriate linkage, the energy equation is still required to be solved alongside with the mass conservation and momentum equations to determine the local temperature T in equation (2.3.6).

2.4 Equations of Motion

To solve the flow physics within the physical domain, CFD requires the subdivision of the domain into a number of smaller, non-overlapping subdomains. This subsequently results in the generation of a mesh (or grid) of cells (elements or control volumes) covering the whole domain. The essential fluid flows that are described in each of these cells are usually solved numerically through fundamental equations governing the fluid dynamics. Discrete values of the flow properties, such as the velocity, pressure, temperature, and other transport parameters of interest, are thereafter determined at each of the respective cells.

The purpose of this section is to derive these governing equations. To obtain the basic equations of fluid motion, the following philosophy is adopted. Firstly, the appropriate fundamental principles from conservation laws of physics are chosen—namely the conservation of mass, Newton's second law for the conservation of momentum, and the first law of thermodynamics for the conservation of energy. Secondly, these physical principles are applied to a suitable model of the fluid flow. Thirdly, the mathematical equations that embody such physical principles are extracted from the model under consideration.

For general applications in fire, the significance of these equations of motion hinges on the basis of a continuum fluid. Let us consider the infinitesimal small fluid element or control volume represented in the flow field with a differential volume ΔV in Figure 2.4. This fluid element can be taken to be infinitesimal and viewed in the same sense as differential calculus. However, it is large enough to encapsulate molecules at macroscopic length scales—say, 1 μm or larger—so that it can be regarded as a continuous medium. The behavior of the fluid can thus be described in terms of macroscopic properties such as the velocity, pressure, density, and temperature as well as their space and time derivatives. These macroscopic properties are not influenced by individual molecules, and this leads immediately to the development of fundamental equations in partial differential form.

2.4.1 Continuity Equation

Consider the following: matter may neither be created nor destroyed, which means that mass must always be conserved. This conservation law is pertinent to the derivation of the continuity equation. Applying to the elemental volume ΔV inside the flow field, as shown in Figure 2.3, the fundamental physical principle for which the conservation of mass is satisfied requires:

$$\begin{array}{ll} \textit{The rate of} & \textit{The net rate at} \\ \textit{increase of mass} & \textit{which mass enters} \\ \textit{within the fluid} = & \textit{the elemental} \\ \textit{element} & \textit{volume} \end{array} \qquad (2.4.1)$$

Let us further consider the enlarged infinitesimal fluid element in a Cartesian coordinate system with a volume ΔV of $\Delta x \, \Delta y \, \Delta z$ that is fixed in space where the *mass conservation* statement applies to the (u,v,w) flow field (see Figure 2.4). Since the mass of the fluid element m is given by $\rho \, \Delta V \, (= \Delta x \, \Delta y \, \Delta z)$, the rate of increase of mass within the fluid element is

$$\frac{\partial m}{\partial t} = \frac{\partial}{\partial t}(\rho \, \Delta x \, \Delta y \, \Delta z) = \frac{\partial \rho}{\partial t} \Delta x \, \Delta y \, \Delta z \qquad (2.4.2)$$

To account for the mass flow across each of the faces of the element, it can be seen from Figure 2.4 that the net rate at which mass enters the elemental volume is given by:

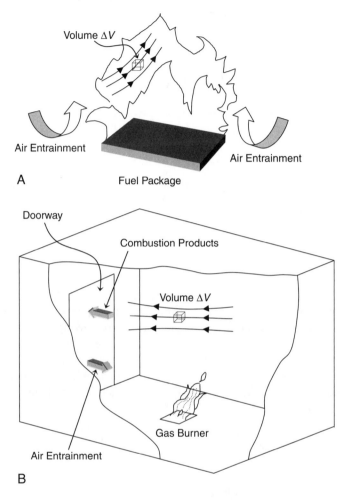

Figure 2.3 The infinitesimal fluid element approach. Representation models of a fluid flow in (a) free-standing fire and (b) single-compartment fire.

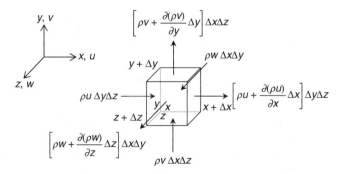

Figure 2.4 The conservation of mass in an infinitesimal fluid element.

$$\underbrace{(\rho u)\Delta y \Delta z - \left[(\rho u) + \frac{\partial(\rho u)}{\partial x}\Delta x\right]\Delta y \Delta z}_{\text{along } x}$$

$$\underbrace{+ (\rho v)\Delta x \Delta z - \left[(\rho v) + \frac{\partial(\rho v)}{\partial y}\Delta y\right]\Delta x \Delta z}_{\text{along } y} \tag{2.4.3}$$

$$\underbrace{+ (\rho w)\Delta x \Delta y - \left[(\rho w) + \frac{\partial(\rho w)}{\partial z}\Delta z\right]\Delta x \Delta y}_{\text{along } z}$$

The rates at which mass enters the control volume through the surfaces perpendicular to x, y, and z are respectively the inflow components of $(\rho u)\,\Delta y\,\Delta z$, $(\rho v)\,\Delta x\,\Delta z$, and $(\rho W)\,\Delta x\,\Delta y$. They are assigned as positive to signify an increase of mass in the element. Since mass needs to be conserved within the element, the rates at which mass leaves the surfaces of $x + \Delta x$, $y + \Delta y$, and $z + \Delta z$ are represented by the negative outflow components as indicated in equation (2.4.3). By equating equations (2.4.2) and (2.4.3), canceling terms and dividing by the constant-size $\Delta x\,\Delta y\,\Delta z$, equation (2.4.1) becomes:

$$\frac{\partial \rho}{\partial t} + \frac{\partial(\rho u)}{\partial x} + \frac{\partial(\rho v)}{\partial y} + \frac{\partial(\rho w)}{\partial z} = 0 \tag{2.4.4}$$

Equation (2.4.4) is essentially the *partial differential form* of the *unsteady, three-dimensional mass conservation* or *continuity equation*.

2.4.2 Momentum Equation

For the derivation of the momentum equation, the concept of substantial derivative is introduced and derived. It is conceivable that the generalization of any variable property ϕ can be expressed in conservative form as:

$$\frac{\partial(\rho\phi)}{\partial t} + \frac{\partial(\rho u\phi)}{\partial x} + \frac{\partial(\rho v\phi)}{\partial y} + \frac{\partial(\rho w\phi)}{\partial z} = 0 \tag{2.4.5}$$

Note that the *partial differential form* for the mass conservation equation (2.4.4) is already expressed in *conservative form*. The preceding formula expresses the rate of change of ϕ per unit volume with the addition of the net flow of ϕ out of the fluid element per unit volume. By applying the chain-rule to equation (2.4.5),

$$\frac{\partial(\rho\phi)}{\partial t} + \frac{\partial(\rho u\phi)}{\partial x} + \frac{\partial(\rho v\phi)}{\partial y} + \frac{\partial(\rho w\phi)}{\partial z} =$$

$$\rho\frac{\partial\phi}{\partial t} + \rho u\frac{\partial\phi}{\partial x} + \rho v\frac{\partial\phi}{\partial y} + \rho w\frac{\partial\phi}{\partial z} + \phi \underbrace{\left[\frac{\partial\rho}{\partial t} + \frac{\partial(\rho u)}{\partial x} + \frac{\partial(\rho v)}{\partial y} + \frac{\partial(\rho w)}{\partial z}\right]}_{=\,0\ by\ definition\ of\ the\ continuity\ equation\ (2.10)} =$$

$$\rho\left[\underbrace{\frac{\partial\phi}{\partial t} + u\frac{\partial\phi}{\partial x} + +v\frac{\partial\phi}{\partial y} + w\frac{\partial\phi}{\partial z}}_{D\phi/Dt}\right]$$

$$(2.4.6)$$

Equation (2.4.6) represents the *non-conservative form* of the rate of change of the variable property ϕ per unit volume. The terms that are represented within the bracket in equation (2.4.6) are essentially the substantial derivative of ϕ with respect to time, designated as $D\phi/Dt$. Both of the conservative and non-conservative forms can be used to express the conservation of a physical quantity. We shall adopt the non-conservative form to derive the next physical law encountered in flow problems, which is the momentum theorem.

In deriving this physical law, let us reconsider the fluid element as described in Figure 2.4 for mass conservation. Newton's second law of motion states that the sum of forces that is acting on the fluid element, as illustrated in Figure 2.5, equals the rate of change of momentum, which is the product of its mass and the acceleration of the element.

$$\begin{matrix} \textit{The rate increase} & & \textit{The sum of forces} \\ \textit{of momentum of} & = & \textit{acting on the} \\ \textit{the fluid element} & & \textit{fluid element} \end{matrix} \qquad (2.4.7)$$

Essentially, there are three scalar relations along the x, y, and z directions of the Cartesian frame of which this particular fundamental law can be invoked. The x component of Newton's second law can be expressed as:

$$\sum F_x = ma_x \qquad (2.4.8)$$

where F_x and a_x are the force and acceleration along the x direction. The acceleration a_x at the right-hand side of the preceding equation is simply the time rate change of u, which is given by the substantial derivative. Thus,

$$a_x = \frac{Du}{Dt} \qquad (2.4.9)$$

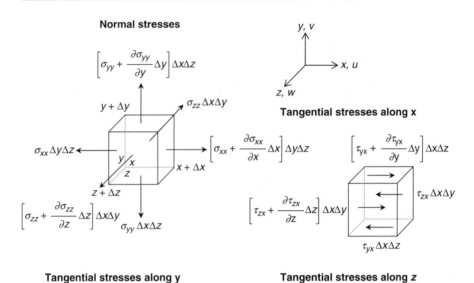

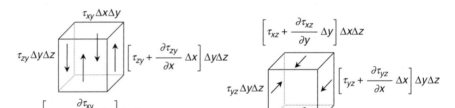

Figure 2.5 Normal and tangential stresses acting on infinitesimal control volumes for velocity components *u*, *v*, and *w* along the Cartesian directions of *x*, *y*, and *z*.

Recalling that the mass of the fluid element *m* is $\rho \Delta V$ ($= \Delta x \, \Delta y \, \Delta z$), the rate of increase of *x*-momentum is

$$\rho \frac{Du}{Dt} \Delta x \Delta y \Delta z \tag{2.4.10}$$

On the left-hand side of equation (2.4.7), two sources of this force that the moving fluid element generally experiences are *body forces* and *surface forces*. These effects are usually incorporated by introducing them as additional source terms into the momentum equations. The surface forces for the velocity component *u*, as seen in Figure 2.5, that deform the fluid element are due to the normal stress σ_{xx} and tangential stresses τ_{yx} and τ_{zx} acting on the surfaces of the fluid element. The net force in the *x* direction is the sum of the force components acting on the fluid element. Considering the velocity component *u* as seen in Figure 2.5, the surface forces are due to the normal stress σ_{xx} and tangential stresses τ_{yx} and τ_{zx} acting on the surfaces of the fluid element. The total net force per unit volume on the fluid due to these surface stresses should be equal to the sum of the normal and tangential forces. Hence, the total net force per unit volume along the x direction is:

$$\left[\frac{\partial \sigma_{xx}}{\partial x} + \frac{\partial \tau_{yx}}{\partial y} + \frac{\partial \tau_{zx}}{\partial z}\right] \Delta x \Delta y \Delta z \qquad (2.4.11)$$

The total net forces per unit volume on the rest of the control volume surfaces along the y direction and z direction can be similarly derived to yield:

$$\left[\frac{\partial \tau_{xy}}{\partial x} + \frac{\partial \sigma_{yy}}{\partial y} + \frac{\partial \tau_{zy}}{\partial z}\right] \Delta x \Delta y \Delta z \qquad (2.4.12)$$

$$\left[\frac{\partial \tau_{xz}}{\partial x} + \frac{\partial \tau_{yz}}{\partial y} + \frac{\partial \sigma_{zz}}{\partial z}\right] \Delta x \Delta y \Delta z \qquad (2.4.13)$$

Combining equation (2.4.11) with the time rate of change of the horizontal velocity component u and body forces, the x-momentum equation after dividing by the control volume $\Delta x\, \Delta y\, \Delta z$ becomes

$$\rho \frac{Du}{Dt} = \frac{\partial \sigma_{xx}}{\partial x} + \frac{\partial \tau_{yx}}{\partial y} + \frac{\partial \tau_{zx}}{\partial z} + \sum F_x^{body\ forces} \qquad (2.4.14)$$

In a similar fashion, the y-momentum and z-momentum equations, using equations (2.4.12) and (2.4.13), can be obtained through

$$\rho \frac{Dv}{Dt} = \frac{\partial \tau_{xy}}{\partial x} + \frac{\partial \sigma_{yy}}{\partial y} + \frac{\partial \tau_{zy}}{\partial z} + \sum F_y^{body\ forces} \qquad (2.4.15)$$

$$\rho \frac{Dw}{Dt} = \frac{\partial \tau_{xz}}{\partial x} + \frac{\partial \tau_{yz}}{\partial y} + \frac{\partial \sigma_{zz}}{\partial z} + \sum F_z^{body\ forces} \qquad (2.4.16)$$

In many fluid flows, a suitable model for the viscous stresses is required. The stresses can usually be expressed as a function of the local deformation rate (or strain rate). It is a common practice to assume that the fluid is Newtonian and that all gases and the majority of liquids are isotropic. The rate of linear deformation on the control volume $\Delta x\, \Delta y\, \Delta z$ caused by the motion of fluid can be expressed in terms of the velocity gradients. Describing the normal stress relationships for σ_{xx}, σ_{yy}, and σ_{zz} appearing in equations (2.4.14), (2.4.15), and (2.4.16) in terms of pressure p and normal viscous stress components τ_{xx}, τ_{yy}, and τ_{zz} acting perpendicular to the control volume:

$$\sigma_{xx} = -p + \tau_{xx} \quad \sigma_{yy} = -p + \tau_{yy} \quad \sigma_{zz} = -p + \tau_{zz} \qquad (2.4.17)$$

the normal and tangential viscous stress components according to *Newton's law of viscosity* can be expressed by

$$\tau_{xx} = 2\mu\frac{\partial u}{\partial x} + \lambda\left[\frac{\partial u}{\partial x} + \frac{\partial v}{\partial y} + \frac{\partial w}{\partial z}\right] \quad \tau_{yy} = 2\mu\frac{\partial v}{\partial y} + \lambda\left[\frac{\partial u}{\partial x} + \frac{\partial v}{\partial y} + \frac{\partial w}{\partial z}\right]$$

$$\tau_{zz} = 2\mu\frac{\partial w}{\partial z} + \lambda\left[\frac{\partial u}{\partial x} + \frac{\partial v}{\partial y} + \frac{\partial w}{\partial z}\right]$$

$$\tau_{xy} = \tau_{yx} = \mu\left(\frac{\partial v}{\partial x} + \frac{\partial u}{\partial y}\right) \quad \tau_{xz} = \tau_{zx} = \mu\left(\frac{\partial w}{\partial x} + \frac{\partial u}{\partial z}\right)$$

$$\tau_{yz} = \tau_{zy} = \mu\left(\frac{\partial w}{\partial y} + \frac{\partial v}{\partial z}\right)$$

$$(2.4.18)$$

The proportionality constants of μ and λ are the (first) dynamic viscosity that relates stresses to linear deformation and the second viscosity that relates stresses to the volumetric deformation, respectively. To this present day, not much is known about the second viscosity. Nevertheless, Stokes hypothesis of $\lambda = -2/3\mu$ is frequently used, and it has been found for gases to be a good working approximation.

Combining equations (2.4.17) and (2.4.18) with equations (2.4.14), (2.4.15), and (2.4.16), the three-dimensional momentum equations for the velocity components u, v, and w can now be rewritten in the following partial differential forms as

$$\rho\frac{Du}{Dt} = -\frac{\partial p}{\partial x} + \frac{\partial}{\partial x}\left[2\mu\frac{\partial u}{\partial x} + \lambda\left(\frac{\partial u}{\partial x} + \frac{\partial v}{\partial y} + \frac{\partial w}{\partial z}\right)\right]$$

$$+ \frac{\partial}{\partial y}\left[\mu\left(\frac{\partial u}{\partial y} + \frac{\partial v}{\partial x}\right)\right] + \frac{\partial}{\partial z}\left[\mu\left(\frac{\partial u}{\partial z} + \frac{\partial w}{\partial x}\right)\right] + \sum F_x^{body\ forces}$$

$$(2.4.19)$$

$$\rho\frac{Dv}{Dt} = -\frac{\partial p}{\partial y} + \frac{\partial}{\partial y}\left[2\mu\frac{\partial v}{\partial y} + \lambda\left(\frac{\partial u}{\partial x} + \frac{\partial v}{\partial y} + \frac{\partial w}{\partial z}\right)\right]$$

$$+ \frac{\partial}{\partial x}\left[\mu\left(\frac{\partial u}{\partial y} + \frac{\partial v}{\partial x}\right)\right] + \frac{\partial}{\partial z}\left[\mu\left(\frac{\partial v}{\partial z} + \frac{\partial w}{\partial y}\right)\right] + \sum F_y^{body\ forces}$$

$$(2.4.20)$$

$$\rho \frac{Dw}{Dt} = -\frac{\partial p}{\partial z} + \frac{\partial}{\partial z}\left[2\mu\frac{\partial w}{\partial y} + \lambda\left(\frac{\partial u}{\partial x} + \frac{\partial v}{\partial y} + \frac{\partial w}{\partial z}\right)\right]$$
$$+ \frac{\partial}{\partial x}\left[\mu\left(\frac{\partial u}{\partial z} + \frac{\partial w}{\partial x}\right)\right] + \frac{\partial}{\partial y}\left[\mu\left(\frac{\partial v}{\partial z} + \frac{\partial w}{\partial y}\right)\right] + \sum F_z^{body\ forces}$$

(2.4.21)

By applying equation (2.4.6), equations (2.4.19), (2.4.20), and (2.4.21) can be re-expressed in their conservative partial differential forms as:

$$\frac{\partial(\rho u)}{\partial t} + \frac{\partial(\rho u u)}{\partial x} + \frac{\partial(\rho v u)}{\partial y} + \frac{\partial(\rho w u)}{\partial z} =$$
$$-\frac{\partial p}{\partial x} + \frac{\partial}{\partial x}\left[2\mu\frac{\partial u}{\partial x} + \lambda\left(\frac{\partial u}{\partial x} + \frac{\partial v}{\partial y} + \frac{\partial w}{\partial z}\right)\right]$$
$$+ \frac{\partial}{\partial y}\left[\mu\left(\frac{\partial u}{\partial y} + \frac{\partial v}{\partial x}\right)\right] + \frac{\partial}{\partial z}\left[\mu\left(\frac{\partial u}{\partial z} + \frac{\partial w}{\partial x}\right)\right] + \sum F_x^{body\ forces}$$

(2.4.22)

$$\frac{\partial(\rho v)}{\partial t} + \frac{\partial(\rho u v)}{\partial x} + \frac{\partial(\rho v v)}{\partial y} + \frac{\partial(\rho w v)}{\partial z} =$$
$$-\frac{\partial p}{\partial y} + \frac{\partial}{\partial y}\left[2\mu\frac{\partial v}{\partial y} + \lambda\left(\frac{\partial u}{\partial x} + \frac{\partial v}{\partial y} + \frac{\partial w}{\partial z}\right)\right]$$
$$+ \frac{\partial}{\partial x}\left[\mu\left(\frac{\partial u}{\partial y} + \frac{\partial v}{\partial x}\right)\right] + \frac{\partial}{\partial z}\left[\mu\left(\frac{\partial v}{\partial z} + \frac{\partial w}{\partial y}\right)\right] + \sum F_y^{body\ forces}$$

(2.4.23)

$$\frac{\partial(\rho w)}{\partial t} + \frac{\partial(\rho u w)}{\partial x} + \frac{\partial(\rho v w)}{\partial y} + \frac{\partial(\rho w w)}{\partial z} =$$
$$-\frac{\partial p}{\partial z} + \frac{\partial}{\partial z}\left[2\mu\frac{\partial w}{\partial z} + \lambda\left(\frac{\partial u}{\partial x} + \frac{\partial v}{\partial y} + \frac{\partial w}{\partial z}\right)\right]$$
$$+ \frac{\partial}{\partial x}\left[\mu\left(\frac{\partial u}{\partial z} + \frac{\partial w}{\partial x}\right)\right] + \frac{\partial}{\partial y}\left[\mu\left(\frac{\partial v}{\partial z} + \frac{\partial w}{\partial y}\right)\right] + \sum F_z^{body\ forces}$$

(2.4.24)

Note that the preceding equations are also commonly known as the Navier-Stokes equations for a Newtonian fluid.

2.4.3 Energy Equation

Based on the consideration of the *first law of thermodynamics,* the energy equation can be derived for a combusting fire system, which states that the rate of change of energy is equal to the net rate of heat addition, plus the heat rate of work done, plus the rate of heat added or removed by the heat source.

The rate increase of energy of the fluid element	=	*The net rate of heat added to the fluid element*	+	*The net rate of work done on the fluid element*	+	*The rate of heat added or removed by heat source on the fluid element*
		$\underbrace{\qquad}$		$\underbrace{\qquad}$		$\underbrace{\qquad}$
		$\sum \dot{Q}$		$\sum \dot{W}$		$\dot{Q}_s \Delta V$

$$(2.4.25)$$

As discussed in section 2.4.2, the time rate of change of any arbitrary variable property ϕ is defined as the product between the density and the substantial derivative of ϕ. The time rate of change of energy for the moving fluid element can thus be given by:

$$\rho \frac{DE}{Dt} \Delta x \Delta y \Delta z \qquad\qquad (2.4.26)$$

Two terms represented by $\sum \dot{Q}$ and $\sum \dot{W}$ in equation (2.4.25) describe the net rate of heat addition to the fluid within the control volume and the net rate of work done by surface forces on the fluid. Referring to Figure 2.6, the rate of work done on the control volume in the x direction is equivalent to the product between the surface forces (caused by the normal viscous stress σ_{xx} and tangential viscous stresses τ_{yx} and τ_{zx}) with the velocity component u. The net rate of work done by these surface forces acting along the x direction is:

$$\left[\frac{\partial(u\sigma_{xx})}{\partial x} + \frac{\partial(u\tau_{yx})}{\partial y} + \frac{\partial(u\tau_{zx})}{\partial z} \right] \Delta x \Delta y \Delta z \qquad\qquad (2.4.27)$$

Work done due to surface stress components along the y direction and z direction can also be similarly derived and these additional rates of work done on the fluid, which are:

$$\left[\frac{\partial(v\tau_{xy})}{\partial x} + \frac{\partial(v\sigma_{yy})}{\partial y} + \frac{\partial(v\tau_{zy})}{\partial z} \right] \Delta x \Delta y \Delta z \qquad\qquad (2.4.28)$$

$$\left[\frac{\partial(w\tau_{xz})}{\partial x} + \frac{\partial(w\tau_{yz})}{\partial y} + \frac{\partial(w\sigma_{zz})}{\partial z} \right] \Delta x \Delta y \Delta z \qquad\qquad (2.4.29)$$

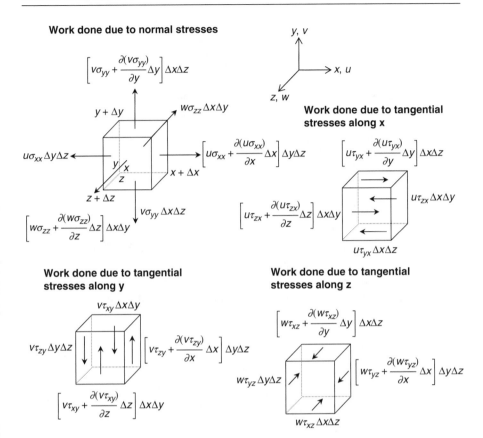

Figure 2.6 Work done due to normal and tangential stresses along the Cartesian directions of x, y, and z.

In addition to the work done by surface forces on the fluid element, it is worth mentioning the possible consideration of work done due to body forces. Generally, this effect is ignored for most practical applications. The net rate work done on the fluid element can thus be comprised of:

$$\sum \dot{W} = \frac{\partial(u\sigma_{xx})}{\partial x} + \frac{\partial(v\sigma_{yy})}{\partial y} + \frac{\partial(w\sigma_{zz})}{\partial z}$$

$$+ \frac{\partial(u\tau_{yx})}{\partial y} + \frac{\partial(u\tau_{zx})}{\partial z} + \frac{\partial(v\tau_{xy})}{\partial x} + \frac{\partial(v\tau_{zy})}{\partial z} + \frac{\partial(w\tau_{xz})}{\partial x} + \frac{\partial(w\tau_{yz})}{\partial y}$$

$$(2.4.30)$$

For heat added, the net rate of heat transfer to the fluid due to the heat flow along the x direction is given by the difference between the heat input at surface at x and heat loss at surface $x + \Delta x$ as depicted in Figure 2.7.

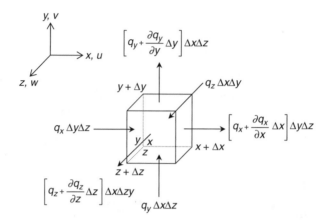

Figure 2.7   Heat added to the fluid along the Cartesian directions of x, y, and z.

Similar considerations are also applied for the net rates of heat transfer along the y direction and z direction. The total rate of heat added to the fluid results in:

$$\sum \dot{Q} = - \left[\frac{\partial q_x}{\partial x} + \frac{\partial q_y}{\partial y} + \frac{\partial q_z}{\partial z} \right] \Delta x \Delta y \Delta z \qquad (2.4.31)$$

The energy fluxes q_x, q_y, and q_z in equation (2.4.31) can be formulated by applying the *Fourier's law of heat conduction* that relates the heat flux to the local temperature gradient:

$$q_x = -k \frac{\partial T}{\partial x} \qquad q_y = -k \frac{\partial T}{\partial y} \qquad q_z = -k \frac{\partial T}{\partial z} \qquad (2.4.32)$$

where k is the thermal conductivity. It is also worth mentioning that besides conduction, reacting flows generally require the consideration of two additional contributions to the heat flux in a combusting fire system. The first is the additional contribution to the heat flux caused by the inter-diffusion process, while the second is Soret and Dufour effects. The latter is essentially based on the Onsager's reciprocal relations for the thermodynamics of irreversible processes, which imply if temperature gives rise to diffusion velocities (the thermal-diffusion effect or Soret effect), concentration gradients must also produce a heat flux. In most cases, the Soret and Dufour effects are rather small, even when thermal diffusion is not negligible, and they are usually omitted in most applications. For the inter-diffusion process, there is no consensus in the literature whether this effect should always be included or excluded totally. Appropriate expressions for this term will nonetheless be presented in the next chapter for completeness. The net rate of heat added to the fluid element without inter-diffusion processes and Soret and Dufour effects becomes:

$$\sum \dot{Q} = \left[\frac{\partial}{\partial x} \left(k \frac{\partial T}{\partial x} \right) + \frac{\partial}{\partial y} \left(k \frac{\partial T}{\partial y} \right) + \frac{\partial}{\partial z} \left(k \frac{\partial T}{\partial z} \right) \right] \Delta x \Delta y \Delta z \qquad (2.4.33)$$

With regards to the last term appearing in equation (2.4.25), heat generally may be added into the system due to a chemical reaction or removed from the system due to radiation heat transfer. For succinctness, the generic description of the source term is retained in this present derivation. Combining all the contributions based on equations (2.4.30) and (2.4.33), and substituting these expressions along with the time rate of change of energy E, from equation (2.4.26) into equation (2.4.25), the equation for the conservation of energy after dividing by the control volume $\Delta x \, \Delta y \, \Delta z$ is given as:

$$\rho \frac{DE}{Dt} = \frac{\partial(u\sigma_{xx})}{\partial x} + \frac{\partial(v\sigma_{yy})}{\partial y} + \frac{\partial(w\sigma_{zz})}{\partial z}$$

$$+ \frac{\partial(u\tau_{yx})}{\partial y} + \frac{\partial(u\tau_{zx})}{\partial z} + \frac{\partial(v\tau_{xy})}{\partial x} + \frac{\partial(v\tau_{zy})}{\partial z} + \frac{\partial(w\tau_{xz})}{\partial x} + \frac{\partial(w\tau_{yz})}{\partial y} \qquad (2.4.34)$$

$$+ \frac{\partial}{\partial x} \left(k \frac{\partial T}{\partial x} \right) + \frac{\partial}{\partial y} \left(k \frac{\partial T}{\partial y} \right) + \frac{\partial}{\partial z} \left(k \frac{\partial T}{\partial z} \right) + \dot{Q}_s$$

By applying the normal stresses described in equation (2.4.17), the preceding energy equation can be alternatively expressed as:

$$\rho \frac{DE}{Dt} = -\frac{\partial(up)}{\partial x} - \frac{\partial(vp)}{\partial y} - \frac{\partial(wp)}{\partial z} + \Phi + \frac{\partial}{\partial x} \left[k \frac{\partial T}{\partial x} \right] + \frac{\partial}{\partial y} \left[k \frac{\partial T}{\partial y} \right]$$

$$+ \frac{\partial}{\partial z} \left[k \frac{\partial T}{\partial z} \right] + \dot{Q}_s$$

$$(2.4.35)$$

The effects due to the viscous stresses in the energy equation are described by the dissipation function Φ that can be shown to be

$$\Phi = \frac{\partial(u\tau_{xx})}{\partial x} + \frac{\partial(u\tau_{yx})}{\partial y} + \frac{\partial(u\tau_{zx})}{\partial z}$$

$$+ \frac{\partial(v\tau_{xy})}{\partial x} + \frac{\partial(v\tau_{yy})}{\partial y} + \frac{\partial(v\tau_{zy})}{\partial z} + \frac{\partial(w\tau_{xz})}{\partial x} + \frac{\partial(w\tau_{yz})}{\partial y} + \frac{\partial(w\tau_{zz})}{\partial z} \qquad (2.4.36)$$

The dissipation function represents a source of energy due to work done deforming the fluid element. This work is extracted from the mechanical energy that causes fluid movement, which is subsequently converted into heat.

The specific energy E of a fluid can often be defined as the sum of the internal energy and kinetic energy. In three dimensions, the specific energy E can be defined as

$$E = \underbrace{e}_{\text{specific internal energy}} + \underbrace{\frac{1}{2}\left(u^2 + v^2 + w^2\right)}_{\text{kinetic energy}} \tag{2.4.37}$$

For compressible flows, equation (2.4.37) is often re-arranged to give an equation for the *enthalpy*. The sensible enthalpy h_s and the total enthalpy h of a fluid can be defined as

$$h_s = e + \frac{p}{\rho} \text{ and } h = h_s + \frac{1}{2}\left(u^2 + v^2 + w^2\right)$$

Combining these two definitions with the specific energy E, we obtain

$$h = e + \frac{p}{\rho} + \frac{1}{2}\left(u^2 + v^2 + w^2\right) = E + \frac{p}{\rho} \tag{2.4.38}$$

Substituting equation (2.4.38) into equation (2.4.34) and after some re-arrangement, the conservative partial differential form of the energy equation in terms of the total enthalpy h is given by

$$\frac{\partial(\rho h)}{\partial t} + \frac{\partial(\rho u h)}{\partial x} + \frac{\partial(\rho v h)}{\partial y} + \frac{\partial(\rho w h)}{\partial z} = \frac{\partial p}{\partial t} + \Phi$$

$$+ \frac{\partial}{\partial x}\left[k\frac{\partial T}{\partial x}\right] + \frac{\partial}{\partial y}\left[k\frac{\partial T}{\partial y}\right] + \frac{\partial}{\partial z}\left[k\frac{\partial T}{\partial z}\right] + \dot{Q}_s \tag{2.4.39}$$

2.4.4 Scalar Equation

The scalar equation can also be similarly formulated based on the derivation of the energy equation described in the previous section. The governing equation of any scalar property can simply be stated as the rate of change of scalar property equal to the net rate of scalar property added plus the rate of creation or destruction caused by external source.

The rate of increase of scalar property of the fluid element	=	The net rate of scalar property added to the fluid element	±	The rate of creation or destruction by external source on the fluid element

$$\underbrace{\qquad\qquad\qquad}_{\dot{R}_s\Delta V}$$

$$(2.4.40)$$

The time rate of change of any scalar property φ is essentially the same expression derived in equation (2.4.6). For the time rate of change of scalar property on a moving fluid element, it can be expressed as:

$$\rho\frac{D\varphi}{Dt}\Delta x\Delta y\Delta z \qquad\qquad (2.4.41)$$

In Figure 2.8, the net rate of scalar transfer to the fluid due to the transport of scalar property in the x direction is given by the difference between the inflow of mass scalar flux at surface at x and outflow of mass scalar flux at surface $x + \Delta x$. Similar considerations can also be applied to obtain the net rates of the transport of scalar property along the y direction and the z direction. The total rate of scalar property added to the fluid results in:

$$-\left[\frac{\partial J_x}{\partial x} + \frac{\partial J_y}{\partial y} + \frac{\partial J_z}{\partial z}\right]\Delta x\Delta y\Delta z \qquad\qquad (2.4.42)$$

Substituting equations (2.4.41) and (2.4.42) into equation (2.4.40), and after dividing by the control volume $\Delta x\,\Delta y\,\Delta z$, yields the equation for the conservation of scalar property—in other words,

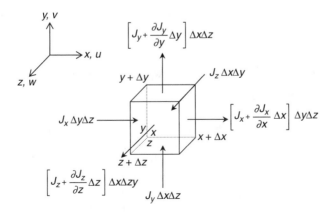

Figure 2.8 Scalar property added to the fluid along the Cartesian directions of x, y, and z.

$$\rho \frac{D\varphi}{Dt} = -\frac{\partial J_x}{\partial x} - \frac{\partial J_y}{\partial y} - \frac{\partial J_z}{\partial z} + \dot{R}_s \tag{2.4.43}$$

To formulate the appropriate scalar mass fluxes of J_x, J_y, and J_z in equation (2.4.42), the *Fick's law of diffusion* can be invoked, which relates the scalar mass flux to the local scalar property gradient:

$$J_x = -\rho D \frac{\partial \varphi}{\partial x} \quad J_y = -\rho D \frac{\partial \varphi}{\partial y} \quad J_z = -\rho D \frac{\partial \varphi}{\partial z} \tag{2.4.44}$$

where D is the mass diffusivity. Note the similarity between the scalar mass fluxes just formulated with the heat fluxes. Just as heat flows in the direction of decreasing temperature, the scalar property also experiences the same effect through diffusion in the direction of decreasing φ. The last term $\dot{R}_s$ appearing in equation (2.4.49) depicts the possibility of the increase or decrease of the scalar property within the fluid element due to some prescribed external sources. For example, the creation or destruction of chemical species in a combusting fire system such as in fire is due to the presence of chemical reactions. Here again, we shall retain the generic description of the source term in this present derivation for succinctness. By applying the scalar mass flux to the local scalar property gradients as described and substitution of these expressions into equation (2.4.44) and applying equation (2.4.6) yields the conservative partial differential form as

$$\frac{\partial(\rho\varphi)}{\partial t} + \frac{\partial(\rho u \varphi)}{\partial x} + \frac{\partial(\rho v \varphi)}{\partial y} + \frac{\partial(\rho w \varphi)}{\partial z} =$$
$$\frac{\partial}{\partial x}\left(\rho D \frac{\partial \varphi}{\partial x}\right) + \frac{\partial}{\partial y}\left(\rho D \frac{\partial \varphi}{\partial y}\right) + \frac{\partial}{\partial z}\left(\rho D \frac{\partial \varphi}{\partial z}\right) + \dot{R}_s \tag{2.4.45}$$

2.5 Differential and Integral Forms of the Transport Equations

Based on the derivation of the fundamental equations just described, let us collate the conservative form of the set of equations, as summarized in Table 2.1, which governs the time-dependent three-dimensional fluid flow and heat transfer of a *compressible* Newtonian fluid in a combusting fire system.

It can be observed from Table 2.1 that there are commonalities between the various equations. If we employ the general variable ϕ as introduced in

Table 2.1 Governing equations of the flow of a compressible Newtonian fluid in Cartesian coordinates.

Mass

$$\frac{\partial \rho}{\partial t} + \frac{\partial (\rho u)}{\partial x} + \frac{\partial (\rho v)}{\partial y} + \frac{\partial (\rho w)}{\partial z} = 0$$

x-Momentum

$$\frac{\partial (\rho u)}{\partial t} + \frac{\partial (\rho u u)}{\partial x} + \frac{\partial (\rho v u)}{\partial y} + \frac{\partial (\rho w u)}{\partial z} = \frac{\partial}{\partial x}\left[\mu \frac{\partial u}{\partial x}\right] + \frac{\partial}{\partial y}\left[\mu \frac{\partial u}{\partial y}\right] + \frac{\partial}{\partial z}\left[\mu \frac{\partial u}{\partial z}\right] + S_{M_x}$$

y-Momentum

$$\frac{\partial (\rho v)}{\partial t} + \frac{\partial (\rho u v)}{\partial x} + \frac{\partial (\rho v v)}{\partial y} + \frac{\partial (\rho w v)}{\partial z} = \frac{\partial}{\partial x}\left[\mu \frac{\partial v}{\partial x}\right] + \frac{\partial}{\partial y}\left[\mu \frac{\partial v}{\partial y}\right] + \frac{\partial}{\partial z}\left[\mu \frac{\partial v}{\partial z}\right] + S_{M_y}$$

z-Momentum

$$\frac{\partial (\rho w)}{\partial t} + \frac{\partial (\rho u w)}{\partial x} + \frac{\partial (\rho v w)}{\partial y} + \frac{\partial (\rho w w)}{\partial z} = \frac{\partial}{\partial x}\left[\mu \frac{\partial w}{\partial x}\right] + \frac{\partial}{\partial y}\left[\mu \frac{\partial w}{\partial y}\right] + \frac{\partial}{\partial z}\left[\mu \frac{\partial w}{\partial z}\right] + S_{M_z}$$

Enthalpy

$$\frac{\partial (\rho h)}{\partial t} + \frac{\partial (\rho u h)}{\partial x} + \frac{\partial (\rho v h)}{\partial y} + \frac{\partial (\rho w h)}{\partial z} = \frac{\partial}{\partial x}\left[k \frac{\partial T}{\partial x}\right] + \frac{\partial}{\partial y}\left[k \frac{\partial T}{\partial y}\right] + \frac{\partial}{\partial z}\left[k \frac{\partial T}{\partial z}\right] + S_b$$

Scalar Property

$$\frac{\partial (\rho \varphi)}{\partial t} + \frac{\partial (\rho u \varphi)}{\partial x} + \frac{\partial (\rho v \varphi)}{\partial y} + \frac{\partial (\rho w \varphi)}{\partial z} = \frac{\partial}{\partial x}\left(\rho D \frac{\partial \varphi}{\partial x}\right) + \frac{\partial}{\partial y}\left(\rho D \frac{\partial \varphi}{\partial y}\right) + \frac{\partial}{\partial z}\left(\rho D \frac{\partial \varphi}{\partial z}\right) + S_\varphi$$

section 2.4.2 and expressing all the fluid flow equations, including equations of enthalpy and scalar quantities, the conservative compressible form of the governing equation can usually be written as

$$\frac{\partial(\rho\phi)}{\partial t} + \frac{\partial(\rho u\phi)}{\partial x} + \frac{\partial(\rho v\phi)}{\partial y} + \frac{\partial(\rho w\phi)}{\partial z} = \frac{\partial}{\partial x}\left[\Gamma_\phi\frac{\partial\phi}{\partial x}\right] + \frac{\partial}{\partial y}\left[\Gamma_\phi\frac{\partial\phi}{\partial y}\right]$$
$$+ \frac{\partial}{\partial z}\left[\Gamma_\phi\frac{\partial\phi}{\partial z}\right] + S_\phi$$

(2.5.1)

Equation (2.5.1) is aptly known as the transport equation for any variable property ϕ. It illustrates the various physical transport processes occurring in the fluid flow: the rate of change of ϕ, which is the *local acceleration* term, accompanied by the *advection* terms on the left-hand side is respectively equivalent to the *diffusion* term (Γ_ϕ is designated as the diffusion coefficient) and the *source* term (S_ϕ) on the right-hand side. In order to bring forth the common features, terms that are not shared between the equations are placed into the source terms. It is noted that the additional source terms in the momentum equations S_{M_x}, S_{M_y} and S_{M_z} comprised of the pressure and non-pressure gradient terms and other possible sources such as gravity that influence the fluid motion.

For *incompressible* and *weakly compressible* flows, it is common practice to transform the energy equation by replacing the heat flux according to the local enthalpy gradient instead of the temperature gradient—in other words,

$$q_x = -\frac{k}{C_p}\frac{\partial h}{\partial x} \qquad q_y = -\frac{k}{C_p}\frac{\partial h}{\partial y} \qquad q_z = -\frac{k}{C_p}\frac{\partial h}{\partial z}$$

(2.5.2)

where C_p is the specific heat of constant pressure. Employing the preceding heat fluxes, the enthalpy equation can be alternatively expressed as

$$\frac{\partial(\rho h)}{\partial t} + \frac{\partial(\rho u h)}{\partial x} + \frac{\partial(\rho v h)}{\partial y} + \frac{\partial(\rho w h)}{\partial z} = \frac{\partial}{\partial x}\left[\frac{k}{C_p}\frac{\partial h}{\partial x}\right] + \frac{\partial}{\partial y}\left[\frac{k}{C_p}\frac{\partial h}{\partial y}\right]$$
$$+ \frac{\partial}{\partial z}\left[\frac{k}{C_p}\frac{\partial h}{\partial z}\right] + S_h$$

(2.5.3)

which now shares the same form as equation (2.5.1). The additional source term S_h in the equation (2.5.3) contains the time derivative of the pressure, dissipation function given by equation (2.4.36), and heat sources or sinks affecting the enthalpy within the reacting flow system. In field modeling, the energy equation in the form of equation (2.5.3) is usually adopted. Note that the kinetic energy $\frac{1}{2}(u^2 + v^2 + w^2)$ in the definition of enthalpy, the pressure work term $\partial p / \partial t$, and the dissipation function that represents the source of energy due to work done deforming the fluid element are usually ignored in most practical applications.

By setting the transport property ϕ equal to 1, u, v, w, h, ϕ and selecting appropriate values for the diffusion coefficient Γ_ϕ and source terms S_ϕ, we obtain the special forms presented in Table 2.2 for each of the partial differential equations for the conservation of mass, momentum, energy, and scalar property. Although we have systematically walked through the derivation of the complete set of governing equations in detail from basic conservation principles, the final general form pertaining to the fluid motion, heat transfer, and so forth conforms simply to the generic form of equation (2.5.1). This equation is of enormous significance, as it allows increasing complexity of physical processes to be accommodated within the CFD framework for solving more complicated problems.

Table 2.2 General form of governing equations for compressible flow in Cartesian coordinates.

ϕ	Γ_ϕ	S_ϕ
1	0	0
u	μ	$-\dfrac{1}{\rho}\dfrac{\partial p}{\partial x} + \dfrac{\partial}{\partial x}\left[\mu\dfrac{\partial u}{\partial x}\right] + \dfrac{\partial}{\partial y}\left[\mu\dfrac{\partial v}{\partial x}\right] + \dfrac{\partial}{\partial z}\left[\mu\dfrac{\partial w}{\partial x}\right]$ $+ \dfrac{\partial}{\partial x}\left[\lambda\left(\dfrac{\partial u}{\partial x}+\dfrac{\partial v}{\partial y}+\dfrac{\partial w}{\partial z}\right)\right] + \sum F_x^{body\ forces}$
v	μ	$-\dfrac{1}{\rho}\dfrac{\partial p}{\partial y} + \dfrac{\partial}{\partial x}\left[\mu\dfrac{\partial u}{\partial y}\right] + \dfrac{\partial}{\partial y}\left[\mu\dfrac{\partial v}{\partial y}\right] + \dfrac{\partial}{\partial z}\left[\mu\dfrac{\partial w}{\partial y}\right]$ $+ \dfrac{\partial}{\partial y}\left[\lambda\left(\dfrac{\partial u}{\partial x}+\dfrac{\partial v}{\partial y}+\dfrac{\partial w}{\partial z}\right)\right] + \sum F_y^{body\ forces}$
w	μ	$-\dfrac{1}{\rho}\dfrac{\partial p}{\partial z} + \dfrac{\partial}{\partial x}\left[\mu\dfrac{\partial u}{\partial z}\right] + \dfrac{\partial}{\partial y}\left[\mu\dfrac{\partial v}{\partial z}\right] + \dfrac{\partial}{\partial z}\left[\mu\dfrac{\partial w}{\partial z}\right]$ $+ \dfrac{\partial}{\partial z}\left[\lambda\left(\dfrac{\partial u}{\partial x}+\dfrac{\partial v}{\partial y}+\dfrac{\partial w}{\partial z}\right)\right] + \sum F_z^{body\ forces}$
h	$\dfrac{k}{C_p}$	$\dot{Q}_s$
ϕ	ρD	$\dot{R}_s$

In order to numerically solve the approximate form of equation (2.5.1), it is convenient to consider the integral form of this generic transport equation over a finite control volume. The advantage of adopting such a form will be covered in more detail in section 2.7. Integration of this equation over a three-dimensional control volume V yields

$$\int_V \frac{\partial(\rho\phi)}{\partial t} dV + \int_V \left\{ \frac{\partial(\rho u\phi)}{\partial x} + \frac{\partial(\rho v\phi)}{\partial y} + \frac{\partial(\rho w\phi)}{\partial z} \right\} dV$$
$$= \int_V \left\{ \frac{\partial}{\partial x}\left[\Gamma\frac{\partial\phi}{\partial x}\right] + \frac{\partial}{\partial y}\left[\Gamma\frac{\partial\phi}{\partial y}\right] + \frac{\partial}{\partial z}\left[\Gamma\frac{\partial\phi}{\partial z}\right] \right\} dV + \int_V S_\phi dV \qquad (2.5.4)$$

By applying Gauss' divergence theorem to the volume integral, the second term on the left-hand side—for example, the spatial derivative along the x direction—can be rewritten as

$$\int_V \frac{\partial(\rho u\phi)}{\partial x} dV = \int_A (\rho u\phi) dA^x \qquad (2.5.5)$$

where dA^x is the elemental projected area along the x direction. The projected area is positive if their outward normal vector from the volume surface is directed in the same direction along the Cartesian coordinate system; otherwise it is negative. Similarly, the first term on the right side for the spatial derivative along the x direction according to Gauss' divergence theorem can be obtained according to

$$\int_V \frac{\partial}{\partial x}\left[\Gamma\frac{\partial\phi}{\partial x}\right] dV = \int_A \left[\Gamma\frac{\partial\phi}{\partial x}\right] dA^x \qquad (2.5.6)$$

The surface integrals along the y and z directions are also attained in exactly the same fashion. Based on the preceding surface integrals, equation (2.5.4) can now be written as:

$$\int_V \frac{\partial(\rho\phi)}{\partial t} dV + \int_A \{ (\rho u\phi) dA^x + (\rho v\phi) dA^y + (\rho w\phi) dA^z \}$$
$$= \int_A \left\{ \left[\Gamma\frac{\partial\phi}{\partial x}\right] dA^x + \left[\Gamma\frac{\partial\phi}{\partial y}\right] dA^y + \left[\Gamma\frac{\partial\phi}{\partial z}\right] dA^z \right\} + \int_V S_\phi dV \qquad (2.5.7)$$

2.6 Physical Interpretation of Boundary Conditions for Field Modeling

The mass, momentum, and energy expressed as the preceding enthalpy and scalar property equations govern the flow and heat transfer of a fluid in a combusting fire system. Boundary conditions, and sometimes initial conditions, strongly dictate the particular solutions to be obtained from the governing equations. This creates particular significance, especially in field modeling, as any numerical solution of the governing equations must result in a strong and compelling numerical representation of the specification of appropriate boundary conditions.

For the velocities, the *no-slip* condition on solid boundaries is described. For this boundary condition, the solid surface can be assumed to have zero relative velocity between the surface and the fluid immediately at the surface. If the surface is stationary, all the velocity components along the Cartesian coordinate directions can be taken to be zero—in other words,

$$u = v = w = 0 \qquad \text{at the surface} \tag{2.6.1}$$

The condition for most flows at *inflow* boundaries requires at least one velocity component to be given for the solution of the governing equations for any transport property ϕ. This can be provided by the *Dirichlet* boundary condition on the velocity—for example, in the x direction, the boundary condition is given by:

$$u = f \quad \text{and} \quad v = w = 0 \qquad \text{at the inflow boundary} \tag{2.6.2}$$

where f can either be specified as a constant value or a velocity profile at the surface. From a computational perspective, *Dirichlet* boundary conditions can be applied rather accurately as long as f is continuous. It is important that outflow boundaries need to be positioned at locations where the flow is approximately unidirectional and where surface stresses can take known values. In a fully developed flow, the velocity component in the direction across the boundary remains unchanged, and by satisfying stress continuity, the shear forces along the surface are taken to be zero to provide the following outflow condition:

$$\frac{\partial u}{\partial n} = \frac{\partial v}{\partial n} = \frac{\partial w}{\partial n} = 0 \qquad \text{at the outflow boundary} \tag{2.6.3}$$

where n is the direction normal to the surface. This condition is commonly known as the *Neumann* boundary condition. Physically, in reference to the continuity equation (2.4.4), it is clear that the appropriate boundary conditions (2.6.1), (2.6.2), and (2.6.3) imposed at any location on the surface walls close the system mathematically and satisfy local and overall mass conservation.

For other transport variables besides velocities, a *no-slip* condition can also be analogously imposed for the rest of the transport variables. Take, for example, the temperature at the wall surface. If the material temperature of the surface is at some temperature designated by T_w, then the temperature fluid layer immediately in contact with the surface should also be T_w. For a given problem where the wall temperature is known, the *Dirichlet* boundary condition applies and the fluid temperature is

$$T = T_w \qquad \text{at the wall} \tag{2.6.4}$$

The *Dirichlet* boundary condition equally applies for the enthalpy and scalar property to yield

$$h = h_w; \quad \varphi = \varphi_w \qquad \text{at the wall} \tag{2.6.5}$$

However, if the wall temperature is not known (e.g., if the temperature is changing as a function of time due to the heat transfer to or from the surface), then Fourier's law of heat condition can be applied to provide the necessary boundary condition at the surface. If we denote the instantaneous wall heat flux as q_w, then, according to Fourier's law,

$$q_w = -\left(k\frac{\partial T}{\partial n}\right)_w \qquad \text{at the wall} \tag{2.6.6}$$

Here, the changing surface temperature T_w is responding to the thermal response of the wall material through the heat transfer to the wall q_w. This condition can also be similarly applied for the enthalpy as

$$q_w = -\left(\frac{k}{C_p}\frac{\partial h}{\partial n}\right)_w \qquad \text{at the wall} \tag{2.6.7}$$

Equations (2.6.6) and (2.6.7), as far as the flow is concerned, are boundary conditions for the temperature and enthalpy gradients at the wall. For the scalar property, Fick's law of diffusion can be applied to provide the necessary boundary condition at the surface. Denoting the instantaneous wall mass flux as J_w, then, according to Fick's law,

$$J_w = -\left(\rho D\frac{\partial \varphi}{\partial n}\right)_w \qquad \text{at the wall} \tag{2.6.8}$$

For the case where there is no heat and mass transfer to the surface, the proper boundary condition comes from equations (2.6.6) and (2.6.7) with $q_w = 0$ and from equation (2.6.8) with $J_w = 0$; hence,

$$\left(\frac{\partial T}{\partial n}\right)_w = \left(\frac{\partial h}{\partial n}\right)_w = \left(\frac{\partial \varphi}{\partial n}\right)_w = 0 \qquad \text{at the wall} \qquad (2.6.9)$$

This condition immediately falls in line with the *Neumann* boundary condition for the velocity at the outflow boundaries. On the inflow and outflow boundaries of the flow domain, it is common to have the variables specified at the inflow boundary and the zero normal gradients to be adopted at the outflow boundary.

Other boundary conditions that are also of importance and often required for field modeling include the open, symmetry, and periodic boundary conditions. For the free-standing fire case, the far free-stream boundary requires the application of an open boundary, which simply states that the normal gradient of any of the transport property ϕ is zero—that is,

$$\left(\frac{\partial \phi}{\partial n}\right)_w = 0 \qquad \text{at the open boundary} \qquad (2.6.10)$$

Gresho and Sani (1991) reviewed the intricacies of open boundary conditions and state that there are some *theoretical concerns* regarding this boundary condition. However, its success in CFD practice left them to recommend it as the simplest and cheapest method when compared with theoretically more satisfying selections. Furthermore, the symmetric boundary condition can be employed to take advantage of special geometrical features of the solution region. This boundary condition can be imposed by prescribing the normal velocity at the surface and the normal gradients of the other velocity components to be zero. *Neumann* boundary condition is subsequently applied for the rest of the variables. For the periodic boundary condition, the transport property of one of the surface ϕ_1 is taken to be equivalent to the transport property of the second surface ϕ_2—that is, $\phi_1 = \phi_2$, depending on which two surfaces of the flow domain experience periodicity. The application of the aforementioned boundary conditions for specific fire cases will be described in more detail in section 2.7.3.

2.7 Numerical Approximations of Transport Equations for Field Modeling

There are four discretisation methods that are currently available in the main stream of CFD: finite difference, finite element, spectral, and finite volume methods.

The finite difference method is believed to be the oldest of the numerical methods. Developed by Euler in 1768, it was used to obtain numerical solutions to partial differential equations by hand calculation. The basic idea of this method stems from the consideration of Taylor series expansions being employed at each nodal point of the grid to generate appropriate finite difference expressions to approximate the partial derivatives of the governing equations. These derivatives, replaced by finite difference approximations, yield an algebraic equation for the flow solution at each grid point. This method is generally more suited for *structured* grids, since it requires a mesh having a high degree of regularity.

The finite element method requires the application of simple piecewise polynomial functions that are employed on local elements to describe the variations of the unknown flow variables. The concept of weighted residuals is introduced to measure the errors associated with the approximate functions, which are later minimized. A set of non-linear algebraic equations for the unknown terms of the approximating functions is solved, hence yielding the flow solution. The finite element method has not enjoyed extensive usage in CFD despite its ability in handling *unstructured* grids of arbitrary geometries. It has generally been found that the finite element method requires greater computational resources and computer processing power than the equivalent finite volume method.

The spectral method employs the same general approach as the finite difference and finite element methods where the unknowns of the governing equations are replaced with a truncated series. The difference is that where the previous two methods employ local approximations, the spectral method uses global approximation that is either by means of a truncated Fourier series or a series of Chebyshev polynomials for the entire flow domain. The discrepancy between the exact solution and the approximation is dealt with by using a weighted residuals concept similar to the finite element method.

The finite volume method, like the finite element method, has the ability of handling arbitrary geometries with ease. It can be applied to *structured* as well as *unstructured* meshes. The latter is gaining in popularity and usage especially in the majority of commercial CFD codes. Be definition, *structured* mesh is a mesh containing cells having either a regular-shaped element with four-nodal corner points in two dimensions or a hexahedral-shaped element with eight-nodal corner points in three dimensions. *Unstructured* mesh can, however, be described as a mesh overlaying with cells in the form of either a triangle-shaped element in two dimensions or a tetrahedron-shaped in three dimensions. More importantly, this method bears many similarities to the finite difference method and is simple to apply. The many advantages it represents, as well as the consistency of the concept of the control volume approach with the finite volume method, favor its application to field modeling. The integrated form of the generic transport equation (2.4.7) through the finite volume method is elaborated in the next section.

2.7.1 Discretisation Methods

2.7.1.1 Steady Flows

The basis for computational procedures in the finite volume method can be illustrated by simplifying equation (2.5.7) into its integral form of a steady transport equation of property ϕ:

$$
\underbrace{\int_A \{(\rho u\phi)dA^x + (\rho v\phi)dA^y + (\rho w\phi)dA^z\}}_{advection}
$$

$$
= \underbrace{\int_A \left\{ \left[\Gamma\frac{\partial\phi}{\partial x}\right]dA^x + \left[\Gamma\frac{\partial\phi}{\partial y}\right]dA^y + \left[\Gamma\frac{\partial\phi}{\partial z}\right]dA^z \right\}}_{diffusion} + \underbrace{\int_V S_\phi dV}_{source}
$$

(2.7.1)

In essence, the finite volume method discretises the integral form of the conservation equations directly in the physical space. Consider the physical geometry to be subdivided into a number of finite contiguous control volumes, where the resulting statements express the exact conservation of relevant properties for each of the control volumes. At the centroid of these volumes, the property ϕ is calculated. The cornerstone of the finite volume method is the *control volume integration*. In a control volume, the bounding surface areas of the element are directly linked to the discretisation of the advection and diffusion terms for ϕ in equation (2.7.1). The discretised form of the advection term from which the surfaces, fluxes are determined at the control volume faces is given by

$$
\int_A \{(\rho u\phi)dA^x + (\rho v\phi)dA^y + (\rho w\phi)dA^z\}
$$

$$
= \sum_{i=1}^{N} (\rho u\phi)_i A_i^x + \sum_{j=1}^{N} (\rho v\phi)_j A_j^y + \sum_{k=1}^{N} (\rho w\phi)_k A_k^z
$$

(2.7.2)

where N is the number of surfaces bounding the control volume. Similarly, the algebraic form of the surface fluxes of the diffusion term is:

$$
\int_A \left\{ \left[\Gamma\frac{\partial\phi}{\partial x}\right]dA^x + \left[\Gamma\frac{\partial\phi}{\partial y}\right]dA^y + \left[\Gamma\frac{\partial\phi}{\partial z}\right]dA^z \right\}
$$

$$
= \sum_{i=1}^{N} \left(\Gamma\frac{\partial\phi}{\partial x}\right)_i A_i^x + \sum_{j=1}^{N} \left(\Gamma\frac{\partial\phi}{\partial y}\right)_j A_j^y + \sum_{k=1}^{N} \left(\Gamma\frac{\partial\phi}{\partial z}\right)_k A_k^z
$$

(2.7.3)

The source term can be approximated accordingly as

$$\int_V S_\phi dV = S_\phi \Delta V \tag{2.7.4}$$

For the purpose of illustration, let us consider the three-dimensional structured grid arrangement as depicted in Figure 2.9 where the centroid of the central control volume indicated by the point P is surrounded by six adjacent control volumes having their respective centroids indicated by the central points: east, E; west, W; north, N; south, S; top, T; and bottom, B. The control volume face between points P and E is separated by the area A_e^x. Subsequently, the rest of the control volume faces are respectively A_w^x, A_n^y, A_s^y, A_t^z, and A_b^z. The surface fluxes along the x direction of the advection term in equation (2.7.2) are:

$$\sum_{i=1}^{6} (\rho u \phi)_i A_i^x = (\rho u \phi)_e A_e^x - (\rho u \phi)_w A_w^x$$
$$+ (\rho u \phi)_n \overset{=0}{A_n^x} - (\rho u \phi)_s \overset{=0}{A_s^x} + (\rho u \phi)_t \overset{=0}{A_t^x} - (\rho u \phi)_b \overset{=0}{A_b^x} \tag{2.7.5}$$

Note that for areas on the control volume sides of w, s, and b, the projected areas are negative, since their outward normal vectors are directed in the opposite directions to the Cartesian coordinate system. The projection areas of A_n^x,

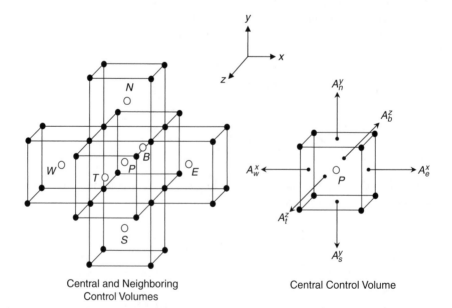

Central and Neighboring   Central Control Volume
Control Volumes

Figure 2.9   Schematic illustration of a three-dimensional structured grid arrangement.

A_s^x, A_t^x, and A_b^x are essentially zero for the structured uniform grid arrangement. For the surface fluxes along the y and z directions in equation (2.51), they are respectively given by:

$$\sum_{j=1}^{6}(\rho v\phi)_j A_j^y = (\rho v\phi)_e^{\nearrow=0} A_e^y - (\rho v\phi)_w^{\nearrow=0} A_w^y$$

$$+(\rho v\phi)_n A_n^y - (\rho v\phi)_s A_s^y + (\rho u\phi)_t^{\nearrow=0} A_t^y - (\rho u\phi)_b^{\nearrow=0} A_b^y \tag{2.7.6}$$

and

$$\sum_{k=1}^{6}(\rho w\phi)_k A_k^z = (\rho v\phi)_e^{\nearrow=0} A_e^z - (\rho v\phi)_w^{\nearrow=0} A_w^z$$

$$+ (\rho u\phi)_n^{\nearrow=0} A_n^z - (\rho u\phi)_s^{\nearrow=0} A_s^z (\rho w\phi)_t A_t^z - (\rho w\phi)_b A_b^z \tag{2.7.7}$$

The discretised form of the advection term for a structured grid arrangement is thus given by:

$$\sum_{i=1}^{6}(\rho u\phi)_i A_i^x + \sum_{j=1}^{6}(\rho v\phi)_j A_j^y + \sum_{k=1}^{6}(\rho w\phi)_k A_k^z =$$
$$(\rho u\phi)_e A_e^x - (\rho u\phi)_w A_w^x + (\rho v\phi)_n A_n^y - (\rho v\phi)_s A_s^y + (\rho w\phi)_t A_t^z - (\rho w\phi)_b A_b^z \tag{2.7.8}$$

Similar to equation (2.7.8), the discretised form of the diffusion term for a structured grid arrangement in equation (2.7.3) can be ascertained as

$$\sum_{i=1}^{6}\left(\Gamma\frac{\partial\phi}{\partial x}\right)_i A_i^x + \sum_{j=1}^{6}\left(\Gamma\frac{\partial\phi}{\partial y}\right)_j A_j^y + \sum_{k=1}^{6}\left(\Gamma\frac{\partial\phi}{\partial z}\right)_k A_k^z$$

$$= \left(\Gamma\frac{\partial\phi}{\partial x}\right)_e A_e^x - \left(\Gamma\frac{\partial\phi}{\partial x}\right)_w A_w^x + \left(\Gamma\frac{\partial\phi}{\partial y}\right)_n A_n^y - \left(\Gamma\frac{\partial\phi}{\partial y}\right)_s A_s^y \tag{2.7.9}$$

$$+ \left(\Gamma\frac{\partial\phi}{\partial z}\right)_t A_t^z - \left(\Gamma\frac{\partial\phi}{\partial z}\right)_b A_b^z$$

As the finite volume method works with control volumes and not the grid intersection points, it has the capacity to accommodate any type of grid. Here, instead of structured grids, unstructured grids that usually comprise of triangular elements in the case of two dimensions or a tetrahedra element in three dimensions can be readily employed. This thereby allows a large number of options for the definition of the shape and location of the control volumes. For such types of grids, computations are now required to evaluate all appropriate projection areas in their respective Cartesian coordinate directions: A_i^x, A_j^y, and A_k^z.

The first order derivatives at the control volume faces in equation (2.7.9) can usually be approximated from the discrete ϕ values of the surrounding elements. For example, in a structured mesh arrangement as shown in Figure 2.9 where the central control volume is surrounded by only one adjacent control volume at each face, the discrete form of the first order derivatives could be obtained by imposing a piecewise linear gradient profile between the central and adjacent nodes. By denoting the distance between points W and P as δx_W and points P and E as δx_E, the diffusive fluxes at the control volume faces e and w along the x direction can be evaluated as

$$\left(\Gamma\frac{\partial\phi}{\partial x}\right)_e A_e^x = \Gamma_e\left(\frac{\phi_E - \phi_P}{\delta x_E}\right)A_e^x; \qquad \left(\Gamma\frac{\partial\phi}{\partial x}\right)_w A_w^x = \Gamma_w\left(\frac{\phi_P - \phi_W}{\delta x_W}\right)A_w^x$$

$$(2.7.10)$$

Other diffusive fluxes at other bounding surfaces of the control volume can also be similarly obtained according to the preceding formulae—in other words,

$$\left(\Gamma\frac{\partial\phi}{\partial y}\right)_n A_n^y = \Gamma_n\left(\frac{\phi_N - \phi_P}{\delta y_N}\right)A_n^y; \qquad \left(\Gamma\frac{\partial\phi}{\partial y}\right)_s A_s^y = \Gamma_s\left(\frac{\phi_P - \phi_S}{\delta y_S}\right)A_s^y$$

$$(2.7.11)$$

$$\left(\Gamma\frac{\partial\phi}{\partial z}\right)_t A_t^z = \Gamma_t\left(\frac{\phi_T - \phi_P}{\delta z_T}\right)A_t^z; \qquad \left(\Gamma\frac{\partial\phi}{\partial z}\right)_b A_b^z = \Gamma_b\left(\frac{\phi_P - \phi_B}{\delta z_B}\right)A_b^z$$

$$(2.7.12)$$

For simplicity, values for the interface diffusion coefficients Γ_e, Γ_w, Γ_n, Γ_s, Γ_t, and Γ_b could be approximated by linear interpolation. If needed, higher-order quadratic profiles may be employed to attain higher *accuracy* for the numerical solution, which nevertheless require the inclusion of more surrounding elemental volumes in addition to the neighboring control volumes presented in Figure 2.9.

The principal problem in the discretisation of the advection term in equation (2.7.8) is the calculation of the interface values of property ϕ at control volume faces and its convective fluxes across these boundaries. For the latter, special treatment is required to evaluate the convective fluxes in order to ensure conservation of mass, which will be elucidated in the next section. A suitable interpolation procedure is usually used to express the variable values at the control volume surface in terms of the central and neighboring values. In hindsight, it appears rather straightforward, but the *stability* of the numerical solution has

been found to be strongly dependent on the flow direction. The various interpolation schemes are explored following.

Commonly-used schemes

Consider the uniform one-dimensional structured grid arrangement as depicted
in Figure 2.10. The interface values of property ϕ at the control volume faces
of e and w are designated by ϕ_w and ϕ_e. As a direct consequence of piecewise
linear gradient profiles applied to approximate the first order gradients in the
diffusion term as just described, it seems rather sensible that linear interpolation could also be realized between the central and neighboring nodes. The
interface values ϕ_w and ϕ_e can be determined as

$$\phi_w = \frac{1}{2}(\phi_W + \phi_P); \qquad \phi_e = \frac{1}{2}(\phi_P + \phi_E) \tag{2.7.13}$$

The preceding approximation is second order accurate, and this interpolation
procedure, commonly known as the *central differencing scheme*, does not exhibit

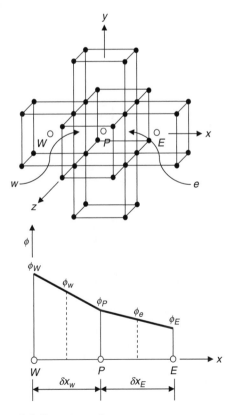

Figure 2.10   The central differencing scheme.

any bias on the flow direction. It has been well documented in literature (Patankar, 1980, Versteeg and Malasekera, 1995) that the inadequacy of this scheme in a strongly convective flow is its inability to identify the flow direction. The preceding treatment usually results in large "undershoots" and "overshoots" in some flow problems, eventually causing the numerical procedure to diverge. In some circumstances, it may yield non-physical solutions. Increasing the mesh resolution for the computational domain with very small grid spacing could possibly overcome the problem. Such an approach, however, usually precludes practical flow calculations to be carried out robustly and effectively in practice.

To overcome the problem due to central differencing, much emphasis has been placed in developing an array of interpolation schemes that accommodate some recognition of the flow direction. Consider for a moment the flow moving across three control volumes from the upstream node W (left) to the downstream node E (right) as illustrated in Figure 2.11. Through the central differencing approximation, the interface values of ϕ are always assumed to be weighted by the influence of the available variables at the neighboring grid

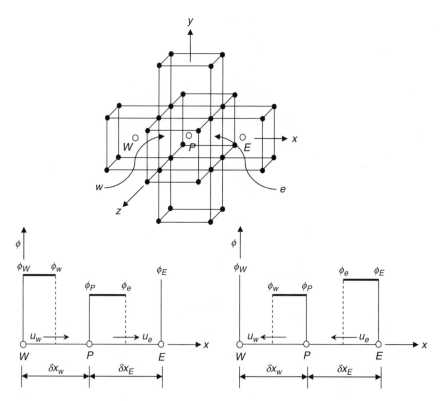

Figure 2.11   The upwind differencing scheme.

nodal points; the downstream values of ϕ_P and ϕ_E are always required during the evaluation of ϕ_w and ϕ_e. These values are usually not known *a priori* in the majority of flow cases. By exerting an unequal weighting influence based on the available variables located at the surrounding nodes, a numerical solution can thus be designed to recognize the direction of the flow in order to appropriately determine the interface values. This is essentially the hallmark of the *upwind* or *donor-cell* concept, which is described as follows. If the interface velocities along the Cartesian x direction are positive: $u_w > 0$ and $u_e > 0$, the interface values ϕ_w and ϕ_e according to the *donor-cell* concept can be approximated according to their upstream neighboring counterparts as

$$\phi_w = \phi_W; \qquad \phi_e = \phi_P \qquad\qquad (2.7.14)$$

Similarly, if the interface velocities are negative: $u_w < 0$ and $u_e < 0$, the interface values ϕ_w and ϕ_e are conversely evaluated by

$$\phi_w = \phi_P; \qquad \phi_e = \phi_E \qquad\qquad (2.7.15)$$

This scheme, known as the *upwind scheme*, promotes numerical stability; satisfies transportiveness (flow direction); boundness (diagonally dominant matrix coefficients ensuring numerical convergence, which will be discussed in the next section); and conservativeness (fluxes that are represented in a consistent manner). Albeit its simplicity, this scheme is only first order accurate.

In order to improve the solution accuracy, Spalding (1972) developed a scheme that combines the central and upwind differencing schemes by employing piecewise formulae based on the local Peclet number Pe. By definition, the Peclet number determines the relative contribution of the convective and diffusive fluxes. This so-called *hybrid differencing scheme* retains a second order accuracy for small Peclet numbers due to central differencing but reverts to the first order upwind differencing for large Peclet numbers. For the case where the interface velocities are positive: $u_w > 0$ and $u_e > 0$, the interface values ϕ_w and ϕ_e according to the hybrid differencing formulae are given by

$$\phi_w = \frac{1}{2}(\phi_W + \phi_P); \quad \phi_e = \frac{1}{2}(\phi_P + \phi_E) \quad \text{for } Pe_w, Pe_e < 2 \qquad (2.7.16)$$

$$\phi_w = \phi_W; \qquad\qquad \phi_e = \phi_P \qquad\qquad \text{for } Pe_w, Pe_e \geq 2 \qquad (2.7.17)$$

where $Pe_w = \dfrac{(\rho u)_w}{\Gamma_w/\delta x_W}$ and $Pe_e = \dfrac{(\rho u)_e}{\Gamma_e/\delta x_E}$

Following equations (2.7.13) and (2.7.15), similar considerations can be obtained for the respective interface values of ϕ_w and ϕ_e when the interface velocities u_w and u_e are in the opposite direction or negative. Like upwind differencing, this scheme is highly stable, satisfies transportiveness, and produces physically realistic solutions. It has been widely used in many computational procedures such as in the commercial CFD code CFX4.4; this scheme is used as the default scheme for flow calculations. Despite the exploitation of the favorable properties of the upwind and central differencing schemes, the accuracy of this scheme is still only first order, since in most cases of real practical flows, such as of burning fires, the majority of the local Peclet numbers will be greater than 2 due to the large flow velocities that exist within the flow system.

Another popular scheme that is considered to yield better results than the hybrid scheme is the *power-law differencing scheme* of Patankar (1980). Here, the upwind differencing becomes effective only when $Pe > 10$. For example, the interface values ϕ_w and ϕ_e according to the power-law differencing formulae for the respective interface velocities $u_w > 0$ and $u_e > 0$ can be determined as

$$\phi_w = (1 - \chi_w)\phi_W + \chi_w\phi_P; \quad \phi_e = (1 - \chi_e)\phi_P + \chi_e\phi_E \text{ for } 0 < Pe_w, Pe_e < 10$$

$$(2.7.18)$$

where $\chi_w = (1 - 0.1Pe_w)^5/Pe_w$ and $\chi_e = (1 - 0.1Pe_e)^5/Pe_e$

$$\phi_w = \phi_W; \quad \phi_e = \phi_P \quad \text{for } Pe_w, Pe_e \geq 10 \tag{2.7.19}$$

The power-law differencing scheme possesses similar properties to the hybrid scheme. It has also enjoyed much extensive usage in practical flow calculations and can be used as an alternative to the hybrid scheme. Commercial CFD code such as FLUENT 6.1.22 uses this scheme as the default scheme for flow calculations. The necessary interface values for property ϕ at the other control volume faces can be similarly obtained according to the preceding evaluation.

The inherent first order accuracy in all of the preceding schemes due to the consideration of the upwind concept makes them prone to unwanted numerical diffusion errors. In order to reduce these numerical errors, high order approximations such as the *second-order upwind differencing scheme* and *third-order QUICK (Quadratic Upstream Interpolation for Convective Kinetics) scheme* of Lenoard (1979) that are widely applied in many CFD problems, have been proposed. More details on these schemes are presented in Appendix A.1.

Other Schemes

The QUICK scheme may yield unnecessary "undershoots" and "overshoots" during numerical calculations. Considerations of some rather mathematically elegant algorithms such as flux limiters by Sweby (1984) and Anderson et al. (1986) in the MINMOD and SUPERBEE schemes or slope limiters due to the Monotone Upwind Scheme for Conservation Laws (MUSCL) family of methods that can be found in Van Leer (1974, 1977a, 1977b, 1979), Godunov

(1959), and approximate Riemann solvers (Toro, 1997) have been proposed to remedy these problems. Amongst these, the total variable diminishing (TVD) schemes have been found to be well suited especially for capturing shock waves in high speed flows; they have nonetheless also proven to be useful for other types of CFD calculations. The design of such schemes is specifically aimed to maintain high accuracy in smooth regions of the flow whilst able to capture sharp non-oscillatory transitions at discontinuities. First-order and second-order upwind schemes as preciously described provide oscillation-free solutions in the vicinity of discontinuities and can be readily shown to obey the TVD condition. Higher-order schemes that belong to the family of TVD algorithms have further shown to be rather successful in minimizing the numerical diffusion caused by lower-order schemes. Some of the most popular and commonly applied schemes are presented in Appendix A.2. For more advanced schemes, the reader is advised to refer in literatures such as Shu and Osher (1988, 1989), Liu et al. (1994), Jiang and Shu (1996), Suresh and Huynh (1997), and Daru and Tenaud (2004) for more in-depth exposition.

2.7.1.2 Unsteady Flows

In order to illustrate the approximate form of the unsteady transport equation of property ϕ, equation (2.5.7) needs to be further augmented with the integration over a finite time step Δt. By changing the order of integration in the time derivative term, we obtain

$$\int_V \left(\int_t^{t+\Delta t} \frac{\partial(\rho\phi)}{\partial t} dt \right) dV + \int_t^{t+\Delta t} \left(\int_A \{(\rho u\phi)dA^x + (\rho v\phi)dA^y + (\rho w\phi)dA^z\} \right) dt$$
$$= \int_t^{t+\Delta t} \left(\int_A \left\{ \left[\Gamma \frac{\partial\phi}{\partial x} \right] dA^x + \left[\Gamma \frac{\partial\phi}{\partial y} \right] dA^y + \left[\Gamma \frac{\partial\phi}{\partial z} \right] dA^z \right\} \right) dt + \int_t^{t+\Delta t} \int_V S_\phi dV dt$$

$$(2.7.20)$$

The discretisation methods as described in the previous section are essentially the same as in steady flows for the treatment of the advection, diffusion, and source terms. The time derivative of equation (2.7.20) for the control volume can also be similarly approximated like the source term yielding

$$\left(\int_t^{t+\Delta t} \frac{\partial(\rho\phi)}{\partial t} dt \right) \Delta V + \int_t^{t+\Delta t} \left(\sum_{i=1}^N (\rho u\phi)_i A_i^x + \sum_{j=1}^N (\rho v\phi)_j A_j^y + \sum_{k=1}^N (\rho w\phi)_k A_k^z \right) dt$$
$$= \int_t^{t+\Delta t} \left(\sum_{i=1}^N \left(\Gamma \frac{\partial\phi}{\partial x} \right)_i A_i^x + \sum_{j=1}^N \left(\Gamma \frac{\partial\phi}{\partial y} \right)_j A_j^y + \sum_{k=1}^N \left(\Gamma \frac{\partial\phi}{\partial z} \right)_k A_k^z \right) dt + \int_t^{t+\Delta t} S_\phi \Delta V dt$$

$$(2.7.21)$$

To solve the preceding equation numerically, suitable methods necessary for time integration are required. In the majority of cases, the time derivative can be obtained by first order approximation as:

$$\int_{t}^{t+\Delta t} \frac{\partial(\rho\phi)}{\partial t}\, dt = \frac{(\rho\phi)^{n+1} - (\rho\phi)^{n}}{\Delta t} \tag{2.7.22}$$

where Δt is the incremental time step and the superscripts n and $n+1$ denote the previous and current time levels, respectively. For the advection, diffusion, and source terms, the integration over time can be generalized by means of introducing a weighting parameter θ between 0 and 1. Using the preceding first order approximation for the time derivative, equation (2.7.21) becomes

$$\left(\frac{(\rho\phi)^{n+1} - (\rho\phi)^{n}}{\Delta t}\right)\Delta V + (1-\theta)\left(\sum_{i=1}^{N}(\rho u\phi)_i A_i^x + \sum_{j=1}^{N}(\rho v\phi)_j A_j^y + \sum_{k=1}^{N}(\rho w\phi)_k A_k^z\right)^n$$

$$\theta\left(\sum_{i=1}^{N}(\rho u\phi)_i A_i^x + \sum_{j=1}^{N}(\rho v\phi)_j A_j^y + \sum_{k=1}^{N}(\rho w\phi)_k A_k^z\right)^{n+1}$$

$$= (1-\theta)\left(\sum_{i=1}^{N}\left(\Gamma\frac{\partial\phi}{\partial x}\right)_i A_i^x + \sum_{j=1}^{N}\left(\Gamma\frac{\partial\phi}{\partial y}\right)_j A_j^y + \sum_{k=1}^{N}\left(\Gamma\frac{\partial\phi}{\partial z}\right)_k A_k^z\right)^n$$

$$+ \theta\left(\sum_{i=1}^{N}\left(\Gamma\frac{\partial\phi}{\partial x}\right)_i A_i^x + \sum_{j=1}^{N}\left(\Gamma\frac{\partial\phi}{\partial y}\right)_j A_j^y + \sum_{k=1}^{N}\left(\Gamma\frac{\partial\phi}{\partial z}\right)_k A_k^z\right)^{n+1}$$

$$+ (1-\theta)S_\phi^n \Delta V + \theta S_\phi^{n+1}\Delta V$$

$$\tag{2.7.23}$$

The exact form of the final discretised equation depends on the appropriate value of θ. For an **explicit** method, θ is set to zero, and all the transport properties of ϕ in the advection, diffusion, and source terms are known at the previous time level n. Assuming the current time level ρ^{n+1} is known, the property ϕ^{n+1} can be immediately evaluated. Nevertheless, if θ is set to unity, this method, commonly referred as the **fully-implicit** procedure, results in the need of calculating all the transport property of ϕ in the time derivative, advection, diffusion, and source terms at the current time level $n+1$. Note that the explicit and fully implicit approaches to equation (2.7.25) are methods of only first order in time. Similar to the first order in space, these methods may also cause unwanted numerical diffusion in time. In order to reduce these numerical errors, second order approximations such as the **explicit Adams-Bashford**, **semi-implicit Crank-Nicolson**, and **second order fully implicit** methods have been proposed, which are further described in Appendix A.3.

2.7.2 Solution Algorithms

A road map of the computational solution procedure for field modeling is illustrated in Figure 2.12. The formulation of appropriate equations of motion and boundary conditions has been carried out in sections 2.4 and 2.6. This represents the first stage of the road map as stipulated in Figure 2.12. With the application of the finite volume method in approximating the transport equations, various discretisation methods have been explored and explained in the previous section 2.6.1. In this section, the array of solution algorithms that are generally required to solve the algebraic form of the transport equations is described. They consist of matrix solvers and pressure-velocity linkage methods.

2.7.2.1 Matrix Solvers

Application of various discretisation methods as described in section 2.7.1 results in a system of algebraic equations either through the steady form of equation (2.7.1) or unsteady form of equation (2.7.20). These governing equations need to be solved by some dedicated numerical solvers, and the degree of complexity depends on the dimensionality and geometry of the physical problem. Whether the equations are linear or non-linear, efficient and robust matrix solvers are required to solve the system of algebraic equations.

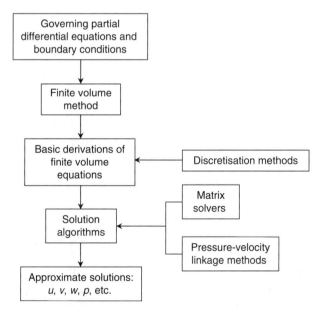

Figure 2.12 Road map of computational solution procedure.

This system of equations can usually be expressed in the form:

$$A\phi = B \tag{2.7.24}$$

where ϕ is the unknown nodal variables of the transport property. Matrix A contains non-zero coefficients of the algebraic equations:

$$A = \begin{bmatrix} A_{11} & A_{12} & A_{13} & \cdots\cdots & A_{1n} \\ A_{21} & A_{22} & A_{23} & \cdots\cdots & A_{2n} \\ A_{31} & A_{32} & A_{33} & \cdots\cdots & A_{3n} \\ \vdots & \vdots & \vdots & \ddots & \vdots \\ A_{n-21} & A_{n-22} & A_{n-23} & \cdots\cdots & A_{n-2n} \\ A_{n-11} & A_{n-12} & A_{n-13} & \cdots\cdots & A_{n-1n} \\ A_{n1} & A_{n2} & A_{n3} & \cdots\cdots & A_{nn} \end{bmatrix} \tag{2.7.25}$$

while B comprises of known values of ϕ, for example, that are given by the boundary conditions or source/sink terms. The diagonal coefficients of the matrix A are represented by the entries of A_{11}, A_{22}, ..., A_{nn}.

For a structured grid arrangement, the resultant matrix of A in equation (2.7.25) via the finite volume method has a regular arrangement (banded) and is a sparse matrix; most of the elements are zero and the non-zero terms are close to the diagonal. In three dimensions, the matrix is:

$$A = \begin{bmatrix} A_P & A_E & 0 & A_N & 0 & 0 & 0 & 0 & 0 & A_T & 0 \\ A_W & A_P & A_E & 0 & A_N & 0 & 0 & 0 & 0 & 0 & \ddots \\ 0 & A_W & A_P & A_E & 0 & A_N & 0 & 0 & 0 & 0 & 0 \\ A_S & 0 & A_W & A_P & A_E & 0 & A_N & 0 & 0 & 0 & 0 \\ 0 & A_S & 0 & A_W & A_P & A_E & 0 & A_N & 0 & 0 & 0 \\ 0 & 0 & A_S & 0 & A_W & A_P & A_E & 0 & A_N & 0 & 0 \\ 0 & 0 & 0 & A_S & 0 & A_W & A_P & A_E & 0 & A_N & 0 \\ 0 & 0 & 0 & 0 & A_S & 0 & A_W & A_P & A_E & 0 & \ddots \\ 0 & 0 & 0 & 0 & 0 & A_S & 0 & A_W & A_P & A_E & 0 \\ A_B & 0 & 0 & 0 & 0 & 0 & A_S & 0 & A_W & A_P & \ddots \\ 0 & \ddots & 0 & 0 & 0 & 0 & 0 & \ddots & 0 & \ddots & \end{bmatrix} \tag{2.7.26}$$

The non-zero coefficients designated by A_B, A_S, A_W, A_P, A_E, A_N, and A_T within each row of the matrix in equation (2.7.26) are typically representative coefficients of a control volume in space of a structured grid arrangement. A comprehensive formulation of these coefficients is presented in Appendix B.1.

For computational economy, iterative matrix solvers are employed out of necessity to solve the set of algebraic equations. One commonly used method is the ADI (Alternating Direction Implicit), introduced by Peaceman and

Rachford (1955), which has been aimed to reduce multi-dimensional problems, whether they are two-dimensional or three-dimensional, to a sequence to be solved as a one-dimensional problem. The resulting matrix for the one-dimensional problem is of a tri-diagonal form; Thomas (1949) algorithm, a special form of a direct Gaussian elimination method, can be applied. This procedure solves the nodal variables for the lines in one direction and repeats for the lines in other directions. Another iterative method for solving multi-dimensional problems is the Strongly Implicit Procedure (SIP) proposed by Stone (1968) or modified SIP by Schneider and Zedan (1981). The basic idea of this method involves approximating the matrix A in the form of equation (2.7.26), by an incomplete LU (Lower-Upper) factorization to yield an iteration matrix M. Unlike other methods, SIP is a rather good iterative solver in its own right. It has been used in some commercial CFD codes as the standard solver for non-linear equations such as in the commercial CFD code CFX4.4.

For an unstructured grid arrangement, the resultant matrix of A does not conform to the sparse matrix for a structured grid arrangement. As such, ADI or SIP matrix solver cannot be directly applied to such system. In many CFD applications, other more sophisticated methods such as conjugate gradient and multigrid methods are being employed with increasing frequency because of their ease in solving large system of algebraic equations formulated especially for unstructured meshes.

The conjugate gradient method is essentially a method that seeks the minimum of a function that belongs to the class of *steepest descent methods*. The basic method in itself converges rather slowly. However, when it is used in conjunction with some *preconditioning* of the original matrix, significant enhancements in its speed of convergence can be realized. For example, this preconditioning technique has been achieved by either applying the *incomplete Cholesky* factorization for symmetric matrices or applying *biconjugate gradients* for asymmetric matrices. Interested readers are encouraged to refer to Concus et al. (1976) and Kershaw (1978) for more details.

For the multigrid method, it can be characterized as being either geometric or algebraic. Geometric multigrid, also known as the FAS (Full Approximation Scheme) multigrid, involves a hierarchy of meshes (cycling between fine and coarse grids) and the discretised equations are evaluated on every level, while in algebraic multigrid, the coarse level equations are generated without any geometry or re-discretisation on the coarse levels—a feature that makes this method particularly attractive for use on unstructured meshes. In algebraic multigrids, once linearization is performed on the system of equations, the non-linear properties are not experienced by the solver until at the fine level where the operator is finally updated. Within each level, simple point-by-point iterative methods such as Jacobi and Gauss-Siedel could be employed for the coarse level equations to determine the immediate values for ϕ. Multigrid approach is more a strategy than a particular method. Details are not presented here in the interest of brevity and keen readers are referred to Wesselling (1995), Timmermann (2000), and Thomas et al. (2003) for the latest trends and developments of multigrid methods. It is noted that conjugate gradient methods and multigrid methods are usually employed in structured grid arrangement to accelerate the iteration process with the goal of achieving quicker convergence for the Poisson equation of pressure (or pressure correction), which will be discussed in the next section.

2.7.2.2 Pressure-Velocity Linkage Methods

For explicit time-marching methods, the marker-and-cell (MAC) method developed by Harlow and Welch (1965) represents the first primitive variable method employing a derived Poisson equation for pressure of which the pressure is used as a mapping parameter to satisfy the continuity equation. The Poisson equation of pressure is formulated by taking the divergence of the momentum equation. Another pressure-based method that is also commonly employed in many CFD applications is the fractional-step procedure (Chorin, 1968 and Yanenko, 1971). Here, the auxiliary velocity field is initially obtained from solving the momentum equation in which the pressure-gradient term is entirely excluded or computed from the pressure in the previous time step. The pressure is later computed through a Poisson equation similar to the MAC method that maps the auxiliary velocity onto a divergence-free velocity field. The MAC method and fractional-step procedure will be further explained in Chapter 5.

In this section, the popular SIMPLE (Semi-Implicit for Method Pressure-Linkage Equation) scheme is described to cater for implicit-type algorithms of steady or unsteady solutions. Pioneered by Patankar and Spalding (1972), this scheme has found widespread application in the majority of commercial CFD codes for practical engineering solutions. Based on the basic philosophy of effectively coupling the pressure with the velocity for an incompressible flow, the pressure is linked to the velocity via the construction of a pressure field to guarantee conservation of mass. The continuity equation becomes now a kinematic constraint on the velocity field rather than a dynamic equation.

By definition, an incompressible flow, which is also applicable to weakly compressible flow, is the approximation of the flow where the flow speed is insignificant compared to the speed of sound of the fluid medium. Mathematically, the incompressible flow formulation poses unique challenges because of its incompressibility requirement. Physically, an incompressible flow is characterized by an elliptic behavior of the pressure waves whereby the speed in a truly incompressible flow is infinite, which imposes stringent requirements on computational algorithms for satisfying incompressibility. The major difference between an incompressible and compressible formulation lies in the mass conservation equation. In general, the incompressible formulation can be viewed as a singular limit of the compressible counterpart where the pressure field in this instance is considered to be represented as part of the solution process. The primary issue in solving the set of governing equations for an incompressible flow is thus to appropriately satisfy the mass conservation equation.

The SIMPLE scheme begins from the discretised forms of the momentum equations. From equations (2.7.24) and (2.7.25), the algebraic x-momentum equation can be expressed as

$$u = \sum \frac{A^u_{ij,j\neq i} u_{nb}}{A^u_{ii}} - \frac{\Delta V}{A^u_{ii}} \frac{\partial p}{\partial x} + B' \tag{2.7.27}$$

where u_{nb} are the neighboring nodes of the u velocity component, A^u_{ij} are A^u_{ii} are the non-diagonal and diagonal coefficients of matrix A for the u velocity component, and B' is the remaining source term after the pressure gradient source term has been removed. The y-momentum and z-momentum equations can also be similarly obtained as

$$v = \sum \frac{A^v_{ij,j\neq i} v_{nb}}{A^v_{ii}} - \frac{\Delta V}{A^v_{ii}} \frac{\partial p}{\partial y} + B' \tag{2.7.28}$$

$$w = \sum \frac{A^w_{ij,j\neq i} w_{nb}}{A^w_{ii}} - \frac{\Delta V}{A^w_{ii}} \frac{\partial p}{\partial z} + B' \tag{2.7.29}$$

The SIMPLE scheme is essentially a "guess-and-correct" procedure for the calculation of pressure through the solution of a pressure correction equation. It is thus an iterative procedure. During the iterative process, the discretised momentum equations can be solved using the guessed pressure field p^* to obtain the guessed velocities u^*, v^*, and w^*:

$$u^* = \sum \frac{A^u_{ij,j\neq i} u^*_{nb}}{A^u_{ii}} - \frac{\Delta V}{A^u_{ii}} \frac{\partial p^*}{\partial x} + B' \tag{2.7.30}$$

$$v^* = \sum \frac{A^v_{ij,j\neq i} v^*_{nb}}{A^v_{ii}} - \frac{\Delta V}{A^v_{ii}} \frac{\partial p^*}{\partial y} + B' \tag{2.7.31}$$

$$w^* = \sum \frac{A^w_{ij,j\neq i} w^*_{nb}}{A^w_{ii}} - \frac{\Delta V}{A^w_{ii}} \frac{\partial p^*}{\partial z} + B' \tag{2.7.32}$$

The corrected velocities u, v, and w with the correct pressure field p may be represented by equations (2.7.27), (2.7.28), and (2.7.29). Subtracting equation (2.7.27) from (2.7.30), equation (2.7.28) from (2.7.31), and equation (2.7.29) from (2.7.32), the following expressions are obtained:

$$u - u^* = \sum \frac{A^u_{ij,j\neq i}(u_{nb} - u^*_{nb})}{A^u_{ii}} - D^u \frac{\partial(p - p^*)}{\partial x} \tag{2.7.33}$$

$$v - v^* = \sum \frac{A^v_{ij,j\neq i}(v_{nb} - v^*_{nb})}{A^v_{ii}} - D^v \frac{\partial(p - p^*)}{\partial y} \tag{2.7.34}$$

$$w - w^* = \sum \frac{A^w_{ij,j\neq i}(w_{nb} - w^*_{nb})}{A^w_{ii}} - D^w \frac{\partial(p - p^*)}{\partial z} \tag{2.7.35}$$

where $D^u = \frac{\Delta V}{A^u_{ii}}$, $D^v = \frac{\Delta V}{A^v_{ii}}$ and $D^w = \frac{\Delta V}{A^w_{ii}}$

The SIMPLE scheme approximates the above equations by the omission of the neighboring nodal terms—in other words,

$$u - u^* = -D^u \frac{\partial p'}{\partial x} \tag{2.7.36}$$

$$v - v^* = -D^v \frac{\partial p'}{\partial y} \tag{2.7.37}$$

$$w - w^* = -D^w \frac{\partial p'}{\partial z} \tag{2.7.38}$$

where p' ($= p - p^*$) is defined as the pressure correction. Since this scheme has been primarily designed to be an iterative procedure, there is no reason why the formula designed to predict the pressure correction p' needs to be physically correct. A formula for p' can be simply constructed as a numerical artifice with the aim to expedite the convergence of the velocity field to a solution that satisfies the continuity equation. By differentiating equation (2.7.36) by the Cartesian x direction, equation (2.7.37) by the Cartesian y direction and equation (2.7.38) by the Cartesian z direction and summing them together yields

$$-\frac{\partial}{\partial x}\left(D^u \frac{\partial p'}{\partial x}\right) - \frac{\partial}{\partial y}\left(D^v \frac{\partial p'}{\partial y}\right) - \frac{\partial}{\partial z}\left(D^v \frac{\partial p'}{\partial z}\right)$$

$$+ \underbrace{\frac{\partial u^*}{\partial x} + \frac{\partial v^*}{\partial y} + \frac{\partial w^*}{\partial z}}_{\text{guessed velocity gradients}} = \underbrace{\frac{\partial u}{\partial x} + \frac{\partial v}{\partial y} + \frac{\partial w}{\partial z}}_{\text{correct velocity gradients}} \tag{2.7.39}$$

The derivation of the pressure correction in equation (2.7.39) represents the Poisson equation for incompressible flow. Note that the corrected velocity gradients equation of the right-hand side in equation (2.7.39) is zero by definition of the continuity equation. An appropriate equation in the form similar to equation (2.7.39) can be obtained after some mathematical manipulation to yield the Poisson form of the pressure correction for weakly compressible flow as:

$$-\frac{\partial}{\partial x}\left(\rho D^u \frac{\partial p'}{\partial x}\right) - \frac{\partial}{\partial y}\left(\rho D^v \frac{\partial p'}{\partial y}\right) - \frac{\partial}{\partial z}\left(\rho D^v \frac{\partial p'}{\partial z}\right)$$

$$+ \underbrace{\frac{\partial(\rho u)^*}{\partial x} + \frac{\partial(\rho v)^*}{\partial y} + \frac{\partial(\rho w)^*}{\partial z}}_{\text{guessed velocity gradients}} = \underbrace{\frac{\partial(\rho u)}{\partial x} + \frac{\partial(\rho v)}{\partial y} + \frac{\partial(\rho w)}{\partial z}}_{\text{corrected velocity gradients}} \tag{2.7.40}$$

Invoking equation (2.4.4), it can be shown herein that the term represented by the source term of the right-hand side for the above equation is the time derivative of the density. Equation (2.7.40) can thus be re-arranged as

$$
\frac{\partial}{\partial x}\left(\rho D^u \frac{\partial p'}{\partial x}\right) + \frac{\partial}{\partial y}\left(\rho D^v \frac{\partial p'}{\partial y}\right) + \frac{\partial}{\partial z}\left(\rho D^v \frac{\partial p'}{\partial z}\right)
$$

$$
= \underbrace{\frac{\partial \rho^*}{\partial t} + \frac{\partial (\rho u)^*}{\partial x} + \frac{\partial (\rho v)^*}{\partial y} + \frac{\partial (\rho w)^*}{\partial z}}_{\textit{mass residual}}
$$

(2.7.41)

The source term appearing in the pressure correction equation (2.7.41), commonly known as the *mass residual*, is normally used in CFD computations as a criterion to terminate the iteration procedure. As the mass residual continues to diminish, the pressure correction p' will be zero, thereby yielding a converged solution of $p^* = p$, $u^* = u$, $v^* = v$ and $w^* = w$.

It is imperative that the mass fluxes $(\rho u)^*$, $(\rho v)^*$, and $(\rho w)^*$, in the *mass residual* of equation (2.7.41) at the respective faces of the control volume are evaluated in a manner of avoiding non-physical solutions (Patankar, 1980, Versteeg and Malasekera, 1995). This so-called "checker-board" effect may cause serious convergence problems due to the spatial oscillations in the pressure and velocity fields. A remedy for this problem is to adopt a *staggered* grid for the velocity components. Consider the two-dimensional structured grid arrangement for the *staggered* grid in Figure 2.13. To conserve mass, the velocities u and v are now evaluated at the control volume faces, while the rest of the variables (such as pressure, enthalpy, etc.) governing the flow-field are stored at the central node of the control volumes (P, W, E, N, S). The discrete values of the velocity component, u, from the x-momentum equation are evaluated and stored at the east, e, and west, w, faces of the control volume. By evaluating the other velocity

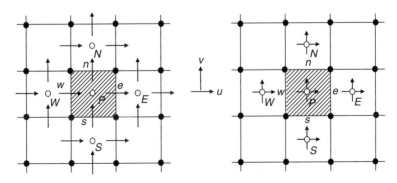

Figure 2.13   (a) Staggered and (b) collocated arrangements of velocity components on a finite volume grid.

components using the y-momentum and z-momentum equations on the rest of the control volume faces, these velocities allow a straightforward evaluation of the mass fluxes that are used in the pressure correction equation (2.7.41).

Alternatively, the *collocated* grid has significant advantages over the *staggered* grid in handling complicated domains especially in the capability of accommodating slope discontinuities or boundary conditions that may be discontinuous. For multigrid methods, the collocated arrangement allows the ease of transfer of information between various grid levels. All flow-field variables including the velocities are stored at the same set of nodal points as illustrated in Figure 2.13. Nonetheless, this grid arrangement suffers from the well-known "checker-board" effect, and it was out of favor for a substantial period because of the difficulties in coupling the pressure with the velocity and the occurrence of oscillations in the pressure. Through significant developments by Rhie and Chow (1983), an interpolation method that provided physically sensible and stable solutions has once again made the *collocated* grid attractive.

Let us take for an example the algebraic x-momentum equation (2.7.27) to briefly illustrate the Rhie-Chow interpolation method, which can be rewritten in the form of

$$u + D^u \frac{\partial p}{\partial x} = A u_{nb} + B^{'} \tag{2.7.42}$$

where $A = \sum \frac{A^u_{ij,j \neq i}}{A^u_{ii}}$. Consider the interpolation to be performed to the east face of a control volume centered at P for the structured grid arrangement as illustrated in Figure 2.13 for the *collocated* grid arrangement. Using equation (2.7.42), the velocity components u_P and u_E obey the algebraic momentum equations:

$$u_P + \left(D^u \frac{\partial p}{\partial x} \right)_P = (A u_{nb})_P + B^{'}_P \tag{2.7.43}$$

$$u_E + \left(D^u \frac{\partial p}{\partial x} \right)_E = (A u_{nb})_E + B^{'}_E \tag{2.7.44}$$

For the east face e, the velocity component u_e also obeys the algebraic momentum equation:

$$u_e + \left(D^u \frac{\partial p}{\partial x} \right)_e = (A u_{nb})_e + B^{'}_e \tag{2.7.45}$$

The prescription of Rhie and Chow is simply a method approximating solutions of equation (2.7.45). The terms on the right hand side of equation (2.7.45) can be assumed to be approximated by linear interpolations, indicated by (¯), between the nodes of P and E of the corresponding terms in equations (2.7.43) and (2.7.44)—in other words,

$$u_e + \overline{\left(D^u \frac{\partial p}{\partial x}\right)_e} = \overline{(Au_{nb})_e} + \overline{B'_e} \tag{2.7.46}$$

Note that $\overline{(Au_{nb})_e} + \overline{B'_e} = \overline{u_e} + \overline{\left(D^u \frac{\partial p}{\partial x}\right)_e}$, equation (2.7.46) can be rewritten according to

$$u_e = \overline{u_e} + \overline{\left(D^u \frac{\partial p}{\partial x}\right)_e} - \left(D^u \frac{\partial p}{\partial x}\right)_e \tag{2.7.47}$$

Assuming that $D^u_e = \overline{D^u_e}$ and $\overline{\left(D^u \frac{\partial p}{\partial x}\right)_e} = \overline{D^u_e}\overline{\left(\frac{\partial p}{\partial x}\right)_e}$, the Rhie-Chow interpolation formula is given by

$$u_e = \overline{u_e} + \overline{D^u_e}\left[\overline{\left(\frac{\partial p}{\partial x}\right)_e} - \left(\frac{\partial p}{\partial x}\right)_e\right] \tag{2.7.48}$$

The last term in equation (2.7.48) is usually approximated according to a first order approximation:

$$\left(\frac{\partial p}{\partial x}\right)_e = \frac{p_E - p_P}{\delta x_E} \tag{2.7.49}$$

where δx_E is the distance between the central nodes P and E.

According to equation (2.7.46), it is also possible to propose another simpler interpolation method of which the current authors have employed with much success. Assuming again for the sake of brevity $D^u_e = \overline{D^u_e}$, the interface velocity u_e can be evaluated as

$$u_e = -\overline{D^u_e}\left(\frac{p_E - p_P}{\delta x_E}\right) + \overline{(Au_{nb})_e} + \overline{B'_e} \tag{2.7.50}$$

In the preceding equation, the interpolated terms of the neighboring nodal velocities and the source at the control volume face e are retained. The pressure gradient is replaced with the first order approximation between the pressures at nodes P and E. The advantage of this prescription is the ability to incorporate the body force due to buoyancy, which is the main driving mechanism in most fires, within the interpolation process. For the other control volume faces, equation (2.7.48) or (2.7.50) can be similarly applied to obtain the necessary interface velocities.

It is now possible to assemble the complete solution procedure for an implicit-type algorithm that is the sequence of which, through repetitive calculations, leads to the final converged solution satisfying all the governing equations involved. It can be summarized as:

1. Initialize all field values by an initial guess
2. Solve the algebraic momentum equations (2.7.30), (2.7.31), and (2.7.32) to obtain u^*, v^*, and w^* based on the guessed pressure p^*. If the *collocated* grid arrangement is adopted, the interface velocities are evaluated employing the Rhie-Chow interpolation method or the simpler procedure as just described
3. Solve the pressure correction equation (2.7.44) to obtain p'
4. Correct the velocities as well as mass fluxes using the expressions (2.7.36), (2.7.37), and (2.7.38) and pressure by $p = p^* + p'$ through the available solution of the pressure correction field
5. Solve additional equations for other property ϕ governing the flow process, if necessary
6. Using the corrected velocities, mass fluxes, and pressure as the prevailing fields for the new iteration cycle, return to step 2.

The sequence of steps 2–5 is repeated until convergence is achieved. The *mass residual* in equation (2.7.41) is normally employed as one of many criteria to terminate the solution procedure. Another way of ascertaining convergence is through the sum of absolute *imbalances* (*residuals*) of the discretised equations at all computational nodes. From equations (2.7.24) and (2.7.25), the *imbalance* of property ϕ at any computational node can be calculated as $R = \left| \sum \dfrac{A^u_{ij,j \neq i} \phi^*_{nb}}{A^u_{ii}} + B' - \phi \right|$. The sum is given by $SR_\phi = \sum\limits_{l=1}^{M} R^n_l$, where M is the total number of nodes and n is the iteration counter. Majority of commercial CFD codes impose their own respective convergence criteria that are generally applicable to a wide range of flow problems. Interested readers may wish to investigate the specified tolerance levels employed by these software packages or refer to books like Fletcher (1991) and Ferziger and Perić, (1999) for more detailed discussion. Appropriate settings of convergence criteria are still nevertheless determined from practical experience and application of CFD methodologies.

Other Pressure-Based Methods

The reader should also be well aware of other types of pressure-velocity coupling algorithms that employ a similar philosophy to the SIMPLE algorithm as just described, which have been employed by many CFD users or adopted in a number of commercial CFD codes. These variant SIMPLE algorithms have been formulated to aid convergence and improve the robustness for numerical computations. We briefly describe a collection of other available popular algorithms and modifications made to the original SIMPLE algorithm.

The SIMPLEC (SIMPLE-Consistent) algorithm by Van Doormal and Raithby (1984) follows the same iterative steps as in the SIMPLE algorithm. The main difference between the SIMPLEC and SIMPLE is that the discretised momentum equations are manipulated so that the SIMPLEC velocity correction formulae omit terms that are less significant than those omitted in

SIMPLE. Another pressure correction procedure that is also commonly employed is the PISO (Pressure Implicit with Splitting of Operators) algorithm proposed by Issa (1986). Originally, this pressure-velocity calculation procedure was developed for non-iterative computation of unsteady compressible flows but has been adapted successfully for the iterative solution of steady state problems. The PISO procedure is simply an extension of SIMPLE by an additional corrector step, which requires the need to solve an additional pressure correction equation to enhance the convergence of the numerical solution. The SIMPLER (SIMPLE-Revised) algorithm developed by Patankar (1980) also falls within the framework of two corrector steps like in PISO. Here, a discretised equation for the pressure provides the intermediate pressure field before the discretised momentum equations are solved. A pressure correction is subsequently solved and the velocities are corrected according to the correction formulae as in the SIMPLE algorithm. Other SIMPLE-like algorithms that readers may also find useful and share the same essence in their derivations are SIMPLEST (SIMPLE-ShorTened) of Spalding (1980), SIMPLEX of Van Doormal and Raithby (1985), and SIMPLEM (SIMPLE-Modified) of Acharya and Moukalled (1989). More details of all the preceding pressure-velocity coupling algorithms are left to the pursuit of keen and interested readers.

2.7.3 Boundary Conditions

Appropriate boundary conditions are essential in attaining meaningful numerical solutions. A number of suitable boundary conditions for field modeling have been presented in section 2.5. For the purpose of illustration, applications of these boundary conditions that mimic the real physical representation of the flow process into a solvable computational problem are described for the cases of a free-standing fire and a compartment fire. Figures 2.14 and 2.15 demonstrate the schematic descriptions of the boundary conditions that are required to be specified for the two fire cases. Note that the many illustrative examples in this chapter as well as in other chapters adopt the boundary conditions as demonstrated in these two schematic drawings.

For the free-standing fire, the computational domain is bounded by a horizontal *inflow* boundary in the middle with a horizontal *solid wall* surrounding the fuel source and five *open* boundaries. For the inflow boundary condition, the inlet velocity can usually be specified according to known values to indicate the fluid entering into the flow domain. Prior knowledge of appropriate values for density, temperature, and other quantities are prescribed similarly to the inlet velocity. By definition, the velocities are zero for the stationary *solid wall* surrounding the fuel source. For density, temperature, and other quantities, they are imposed at either fixed known values or given fluxes. At the *open* boundaries, care should always be exercised in defining these boundaries, as they have to be treated in a manner whereby they need to be sufficiently placed far away from the region of interest within the solution domain in order that

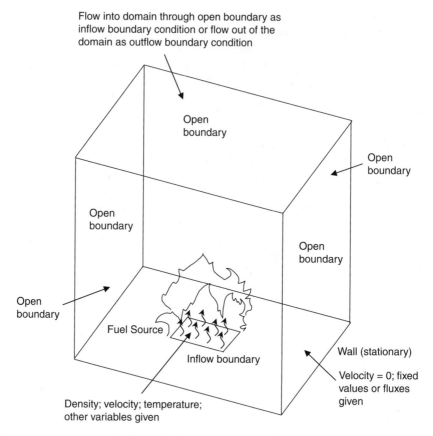

Flow into domain through open boundary as
inflow boundary condition or flow out of the
domain as outflow boundary condition

Open
boundary

Open
boundary

Open
boundary

Open
boundary

Open
boundary

Fuel Source

Inflow boundary

Wall (stationary)

Velocity = 0; fixed
values or fluxes
given

Density; velocity; temperature;
other variables given

Figure 2.14   Boundary conditions for the case of a free-standing fire.

physically meaningful results are realizable. For the velocity, boundary values
can be approximated by extrapolation from adjacent nodes. Boundary conditions for density, temperature, and other quantities are usually, however, set
according to the flow direction. Values at ambient conditions are usually prescribed for *inflow*, whilst otherwise extrapolated from upstream for *outflow*.

   To take advantage of special geometrical features that the solution region
possess for the compartment fire, a *symmetric* boundary condition can be
employed as illustrated in Figure 2.15 to speed up computations and enhance
computational accuracy by the additional number of cells that could be further
accommodated into the simplified geometry. The conditions that apply to the
plane of symmetry are no cross-flow for the velocity (zero convective flux) and zero
diffusion flux for the other dependant variables in the normal direction. For
the *inflow* boundary of the fuel source in the middle of the enclosure and *solid
walls* bounding the computational domain, similar boundary conditions can be
applied as previously described for the free-standing fire case. The *open*

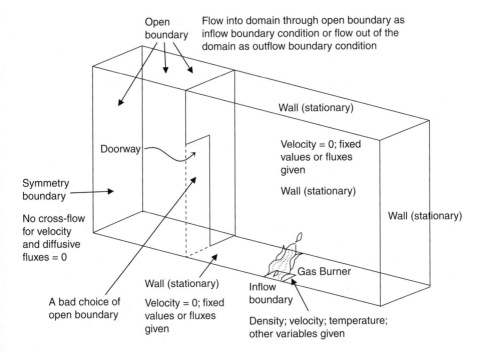

Figure 2.15   Boundary conditions for the case of a compartment fire.

boundary should also be placed far downstream from the region of interest. For this particular case, the *open* boundary defined at the doorway represents a bad choice because of the possible accumulation errors in estimating the boundary condition that may propagate into the upstream flow within the compartment yielding inaccurate solutions. In Figure 2.15, the *open* boundaries are defined at locations away from the doorway, assumed at ambient conditions.

2.8 Summary

The mathematical basis for a comprehensive general purpose model of fluid flow and heat transfer for field modeling is formulated from the basic principles of conservation of mass, momentum, and energy and other concepts to attain additional equation for any scalar property. Through the consideration of an infinitesimal small control volume, the governing equations are derived. In the momentum equations, the Newtonian model of viscous stresses is employed to close the system of equations.

Through the control volume approach, significant commonalities between the conservation equations have lead to the formulation of generic forms of the governing equations. The consistency of the control volume approach with the finite volume method allows the immediate discretisation of the generic transport equation into its algebraic form. Through the basic derivations of these finite volume equations, discretisation methods that yield physical solutions are employed to sufficiently approximate the terms representing the local acceleration, advection, diffusion, and sources. Application of suitable matrix solvers and pressure-velocity linkage methods together with appropriate boundary conditions complete the fourth stage of the road map of computational solution for field modeling as exemplified in Figure 2.12.

The governing equations described in Part 1 of this chapter could at best only provide a description of the fire dynamics associated with a laminar combusting flow system. To adequately resolve practical fires, additional models to better capture all the physical processes involving turbulence, combustion, radiation, smoke movement and production, and solid pyrolysis are generally needed to predict the fire characteristics occurring in real fires. These models are incorporated within field modeling before computing the approximate solutions of the velocity, pressure, temperature, and so on. Figure 2.16 illustrates an extract of the fifth stage of the road map in Figure 2.12. In the context of fire engineering, most practical fires are turbulent in nature. The many aspects of turbulence modeling will be expounded in Part 2 of this chapter. Relevant models pertaining to combustion and radiation modeling of fires are explored in Chapter 3, while the considerations of smoke (soot) movement and production and solid pyrolysis as supplementary models for fire investigations are subsequently discussed in the proceeding Chapter 4. It will be demonstrated later that the additional equations governing the processes of turbulence, combustion, radiation, smoke (soot) movement and production, and solid pyrolysis are simply forms of the same generic transport equation. Increasing model complexities can thus be accommodated with relative ease for the sequence of numerical calculations to be performed in the field modeling approach for practical fires. For the benefit of code developers or even code users, a flow chart detailing the overall computational procedures on the application of these models in practice can be found in Appendix B.

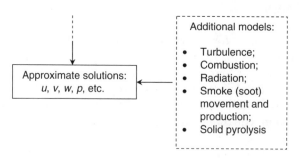

Figure 2.16 Additional modeling considerations to the road map of computational solution procedure.

PART II TURBULENCE

2.9 What Is Turbulence?

Turbulent flows can generally be viewed as the motion of a fluid becoming intrinsically unstable and unsteady so the final state of the fluid behaves in a random and chaotic manner. They always have a three-dimensional spatial character, and rapid fluctuations are generally exhibited in such flows. Numerous visualizations of turbulent flows have clearly revealed the presence of rotational flow structures, so-called turbulent eddies, spanning a wide range of length and velocity scales, commonly known as turbulent scales. As an example, Figure 2.17 exemplifies a typical turbulent free jet that has been observed during experiments.

The largest eddies in the fluid can usually be described as having a characteristic velocity and a characteristic length of the same order as the velocity scale and length scale of the mean flow. This suggests that for turbulent flows, the largest eddies whose scales are comparable with the mean flow are dominated primarily by inertia effects rather than viscous effects. The large eddies are therefore effectively inviscid. By the transport of eddies, energy is extracted from the mean flow by a process called vortex stretching. Angular momentum is conserved during vortex stretching, and the stretching work done by the mean flow on the large eddies provides the energy that maintains the turbulence. These larger eddies will have a tendency to breed new instabilities within the flow thereby creating smaller eddies.They are transported mainly by vortex stretching from the larger eddy rather than from the mean flow. Energy is subsequently transferred from the larger eddy to the smaller eddy until the turbulent eddies become so small that viscous effects dominate. Work is performed against the action of the viscous stresses, so that the energy associated with the eddy motions is dissipated and

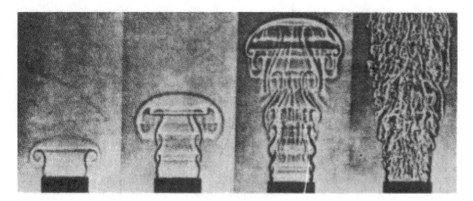

Figure 2.17 Schlieren photographs of a free jet highlighting the formation of separate flow vortices at initial stages and highly turbulent flow in the later stages (after Garside et al., 1943).

converted into thermal internal energy. The continual transfer of energy from the larger eddy to smaller eddy is termed as the energy cascade.

Larger eddies are flow dependent, as they are generated from mean flow characteristics. Their turbulent scales are large compared with the viscosity causing the structure of the eddy to be highly *anisotropic* (that is, varying in all directions). Small eddies have much smaller turbulent scales compared with viscosity, causing the flow to be *isotropic*, since the diffusive effects of viscosity dominates and smears out the directionality of the flow structure.

Some important characteristics of turbulent flows according to Tennekes and Lumley (1972) can be summarized as:

- *Irregular.* All turbulent flows are random, or irregular. Statistical methods present the only viable way of quantifying the characteristics of turbulence.
- *Diffusivity.* An important feature of all turbulent flows is the diffusivity of turbulence. The presence of rigorous mixing increases the rates of momentum, heat, and mass transfer in such flows.
- *Large Reynolds Number.* By definition, Reynolds number, a non-dimensional parameter designated as *Re*, describes the ratio between the inertia force and the friction force:

$$Re \equiv \frac{\rho U^* L^*}{\mu} = \frac{\text{inertia force}}{\text{friction force}} \tag{2.9.1}$$

where U^* and L^* represent the reference velocity and length scale. In turbulent flows, the inertia force tends to be much larger than the friction force at high Reynolds numbers. Turbulence also often originates as an instability of laminar flows, which is related to interaction of viscous and inertia terms in the equation of motion.

- *Three-Dimensional Vorticity Fluctuations.* Turbulence is rotational and three-dimensional. The dynamics of vorticity play an important role in the description of turbulent flows.
- *Dissipation.* Turbulent flows are always dissipative. Work done due to viscous shear stresses increases the internal energy of the fluid at the expense of the kinetic energy of the turbulence. Turbulence needs a continuous supply of energy to compensate for these viscous losses.
- *Continuum.* Turbulence is a continuum phenomenon. It is thus governed by the equations of fluid mechanics. Even the smallest turbulent scales are ordinarily much larger than any of the molecular length scale.

2.10 Overview of Turbulence Modeling Approaches

As described in the previous section, turbulence is associated with the existence of random fluctuations in the fluid. This is best exemplified by an illustration of a temporal variation of a transport property ϕ in Figure 2.18. Even with the current computational technology, the random nature of the fluid flow still precludes computations based on the equations that describe the fluid motion to be carried out to the required accuracy. It is thus more preferable that there

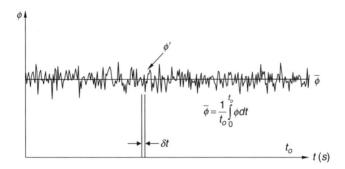

Figure 2.18   Transport property ϕ fluctuating with time at some point in a turbulent flow.

be some means of practically resolving the random transient distribution of the property ϕ with time in Figure 2.18. By decomposing the instantaneous property ϕ as a steady mean motion $\overline{\phi}$ and a fluctuating, or eddy, motion ϕ' as:

$$\phi = \overline{\phi} + \phi' \quad and \quad \overline{\phi} = \frac{1}{t_o} \int_0^{t_o} \phi\, dt \tag{2.10.1}$$

where t_o is a large enough time interval exceeding the time scales of the slowest variations (due to largest eddies) as shown in Figure 2.18. This approach presents an attractive way to characterize a turbulent flow by the mean values of flow properties ($\overline{u}$, $\overline{v}$, $\overline{w}$, $\overline{p}$, etc.) with its corresponding statistical fluctuating property (u', v', w', p', etc.). It is noted that the time averaged of the fluctuating component ϕ' is, by definition, zero:

$$\overline{\phi'} = \frac{1}{t_o} \int_0^{t_o} \phi'\, dt \equiv 0 \tag{2.10.2}$$

In many engineering applications, the knowledge of mean values of flow properties is usually of greater significance than their fluctuating components. The introduction of these definitions into relevant conservation equations produces a more desirable form of these equations, which is the *time-averaged* form.

Through equation (2.10.1), instantaneous density, velocities, enthalpy, scalar property, and so on can be expressed in terms of their mean and fluctuating quantities. Substituting into the governing equations and taking the time average, a system of equations commonly known as the **Reynolds-Averaged Navier-Stokes** equations can be derived, and they are expressed in compact form in Table 2.3. These conservation equations have been formulated based on the application of the Reynolds averaging rules:

Table 2.3 Reynolds-Averaged Navier-Stokes equations in Cartesian coordinates.

Time-Averaged Mass

$$\frac{\partial \overline{\rho}}{\partial t} + \frac{\partial}{\partial x_j}(\overline{\rho}\overline{u}_j + \overline{\rho'u_j'}) = 0 \qquad j = 1, 2, 3$$

Time-Averaged Momentum

$$\frac{\partial}{\partial t}(\overline{\rho}\overline{u}_i + \overline{\rho'u_i'}) + \frac{\partial}{\partial x_j}(\overline{\rho}\overline{u}_i\overline{u}_j + \overline{\rho}\overline{u_i'u_j'} + \overline{u}_i\overline{\rho'u_j'} + \overline{u}_j\overline{\rho'u_i'} + \overline{\rho'u_i'u_j'}) = -\frac{\partial \overline{\sigma}_{ij}}{\partial x_j} + \overline{S}_{u_i}$$

where

$$\overline{\sigma}_{ij} = \overline{p}\delta_{ij} - \mu\left(\frac{\partial \overline{u}_i}{\partial x_j} + \frac{\partial \overline{u}_i}{\partial x_j}\right) + \frac{2}{3}\mu\frac{\partial \overline{u}_i}{\partial x_j}\delta_{ij} \qquad i, j = 1, 2, 3$$

Time-Averaged Enthalpy

$$\frac{\partial}{\partial t}(\overline{\rho}\overline{h} + \overline{\rho'h'}) + \frac{\partial}{\partial x_j}(\overline{\rho}\overline{u}_j\overline{h} + \overline{\rho}\overline{u_j'h'} + \overline{u}_j\overline{\rho'h'} + \overline{h}\overline{\rho'u_j'} + \overline{\rho'u_j'h'})$$

$$= \frac{\partial}{\partial x_j}\left[\frac{k}{C_p}\frac{\partial \overline{h}}{\partial x_j}\right] + \overline{S}_h \qquad j = 1, 2, 3$$

Time-Averaged Scalar Property

$$\frac{\partial}{\partial t}(\overline{\rho}\overline{\varphi} + \overline{\rho'\varphi'}) + \frac{\partial}{\partial x_j}(\overline{\rho}\overline{u}_j\overline{\varphi} + \overline{\rho}\overline{u_j'\varphi'} + \overline{u}_j\overline{\rho'\varphi'} + \overline{\varphi}\overline{\rho'u_j'} + \overline{\rho'u_j'\varphi'})$$

$$= \frac{\partial}{\partial x_j}\left[\overline{\rho}D\frac{\partial \overline{\varphi}}{\partial x_j}\right] + \frac{\partial}{\partial x_j}\left[D\rho\overline{\frac{\partial \varphi'}{\partial x_j}}\right] + \overline{S}_\varphi \qquad j = 1, 2, 3$$

Note: $(u_1, u_2, u_3) \equiv (u, v, w)$; $(x_1, x_2, x_3) \equiv (x, y, z)$

$$\overline{\overline{\alpha}} = \overline{\alpha}; \quad \overline{\alpha + \beta} = \overline{\alpha} + \overline{\beta}; \quad \overline{\overline{\alpha}\beta} = \overline{\alpha}\overline{\beta}$$

$$\overline{\alpha\beta} = \overline{\alpha}\overline{\beta} + \overline{\alpha'\beta'}; \quad \overline{\frac{\partial \alpha}{\partial s}} = \frac{\partial \overline{\alpha}}{\partial s}; \quad \int \alpha ds = \int \overline{\alpha} ds \qquad (2.10.3)$$

where α and β are two dependent variables and s denotes any one of the independent variables x, y, z, and t. In Table 2.3, the Kronecker delta, δ_{ij} is given by $\delta_{ij} = 1$ if $i = j$ and $\delta_{ij} = 0$ if $i \neq j$. It is worthwhile noting that the fluctuations in the dynamics viscosity μ, thermal conductivity k, specific heat C_p, and mass diffusivity D are usually neglected in majority of practical applications.

The Favre-averaged approach (Favre, 1965, 1969) is described for the derivation of the conservation equations. If we define a mass-weighted mean property ϕ as

$$\tilde{\phi} = \frac{\overline{\rho\phi}}{\overline{\rho}} \qquad (2.10.4)$$

the instantaneous property ϕ may now be written according to

$$\phi = \tilde{\phi} + \phi'' \tag{2.10.5}$$

where ϕ'' is the superimposed velocity fluctuation. Multiplying equation (2.10.5) by density ρ, we obtain

$$\rho\phi = \rho(\tilde{\phi} + \phi'') = \rho\tilde{\phi} + \rho\phi''$$

By time-averaging the preceding equation,

$$\overline{\rho\phi} = \overline{\rho\tilde{\phi}} + \overline{\rho\phi''} \tag{2.10.6}$$

From the definition of equation (2.10.4), it follows that

$$\overline{\rho\phi''} = 0 \tag{2.10.7}$$

Substituting the instantaneous density, velocities, enthalpy, scalar property, and so on expressed in terms of their mass-weighted mean and fluctuating quantities in the form of equation (2.10.5) into the governing equations and taking the time average, the system of equations known as the **Favre-Averaged Navier-Stokes** equations can be alternatively expressed in compact form as shown in Table 2.4.

Table 2.4 Favre-Averaged Navier-Stokes equations in Cartesian coordinates.

Favre-Averaged Mass

$$\frac{\partial \overline{\rho}}{\partial t} + \frac{\partial}{\partial x_j}(\overline{\rho}\tilde{u}_j) = 0 \qquad j = 1, 2, 3$$

Favre-Averaged Momentum

$$\frac{\partial}{\partial t}(\overline{\rho}\tilde{u}_i) + \frac{\partial}{\partial x_j}(\overline{\rho}\tilde{u}_i\tilde{u}_j + \overline{\rho u_i'' u_j''}) = -\frac{\partial \overline{\sigma}_{ij}}{\partial x_j} + \overline{S}_{u_i}$$

where

$$\overline{\sigma}_{ij} = \overline{p}\delta_{ij} - \mu\left(\frac{\partial \tilde{u}_i}{\partial x_j} + \frac{\partial \tilde{u}_j}{\partial x_i}\right) + \frac{2}{3}\mu\frac{\partial \tilde{u}_i}{\partial x_j}\delta_{ij} \qquad i, j = 1, 2, 3$$

Favre-Averaged Enthalpy

$$\frac{\partial}{\partial t}(\overline{\rho}\tilde{h}) + \frac{\partial}{\partial x_j}(\overline{\rho}\tilde{u}_j\tilde{h} + \overline{\rho u_j'' h''}) = \frac{\partial}{\partial x_j}\left[\frac{k}{C_p}\frac{\partial \tilde{h}}{\partial x_j}\right] + \overline{S}_h \qquad j = 1, 2, 3$$

Favre-Averaged Scalar Property

$$\frac{\partial}{\partial t}(\overline{\rho}\tilde{\varphi}) + \frac{\partial}{\partial x_j}(\overline{\rho}\tilde{u}_j\tilde{\varphi} + \overline{\rho u_j'' \varphi''}) = \frac{\partial}{\partial x_j}\left[\overline{\rho}D\frac{\partial \tilde{\varphi}}{\partial x_j}\right] + \frac{\partial}{\partial x_j}\left[D\rho\overline{\frac{\partial \varphi''}{\partial x_j}}\right] + \overline{S}_\varphi \qquad j = 1, 2, 3$$

Note: $(u_1, u_2, u_3) \equiv (u, v, w)$; $(x_1, x_2, x_3) \equiv (x, y, z)$

By comparing Tables 2.3 and 2.4, it is evident that time-averaging results is a completely different system of equations from Favre-averaging. The equations through Favre-averaging are much simpler by the mere presence of only the Reynolds and scalar stress terms: $\overline{\rho u_i'' u_j''}$, $\overline{\rho u_i'' h''}$, and $\overline{\rho u_i'' \varphi''}$. Since most experimental sampling probes measure values that approximate mass-weighted concentrations rather than time-averaged concentrations, Favre-averaged equations are very amenable to obtain practical solutions in many numerically related fire investigations. To solve the system of equations, the Reynolds and scalar stresses must be related to the mean quantities of the flow field. In the next section, the approach based on the eddy or turbulent viscosity concept is introduced to determine the appropriate relationships of these stresses.

2.11 Additional Equations for Turbulent Flow—Standard *k-ε* Turbulence Model

Boussinesq (1877) first introduced the eddy or turbulent viscosity concept by suggesting the possibility of linking Reynolds stresses to the mean rates of eddy deformation. In the Favre-averaged form of the momentum equation, the hypothesis effectively expresses the Reynolds stresses to the mean velocity gradients through the relation:

$$-\overline{\rho u_i'' u_j''} = -\overline{\rho \widetilde{u_i'' u_j''}} = \mu_T \left(\frac{\partial \tilde{u}_i}{\partial x_j} + \frac{\partial \tilde{u}_j}{\partial x_i} \right) - \frac{2}{3} \left(\mu_T \frac{\partial \tilde{u}_i}{\partial x_j} + \overline{\rho} k \right) \delta_{ij} \qquad (2.11.1)$$

where μ_T is the turbulent or eddy viscosity. The right-hand side of equation (2.11.1) is analogous to *Newton's law of viscosity*, which applies to a laminar flow, except for the appearance of the turbulent viscosity μ_T—a function of the flow rather than of the fluid—and turbulent kinetic energy k. Similarly, the scalar stress for the enthalpy may also be taken to be proportional to the gradient of the mean value of the transported quantity. In order words,

$$-\overline{\rho u_i'' h''} = -\overline{\rho \widetilde{u_i'' h''}} = \Gamma_h^T \frac{\partial \tilde{h}}{\partial x} \qquad (2.11.2)$$

where Γ_h^T is the turbulent diffusivity for enthalpy. Since the turbulent transport of momentum and heat is due to the same mechanisms—eddy mixing—it is conceivable that the value of the turbulent viscosity in equation (2.11.2) can be taken to be close to that of turbulent viscosity μ_T. Based on the definition of the laminar Prandtl number given as

$$Pr \equiv \frac{\mu C_p}{k} = \frac{\text{Molecular diffusivity of momentum}}{\text{Molecular diffusivity of heat}} \qquad (2.11.3)$$

the turbulent Prandtl number Pr_T may be similarly defined as

$$Pr_T = \frac{\mu_T}{\Gamma_h^T} \qquad (2.11.4)$$

For the scalar property, the scalar stress for the scalar property may also be expressed in terms of mean quantity like the enthalpy as

$$-\overline{\rho u_i'' \varphi''} = -\overline{\rho} \widetilde{u_i'' \varphi''} = \Gamma_\varphi^T \frac{\partial \tilde{\varphi}}{\partial x} \qquad (2.11.5)$$

where Γ_φ^T is the turbulent diffusivity for scalar property. The turbulent transport of the scalar property should behave in an analogous manner to the turbulent transport of momentum and heat by eddy mixing. Through the definition of the laminar Schmidt number:

$$Sc \equiv \frac{\mu \rho}{D} = \frac{\text{Molecular diffusivity of momentum}}{\text{Molecular diffusivity of mass}} \qquad (2.11.6)$$

the turbulent Schmidt number Sc_T may also be similarly obtained as

$$Sc_T = \frac{\mu_T}{\Gamma_\varphi^T} \qquad (2.11.7)$$

To satisfy dimensional requirements, at least two scaling parameters are required to relate the Reynolds stresses to the rate of deformation. In most engineering flow problems, the complexity of turbulence precludes the use of any simple formulae. A feasible choice is the turbulent quantity k and other turbulent quantities, one of which is the rate of dissipation of turbulent energy ε. A typical two-equation turbulence model commonly used in handling many turbulent fluid engineering problems is the *standard k-ε model* by Launder and Spalding (1974).

The local turbulent viscosity μ_T can be either obtained from dimensional analysis or from analogy to the laminar viscosity as $\mu_T \propto \overline{\rho} v_t l$. Based on the characteristic velocity v_t defined as $k^{1/2}$, and the characteristic length l as $k^{3/2}/\varepsilon$, the turbulent viscosity μ_T is given as:

$$\mu_T = \overline{\rho} C_\mu \frac{k^2}{\varepsilon} \qquad (2.11.8)$$

where C_μ is an empirical constant. The turbulent kinetic energy k and the rate of dissipation of turbulent energy ε are respectively defined as $k = \frac{1}{2}\widetilde{u_i'' u_i''}$ and $\varepsilon = \frac{\mu_T}{\rho}\overline{\left(\frac{\partial u_i''}{\partial x_j}\right)\left(\frac{\partial u_i''}{\partial x_j}\right)}$. In order to evaluate the turbulent viscosity in equation (2.11.8), the values of k and ε must be known, which are generally obtained through solution of their respective transport equations. The derivation of the governing equations for k and ε involves substantial mathematical manipulation of the instantaneous Navier-Stokes equations (2.4.22)–(2.4.24) alongside with their averaged counterparts in Table 2.4. It is not our intention to burden the reader with detailed developments of these equations. There are many established literatures and books—for example, Tennekes and Lumley (1972) and Versteeg and Malasekera (1995)—where the treatises of these equations have been carried out with much vigor. The reader may wish to refer to them for a more in-depth understanding of the specific considerations in formulating these transport equations. Physically, the transport equations bear many similarities with the generic transport equation. In other words,

$$
\begin{array}{c}
\textit{The local} \\
\textit{rate of} \\
\textit{change of} \\
\textit{k or } \varepsilon
\end{array}
+
\begin{array}{c}
\textit{The transport} \\
\textit{of k or } \varepsilon \\
\textit{by advection}
\end{array}
=
\begin{array}{c}
\textit{The transport of} \\
\textit{k or } \varepsilon \textit{ by} \\
\textit{diffusion}
\end{array}
+
\begin{array}{c}
\textit{The net source} \\
\textit{rate of k or } \varepsilon
\end{array}
\qquad (2.11.9)
$$

The net source rate of k or ε in equation (2.11.9) includes the rate of production as well as the rate of destruction of k or ε. Additional transport equations for turbulent flow expressed in compact form for the *standard k-ε model* are given by:

$$
\frac{\partial}{\partial t}(\overline{\rho}k) + \frac{\partial}{\partial x_j}(\overline{\rho}u_j k) = \frac{\partial}{\partial x_j}\left[\frac{\mu_T}{\sigma_k}\frac{\partial k}{\partial x_j}\right] \underbrace{- \overline{\rho u_i'' u_j''}\frac{\partial \tilde{u}_i}{\partial x_j}}_{production} - \underbrace{\overline{\rho}\varepsilon}_{destruction}
$$

$$
\frac{\partial}{\partial t}(\overline{\rho}\varepsilon) + \frac{\partial}{\partial x_j}(\overline{\rho}u_j \varepsilon) = \frac{\partial}{\partial x_j}\left[\frac{\mu_T}{\sigma_\varepsilon}\frac{\partial \varepsilon}{\partial x_j}\right] \underbrace{- C_{\varepsilon 1}\frac{\varepsilon}{k}\left(\overline{\rho u_i'' u_j''}\frac{\partial \tilde{u}_i}{\partial x_j}\right)}_{production} - \underbrace{C_{\varepsilon 2}\overline{\rho}\frac{\varepsilon^2}{k}}_{destruction}
$$

Substituting equation (2.11.1) into the above two equations yields

$$
\frac{\partial}{\partial t}(\overline{\rho}k) + \frac{\partial}{\partial x_j}(\overline{\rho}u_j k) = \frac{\partial}{\partial x_j}\left[\frac{\mu_T}{\sigma_k}\frac{\partial k}{\partial x_j}\right]
$$

$$
\underbrace{+ \mu_T\frac{\partial \tilde{u}_i}{\partial x_j}\left(\frac{\partial \tilde{u}_i}{\partial x_j} + \frac{\partial \tilde{u}_j}{\partial x_i}\right) - \frac{2}{3}\frac{\partial \tilde{u}_i}{\partial x_j}\left(\mu_T\frac{\partial \tilde{u}_i}{\partial x_j} + \overline{\rho}k\right)\delta_{ij}}_{production} - \underbrace{\overline{\rho}\varepsilon}_{destruction}
$$

$$
(2.11.10)
$$

$$\frac{\partial}{\partial t}(\bar{\rho}\varepsilon) + \frac{\partial}{\partial x_j}(\bar{\rho}\bar{u}_j\varepsilon) = \frac{\partial}{\partial x_j}\left[\frac{\mu_T}{\sigma_\varepsilon}\frac{\partial \varepsilon}{\partial x_j}\right]$$

$$+ \underbrace{C_{\varepsilon 1}\frac{\varepsilon}{k}\left[\mu_T\frac{\partial u_i}{\partial x_j}\left(\frac{\partial u_i}{\partial x_j}+\frac{\partial u_j}{\partial x_i}\right) - \frac{2}{3}\frac{\partial u_i}{\partial x_j}\left(\mu_T\frac{\partial u_i}{\partial x_j}+\bar{\rho}k\right)\delta_{ij}\right]}_{\text{production}} - \underbrace{C_{\varepsilon 2}\bar{\rho}\frac{\varepsilon^2}{k}}_{\text{destruction}}$$

$$(2.11.11)$$

The constants for the *standard k-ε model* have been arrived through comprehensive data fitting for a wide range of turbulent flows (see Launder and Spalding, 1974):

$$C_\mu = 0.09, \quad \sigma_k = 1.0, \quad \sigma_\varepsilon = 1.3, \quad C_{\varepsilon 1} = 1.44, \quad C_{\varepsilon 2} = 1.92.$$

The production and destruction of turbulent kinetic energy are always closely linked in the k-equation (2.11.10). When the dissipation rate ε is large, the production of k is also large. Equation (2.11.11) assumes that the production and destruction terms are proportional to the production and destruction terms of the k-equation. Adoption of such terms ensures that ε increases rapidly if k increases rapidly and that it decreases sufficiently fast to avoid non-physical (negative) values of turbulent kinetic energy if k decreases. The factor ε / k in the production and destruction terms in equation (2.11.11) makes these terms dimensionally correct in the ε-equation.

On the basis of equations (2.11.1), (2.11.4) and (2.11.7), the eddy viscosity hypothesis provides closure for the system of **Favre-Averaged Navier-Stokes** equations derived in Table 2.4, which in terms of the turbulent viscosity μ_T is summarized in Table 2.5. Note that the term k / C_p appearing in the Favre-averaged enthalpy equation is usually replaced by an alternative expression in terms of the laminar viscosity μ and laminar Prandtl number Pr ($= \mu\, C_p / k$) and the term $\frac{\partial}{\partial x_j}\left[\overline{D\rho\frac{\partial \varphi''}{\partial x_j}}\right]$ appearing in the left-hand side of the Favre-averaged scalar property equation is omitted in most cases. It is also evidently clear that the system of equations in Table 2.5 corresponds to the generic form of the transport equation.

2.12 Other Turbulence Models

Turbulent states are encountered across the whole range of fluid flows. They are generally very rich, complex, and varied. No single turbulence model can thus far be readily employed to span these states, since none is expected to be universally valid for all types of flows. The Favre-averaged approach as

Table 2.5 Favre-Averaged Navier-Stokes equations in Cartesian coordinates based on the eddy viscosity concept.

Favre-Averaged Mass

$$\frac{\partial \overline{\rho}}{\partial t} + \frac{\partial}{\partial x_j}(\overline{\rho}\tilde{u}_j) = 0 \qquad j = 1,2,3$$

Favre-Averaged Momentum

$$\frac{\partial}{\partial t}(\overline{\rho}\tilde{u}_i) + \frac{\partial}{\partial x_j}(\overline{\rho}\tilde{u}_i\tilde{u}_j) = \frac{\partial}{\partial x_j}\left[(\mu + \mu_T)\frac{\partial \tilde{u}_i}{\partial x_j}\right] + \overline{S}'_{u_i}$$

where

$$\overline{S}'_{u_i} = -\frac{\partial \overline{p}}{\partial x_j}\delta_{ij} + \frac{\partial}{\partial x_j}\left[(\mu + \mu_T)\left(\frac{\partial \tilde{u}_i}{\partial x_j} + \frac{\partial \tilde{u}_j}{\partial x_i}\right)\right]$$

$$-\frac{\partial}{\partial x_j}\left[\frac{2}{3}(\mu + \mu_T)\frac{\partial \tilde{u}_i}{\partial x_j}\delta_{ij} + \overline{\rho}k\delta_{ij}\right] + \sum F_{x_i}^{body\ forces} \qquad i,j = 1,2,3$$

Favre-Averaged Enthalpy

$$\frac{\partial}{\partial t}(\overline{\rho}\tilde{h}) + \frac{\partial}{\partial x_j}(\overline{\rho}\tilde{u}_j\tilde{h}) = \frac{\partial}{\partial x_j}\left[\left(\frac{\mu}{Pr} + \frac{\mu_T}{Pr_T}\right)\frac{\partial \tilde{h}}{\partial x_j}\right] + \overline{S}_h \qquad j = 1,2,3$$

Favre-Averaged Scalar Property

$$\frac{\partial}{\partial t}(\overline{\rho}\tilde{\varphi}) + \frac{\partial}{\partial x_j}(\overline{\rho}\tilde{u}_j\tilde{\varphi}) = \frac{\partial}{\partial x_j}\left[\left(\overline{\rho}D + \frac{\mu_T}{Sc_T}\right)\frac{\partial \tilde{\varphi}}{\partial x_j}\right] + \overline{S}_\varphi \qquad j = 1,2,3$$

Note: $(u_1,u_2,u_3) \equiv (u,v,w)$; $(x_1,x_2,x_3) \equiv (x,y,z)$

described in the previous section to turbulent flow results in the formulation of the *standard k-ε model* as proposed by Launder and Spalding (1974). For most engineering purposes, this model has been demonstrated to yield sensible solutions to most industrially relevant flows. It has been widely validated and is well established in the CFD community.

For compartment fires, the *standard k-ε model* has been employed with remarkable success in handling recirculating and confined flows. A survey of currently available turbulence models in commercial CFD codes as well as in literature reveals, however, other turbulence models that could also be employed in field modeling. As an alternative to the *standard k-ε model*, other eddy viscosity models, such as *RNG k-ε model* and *realizable k-ε model* proposed by Yakhot et al. (1992) and Shih et al. (1995), are possible recommendations. The improved features of these models have shown to be aptly applicable to predict important flow cases having flow separation, flow re-attachment, flow recovery, and some unconfined flows (e.g., free shear jet) of which may suffice in some fire engineering investigations.

The *standard k-ε model* that is a consequence of the eddy viscosity hypothesis assumes that the turbulent stresses are linearly related to the rate of strain by a scalar turbulent viscosity. The principal strain directions are always aligned to the principal stress directions; they behave in an *isotropic* manner. In some flow cases, secondary flows that exist within the geometry are driven by *anisotropic* normal Reynolds stresses. As such, the assumption of *isotropic* normal Reynolds stresses is unrealistic for such kinds of flows, and a more complicated approach by evaluating each Reynolds stresses $u_i''u_j''$ is required. The *Reynolds Stress Model*, also called the second-moment closure model, determines the turbulent stresses directly by solving a transport equation for each stress component. These equations represent the turbulent transport, generation, dissipation, and redistribution of Reynolds stresses. An additional equation for the dissipation ε is solved to provide a length scale determining quantity. Interested readers can consult relevant texts by Launder et al. (1989) and Rodi (1993) for the description of this model. There is no doubt that the *Reynolds stress model* has a greater potential to represent the turbulent flow phenomena more correctly than the *standard k-ε model*. This type of model can handle complex strain and, in principle, can cope with non-equilibrium flows.

For wall-attached boundary layers such as those prevalent in compartment fires, turbulent fluctuations are suppressed adjacent to the wall, and the viscous effects become prominent in this region known as the *viscous sub-layer*. Consider the fire plume interacting with a ceiling within a section of the compartment housing the fire source as shown in Figure 2.19. The modified turbulent structure of near-wall ceiling flow generally precludes the application of the two-equation models such as *standard k-ε model*, *RNG k-ε model*, and *realizable k-ε model* or even the *Reynolds Stress Model* at the near-wall region. One common approach is to adopt the so-called wall-function method; the near-wall region is bridged with *wall functions* to avoid resolving the *viscous*

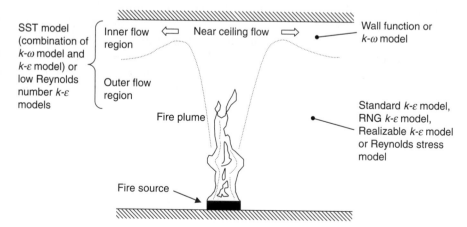

Figure 2.19 A schematic illustration of the fire plume and its interaction with the ceiling of a compartment and appropriate models to resolve the turbulent flow characteristics.

sub-layer. More discussions on the use of *wall functions* will be provided in the next section. It is also possible to totally resolve the *viscous sub-layer* by the application of *low Reynolds number turbulence models*. Here, the *standard k-ε model* is modified by the introduction of wall damping functions to ensure that viscous stresses dominate over the turbulent Reynolds stresses at low Reynolds numbers and in the viscous sub-layer adjacent to solid walls. Different version of the *low Reynolds number k-ε model* by Johns and Launder (1972), Chien (1980), and Lam and Bremhorst (1981) are models that have achieved some considerable success in resolving wall-bounded flows. Another model developed by Wilcox (1998), the *standard k-ω model*, where ω is a frequency of the large eddies, has also shown to perform splendidly close to walls in boundary layer flows. The *standard k-ω model* is nevertheless very sensitive to the free-stream conditions, and unless great care is exercised, spurious results are obtained in flow regions away from the solid walls. To overcome such problems, the *SST (Shear Stress Transport)* variation of *Menter's model* (1993, 1996) was developed with the aim of combining the favorable features of the *standard k-ε model* with the *standard k-ω model* in order that the inner region of the boundary layer is adequately resolved by the latter while the former is employed to obtain numerical solutions in the outer part of the boundary layer. This model is increasing being employed and works exceptionally well in handling non-equilibrium boundary layer regions such as flow separation.

The formulations of the two-equation eddy viscosity models such as *RNG k-ε model*, *realizable k-ε model*, *low Reynolds number turbulence models*, and *SST model*, as well as the *Reynolds Stress Model* are briefly discussed in the subsequent sections below. More details on these respective turbulent models can be found in the previously indicated reference texts.

2.12.1 *Variant of Standard k-ε Turbulence Models*

Some pertinent differences of the *RNG k-ε model* and *realizable k-ε model* in comparison to the *standard k-ε model* are highlighted. The *RNG k-ε model* is based on the renomalization group theory (a rigorous statistical technique) analysis of the Navier-Stokes equations. The transport *k*-equation remains the same as in the *standard k-ε model* except for model constants. Modifications are nevertheless made to the ε-equation whereby an additional term R_ε is introduced into the source term according to:

$$\frac{\partial}{\partial t}(\bar{\rho}\varepsilon) + \frac{\partial}{\partial x_j}(\bar{\rho}\tilde{u}_j\varepsilon) = \frac{\partial}{\partial x_j}\left[\frac{\mu_T}{\sigma_\varepsilon}\frac{\partial \varepsilon}{\partial x_j}\right]$$

$$+ \underbrace{C_{\varepsilon 1}\frac{\varepsilon}{k}\left[\mu_T\frac{\partial \tilde{u}_i}{\partial x_j}\left(\frac{\partial \tilde{u}_i}{\partial x_j}+\frac{\partial \tilde{u}_j}{\partial x_i}\right) - \frac{2}{3}\frac{\partial \tilde{u}_i}{\partial x_j}\left(\mu_T\frac{\partial \tilde{u}_i}{\partial x_j}+\bar{\rho}k\right)\delta_{ij}\right]}_{production} - \underbrace{C_{\varepsilon 2}\bar{\rho}\frac{\varepsilon^2}{k} - R_\varepsilon}_{destruction}$$

$$(2.12.1)$$

In the *standard k-ε model*, the rate of strain term R_ε in the preceding equation is absent. This R_ε term according to Yakhot and Orzag (1986) is formulated as:

$$R_\varepsilon = \frac{C_\mu \bar{\rho} \eta^3 (1 - \eta/\eta_o)}{1 + \beta \eta^3} \frac{\varepsilon^2}{k} \qquad (2.12.2)$$

where

$$\eta = S\frac{k}{\varepsilon}; \qquad S = \sqrt{2 S_{ij} S_{ij}}; \qquad S_{ij} = \frac{1}{2}\left(\frac{\partial u_i}{\partial x_j} + \frac{\partial u_j}{\partial x_i}\right)$$

and β and η_o are constants with values of 0.015 and 4.38. In flow regions where $\eta < \eta_o$, the R_ε term makes a positive contribution. For weakly to moderately strained flows, the *RNG k-ε model* tends to yield numerical results that are largely comparable to the *standard k-ε model*. In flow regions where $\eta > \eta_o$, the R_ε term makes, however, a negative contribution. This is rather significant, since for rapidly strained flows, the *RNG k-ε model* yields a lower turbulent viscosity than the *standard k-ε model* due to the term R_ε compensating the destruction of ε in the source term (see equation (2.12.1)). Hence, the effects of rapid strain and streamline curvature are better accommodated through the *RNG k-ε model* for a wider class of flows than the *standard k-ε model*. According to renomalization group theory, the constants in the turbulent transport equations are given by:

$$C_\mu = 0.0845, \quad \sigma_k = 0.718, \quad \sigma_\varepsilon = 0.718, \quad C_{\varepsilon 1} = 1.42, \quad C_{\varepsilon 2} = 1.68.$$

It is worthwhile noting that the value of C_μ is very close to the empirically determined value of 0.09 in the *standard k-ε model*.

For the *realizable k-ε model*, the term *realizable* means that the model satisfies certain mathematical constraints on the normal Reynolds stresses, consistent with the physics of turbulent flows. The model's core aspect in ensuring realizability (positive of normal stresses) is to purposefully make C_μ variable by sensitizing it to the mean flow (mean deformation) and the turbulence quantities (k, ε). This involves the formulation of a new eddy-viscosity formula for the variable C_μ in the turbulent viscosity relationship. The model also differs in the changes imposed to the transport ε-equation (based on the dynamic equation of the mean-square vorticity fluctuation) where the source term is now solved according to:

$$\frac{\partial}{\partial t}(\bar{\rho}\varepsilon) + \frac{\partial}{\partial x_j}(\bar{\rho}\bar{u}_j \varepsilon) = \frac{\partial}{\partial x_j}\left[\frac{\mu_T}{\sigma_\varepsilon}\frac{\partial \varepsilon}{\partial x_j}\right] + \underbrace{C_1 \bar{\rho} S \varepsilon}_{production}$$

$$\underbrace{- C_2 \bar{\rho}\frac{\varepsilon^2}{k + \sqrt{(\mu_T/\bar{\rho})\varepsilon}}}_{destruction}, \quad S = \sqrt{2 S_{ij} S_{ij}}, \quad S_{ij} = \frac{1}{2}\left(\frac{\partial u_i}{\partial x_j} + \frac{\partial u_j}{\partial x_i}\right) \qquad (2.12.3)$$

and the variable constant C_1 is expressed as:

$$C_1 = \max\left[0.43, \frac{\eta}{\eta + 5}\right]; \quad \eta = S\frac{k}{\varepsilon}$$

The variable C_μ, no longer a constant, is evaluated from:

$$C_\mu = \frac{1}{A_o + A_s \frac{kU^*}{\varepsilon}} \tag{2.12.4}$$

Consequently, model constants A_o and A_s are determined as:

$$A_o = 4.04, \quad A_s = \sqrt{6}\cos\Theta, \quad \Theta = \frac{1}{3}\cos^{-1}(\sqrt{6}W), \quad W = \frac{S_{ij}S_{jk}S_{ki}}{\tilde{S}^3},$$

$$\tilde{S} = \sqrt{S_{ij}S_{ij}}$$

while the parameter U^* is given by:

$$U^* \equiv \sqrt{S_{ij}S_{ij} + \tilde{\Omega}_{ij}\tilde{\Omega}_{ij}}, \quad \tilde{\Omega}_{ij} = \Omega_{ij} - 2e_{ijk}\omega_k, \quad \Omega_{ij} = \tilde{\Omega}_{ij} - e_{ijk}\omega_k$$

where $\tilde{\Omega}_{ij}$ is the mean rate-of rotation viewed in a rotating frame with the angular rotation vector ω_k; $e_{ijk} = +1$ if i, j, and k are different and in cyclic order, $e_{ijk} = -1$ if i, j, and k are different and in anti-cyclic order and $e_{ijk} = 0$ if any two indices are the same. Other constants in the turbulent transport equations for this model are $C_2 = 1.9$, $\sigma_\kappa = 1.0$, and $\sigma_\varepsilon = 1.2$, respectively. Here, the transport k-equation in *realizable k-ε model* is the same as that in the *standard k-ε model* except for model constants. One noteworthy feature of this model is that the production term in the ε-equation is different from those of the *standard k-ε model* and *RNG k-ε model*. It is believed the form suggested as in equation (2.12.3) represented better spectral energy transfer in the turbulent flow. Another important feature is that the destruction term does not have any singularity (i.e., its denominator never vanishes), even if the turbulent kinetic energy k vanishes or becomes smaller than zero. The *realizable k-ε model* has been found to be superior in accurately predicting a wide range of flows including free flows including jets and mixing layers, channel and boundary layer flows, and separated flows.

In order to allow calculation of turbulent flows at low Reynolds number, probably in the range 5000 to 30000, modifications are introduced to the *standard k-ε model* to cope with such flows. This model commonly known as *low Reynolds number k-ε model* involves a damping of the eddy viscosity when the local turbulent Reynolds number is low, a modified definition of the dissipation ε that either goes to zero or a prescribed zero normal gradient at the wall and specific modifications of the source terms in the transport ε-equation. The equations of the *low Reynolds number k-ε model* become:

$$\mu_T = \bar{\rho} C_\mu f_\mu \frac{k^2}{\varepsilon} \tag{2.12.5}$$

$$\frac{\partial}{\partial t}(\bar{\rho}k) + \frac{\partial}{\partial x_j}(\overline{\rho u_j}k) = \frac{\partial}{\partial x_j}\left[\mu + \frac{\mu_T}{\sigma_k}\frac{\partial k}{\partial x_j}\right]$$

$$\underbrace{+\mu_T \frac{\partial \tilde{u}_i}{\partial x_j}\left(\frac{\partial \tilde{u}_i}{\partial x_j} + \frac{\partial \tilde{u}_j}{\partial x_i}\right) - \frac{2}{3}\frac{\partial \tilde{u}_i}{\partial x_j}\left(\mu_T \frac{\partial \tilde{u}_i}{\partial x_j} + \bar{\rho}k\right)\delta_{ij}}_{production} - \underbrace{\bar{\rho}\varepsilon}_{destruction} + D$$

$$\tag{2.12.6}$$

$$\frac{\partial}{\partial t}(\bar{\rho}\varepsilon) + \frac{\partial}{\partial x_j}(\overline{\rho u_j}\varepsilon) = \frac{\partial}{\partial x_j}\left[\mu + \frac{\mu_T}{\sigma_\varepsilon}\frac{\partial \varepsilon}{\partial x_j}\right]$$

$$\underbrace{+C_{\varepsilon 1}f_1 \frac{\varepsilon}{k}\left[\mu_T \frac{\partial \tilde{u}_i}{\partial x_j}\left(\frac{\partial \tilde{u}_i}{\partial x_j} + \frac{\partial \tilde{u}_j}{\partial x_i}\right) - \frac{2}{3}\frac{\partial \tilde{u}_i}{\partial x_j}\left(\mu_T \frac{\partial \tilde{u}_i}{\partial x_j} + \bar{\rho}k\right)\delta_{ij}\right]}_{production} - \underbrace{C_{\varepsilon 2}f_2\bar{\rho}\frac{\varepsilon^2}{k}}_{destruction} + E$$

$$\tag{2.12.7}$$

Here, the modifications are made by including the viscous contribution (laminar viscosity) in the diffusion terms in both the k-equation and ε-equation and an additional term D in the k-equation and an additional term E in the ε-equation. Wall-damping functions f_μ, f_1 and f_2 are also introduced, and they are incorporated into the eddy or turbulent viscosity expression and ε-equation, respectively. Based on the various versions of the *low Reynolds number k-ε models* as aforementioned, the model constants, wall-damping functions, and additional terms D and E are formulated as:

Low Reynolds number k-ε models of Johns and Launder (1972)

$$C_\mu = 0.09, \quad \sigma_k = 1.0, \quad \sigma_\varepsilon = 1.3, \quad C_{\varepsilon 1} = 1.44, \quad C_{\varepsilon 2} = 1.92.$$

$$f_\mu = \exp\left(-\frac{2.5}{1 + Re_t/50}\right), \quad f_1 = 0, \quad f_2 = 1 - 0.3\exp(-Re_t^2),$$

$$D = -2\mu\left[\left(\frac{\partial\sqrt{k}}{\partial x}\right)^2 + \left(\frac{\partial\sqrt{k}}{\partial y}\right)^2 + \left(\frac{\partial\sqrt{k}}{\partial z}\right)^2\right],$$

$$E = -2\mu\mu_T\left[\left(\frac{\partial^2 u}{\partial y^2}\right)^2 + \left(\frac{\partial^2 v}{\partial x^2}\right)^2 + \left(\frac{\partial^2 u}{\partial z^2}\right)^2 + \left(\frac{\partial^2 w}{\partial x^2}\right)^2 + \left(\frac{\partial^2 v}{\partial z^2}\right)^2 + \left(\frac{\partial^2 w}{\partial y^2}\right)^2\right]$$

Low Reynolds number k-ε models of Chien (1980)

$$C_\mu = 0.09, \quad \sigma_k = 1.0, \quad \sigma_\varepsilon = 1.3, \quad C_{\varepsilon1} = 1.35, \quad C_{\varepsilon2} = 1.8.$$

$$f_\mu = 1 - \exp(-0.0115y^+), \quad f_1 = 0, \quad f_2 = 1 - \frac{2}{9}\exp\left(-(Re_t/6)^2\right)$$

$$D = -2\mu\frac{k}{d_n^2}, \quad E = -2\mu\frac{\varepsilon}{d_n^2}\exp(-0.5y^+)$$

Low Reynolds number k-ε models of Lam and Bremhorst (1981)

$$C_\mu = 0.09, \quad \sigma_k = 1.0, \quad \sigma_\varepsilon = 1.3, \quad C_{\varepsilon1} = 1.44, \quad C_{\varepsilon2} = 1.92$$

$$f_\mu = [1 - \exp(-0.0165Re_y)]^2\left(1 + \frac{20.5}{Re_t}\right), \quad f_1 = \left(1 + \frac{0.05}{f_\mu}\right)^3,$$

$$f_2 = 1 - \exp(-Re_t^2)$$

$$D = 0, \ E = 0$$

In the preceding relationships, the following dimensionless variables are defined as: $y^+ = \overline{\rho}d_n u_\tau/\mu$, $Re = \overline{\rho}k^2/(\mu\varepsilon)$, and $Re_y = \overline{\rho}k^{1/2}d_n/\mu$, where d_n is the distance closest to a fixed wall and $u_\tau = (\tau_w/\overline{\rho})^{1/2}$ is the so-called friction velocity.

As an alternative to the *low Reynolds number k-ε models*, the *standard k-ω model* by Wilcox (1998) represents another useful model for the near wall treatment for low Reynolds number flow computations. The model does not involve complex non-linear damping functions such as that described in the *low Reynolds number k-ε models*, and it should therefore be construed as being more robust and, in some circumstances, more accurate. The *standard k-ω model* is an empirical model that is based on the transport equations of the turbulent kinetic energy k and the turbulent frequency ω, considered as the ratio of ε to k—that is, $\omega = \varepsilon / k$. To formulate the *SST model*, the *standard k-ε model* is required to be transformed into a form consistent with the *k-ω* formulation. A blending function F_1 is introduced whereby the *standard k-ω model* is multiplied by this function F_1 and the transformed *k-ε model* by a function $1 - F_1$. At the boundary layer edge, and outside the boundary layer, the *standard k-ε model* is recovered when $F_1 = 0$. The equations of the *SST model* are given as:

$$\mu_T = \frac{\overline{\rho} a_1 k}{\max(a_1 \omega, SF_2)}, \quad S = \sqrt{2S_{ij}S_{ij}}, \quad S_{ij} = \frac{1}{2}\left(\frac{\partial u_i}{\partial x_j} + \frac{\partial u_j}{\partial x_i}\right) \tag{2.12.8}$$

$$\frac{\partial}{\partial t}(\overline{\rho} k) + \frac{\partial}{\partial x_j}(\overline{\rho} \overline{u}_j k) = \frac{\partial}{\partial x_j}\left[\mu + \frac{\mu_T}{\sigma_{k3}}\frac{\partial k}{\partial x_j}\right]$$

$$\underbrace{+\mu_T \frac{\partial \tilde{u}_i}{\partial x_j}\left(\frac{\partial \tilde{u}_i}{\partial x_j} + \frac{\partial \tilde{u}_j}{\partial x_i}\right) - \frac{2}{3}\frac{\partial \tilde{u}_i}{\partial x_j}\left(\mu_T \frac{\partial \tilde{u}_i}{\partial x_j} + \overline{\rho} k\right)\delta_{ij}}_{\text{production}} - \underbrace{\overline{\rho}\beta' k\omega}_{\text{destruction}} \tag{2.12.9}$$

$$\frac{\partial}{\partial t}(\overline{\rho}\omega) + \frac{\partial}{\partial x_j}(\overline{\rho}\overline{u}_j\omega) = \frac{\partial}{\partial x_j}\left[\mu + \frac{\mu_T}{\sigma_{\omega3}}\frac{\partial \omega}{\partial x_j}\right] + \underbrace{2\overline{\rho}(1 - F_1)\frac{1}{\sigma_{\omega2}\omega}\frac{\partial k}{\partial x_j}\frac{\partial \omega}{\partial x_j}}_{\text{cross-diffusion modification}}$$

$$\underbrace{+\alpha_3 \frac{\omega}{k}\left[\mu_T \frac{\partial \tilde{u}_i}{\partial x_j}\left(\frac{\partial \tilde{u}_i}{\partial x_j} + \frac{\partial \tilde{u}_j}{\partial x_i}\right) - \frac{2}{3}\frac{\partial \tilde{u}_i}{\partial x_j}\left(\mu_T \frac{\partial \tilde{u}_i}{\partial x_j} + \overline{\rho} k\right)\delta_{ij}\right]}_{\text{production}} - \underbrace{\overline{\rho}\beta_3\omega^2}_{\text{destruction}}$$

$$\tag{2.12.10}$$

where

$$\sigma_{k3} = F_1\sigma_{k1} + (1 - F_1)\sigma_{k2}, \quad \sigma_{\omega3} = F_1\sigma_{\omega1} + (1 - F_1)\sigma_{\omega2},$$

$$\alpha_3 = F_1\alpha_1 + (1 - F_1)\alpha_2, \quad \beta_3 = F_1\beta_1 + (1 - F_1)\beta_2.$$

The success of this model hinges on the use of appropriate blending functions of F_1 and F_2. For function F_1, it is given by

$$F_1 = \tanh(\Phi_1^4) \tag{2.12.11}$$

with

$$\Phi_1 = \min\left[\max\left(\frac{\sqrt{k}}{0.09\omega d_n}, \frac{500\mu}{\overline{\rho}\omega d_n^2}\right), \frac{4\overline{\rho}k}{D_\omega^+ \sigma_{\omega2} d_n^2}\right] \tag{2.12.12}$$

The variable D_ω^+ appearing in equation (2.12.12) is evaluated according to

$$D_\omega^+ = \max\left(2\overline{\rho}\frac{1}{\sigma_{\omega2}\omega}\frac{\partial k}{\partial x_j}\frac{\partial \omega}{\partial x_j}, 10^{-10}\right) \tag{2.12.13}$$

Note the use of the cross-diffusion modification term without the function $(1 - F_1)$ in the transport ω-*equation* in determining the condition in equation (2.12.13). For function F_2, it can be similarly expressed as

$$F_2 = \tanh(\Phi_2^2) \tag{2.12.14}$$

with

$$\Phi_2 = \max\left(\frac{2\sqrt{k}}{0.09\omega d_n}, \frac{500\mu}{\bar{\rho}\omega d_n^2}\right) \tag{2.12.15}$$

As in the *low Reynolds number k-ε models*, the *SST model* also requires d_n, the distance to the nearest wall, as stipulated in equations (2.12.12) and (2.12.15) to calculate the blending functions in order to appropriately switch between the *k-ω* and *k-ε* models. The model constants in the transport equations of the *SST model* are given by:

$$C_\mu = 0.09, \quad \sigma_{k1} = 1.176, \quad \sigma_{\omega1} = 2.0, \quad \alpha_1 = 5.0/9.0, \quad \beta_1 = 0.075.$$

$$a_1 = 0.31, \quad \sigma_{k2} = 1.0, \quad \sigma_{\omega2} = 1.168, \quad \alpha_2 = 0.44, \quad \beta_2 = 0.0828.$$

2.12.2 Reynolds Stress Models

These more complicated models aim to circumvent a number of major drawbacks experienced by the two-equation *k-ε* models in the prediction of fluid flows with complex strain fields or significant body forces. Under such conditions, the individual Reynolds stresses are poorly represented by the eddy or turbulent viscosity in equation (2.11.1). In order to better accommodate the prevailing anisotropy nature of these stresses, solutions to the exact Reynolds stress transport equations are required.

The transport equation for each of the Reynolds stresses $\widetilde{u_i'' u_j''}$ can be expressed in accordance with the generic transport equation as:

$$\begin{array}{llll}
\textit{The rate} & \textit{The transport} & \textit{The transport} & \\
\textit{of change} + \textit{of } \widetilde{u_i'' u_j''} \textit{ by} & = \textit{of } \widetilde{u_i'' u_j''} \textit{ by} & + \textit{The net source} \\
\textit{of } \widetilde{u_i'' u_j''} & \textit{advection} & \textit{diffusion} & \textit{rate of } \widetilde{u_i'' u_j''}
\end{array}$$

$$\tag{2.12.16}$$

Equation (2.12.16) describes six partial differential equations: one for the transport of each of the six independent Reynolds stresses ($\widetilde{u_1'' u_1''}$, $\widetilde{u_2'' u_2''}$, $\widetilde{u_3'' u_3''}$, $\widetilde{u_1'' u_2''}$, $\widetilde{u_1'' u_3''}$ and $\widetilde{u_2'' u_3''}$, since $\widetilde{u_2'' u_1''} = \widetilde{u_1'' u_2''}$, $\widetilde{u_3'' u_1''} = \widetilde{u_1'' u_3''}$ and $\widetilde{u_3'' u_2''} = \widetilde{u_2'' u_3''}$). Based on the modeling strategy from the original work of Launder et al. (1975), the net source rate of $\widetilde{u_i'' u_j''}$ consists primarily of the rate of production of $\widetilde{u_i'' u_j''}$, rate of destruction of $\widetilde{u_i'' u_j''}$, transport of $\widetilde{u_i'' u_j''}$ due to turbulent pressure-strain interactions, and transport of $\widetilde{u_i'' u_j''}$ due to rotation. The exact equation for the transport of $\widetilde{u_i'' u_j''}$ takes therefore the following form as:

$$\frac{\partial}{\partial t}(\overline{\rho}\,\widetilde{u_i'' u_j''}) + \frac{\partial}{\partial x_k}(\overline{\rho}\,\widetilde{u_k}\,\widetilde{u_i'' u_j''}) = \underbrace{D_{ij}}_{diffusion} + \underbrace{P_{ij}}_{production} + \underbrace{\Pi_{ij}}_{pressure-strain} + \underbrace{\Omega_{ij}}_{rotation} + \underbrace{\varepsilon_{ij}}_{destruction}$$

(2.12.17)

For the transport of $\widetilde{u_i'' u_j''}$ by diffusion, the term represented by D_{ij}, the rate of transport is assumed to be proportional to the gradients of Reynolds stresses. This gradient diffusion idea recurs throughout the turbulence modeling as exemplified by the two-equation turbulent models. The term D_{ij} can be expressed as:

$$D_{ij} = \frac{\partial}{\partial x_k}\left[\left(\mu + \overline{\rho}\,C_s\,\frac{k}{\varepsilon}\,\widetilde{u_k'' u_l''}\right)\frac{\partial\widetilde{u_i'' u_j''}}{\partial x_l}\right], \qquad k,l = 1,2,3 \qquad (2.12.18)$$

Concerning the rate or production of $\widetilde{u_i'' u_j''}$, the exact form of P_{ij}, derived after rigorous mathematical manipulation, is:

$$P_{ij} = -\left(\overline{\rho}\,\widetilde{u_j'' u_k''}\,\frac{\partial\widetilde{u_i}}{\partial x_k} + \overline{\rho}\,\widetilde{u_i'' u_k''}\,\frac{\partial\widetilde{u_j}}{\partial x_k}\right) \qquad (2.12.19)$$

The rate of destruction of $\widetilde{u_i'' u_j''}$ as indicated by the dissipation rate ε_{ij} is modeled assuming isotropic small-scale turbulence at high Reynolds number. The modeled expression for this term is:

$$\varepsilon_{ij} = \frac{2}{3}\overline{\rho}\varepsilon\delta_{ij} \qquad (2.12.20)$$

The transport of $\widetilde{u_i'' u_j''}$ due to turbulent pressure-strain interactions represents the most important part of the model, since it governs the level of isotropy of the Reynolds stresses. Their effects on the Reynolds stresses are twofold. On one physical process, pressure fluctuations due to two turbulent eddies can interact with each other, while on the other, pressure fluctuations interaction of a turbulent eddy with a region of flow can result in different mean velocity. These rather two distinct physical processes have an overall effect of causing the pressure-strain term to redistribute energy so as to make the normal Reynolds stresses more isotropic and to reduce the influence of the Reynolds shear stresses. This linear-pressure term Π_{ij} can be modeled in two parts:

$$\Pi_{ij} = \phi_{ij,1} + \phi_{ij,2} \tag{2.12.21}$$

The first term $\phi_{ij,1}$ that is the slow pressure-strain term, also known as the return-to-isotropy term, in equation (2.12.21) is modeled according to Rotta (1951), and it represents a trend toward isotropy at the rate of the turbulent time scale. It is modeled as:

$$\phi_{ij,1} = -C_1 \frac{\varepsilon}{k} \left(\widetilde{u_i'' u_j''} - \frac{2}{3} k \delta_{ij} \right) \tag{2.12.22}$$

The second term, called the rapid pressure-strain term, in the same equation is modeled according to Launder et al. (1975) as

$$\phi_{ij,2} = -C_2 \left(P_{ij} - \frac{1}{3} P_{ii} \delta_{ij} \right) \tag{2.12.22}$$

where P_{ij} is given in equation (2.12.19) and $P = P_{ii}/2$. The preceding expression is the counterpart of Rotta's proposal for $(\phi_{ij,2} + \phi_{ji,2})$ that tends to isotropize the turbulence production. Finally, the rotational term is given by

$$\Omega_{ij} = -2\omega_k(\widetilde{u_i'' u_l''} e_{jkl} + \widetilde{u_j'' u_l''} e_{ikl}) \tag{2.12.23}$$

Here again ω_k denotes the angular rotation vector, which is also considered in the *realizable k-ε model* for the purpose of evaluating the mean rate-of-rotation $\tilde{\Omega}_{ij}$. Turbulent kinetic energy k as required in the preceding formulae can be determined by adding the normal Reynolds stresses together:

$$k = \frac{1}{2} (\widetilde{u_1'' u_1''} + \widetilde{u_2'' u_2''} + \widetilde{u_3'' u_3''}) \tag{2.12.24}$$

The dissipation *ε-equation* used with the Reynolds stresses model is the same as that of the two-equation turbulent models, with the exception that the transport term is modeled in terms of the stresses as:

$$\frac{\partial}{\partial t}(\bar{\rho}\varepsilon) + \frac{\partial}{\partial x_j}(\bar{\rho}\tilde{u}_j \varepsilon) = \frac{\partial}{\partial x_j}\left[\left(\mu + \bar{\rho} C_\varepsilon \frac{k}{\varepsilon} \widetilde{u_k'' u_l''} \right) \frac{\partial \varepsilon}{\partial x_l} \right] + \underbrace{\frac{1}{2} C_{\varepsilon 1} \frac{\varepsilon}{k} P_{ij}}_{\text{production}} - \underbrace{C_{\varepsilon 2} \bar{\rho} \frac{\varepsilon^2}{k}}_{\text{destruction}}$$
$$\tag{2.12.25}$$

The Reynolds stress equations derived from the preceding can be readily solved for fluids flows away from the proximity of solid walls. For near-wall flows, measurements have indicated that wall effect increases the anisotropy of the normal Reynolds stresses by damping out the fluctuations in the directions normal to the wall and decreases the magnitude of the Reynolds shear stresses. Corrections are thus needed to account for the influence of wall proximity on the pressure-strain terms. The wall-reflection term is generally considered in addition to the terms in equation (2.12.21). This term that tends to damp the normal stresses perpendicular to the wall while enhancing the stresses parallel to the wall is modeled according as:

$$
\begin{aligned}
\phi_{ij,w} = {} & C_1' \frac{\varepsilon}{k} \left(\widetilde{u_k'' u_m''} n_k n_m \delta_{ij} - \frac{3}{2} \widetilde{u_i'' u_k''} n_j n_k - \frac{3}{2} \widetilde{u_j'' u_k''} n_i n_k \right) \frac{k^{3/2}}{C_l \varepsilon d_n} \\
& + C_2' \left(\phi_{km,2} n_k n_m \delta_{ij} - \frac{3}{2} \phi_{ik,2} n_j n_k - \frac{3}{2} \phi_{jk,2} n_i n_k \right) \frac{k^{3/2}}{C_l \varepsilon d_n}
\end{aligned}
\tag{2.12.26}
$$

where n_k is the x_k component of the unit normal to the wall, d_n is the distance to the nearest wall, and $C_l = C_\mu^{3/2}/\kappa$ of which κ is the von Kármán constant. Quadratic correlations to the pressure-strain interactions may also be adopted if needed for improved accuracy. A model proposed by Speziale, Sarkar, and Gatski (1991) has been shown to give superior performance in a range of basic fluid flows. Interested readers are encouraged to refer to the literature for more in-depth analysis. The model constants in the transport equations of the Reynolds stress models are given by:

$$
C_s = 0.22, \quad \sigma_k = 0.82, \quad \sigma_\varepsilon = 1.0, \quad C_{\varepsilon 1} = 1.44, \quad C_{\varepsilon 2} = 1.92,
$$

$$
C_\varepsilon = 0.15, \quad C_1 = 1.8, \quad C_2 = 0.6, \quad C_1' = 0.5, \quad C_2' = 0.3
$$

The full Reynolds stress model dispenses with the notion of turbulent viscosity. Besides equations (2.12.7) and (2.12.26), additional transport equations for the scalar stresses such as $\overline{\rho u_i'' h''}$ and $\overline{\rho u_i'' \varphi''}$ are required to evaluate the diffusion term of the Favre-averaged equations of the enthalpy and scalar property (see Table 2.4). This therefore results in the increasing complexity of the modeling requirements of turbulence flows. The full Reynolds stress model has nonetheless been found to be very susceptible to unwanted numerical instabilities. To remedy such a problem, it is common practice to invoke the use of turbulent viscosity to promote numerical stability in contrast to employing the full Reynolds stress model for turbulent flows. For the Reynolds stress transport equation, the diffusion term can be simplified in the form of a scalar turbulent viscosity as:

$$
D_{ij} = \frac{\partial}{\partial x_k} \left[\left(\mu + \frac{\mu_T}{\sigma_k} \right) \frac{\partial \widetilde{u_i'' u_j''}}{\partial x_k} \right]
\tag{2.12.27}
$$

The turbulent viscosity μ_T is computed similar to the *standard k-ε model*, which is $\mu_T = \bar{\rho} C_\mu (k^2/\varepsilon)$, where $C\mu = 0.09$. Similar to the transport of $\widetilde{u_i'' u_j''}$ by diffusion, the diffusion process in the *ε-equation* is alternatively expressed in terms of a scalar turbulent viscosity as:

$$D_{ij}^\varepsilon = \frac{\partial}{\partial x_k} \left[\left(\mu + \frac{\mu_T}{\sigma_k} \right) \frac{\partial \varepsilon}{\partial x_k} \right] \tag{2.12.28}$$

For the scalar stresses $\widetilde{\rho u_i'' h''}$ and $\widetilde{\rho u_i'' \varphi''}$ in the diffusion term of the Favre-averaged enthalpy and scalar property transport equations, expressions for the scalar stresses in the form similar to equation (2.12.7) remove the need to solve additional transport equations. This simplification greatly reduces the computational burden and better promotes more robust numerical calculations.

2.13 Near-Wall Treatments

Near-wall modeling significantly impacts the reliability of the predicted numerical solutions. Consider the near-wall ceiling flow of the fire scenario depicted in Figure 2.19. In order to predict the wall-bounded turbulent flows with sufficient accuracy, appropriate near-wall models need to be employed. As exemplified in the previous section, it is possible to fully resolve the flow right up to the wall through the use of either different versions of the *low Reynolds number k-ε model* or *k-ω model*. The advantage of employing such models is that no additional assumptions are required concerning the variation of the variables near the wall. Nevertheless, the downside of such models is that they generally require a very fine near-wall resolution. For the *k-ω model*, a wall distance $y^+ \sim 2$ at all the wall nodes is required to sufficiently resolve the fluid flow adjacent to the solid wall. The *low Reynolds number k-ε model*, however, needs an even finer resolution with a wall distance $y^+ \sim 0.2$ at all the wall nodes. Such prerequisites are usually rather difficult to be achieved especially for large full-scale flow problems such as those existing in building fires.

One possible approach to overcome the difficulty of modeling the near-wall region is through the prescription of *wall functions*. The use of these functions can be found practically in every CFD commercial and in-house computer codes and is prevalent in many industrial practices. Through this approach, the difficult near-wall region is not explicitly resolved within the numerical model but rather is bridged using such functions (Launder and Spalding, 1974). To construct these functions, the region close to the wall can be characterized by considering the dimensionless velocity u^+ and wall distance y^+ with respect to the local conditions at the wall. Letting $\tilde{U}$ to be some Favre-averaged velocity parallel to the wall, the dimensionless velocity u^+ can be expressed in the form as $\tilde{U}/u_\tau$, where u_τ is the wall friction velocity defined with respect to the wall shear stress τ_w as $\sqrt{\tau_w/\rho}$.

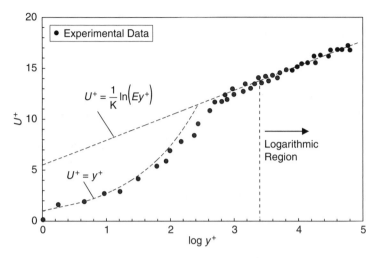

Figure 2.20 The turbulent boundary layer: respective dimensionless velocity profile as a function of the wall distance in comparison to experimental data.

Figure 2.20 illustrates the universal wall function for the velocity. For a wall distance of $y^+ < 5$, the boundary layer is predominantly governed by viscous forces that produce the no-slip condition. This region is subsequently called the *viscous sub-layer*. By assuming that the shear stress is approximately *constant* and equivalent to the wall shear stress τ_w, a linear relationship between the time-averaged velocity and the distance from the wall can be obtained yielding

$$u^+ = y^+ \quad \text{for} \quad y^+ < y_0^+ \tag{2.13.1}$$

With increasing wall distance y^+, turbulent diffusion effects begin to dominate outside the *viscous sub-layer*. It is common to employ a logarithmic relationship; the profile is expressed as:

$$u^+ = \frac{1}{\kappa}\ln(Ey^+) \quad \text{for} \quad y^+ > y_0^+ \tag{2.13.2}$$

The preceding relationship is often called the *log-law* and the layer where the wall distance y^+ lies between the range of $30 < y^+ < 500$ is known as the *log-law layer*. The values for κ (~0.4) and E (~9.8) are universal constants valid for all turbulent flows past smooth walls at high Reynolds numbers. For rough surfaces, the constant E in equation (2.13.2) is usually reduced. The law of the wall can be modified by scaling the normal wall distance d_n on the equivalent roughness height, h_0 (i.e., y^+ is replaced by d_n/h_0), and appropriate values must be selected from data or literature. The cross-over point y_0^+ can be ascertained by computing the intersection between the viscous sub-layer and the logarithmic region based on the upper root of

$$y_0^+ = \frac{1}{\kappa} \ln(E y_0^+) \tag{2.13.3}$$

A similar universal, non-dimensional function can also be constructed for heat transfer. By Reynolds' analogy, the treatment follows the same law-of-the-wall for mean velocity of which law-of-the-wall for enthalpy comprises of:

- Linear law for the thermal conduction in the sub-layer where conduction is important
- Logarithmic law for the turbulent region where effects of turbulence dominate over conduction

The enthalpy in the wall layer is assumed to be:

$$\begin{aligned} h^+ &= Pr y^+ & \text{for } y^+ < y_h^+ \\ h^+ &= \frac{Pr_T}{\kappa} \ln(F_h y^+) & \text{for } y^+ > y_h^+ \end{aligned} \tag{2.13.4}$$

where F_h is determined by using the formula of Jayatilleke (1969):

$$F_h = E \exp\left\{ 9.0\kappa \left[\left(\frac{Pr}{Pr_T} \right)^{0.75} - 1 \right] \left[1 + 0.28 \exp\left(-0.007 \frac{Pr}{Pr_T} \right) \right] \right\} \tag{2.13.5}$$

and the dimensionless enthalpy h^+ is defined as

$$h^+ = \frac{(\tilde{h}_w - \tilde{h}) \bar{\rho} C_\mu^{0.25} k^{0.5}}{J_h} \tag{2.13.6}$$

In equation (2.13.6), the diffusion flux J_h is equivalent to the normal gradient of the enthalpy $(\partial h / \partial n)_w$ perpendicular to the wall and $\tilde{h}_w$ is the value of enthalpy at the wall. The thickness of the thermal conduction layer is usually different from the thickness of the viscous sub-layer, and changes from fluid to fluid. As demonstrated in equation (2.13.3), the cross-over point y_h^+ can also be similarly computed through the intersection between the thermal conduction layer and the logarithmic region based on the upper root of

$$Pr y_h^+ = Pr_T \frac{1}{\kappa} \ln(F_h y_h^+) \tag{2.13.7}$$

Analogous to the heat transfer, wall functions for the scalar property can also be formulated according to linear law for the viscous sub-layer with laminar

Schmidt number and logarithmic law for the turbulent region with turbulent Schmidt number as:

$$\varphi^+ = Scy^+ \qquad \text{for } y^+ < y_\varphi^+$$
$$\varphi^+ = \frac{Sc_T}{\kappa}\ln(F_\varphi y^+) \quad \text{for } y^+ > y_\varphi^+ \qquad\qquad (2.13.8)$$

The dimensionless scalar property φ^+ is expressed in the similar form of equation (2.13.6) given by

$$\varphi^+ = \frac{(\tilde{\varphi}_w - \tilde{\varphi})\overline{\rho}\,C_\mu^{0.25}k^{0.5}}{J_\varphi} \qquad\qquad (2.13.9)$$

where J_φ is the diffusion flux based on the normal wall gradient $(\partial\tilde{\varphi}/\partial n)_w$. It should be noted that the formula F_φ and cross-over point y_φ^+ are calculated in a similar way as F_h and y_h^+ with differences being that the Prandtl numbers are replaced by the corresponding Schmidt numbers.

Strictly speaking, the universal profiles derived from above have been based on an attached two-dimensional Couette flow configuration with *small pressure gradients*, *local equilibrium of turbulence* (production rate of k equals to its destruction rate) and *a constant near-wall stress layer*. For some applications, applying such wall functions may lead to significant inaccuracies in the modeling of wall-bounded turbulent flows. In order to remove some of the limitations imposed by these standard wall functions, *non-equilibrium wall functions* and *enhanced wall treatment* that combines a two-layer model with enhanced wall functions are also available in practice.

Based on the conceptual development of Kim and Choudbury (1995), the key elements of the *non-equilibrium wall functions* are that the log-law is now sensitized to pressure gradient effects and the two-layer-based concept is adopted to calculate the cell-averaged turbulence kinetic energy production and destruction in wall-adjacent cells. Based on the latter aspect, the turbulence kinetic energy budget for the wall-adjacent cells is sensitized to the proportions of the *viscous sub-layer* as well as the *fully turbulent layer*, which can significantly vary from cell to cell in highly non-equilibrium flows. This effectively relaxes the *local equilibrium of turbulence* that is adopted by the standard wall functions. Another near-wall modeling methodology that has also proven to be rather useful is the *enhanced wall treatment*. Here, a single wall law is formulated for the entire wall region. A blending function is introduced to allow a smooth transition between the linear and logarithmic laws. This turbulent law always guarantees the correct asymptotic behavior for large and small values of the wall distance y^+ and provides reasonable representation of the velocity profiles in cases where y^+ lies insides the wall buffer region $(3 < y^+ < 10)$. More details of this approach can be referred in Kader (1993). *Non-equilibrium wall functions* and *enhanced wall treatment* are recommended for complex flows that may involve flow separation, flow re-attachment and flow impingement. Significant improvements have been obtained especially in the prediction of wall shear and heat transfer.

2.14   Setting Boundary Conditions

In many CFD problems, some sensible engineering judgment usually needs to be exercised for the specification of appropriate turbulence quantities at the boundary walls of the flow domain. Some useful guidelines in setting proper boundary conditions for turbulence and in handling various types of boundaries are provided in this section.

For solid walls, boundary conditions for k and ε or ω are substantially different depending on whether turbulence models catered for low Reynolds number effects or the wall function methods are employed.

Consider the schematic illustrations of two different approaches to wall modeling in Figure 2.21 in resolving the near-wall ceiling flow of the fire scenario depicted in Figure 2.19. Let us initially focus on the low Reynolds turbulence models, which require resolution of the mesh right up to the solid wall. For the application of *low Reynolds number k-ε models*, Johns and Launder (1972) and Chien (1980) impose the following boundary conditions:

$$k_{wall} = \varepsilon_{wall} = 0 \qquad\qquad (2.14.1)$$

In Lam and Bremhorst (1981), they also employ a zero value for the wall turbulent kinetic energy. However, a zero normal gradient of the dissipation ε at the wall is prescribed. The boundary conditions are thus:

$$k_{wall} = \left.\frac{\partial \varepsilon}{\partial n}\right|_{wall} = 0 \qquad\qquad (2.14.2)$$

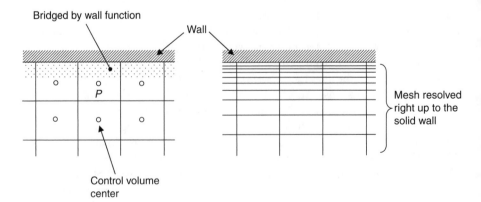

Figure 2.21   Schematic illustrations of two different approaches to wall modeling.

For the *k-ω model*, the rough-wall method by Wilcox (1998) allows the possibility of determining the surface value for ω according to:

$$\omega_{wall} = \frac{\overline{\rho} u_\tau}{\mu} S_R \qquad (2.14.3)$$

where u_τ is the friction velocity and S_R is a non-dimensional function determining the degree of surface roughness of the wall. It is worthwhile noting that the successful application of these models is greatly influenced by the grid spacing of the overlay mesh near the wall. Specific meshing guidelines will be further expounded in the next section.

For wall modeling through the use of wall functions, this simpler approach allows the flexibility of relating the flow variables to the first computational mesh point as shown in Figure 2.21 through universal wall functions as described in the previous section, which thereby removes the requirement to totally resolve the flow structure within the *viscous sub-layer*. When the law-of-the-wall type boundary conditions are employed, a zero normal gradient of k is prescribed:

$$\left.\frac{\partial k}{\partial n}\right|_{wall} = 0 \qquad (2.14.4)$$

On the basis of the *local equilibrium of turbulence* assumption, the dissipation ε is not applied at the wall but is determined at the first computational mesh point. At the central point of the control volume, ε_P is evaluated according to:

$$\varepsilon_P = \frac{C_\mu^{3/4} k_P^{3/2}}{\kappa d_P} \qquad (2.14.5)$$

where d_P denotes the first computational mesh point normal to the wall. In cases where the non-equilibrium wall function is used instead, ε_P can be evaluated as:

$$\varepsilon_P = \frac{1}{2n_P}\left[\frac{2\mu}{\rho y_v} + \frac{C_\mu^{3/4} k_P}{\kappa}\ln\left(\frac{2n_p}{y_v}\right)\right] k_P \qquad (2.14.6)$$

where y_v is the physical viscous sub-layer thickness computed from $y^*\mu/\rho C_\mu^{1/4} k_P^{1/2}$ and y^* is usually set at a value of 11.225. More details considering the formulation of equation (2.14.6) can be found in Kim and Choudbury (1995).

At inlet boundaries, experimentally verified quantities, if possible, should always be applied as inlet boundary conditions for the turbulent kinetic energy

k and its dissipation ε. Such readily accessible measurements of k_{inlet} and ε_{inlet} is, however, very rare in practice; these quantities are particularly unknown at the surface influx of the fuel into the surrounding air (see the schematic illustrations of the free-standing and compartment fires described in section 2.7.3). Through some sensible engineering assumptions, specification of the inlet turbulent kinetic energy k_{inlet} can nevertheless be realized by relating the inlet turbulence to the turbulence intensity I, defined as the ratio of the fluctuating component of the velocity to the mean velocity, as well as the upstream flow conditions. Approximate values for k_{inlet} can be determined according to the following relationship as:

$$k_{inlet} = \frac{3}{2}(u_{inlet}I)^2 \tag{2.14.7}$$

where u_{inlet} is the inlet mean velocity. Similarly, the specification of the dissipation ε_{inlet} can be approximated by the following assumed form:

$$\varepsilon_{inlet} = C_\mu^{3/4} \frac{k_{inlet}^{3/2}}{l} \tag{2.14.8}$$

where l appearing in equation (6.19) is the characteristic length scale. If the k-ω model is employed, ω_{inlet} can be approximated by:

$$\omega_{inlet} = \frac{k_{inlet}^{1/2}}{C_\mu^{1/4}l} \tag{2.14.9}$$

As a guide, the turbulence intensity level I in equation (2.14.6) can be typically set at 0.3% for external flows, while the turbulence level between 5% and 10% is deemed to be appropriate for internal flows. For the length scale l in equations (2.14.8) and (2.14.9), a constant value of length scale derived from a characteristic geometrical feature such as 1% to 10% of the inlet hydraulic diameter can be employed for internal flows. A value determined from the assumption that the ratio of turbulent and molecular viscosity between 1 and 10 is nonetheless a reasonable guess for external flows remote from the boundary layers. For the Reynolds stress model, each stress components ($u_1''u_1''$, $u_2''u_2''$, $u_3''u_3''$, $u_1''u_2''$, $u_1''u_3''$, and $u_2''u_3''$) are required to be properly specified. If these are unavailable, as often the case, the diagonal components ($u_1''u_1''$, $u_2''u_2''$, and $u_3''u_3''$) are taken to be equal to $\frac{2}{3}k$, whereas the extra-diagonal components ($u_1''u_2''$, $u_1''u_3''$, and $u_2''u_3''$) are set to zero (assuming isotropic turbulence). In cases where problems arise in specifying appropriate turbulence quantities, *the inflow boundary for the application of all turbulence models should be moved sufficiently far away from the region of interest so that the inlet boundary layer and subsequently the turbulence are allowed to be developed naturally.*

The *Neumann* boundary conditions, as described in 2.5, can be aptly applied at the outlet, open, or symmetry boundaries. For all the turbulence models presented, the boundary conditions are:

$$\left.\frac{\partial k}{\partial n}\right|_{wall} = \left.\frac{\partial \varepsilon}{\partial n}\right|_{wall} = \left.\frac{\partial \widetilde{u_i'' u_j''}}{\partial n}\right|_{wall} = 0 \qquad (2.14.10)$$

2.15 Guidelines for Setting Turbulence Models in Field Modeling

A range of turbulence models has been presented and discussed. It is clear that no general turbulence model can be used to span all turbulent states, since none can be universally valid for all types of flows. Since different types of turbulent flows require the application of different turbulence models, *how does one viably choose the appropriate turbulence model in the context of field modeling?*

A number of practical guidelines in setting suitable turbulence models to resolve a range of CFD fire problems are presented in this section. In the event where insufficient knowledge precludes the selection of an appropriate turbulence model, the authors strongly encourage the use of the *standard k-ε model* as a starting point for turbulent analysis. In comparison to other sophisticated turbulence models, the *standard k-ε model* offers the simplest level of closure, since it has no dependence on the geometry or flow regime input. This model is robust and stable, and is as good as any other more sophisticated turbulence models in some applications. In most in-house and commercial codes, this model is a default option for handling flows that are turbulent. It is therefore not entirely surprising that it has been a de facto standard in industrial applications and still remains the work-horse of practical computations.

Should the numerical solutions attained through the use of the *standard k-ε model* be unsatisfactory, the palliative action would be to trial more advanced turbulent model. These advisory actions should not be construed as definitive cures but rather recommendations whereby possible alternatives to the *standard k-ε model* can be investigated to improve the numerical predictions. The authors strongly recommend an incremental step-by-step approach of systematically introducing higher levels of model sophistication into the numerical calculations. Through this, the field modeler will benefit immensely from over prescribing the solution with unnecessary complexities. For example, the use of *RNG k-ε model* and *realizable k-ε model* may suffice in possibly yielding the required numerical solutions for most turbulent flows but if all fails, the option to select Reynolds stress model could be exercised. In short, the careful validation of increasingly complex turbulence models is the best-practice method of finding the simplest turbulence model without being too simple as to attain unphysical results. This is the age-old engineering compromise between accuracy and efficiency. The numerical modeler must balance numerical solution time with (field-modeling) solution accuracy.

For near-wall modeling, it is imperative that the lower limit of y^+ through the use of wall functions must be carefully placed so that it does not fall within the *viscous sub-layer*. In such circumstance, the meshing should be arranged in order that the values of y^+ at all the wall-adjacent integration points are set above the recommended limit, typically between 20 and 30. Since the whole emphasis of purposefully adopting the wall functions is to remove the resolution of the *viscous sub-layer*, this procedure offers the best opportunity to practically resolve the turbulent portion of the boundary layer. Besides checking the lower limit of y^+, it is also rather important that the upper limit of y^+ is also investigated during the computational calculation. A flow with moderate Reynolds number could, for example, have a boundary layer that extends up to y^+ between 300 and 500. If the first integration point is placed at a value of $y^+ = 100$, then this will certainly yield an impaired solution due to an insufficient resolution of the region. Adequate boundary layer resolution generally requires at least 8–10 grid nodal points in the layer. The authors always recommend that a post-analysis of the numerical solution be undertaken to determine whether the degree of resolution is achieved or the flow calculation should be subsequently performed with a finer mesh.

There are nonetheless some flow conditions where the universal near-wall behavior over a practical range of y^+ may not be realizable. For such flows, the wall function concept breaks down and its continual use during numerical calculations will lead to significant errors. It is therefore preferable that the flow is fully resolved right up to the wall through the application of low Reynolds number turbulence models. In order to resolve the viscous sub-layer inside the turbulent boundary layer, y^+ at the first node adjacent to the wall should be set preferably less than unity for the *low Reynolds number k-ε models* ($y^+ \sim 0.2$). However, a higher y^+ is acceptable for the standard *k-ω model* ($y^+ \sim 2$) or even larger in some flow cases so long as it is still well within the *viscous sub-layer* ($y^+ = 4$ or 5). Depending on the Reynolds number, it is always important to ensure that there are 5–10 grid nodal points between the wall and the location where $y^+ = 20$ to resolve the mean velocity and turbulent quantities that lie within the viscosity-affected near-wall region. This probably results in 30–60 grid nodal points inside the boundary layer in order to achieve adequate boundary layer resolution. It should be noted that the cost of the numerical solution is around an order of magnitude greater than the use of wall functions because of the additional grid nodal points involved.

2.16   Worked Examples on the Application of Turbulence Models in Field Modeling

2.16.1   Single-Room Compartment Fire

This particular worked example aims to demonstrate the application of the fundamental transport equations of mass, momentum, and energy to resolve the fluid flow alongside with different turbulent models to characterize the effect of turbulence. Based on the eddy viscosity concept, the standard two-equation k-ε model assumes isotropy of the turbulent stresses. Nevertheless, the Reynolds Stress model that solves six components of the turbulent stresses result in a more sophisticated turbulence model in order to better accommodate the possible anisotropy of the flow behavior. Both of these models are assessed against the full-scale single compartment fire experiment performed by Steckler et al. (1984). This particular fire case has been specifically chosen because it represents a *benchmark fire case* that has been used by many field modelers for the primary purpose to test the capability of fire models in predicting the temperature and flow distributions in a compartment subjected to a steady non-spreading fire.

Figure 2.22 shows the schematic drawing of the particular geometry of the compartment. The non-spreading fire was fueled by commercial grade methane, having a circular gas burner diameter D of 0.3 m centrally located in the room on a square enclosure of side 2.8 m and height 2.18 m. Air was drawn into the burn room through a doorway opening of 1.83 m high and 0.74 m wide located in one of the walls as depicted in Figure 2.22. Compartment walls and ceiling were covered with ceramic fiber board insulation to establish near steady state conditions within 30 minutes. Detailed measurements of temperature, using aspirated thermocouples, and velocity by bi-directional probes reported in Steckler et al. (1984) at the doorway are employed to validate the model predictions. Numerical simulations are performed through an in-house computer code FIRE3D and the widely used ANSYS Inc., CFX, commercial CFD code. A heat release rate $\dot{Q}$ of 62.9 kW was selected for the validation exercise. For this worked example, the problem is solved by representing the fire as a volumetric heat source with a specified volume of $0.3 \times 0.3 \times 0.3$ m^3.

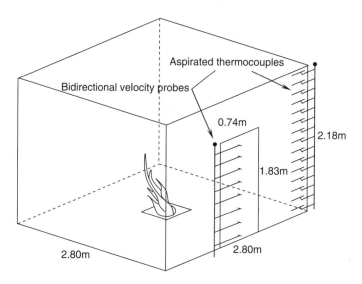

Figure 2.22   Schematic drawing Steckler's burn room.

Numerical features: To ascertain the type of fire that is emanating from the circular gas burner, the Froude number (*Fr*) may be used to as a means of classifying the relative importance of momentum (or inertia) and buoyancy in the flame. By definition, *Fr* can be expressed as

$$Fr \equiv \frac{U^{*2}}{gL^*} = \frac{\text{inertia}}{\text{buoyancy}} \tag{2.16.1}$$

where g is the acceleration due to gravity. The characteristic length scale L^* for this particular problem can be taken to be equivalent to the gas burner diameter while the velocity length scale U^* may be derived from the rate of heat release as

$$U^* = \frac{\dot{Q}}{\Delta H_c \rho_{fuel}(\pi D^2/4)} = \frac{62.9 \times 10^3}{\underbrace{50 \times 10^6}_{\Delta H_c} \times \underbrace{0.65}_{\rho_{fuel}} \times \underbrace{0.0707}_{\pi 0.3^2/4}} = 0.0274 \tag{2.16.2}$$

where ρ_{fuel} is the fuel density of methane at 300 K and ΔH_c is the heat of combustion of methane, which is obtained from Table B.1 of Appendix B. Further discussion on the concept of heat of combustion will be presented in the next chapter. Using equation (2.16.1), the Froude number yields a dimensionless value of 2.55×10^{-4}.

As indicated in Drysdale (1999), the fire clearly corresponds to a buoyancy-driven turbulent diffusion flame on the basis of a very small *Fr* number. It is worthwhile noting that buoyant flow generally requires additional modeling effort. Firstly, the dominant body force resulting from buoyancy should be included in the momentum equation. For this particular example, the gravity acts in the y direction; hence, the additional source term due to buoyancy needs to be incorporated in the *y-momentum* equation. The Favre-averaged *y-momentum* equation becomes

$$\frac{\partial}{\partial t}(\bar{\rho}\tilde{v}) + \frac{\partial}{\partial x_j}(\bar{\rho}\tilde{u}_j\tilde{v})$$

$$= \frac{\partial}{\partial x_j}\left[(\mu + \mu_T)\frac{\partial\tilde{v}}{\partial x_j}\right] - \frac{\partial\bar{p}}{\partial y} + \frac{\partial}{\partial x_j}\left[(\mu + \mu_T)\left(\frac{\partial\tilde{u}_i}{\partial x_j} + \frac{\partial\tilde{u}_j}{\partial x_i}\right)\right] \tag{2.16.3}$$

$$- \frac{\partial}{\partial x_j}\left[\frac{2}{3}(\mu + \mu_T)\frac{\partial\tilde{u}_i}{\partial x_j}\delta_{ij} + \bar{\rho}k\delta_{ij}\right] - g(\rho - \rho_{ref})$$

The additional source term $-g(\rho - \rho_{ref})$ in equation (2.16.3) represents the buoyancy term with the reference density denoted by the variable ρ_{ref}. Secondly, appropriate modifications to the respective turbulence models are also required besides the buoyancy term incorporated within the *y-momentum* equation. For the standard k-ε model, the additional source term for the transport equation of k characterized by the generation term G_{buoy} due to buoyancy along the y direction can be formulated as:

$$S_k \equiv G_{buoy} = g \frac{\mu_T}{\sigma_T} \frac{1}{\rho} \frac{\partial \rho}{\partial y} \tag{2.16.4}$$

For the dissipation of turbulence kinetic energy ε, the additional source term is nevertheless modeled according to

$$S_\varepsilon \equiv \frac{\varepsilon}{k} C_{\varepsilon 1} C_{\varepsilon 3} \max(G_{buoy}, 0) \tag{2.16.5}$$

From the above equations (2.16.4) and (2.16.5), the adjustable constants σ_T and $C_{\varepsilon 3}$ are usually specified as having values of 0.9 and 1.0 respectively. For the Reynolds stresses transport equations, the buoyancy generation term is modeled as:

$$S_{\widetilde{u_i u_j}} \equiv G_{ij} = \frac{\mu_T}{\sigma_T} \frac{1}{\rho} \left(g_i \frac{\partial \rho}{\partial x_j} + g_j \frac{\partial \rho}{\partial x_i} \right) \tag{2.16.6}$$

The buoyancy source term in the dissipation transport equation is the same according to the expression derived in equation (2.16.5). Logarithmic wall functions are employed for the standard k-ε and Reynolds Stress models to bridge the wall and the fully developed turbulent flow. Thirdly, the power density based on the prescribed volume of $0.3 \times 0.3 \times 0.3$ m^3 represents the additional source required for the Favre-averaged enthalpy equation which is: $\bar{S}_h = (62.9 \times 10^3/0.027) = 2.33 \times 10^6$ W/m^3.

For the in-house computer code, the computational geometry was constructed as depicted in Figure 2.23. Since the room configuration is symmetrical about the vertical plane bisecting the doorway and burner, mesh generation for the compartment fire was only carried out on half of the room, thus improving the resolution of the flow and thermal fields. A uniform rectangular mesh is generated spanning the length, width and height of the reduced configuration. In order to eliminate any errors that could arise due to mesh generation, exact mesh distribution comprising of a total of 83160 grid nodes is used for the numerical calculations performed in both the in-house and commercial computer codes. As the doorway temperature and velocity distributions are of significant interest for model assessments, a large extended region away from the doorway is constructed to reduce the end effects of the extended boundaries affecting the flow and thermal characteristics at the doorway. A fixed pressure boundary condition is imposed on all the external boundaries (open boundaries). All the compartment walls are taken to be adiabatic.

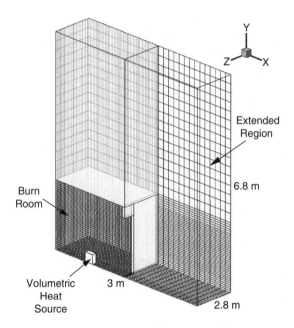

Figure 2.23 Schematic drawing Steckler's burn room.

The governing equations in both of the in-house and commercial computer codes are discretised via the finite volume method and all variables are stored in a co-located grid arrangement. For the in-house computer code, the algorithm for the solution of the governing equations relies on the implicit segregated velocity-pressure formulation such as the SIMPLE scheme along with the *hybrid differencing scheme* with only the standard k-ε model. A Poisson equation for the pressure correction is solved through a default iterative solver (preconditioned conjugate gradient). To avoid non-physical oscillations of the pressure field and the associated difficulties in obtaining a converged solution, the Rhie and Chow (1983) interpolation scheme is employed. For the ANSYS Inc., CFX computer code, a coupled treatment to the discretised form of equations is adopted instead. The advection terms are approximated by the *high resolution scheme*. Standard k-ε and Reynolds Stress turbulence models are applied. A multigrid accelerated Incomplete Lower Upper (ILU) factorization technique is employed to solve the set of linear equations. A strategy similar to the Rhie and Chow (1983) interpolation scheme employed in the in-house computer code to overcome the possible occurrence of spurious velocity and pressure is also adopted in ANSYS Inc., CFX. In both computer codes, numerical simulations are performed in transient mode with a time step of 0.1 s for the solutions to models employed the standard k-ε turbulence model, while a smaller time step of 0.05 s is employed for the Reynolds Stress model to ensure computational stability. Steady state solutions were obtained after a total time of 500 s has elapsed.

Numerical results: One useful way of purposefully illustrating the comparison between measured doorway temperature and velocity profiles of Steckler et al. (1984) and numerical predictions obtained through FIRE3D and ANSYS Inc., CFX computer codes for different turbulent models are by line graphs as shown in Figures 2.24 and 2.25. Through these two figures, two important modeling aspects are examined: the validation between the numerical results and experimental data, and the verification within model predictions.

For validation, model predictions of the temperature at the doorway, whether using the standard k-ε or Reynolds Stress model, are seen to be over-predicted by a significant margin just below the soffit. Principally, this discrepancy is due to the absence of radiant heat loss not considered within the fire models. Numerical simulations carried out by Lewis et al. (1997) on the same fire problem have essentially confirmed the importance of accounting radiative heat loss in their numerical analyses in order to markedly improve the prediction of the upper layer temperature. The hydraulic behavior at the doorway is nevertheless not significantly affected by the temperature; model predictions of the velocity yield substantially better agreement with the experimental profile than the temperature. For this centrally located fire within the compartment, the two-layer structure at the doorway is generally rather well defined. The ability of the prescriptive approach based on the volumetric heat source to preserve the two-layer structure, a prevalent feature identifying compartment fires as illustrated earlier in Figure 2.2, clearly indicates the required predicted transition between the cold air inflow an dilute hot air outflow as typified by the model results reported in Figure 2.24.

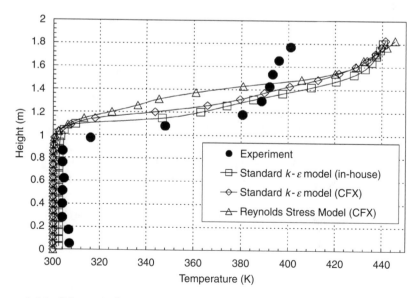

Figure 2.24   Schematic drawing Steckler's burn room.

For the model verification between FIRE3D and ANSYS Inc., CFX computer codes, numerical results obtained for the doorway temperature and velocity profiles using the standard k-ε model from the former are seen to agree rather well against those of the latter with the same turbulence model. In spite of the different numerical algorithms adopted within these two computer codes, the cross model verification instills confidence primarily of the numerical models that have been developed in the in-house computer code FIRE3D. It is nonetheless noted that the model prediction of the velocity via FIRE3D yields a better agreement of the air entrainment into the burn at the lower part of the compartment. During the course of obtaining the numerical results, one the greatest shortcoming of the Reynolds Stress model is the very large computational costs incurred. Since six transport equations for the stress components plus an equation for the dissipation were required to be solved, the computational time increases dramatically to almost three times more than the standard k-ε model. In spite of its supposedly greater accuracy, the Reynolds Stress model did not reveal, however, the significant improvement to the numerical predictions when compared to the standard k-ε model as shown in Figures 2.24 and 2.25. One of the many pertinent guidelines in section 2.15 that has been recommended is to deploy the standard k-ε model as a starting point for any turbulent analysis. In this worked example, the use of the standard k-ε model for the particular fire problem investigated appears to be adequate. The standard k-ε model offers the simplest level of closure, is robust and stable, and is as good as any other more sophisticated turbulence models in majority of practical applications. It is nonetheless noted that there is no single turbulence model that can span the turbulent states that are rich, complex, and varied,

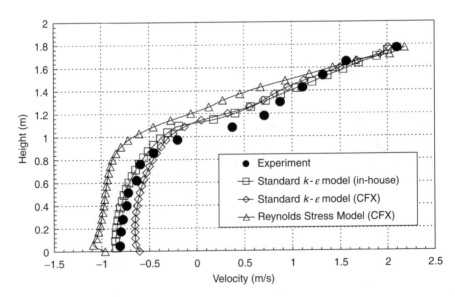

Figure 2.25   Schematic drawing Steckler's burn room.

since none is expected to be universally valid for all flows. Unless the standard k-ε model fails to yield any sensible solutions, more sophisticated turbulence models such as the Reynolds Stress model can be adopted, since it has a greater potential to represent the turbulent flow phenomena more correctly than the standard k-ε model especially for handling non-equilibrium fluid flows or flows having complex strains.

Figures 2.26 and 2.27 illustrate the line temperature contours and velocity vectors at the vertical symmetry plane of the compartment. In CFD, contour plots are one of the most commonly found graphical representations of data while vector plots provide the visual depiction of vector quantities that are displayed at discrete points (usually velocity) with arrows whose orientation indicates direction and whose size indicates magnitude. Majority of commercial CFD codes often possess their own post-processing visualization tools to graphically view the CFD results as illustrated in Figures 2.26b, 2.26c, 2.27b, and 2.27c. The in-house simulation results are however plotted using TEC-PLOT, a commercial package designed specifically for post-processing CFD results. Qualitatively, all the temperature contours show similar characteristic of a deflected fire plume due to the induced incoming fluid flow. An inclination angle with respect to the horizontal plane of about 60° is attained of which in comparison to the experiment performed by Quintiere et al. (1981) of similar heat release rate, this particular flame angle is found to be substantially larger than the observed angle between 33° and 43°. The prescribed heat source approach is certainly not without fault. In comparison to accounting the combustion chemistry through suitable models, which will be further discussed in the next chapter, the energy release at this instance is not allowed to be spatially distributed and coupled to the flow field. Evidently, this contributes to the poor prediction of the structure of the fire source and plume.

Conclusions: The simple approach based on characterizing the fire as a volumetric heat source is presented in this worked example. As demonstrated above, the model via a proper prescription of the fire size (volume) alongside with a suitable two-equation turbulence model is capable of predicting satisfactory post-combustion temperature and velocity distributions at the doorway. This approach, however, depends heavily on the *a priori* knowledge of the shape and size of the volumetric heat source. On the basis of this stringent prerequisite, careful consideration of the model's applicability should always be frequently exercised in order to duly avoid an unrealistic prediction of the fire phenomena due to incorrect prescription of the energy release.

2.16.2 Influence of Gaps of Fire Resisting Doors on Smoke Spread

This worked example illustrates the approach based on the prescriptive volumetric heat source to purposefully investigate the influence of gap sizes affecting the smoke spread from a burn room (Cheung et al., 2006). In practice, fire resisting wall provides one of the simplest ways for the confinement of smoke spread and is therefore commonly employed. This inevitably results in a

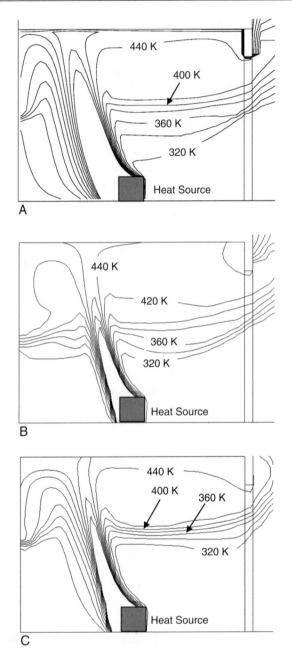

Figure 2.26   Temperature contours: (a) Standard k-ε model (in-house), (b) Standard k-ε model (CFX), and (c) Reynolds Stress model (CFX).

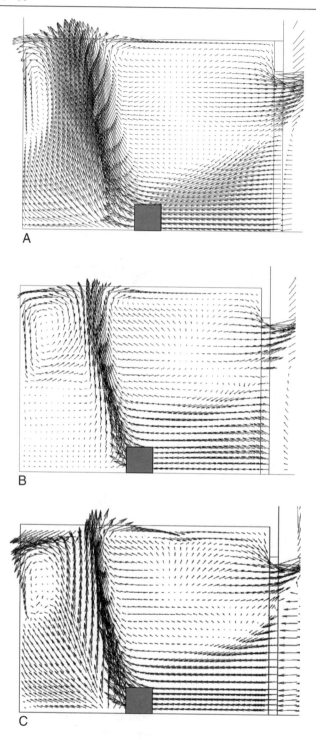

Figure 2.27 Velocity vectors: (a) Standard k-ε model (in-house), (b) Standard k-ε model (CFX), and (c) Reynolds Stress model (CFX).

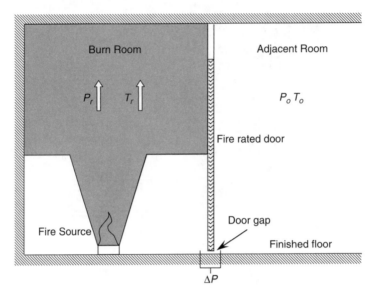

Figure 2.28   Mechanism of smoke spread through the gap between the fire rated door panel and top of finished floor.

doorway being constructed, which in turn becomes an effective passage for the likelihood of smoke spreading in the event of a fire through the gaps.

Figure 2.28 illustrates a schematic diagram illustrating the mechanism of smoke spread through the gap of the fire rated door within the burn and adjacent rooms. A fire source that resides in the burn room is enclosed by the surrounding walls and separated from the adjacent room by a fire rated door. A small gap exists between the door panel and the top of finished floor. As the fire develops and continues to burn, the accumulation of heat causes the air temperature T_r to rise. The static pressure of the room P_r also increases due to the increasing air temperature. If the temperature and pressure of the adjacent room (i.e. T_o and P_o) remain constant at their respective initial conditions, positive pressure difference ΔP between the two rooms can become sufficiently large in eventually driving the hot smoke through the small door gap away from the burn room and spreads into the adjacent room.

Numerical simulations are performed to parametrically investigate four different heights of the door gap: 3 mm (Case I), 5 mm (Case II), 7 mm (Case III) and 10 mm (Case IV). A designed fire is considered to be burning on a square fuel bed centrally located in the single compartment with a 10 MJ total heat content as schematically shown in Figure 2.29. The ANSYS Inc., CFX, commercial CFD code is adopted. This worked example aims to (i) demonstrate the application of transport equations governing mass, momentum and energy alongside with a suitable turbulence model in order to better understand the phenomenological behavior of the smoke egress through the various heights

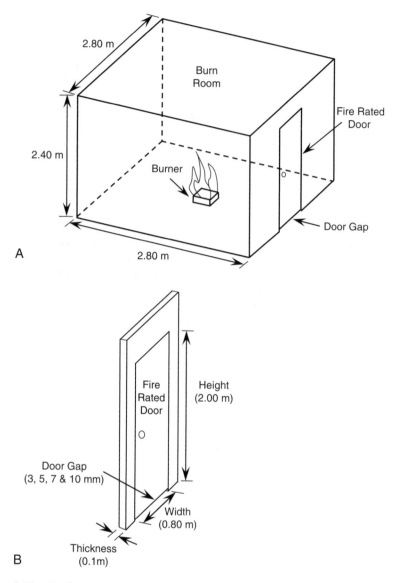

Figure 2.29 Single-compartment fire used in this study: (a) Layout of the generic case and (b) Detail configurations of the fire rated door gap.

of the door gap and (ii) determine which door gap size best impedes the smoke passage.

Numerical features: From a numerical perspective, simulating the fire development and smoke spread behavior through a narrow door gap is challenging. The problem involves a fire burning in a fully confined compartment and without the entrainment of any fresh air into the compartment the fire will burn like a "ghosting flame" and eventually reach a point where it is extinguished. A pragmatic approach to investigating the smoke spread behavior through the door gap can be achieved by simply characterizing the "ghosting flame" as a volumetric heat source. This approach allows, as a first step to capture the essence of the overall effect of the fire behaving in an under-ventilated condition. It therefore provides the means of prescribing the growing heat release rate distribution according to some fire growth characteristic such as a T-square fire as well as the decaying heat release rate that mimics the fire being extinguished as the amount of oxygen depletes within the enclosed environment of the burn room. Figure 2.30 illustrates the heat release rate profile of the designed fire employed in the current study. The transient heat release rate of the fire is uniformly distributed within a volumetric heat source of 0.3 m in width, 0.3 m in length and 1.1 m in height. The height of the heat source is calculated using an empirical fire plume equation that correlates the flame height to its maximum heat release rate and size of the fuel bed. This empirical form of the plume equation can be found in the SFPE handbook (1996) and the reference therein.

The compartment configuration is modeled according to the computational geometry shown in Figure 2.31. Inherent symmetry about the vertical middle plane cutting across the burner and the fire rated door allows the feasibility of only solving half of the burn room in order to improve the resolution of the flow and thermal fields. An extended region of 3.0 m in length, 1.4 m in

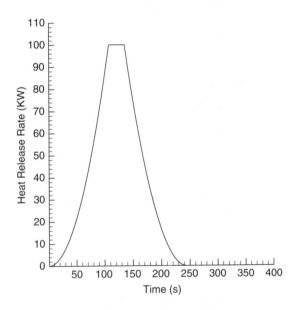

Figure 2.30   Heat release rate profile of the designed fire.

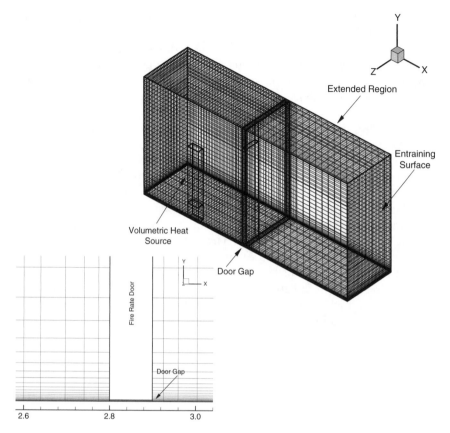

Figure 2.31 Mesh distribution adopted for the single compartment fire and at the door gap and its vicinity locations.

width and 2.4 m in height is included in the simulation model to properly model the flow characteristic passing through the door gap. The domain is filled with a non-uniform grid of $66 \times 49 \times 33$ grid nodal points (totaling 103,740 control volumes). Moreover, a highly concentrated mesh is generated to resolve the smoke egress through the door gap. For the purpose of illustration, the mesh distribution for the 3mm door gap (i.e., Case I) is also demonstrated in Figure 2.30. No-slip wall condition is specified for all the wall surfaces by setting all the velocities to zero. Adiabatic condition is imposed for the wall temperature calculation. For the temperatures, Neumann boundary condition (i.e., the normal gradient being zero) is employed. At the fire source, the transient heat release rate profile is specified and uniformly distributed. It is reduced by 20% to account for radiation loss from the fire (Markatos et al., 1986). For the extended boundaries, the vertical middle plane is treated as entraining surface in the present study. Other surfaces are treated as a solid wall boundary condition.

The relevant numerical features of the ANSYS Inc., CFX computer code have been described in the previous worked example. For this particular problem, numerical simulations are performed in transient mode employing the standard k-ε turbulence model with a time step of 0.1 s for a total time of 400 s.

Numerical results: The variation of the smoke spread or smoke leakage rates with respect to time is illustrated in Figure 2.32. As observed, the narrow 3 mm door gap (Case I) result yields the lowest smoke spread rates. The spreading or leakage rate for this case reaches a maximum value of 0.148 m^3/s at about 130 seconds corresponding to the same time when the fire peaks at its maximum heat release rate. After 136 seconds, the spreading or leakage rates for door gap configurations decline, closely following the behavior of the fire decaying before it extinguishes at 244 seconds. The difference in the maximum spreading or leakage rates displayed in Figure 2.32 is mainly attributed to the rate at which the smoke takes its course to transgress through the door gap since the same fire source in all four cases would generate the same total volume of smoke. Nevertheless, the maximum spreading or leakage rate is not greatly affected by the height of the door gap.

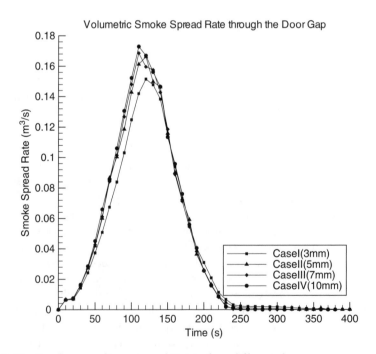

Figure 2.32 Smoke spreading rates subject to four different door gaps.

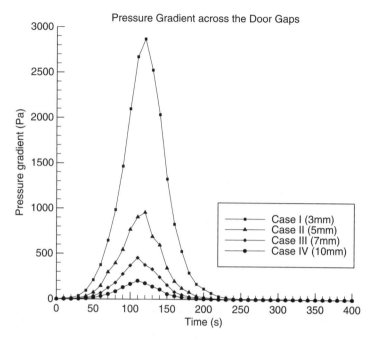

Figure 2.33   Transient pressure gradients across different door gaps.

The transient profiles of the pressure gradients across the door gaps are illustrated in 2.33. It is not surprising that Case I yields the highest pressure gradient because of the greater resistance to the hot gases spread by the 3 mm door gap. This is also confirmed by the lowest hot gases spread rates predicted in Figure 2.33. By increasing the door gap heights, the maximum pressure gradient across the door gap dramatically reduces thus resulting in a lower pressure residing in the burn room. It is not entirely surprising that the pressure in the burn room is directly affected by the configuration of the door gap. Overall, the pressure gradient profiles are rather similar to the profiles of the smoke spread and fire heat release rates.

Figure 2.34 illustrates the averaged temperature profiles of hot gases or smoke passing through the door gaps. For all four cases, the smoke takes more than 90 seconds before spreading or leaking into the surrounding. Before this, the continuous stream of cold air through the door gaps actually contains the smoke within the burn room. As the pressure and temperature increase in the burn room, the positive pressure difference that is created eventually forces the hot gases through the door gaps. Comparing the temperatures predicted by each case, Case IV predicts the highest gas temperature passing through the gap while Case I predicts the lowest temperature. Increasing the height of the gap allows more hot gases to pass through and eventually increases the

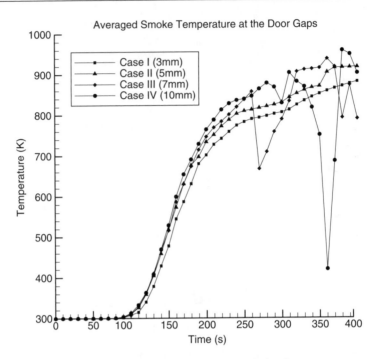

Figure 2.34 Averaged gas temperatures passing through the door gaps.

average temperature at the door gap. Most interestingly, significant temperature fluctuations are evident for Cases III and IV after 250 seconds. For Cases I and II, hot gases are continually driven out of the burn room because of the existence of substantial pressure gradients operating across the door gaps as depicted in Figure 2.34. However, without substantial pressure gradients such as experienced in Cases III and IV, the hot gases may become stagnant in the door gap and mix with the incoming cold air from the surrounding. This could explain the temperature fluctuations predicted in the two cases.

Conclusions: The rise of the burn room pressure incurs the likelihood of the smoke spreading through the door gaps. As demonstrated through the numerical simulations, the various door gaps of different heights fail to impede the spread of smoke. It is also observed that the large door gap (10mm) allows a higher temperature of smoke spreading into the surrounding but is subjected to a lower-pressure gradient. This low pressure gradient may permit the stream of cold air to be re-entrained into the burn room. In an under-ventilated condition, a "ghosting flame" exists due to the limited oxygen content. With depleting supply of oxygen/oxidant in the burn room, the fire will be subsequently extinguished. For the case of the large door gap with the possibility where cold air could be re-entrained back into the burn room, the re-entrainment of cold air may continue to sustain the fire and incur the possible *back-draft* phenomenon where the door is assessed by fire fighters. On the basis of these numerical predictions, a fire rated door with a 3 mm door gap height is demonstrated to represent the best measure for impeding the smoke spread while maintaining reasonable smoothness for the door movement.

2.17 Summary

Practical fires are invariably turbulent in nature. The flow fluctuations associated with turbulence create additional transfer of momentum, heat and mass in the form of extra stresses on the mean flow. These extra stresses must be modeled for field modeling of fires. What actually makes the prediction of the effects of turbulence difficult is that no one single turbulence model can be readily employed to span all the length and time scales since none is expected to be universally applicable for all categories of turbulent flows. The two-equation k-ε model is widely used in practice. It is still valued for its robustness and comes highly recommended for practicable computation times and are only marginally less accurate in comparison to more sophisticated turbulence models. The formulation of other two-equation eddy viscosity models such as RNG k-ε model, realizable k-ε model, low Reynolds number turbulence models and SST model are noted for their relevance in handling a range of industrial external and internal flow computations. Reynolds Stress Model, which allows the possibility of capturing the anisotropic nature of turbulence, is gaining in recognition and in some circles, strongly argued as the only viable way forward toward a general purpose classical turbulence model.

On the basis of the application of the standard k-ε model and Reynolds Stress Model in characterizing the turbulence in a single-room compartment fire via the volumetric heat source approach, the verdict on these models remains open on the applicability of these models in fire problems as demonstrated in the first worked example. The approach of representing the fire as a volumetric heat source should always be adopted with caution as such prescriptive designs depend heavily on the knowledge of the shape and size to be specified within the computational domain. Nevertheless, if the shape and size of the fire volume can be a priori determined, the volumetric heat source approach has been demonstrated as a rather effective method, as exemplified in the second worked example, in studying the influence of gap sizes affecting the smoke spread from a burn room especially through possible narrow spaces between the door panel and surrounding doorframe.

All turbulence models contain adjustable constants that have been pre-determined from experiments. For any CFD calculations of *new* turbulent flows to be fully accepted especially in field modeling investigations, they should always be verified or validated through available experimental or numerical data.

Review Questions

2.1. CFD covers three major disciplines. What are they?

2.2. What examples can the reader list from which CFD is being employed in traditional and non-traditional fluid engineering applications?

2.3. What are some of the advantages of applying CFD?

2.4. What are the limitations and disadvantages of applying CFD?

2.5. In fire engineering, what is field modeling?

2.6. What is the difference between a free-standing fire and a compartment fire?

2.7. A system experiences a state of thermodynamic equilibrium. How is it achieved?

2.8. What are intensive and extensive properties? What is the difference between them?

2.9. What is the difference between compressible and incompressible flows?

2.10. Weakly compressible assumption is generally invoked in modeling fires. How is this achieved?

2.11. What are the basic equations of fluid motion? Explain the basis of a continuum fluid in the development of these equations.

2.12. What is the basic law of the conservation of mass?

2.13. What is the difference between the conservative and non-conservative forms of the governing equations?

2.14. What is Newton's second law of motion?

2.15. Derive a force balance equation for all the forces acting on a differential control volume.

2.16. What is a Newtonian fluid?

2.17. Explain the first law of thermodynamics in deriving the energy equation.

2.18. What is *Fourier's law of heat conduction*?

2.19. What are additional contributions to the heat flux in a combusting fire system?

2.20. In the scalar equation, what is *Fick's law of diffusion*?

2.21. What types of boundary conditions can be imposed for field modeling?

2.22. There are four discretisation methods in the main stream of CFD. What are they?

2.23. What are the main advantages and disadvantages of discretisation of the governing equations through the finite difference and finite volume methods?

2.24. What is the main difference between a structured and an unstructured mesh?

2.25. Is finite difference or finite volume method more suited for structured or unstructured mesh geometries? Why?

2.26. For the figure below, show that the one-dimensional steady state diffusion term $\frac{\partial}{\partial x}\left(\Gamma\frac{\partial\phi}{\partial x}\right)$ can be discretised according to $\left(\Gamma\frac{\partial\phi}{\partial x}\right)_e A_e - \left(\Gamma\frac{\partial\phi}{\partial x}\right)_w A_w$ for the central grid nodal point P.

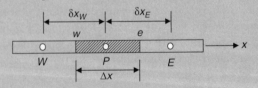

2.27. Based also on the preceding figure, show that the one-dimensional steady state advection term $\dfrac{\partial(\rho u \phi)}{\partial x}$ can be discretised according to $(\rho u \phi)_e A_e - (\rho u \phi)_w A_w$ for the central grid nodal point P.

2.28. Why are upwind schemes important for strongly convective flow?

2.29. Besides the first order upwind scheme, what are other useful schemes can be employed to enhance the accuracy of the numerical solution?

2.30. What are TVD schemes?

2.31. For unsteady flows, what is the difference between explicit and implicit time-marching approaches?

2.32. What are common iterative matrix solvers that can be employed to solve the set of algebraic equations for a structured or an unstructured grid arrangement?

2.33. What is the purpose of the SIMPLE scheme? Is it a direct or iterative method?

2.34. Derive the Poisson form of the pressure correction for weakly compressible flow.

2.35. Collocated grid arrangement suffers from a well-known effect. What is it? How is it remedy?

2.36. Besides SIMPLE, what are other variant SIMPLE algorithms?

2.37. What is the energy cascade process in turbulence?

2.38. Why do large eddies tend to be anisotropic in nature? On the other hand, why are small-scaled eddies isotropic in nature?

2.39. What are the important characteristics of turbulent flows?

2.40. The Reynolds number is a ratio of two forces. What are they?

2.41. What is the difference between Reynolds-Averaged and Favre-Averaged Navier-Stokes equations?

2.42. What is Newton's law of viscosity?

2.43. The Prandtl number is a ratio of two fluid properties. What are they?

2.44. The Schmidt number is a ratio of two fluid properties. What are they?

2.45. Besides the standard k-ε model, what other turbulence models that can be feasibly applied to characterize the turbulent flow in compartment fires?

2.46. Why near-wall modeling is required for turbulence modeling? What are possible approaches that can be adopted to overcome the difficulty of modeling the near-wall region?

3 Additional Considerations in Field Modeling

Abstract

Combustion and radiation are characteristic and inseparable features of fires. In order to gain a better understanding of the flaming behavior of all fires, the inclusion of combustion and radiation characteristics into field modeling represents a fundamental addition to the complex numerical modeling of fires.

The necessity to adopt relevant models in specifically handling the combustion chemistry and radiation heat transfer in fires is exemplified in this chapter. For combustion, the principal knowledge of whether the process is governed by chemical kinetics or turbulent mixing determines the appropriate selection of suitable models. For radiation, the consideration of appropriate models depends on the level of simplification in the radiative properties of absorbing gases, and in the level of sophistication of the radiative transfer and total heat transfer modeling. A range of models is described to demonstrate their applicability in meeting the various challenging aspects of aptly simulating the combustion and radiation processes associated with practical fires.

PART III COMBUSTION

3.1 Turbulent Combustion in Fires

Subject to the availability of three basic elements as illustrated in Figure 3.1 (fuel, oxygen, and heat), favorable flammable or combustion conditions will occur when adequate supplies of fuel and oxygen react in an environment with sufficient heat. A *flaming fire* is categorically a rapid oxidation process. The chemical reaction itself is the convergent result of the three elements of heat, oxygen, and fuel. To create a self-sustaining combustion process, it is the chemical reaction that actually feeds the fire more heat and allows it to continue. A burning fire can therefore be regarded as a manifestation of the chemical reaction. It is nonetheless recognized that the mode of burning may depend somewhat on the physical condition and distribution of the fuel and its surrounding environment than on its chemical nature.

In a state of combustion, a flame is the product of a highly exothermic reaction and can be regarded as a body of gaseous material consisting of reacting gases and finely dispersed carbonaceous particles (soot). The soot particles emit specific bands of electromagnetic (EM) wavelengths, depending on the

Computational Fluid Dynamics in Fire Engineering
Copyright © 2009 by Academic Press. Inc. All rights of reproduction in any form reserved.

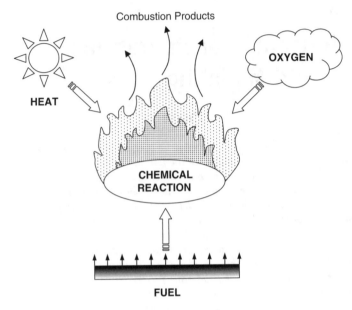

Figure 3.1   A schematic representation of four essential elements for a flaming fire.

combustion chemistry on the fuel involved. Consider a naturally flaming process as described by a burning candle in Figure 3.2. This common flame is a classical example of a *diffusion flame*. Another example is a Bunsen burner with a closed oxygen valve. The principal characteristic of the diffusion flame is that the fuel and oxidizer (oxygen) are initially separated and combustion occurs in the zone where the gases are mixed. A candle flame operates through

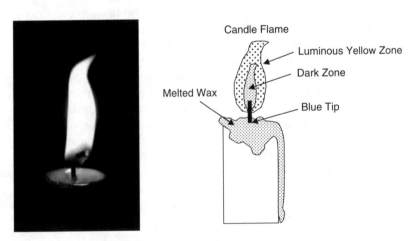

Figure 3.2   An example of a naturally flaming process: a burning candle.

the evaporation of the fuel rises in a laminar flow of hot gas, which reacts with the oxygen diffusing into the flame from the surrounding air. The identifiable luminous yellow zone of the candle flame represents the soot being produced and becomes incandescent from the heat of the chemical reaction. In the candle flame, a zone of blue tip exists because of the occurrence of some premixing of the fuel and oxidizer at the bottom edge of the wick where the flame is quenched, which incidentally has a similar appearance to a premixed flame. In a *premixed flame*, the structure of the flame is rather different. A fully aerated Bunsen burner is the simplest example in developing such a flame where the oxygen supply is premixed with the fuel prior to combustion. A blue-colored flame demonstrates the efficiency of the combustion process and results in less radiative soot being produced.

Laminar flames have low flame speeds. The slow diffusion process of a burning candle represents a typical laminar flame. Most practical fires occur, however, at high flame speeds. The combustion process is now characterized by turbulent flows. The *eddy mixing* process strongly governs the turbulent combustion between the fuel and the oxidizer. The size of the turbulent eddies that move back and forth randomly and across adjacent fluid layers significantly influence the flame-front thickness. According to Damköhler's (1940, 1947) phenomenological investigations, sufficiently large-scale turbulence, of which the localized turbulent eddies are much larger than the flame-front thickness, is seen to merely wrinkle the laminar flame without significantly modifying its internal structure as illustrated by the near-turbulent flame structure in the left-hand pictures of Figure 3.3. On the other hand, sufficiently small-scale turbulence alters the effective transport coefficients in the flame without

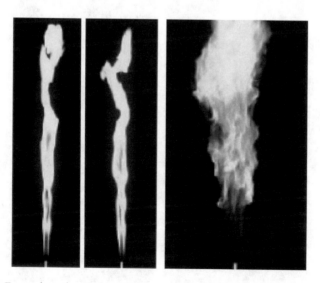

Figure 3.3 Examples of turbulent flame structures of methane: near-turbulent (left) and fully turbulent (right) (both after Lakshminarasimhan et al., 2006)

significantly wrinkling the flame front. Under these circumstances, transport of heat and chemical species is due to the *turbulent diffusivity* but not the *molecular diffusivity*. Further insights into the behavior of turbulent flames have nevertheless led to the proposal of a series of mechanisms of a distributed reaction zone describing the effect of turbulence on the combustion zone in order to overcome the failure of the wrinkled-flame concept developed initially by Damköhler. Three mechanisms were identified by Kovasznay (1956): (1) weak turbulence merely wrinkles the laminar flame front; (2) stronger turbulence disrupts the flame front; and (3) still stronger turbulence results in homogeneous reaction mixtures, in the limit sometimes called the *continuously stirred reactor*. At very high flame speeds—in other words, very high Reynolds numbers—the flame surface becomes very complex as demonstrated by the irregular structure of the flame envelope in the right-hand picture of Figure 3.3. Turbulent flames, unlike laminar flames, are often accompanied by noise and rapid fluctuations of the flame envelope. Turbulence first appears at the tip of the flame and thereafter extends further down toward the burner nozzle as the jet velocity increases. For fully developed turbulent flames, the oxidation process is dominated by *eddy mixing* due to the increased entrainment of air into the reaction zone, which subsequently results in more efficient combustion.

Most practical fires are naturally flaming processes. Flames from natural fires such as the burning of condensed fuels (for example, a campfire depicted in the left-hand picture of Figure 3.4), are considerably different from jet

Figure 3.4   Flames from a campfire (left) and a porous bed gas burner (right).

flames as seen in Figure 3.3. Natural flames are dominated by *buoyancy*, and the momentum (or inertia) of the volatiles rising from the surface is generally very low. In contrast, the momentum of the fuel in jet flames largely determines the behavior and structure of the flames. This chapter deals principally with the combustion of flames from burning solids, although much knowledge has been gained from numerous experimental studies of fires on flat, porous bed gas burners employing hydrocarbon fuels such as those by McCaffrey (1979) and Cox and Chitty (1980) (for example, the flame shown in the right-hand picture of Figure 3.4). In most compartment fire investigations, flat, porous bed gas burners are generally used. For fuel beds that are less than 0.05 m in diameter, the flames are generally considered as laminar. Buoyant diffusion flames with fully developed turbulence are observed for fuel beds having diameters greater than 0.3 m. In spite of the different flow characteristics, the combustion chemistry within the jet and buoyant diffusion flames is in effect identical; there are some commonalities in understanding the process of combustion or burning in both types of flames. Like fully developed turbulent jet flames, the oxidation process of buoyant diffusion flames with fully developed turbulence is also dominated by *eddy mixing*. In the next section, the process of combustion or burning of fires is described via the understanding of the chemical kinetics. In particular, the theories of turbulent combustion and assumptions made therein are discussed.

3.2 Detailed Chemistry versus Simplified Chemistry

In retrospect, combustion or burning is usually associated with the occurrence of a complex sequence of exothermic chemical reactions between a fuel and an oxidant. During such a process, combustion products are generated accompanied by the release of heat or both heat and light in the form of flames. Some chemical reactions can occur very rapidly, while others can occur very slowly. Most chemical reactions develop rapidly as the temperature increases. In general, combustion of practical fires, as will be explained in this section, is full of complex chemical reactions and is greatly influenced by turbulence. Similar to many combustion studies of flames, understanding the chemical kinetics in fires is a branch of chemical science that quantitatively studies the rates of chemical reactions and, more importantly, is the part of chemical science that deals with the interpretation of the empirical kinetic laws in terms of reaction mechanisms. All chemical reactions take place at a definite rate. They are strongly influenced by a number of prevailing conditions that exist within the system. Some of the important conditions are concentrations of the chemical compounds, surrounding pressure, temperature, presence of a catalyst or inhibitor, and the effects of radiation heat transfer.

In laminar combustion, the rate of reaction of a chemical reaction can usually be defined as the rate at which one of the reactants disappears to form products—in other words, the rate at which the products are formed. In a state of complete combustion, the reaction mechanism can be expressed in a single-step fashion as

$$\text{Fuel} + \text{Oxidant} \rightarrow \text{Products} \tag{3.2.1}$$

For a hydrocarbon such as methane, the products would comprise only of carbon dioxide and water vapor in the stoichiometric equation of

$$CH_4 + 2O_2 \rightarrow CO_2 + 2H_2O \tag{R1}$$

The preceding chemical reaction depicts the fuel and the oxygen that are in exactly equivalent or stoichiometric proportions. Equation (R1) illustrates that 1 mole of methane reacts with 2 moles of oxygen to form 1 mole of carbon dioxide and 2 moles of water vapor. For combustion occurring in air, equation (R1) may be modified to include the nitrogen complement as

$$CH_4 + 2O_2 + 2 \times \left(\frac{79}{21}\right)N_2 \rightarrow CO_2 + 2H_2O + 2 \times \left(\frac{79}{21}\right)N_2 \tag{R2}$$

Equation (R2) demonstrates that 1 mole of methane requires 9.524 ($= 2 + (79/21) \times 2$) moles of air. Introducing the molecular weights of methane and air (16 g and 28.95 g, respectively), 1 g of methane requires 17.23 g of air for stoichiometric burning to carbon dioxide and water vapor. More generally,

$$1(\text{g or kg}) \text{ fuel} + r \text{ (g or kg) air} \rightarrow (1 + r) \text{ (g or kg) products} \tag{3.2.2}$$

where r is the stoichiometric air requirement for the fuel in question.

All combustion reactions occur with the release of energy. This may be quantified by defining the heat of combustion (ΔH_c) as the total amount of heat released when a unit quantity of a fuel is oxidized completely. It is noted that heats of combustion are normally determined at constant volume in a *bomb calorimeter* in which a known mass of fuel is burnt completely in an atmosphere of pure oxygen (Moore, 1972). Assuming that there is no heat loss (*adiabatic*), the quantity of heat released is calculated from the temperature rise of the calorimeter and its contents, whose thermal capacities are known. The use of pure oxygen ensures complete combustion, and the result yields the heat release at constant volume. Values of heat of combustion for a range of gases, liquids, and solids are given in Table D.1 of Appendix D. Bomb calorimetry also provides the means by which standard heats of formation $\left(\Delta H_f^o\right)$ of many chemical compounds that may be determined. Values of various standard heats of formation of selected species are given in Table D.2 of Appendix D. More complete information can be found either from *JANAF Thermochemical Tables* or the *Handbook of Chemistry and Physics*. In practice, the heat of combustion may also be expressed in terms of air consumed. It is convenient for many purposes to assume that air consists only of oxygen (21%) and nitrogen (79%) as the main constituent gases. The ratio of nitrogen to oxygen in air is approximately $79/21 = 3.76$. Values of heat of combustion in air for a range of gases, liquids, and solids are also given in Table D.1.

Chemical equations such as those represented by equations (R1) or (R2) define the stoichiometry of a complete chemical reaction. The rate-controlling mechanism can be postulated to involve the collision or interaction of a single molecule of fuel with a single molecule of oxygen. In such a case, the rate of reaction is proportional to the collision of the fuel and oxygen. At a given temperature, the number of collisions according to the *law of mass action*[1] is proportional to the products of the local concentrations of reactants in the mixture. The reaction rate of the fuel is

$$R_{fu} = kC_{fu}C_{ox} \qquad (3.2.3)$$

where k is the proportionality constant called the *specific reaction rate constant*. For a given chemical reaction, k is independent of the concentrations of the fuel and oxidant—C_{fu} and C_{ox}—and depends only on the temperature. The influence of temperature on the rate of reaction may be expressed in terms of the Boltzmann factor—$\exp(-E_a/R_u T)$—specifying the fraction of collisions that have an energy greater than the activation energy E_a, as suggested by Arrhenius (1889). The equation for k to calculate chemical reactions rates, which incidentally is also called the Arrhenius law, may be expressed as

$$k = A_o \exp\left(-\frac{E_a}{R_u T}\right) \qquad (3.2.4)$$

Here, A_o is assumed to include the effect of the collisions of molecules, the steric factor associated with the orientation of the colliding molecules, and the mild temperature dependence of the pre-exponential factor. This parameter A_o, which represents the *collision frequency*, is usually expressed as a function of temperature in the form of AT^n. The appropriate values of A, n, and E_a are based on the nature of the elementary reaction. It should be noted that these parameters are neither functions of the concentrations nor of temperature for given chemical changes.

Combustion processes are never perfect or complete in reality. For example, the gas-phase oxidation mechanism for methane usually involves a series of elementary steps, as shown by the detailed reaction mechanism in Table 3.1. This table is an extension of the original table of reaction mechanism for C_1/C_2 hydrocarbon combustion formulated in Leung et al. (1991), with the consideration of additional reaction steps of methanol (CH_3OH) (between reaction

---

[1]The origin of the law of mass action and its relation to first principles can be found in Penner (1955).

TABLE 3.1 Reaction mechanism for C_1/C_2 hydrocarbon combustion. Rate coefficient in the form $k_f = AT^n \exp(-E_a/R_u T)$

	Reaction		A	n	E
1.	$H + O_2$	$\rightarrow$ $OH + O$	2.00E + 14	0.0	7.03E + 04
2.	$O + H_2$	$\rightarrow$ $OH + H$	5.12E + 04	2.67	2.63E + 04
3.	$OH + H_2$	$\rightarrow$ $H_2O + H$	1.00E + 08	1.6	1.38E + 04
4.	$OH + OH$	$\rightarrow$ $H_2O + O$	1.50E + 09	1.14	4.16E + 02
5.	$2H + M$	$\rightarrow$ $H_2 + M$	9.80E + 13	− 0.6	0.0
6.	$H + OH + M$	$\rightarrow$ $H_2O + M$	2.20E + 19	− 2.0	0.0
7.	$H + O_2 + M$	$\rightarrow$ $HO_2 + M$	2.30E + 15	− 0.8	0.0
8.	$H + HO_2$	$\rightarrow$ $2OH$	1.68E + 14	0.0	3.66E + 03
9.	$H + HO_2$	$\rightarrow$ $H_2 + O_2$	4.30E + 13	0.0	5.90E + 03
10.	$O + HO_2$	$\rightarrow$ $OH + O_2$	3.20E + 13	0.0	0.0
11.	$OH + HO_2$	$\rightarrow$ $H_2O + O_2$	2.90E + 13	0.0	2.08E + 03
12.	$CO + OH$	$\rightarrow$ $CO_2 + H$	4.40E + 06	1.5	− 3.08E + 03
13.	$CH_4 + H$	$\rightarrow$ $CH_3 + H_2$	1.32E + 04	3.0	3.36E + 04
14.	$CH_4 + O$	$\rightarrow$ $CH_3 + OH$	6.92E + 08	1.56	3.55E + 04
15.	$CH_4 + OH$	$\rightarrow$ $CH_3 + H_2O$	1.56E + 07	1.83	1.16E + 04
16.	$CH_3 + H$	$\rightarrow$ CH_4	1.90E + 36	− 7.0	3.79E + 04
17.	$CH_3 + O$	$\rightarrow$ $CH_2O + H$	8.43E + 13	0.0	0.0
18.	$CH_3 + OH$	$\rightarrow$ $CH_2O + H_2$	8.00E + 12	0.0	0.0
19.	$CH_3 + CH_3$	$\rightarrow$ C_2H_6	1.70E + 53	−12.0	8.12E + 04
20.	$CH_3 + CH_3$	$\rightarrow$ $C_2H_5 + H$	8.00E + 14	0.0	1.11E + 05
21.	$CH_3 + CH_2$	$\rightarrow$ $C_2H_4 + H$	4.00E + 13	0.0	0.0
22.	$CH_2 + H$	$\rightarrow$ $CH + H_2$	6.00E + 12	0.0	− 7.50E + 03
23.	$CH_2 + O$	$\rightarrow$ $CO + 2H$	1.20E + 14	0.0	0.0
24.	$CH_2 + O_2$	$\rightarrow$ $CO_2 + 2H$	3.13E + 13	0.0	0.0
25.	$CH_2 + C_2H_2$	$\rightarrow$ $C_3H_3 + H$	1.80E + 12	0.0	0.0
26.	$CH_2 + C_2HO$	$\rightarrow$ $C_2H_3 + CO$	2.00E + 13	0.0	0.0
27.	$CH_2O + H$	$\rightarrow$ $CHO + H_2$	2.30E + 10	1.05	1.37E + 04
28.	$CH_2O + O$	$\rightarrow$ $CHO + OH$	4.15E + 11	0.57	1.16E + 04
29.	$CH_2O + OH$	$\rightarrow$ $CHO + H_2O$	3.43E + 09	1.18	− 1.87E + 03
30.	$CH + O$	$\rightarrow$ $CO + H$	4.00E + 13	0.0	0.0
31.	$CH + O_2$	$\rightarrow$ $CO + OH$	3.30E + 13	0.0	0.0
32.	$CH + C_2H_2$	$\rightarrow$ C_3H_3	1.90E + 13	0.0	0.0
33.	$CHO + H$	$\rightarrow$ $CO + H_2$	9.00E + 13	0.0	0.0

Continued

TABLE 3.1 Reaction mechanism for C_1/C_2 hydrocarbon combustion. Rate coefficient in the form $k_f = AT^n \exp(-E_a/R_u T)$.—Cont'd

	Reaction		A	n	E
34.	$CHO + O$	$\rightarrow$ $CO + OH$	3.00E + 13	0.0	0.0
35.	$CHO + O$	$\rightarrow$ $CO_2 + H$	3.00E + 13	0.0	0.0
36.	$CHO + OH$	$\rightarrow$ $CO + H_2O$	1.02E + 14	0.0	0.0
37.	$CHO + O_2$	$\rightarrow$ $CO + HO_2$	3.00E + 12	0.0	0.0
38.	$CHO + M$	$\rightarrow$ $CO + H + M$	2.50E + 14	0.0	7.03E + 04
39.	$CH_3OH + H$	$\rightarrow$ $CH_3O + H_2$	4.00E + 13	0.0	2.55E + 04
40.	$CH_3OH + O$	$\rightarrow$ $CH_3O + OH$	1.00E + 13	0.0	1.96E + 04
41.	$CH_3OH + OH$	$\rightarrow$ $CH_3O + H_2O$	1.00E + 13	0.0	7.10E + 03
42.	$CH_3O + H$	$\rightarrow$ $CH_2O + H_2$	2.00E + 13	0.0	0.0
43.	$CH_3O + O_2$	$\rightarrow$ $CH_2O + HO_2$	1.00E + 13	0.0	3.00E + 04
44.	$CH_3O + M$	$\rightarrow$ $CH_2O + H + M$	1.00E + 14	0.0	1.05E + 05
45.	$C_2H_6 + H$	$\rightarrow$ $C_2H_5 + H_2$	1.44E + 09	1.5	3.10E + 04
46.	$C_2H_6 + O$	$\rightarrow$ $C_2H_5 + OH$	1.00E + 09	1.5	2.43E + 04
47.	$C_2H_6 + OH$	$\rightarrow$ $C_2H_5 + H_2O$	7.23E + 06	2.0	3.62E + 03
48.	$C_2H_5 + O$	$\rightarrow$ $C_2H_4O + H$	8.43E + 13	0.0	0.0
49.	$C_2H_5 + O_2$	$\rightarrow$ $C_2H_4 + HO_2$	1.02E + 10	0.0	− 9.15E + 03
50.	C_2H_5	$\rightarrow$ $C_2H_4 + H$	1.00E + 46	− 9.1	2.44E + 05
51.	$C_2H_4 + H$	$\rightarrow$ $C_2H_3 + H_2$	5.42E + 14	0.0	6.28E + 04
52.	$C_2H_4 + O$	$\rightarrow$ $CHO + CH_3$	3.50E + 06	2.08	0.0
53.	$C_2H_4 + OH$	$\rightarrow$ $C_2H_3 + H_2O$	7.00E + 13	0.0	1.26E + 04
54.	$C_2H_4 + OH$	$\rightarrow$ $CH_3 + CH_2O$	1.99E + 12	0.0	4.02E + 03
55.	$C_2H_4O + H$	$\rightarrow$ $C_2H_3O + H_2$	4.09E + 09	1.16	1.00E + 04
56.	$C_2H_4O + O$	$\rightarrow$ $C_2H_3O + OH$	5.80E + 12	0.0	7.60E + 03
57.	$C_2H_4O + OH$	$\rightarrow$ $C_2H_3O + H_2O$	2.35E + 10	0.73	− 4.65E + 03
58.	$C_2H_3 + H$	$\rightarrow$ $C_2H_2 + H_2$	1.20E + 13	0.0	0.0
59.	$C_2H_3 + O$	$\rightarrow$ $C_2H_2O + H$	3.00E + 13	0.0	0.0
60.	$C_2H_3 + O_2$	$\rightarrow$ $CH_2O + CHO$	5.40E + 12	0.0	0.0
61.	C_2H_3	$\rightarrow$ $C_2H_2 + H$	5.30E + 31	− 5.5	1.94E + 05
62.	$C_2H_3O + M$	$\rightarrow$ $CH_3 + CO + M$	1.00E + 15	0.0	3.94E + 04
63.	$C_2H_2 + H$	$\rightarrow$ $C_2H + H_2$	6.00E + 13	0.0	1.16E + 05
64.	$C_2H_2 + O$	$\rightarrow$ $CH_2 + CO$	2.17E + 04	2.8	2.08E + 03
65.	$C_2H_2 + O$	$\rightarrow$ $C_2HO + H$	2.17E + 04	2.8	2.08E + 03
66.	$C_2H_2 + OH$	$\rightarrow$ $C_2H_2O + H$	6.00E + 13	0.0	5.40E + 04
67.	$C_2H_2 + OH$	$\rightarrow$ $C_2H + H_2O$	6.00E + 13	0.0	5.40E + 04

Continued

TABLE 3.1 Reaction mechanism for C_1/C_2 hydrocarbon combustion. Rate coefficient in the form $k_f = AT^n \exp(-E_a/R_u T)$.—Cont'd

	Reaction		A	n	E
68.	$C_2H_2O + H$	$\rightarrow$ $CH_3 + CO$	1.80E + 13	0.0	1.40E + 04
69.	$C_2H_2O + O$	$\rightarrow$ $CHO + CHO$	2.30E + 12	0.0	5.70E + 03
70.	$C_2H_2O + O$	$\rightarrow$ $CO + CH_2O$	1.00E + 13	0.0	0.0
71.	$C_2H_2O + OH$	$\rightarrow$ $CH_2O + CHO$	1.00E + 13	0.0	0.0
72.	$C_2H_2O + M$	$\rightarrow$ $CH_2 + CO + M$	1.00E + 16	0.0	2.48E + 05
73.	$C_2H + O_2$	$\rightarrow$ $CO + CHO$	5.00E + 13	0.0	6.30E + 03
74.	$C_2H + C_2H_2$	$\rightarrow$ $C_4H_2 + H$	3.50E + 13	0.0	0.0
75.	$C_2HO + H$	$\rightarrow$ $CH_2 + CO$	3.00E + 13	0.0	0.0
77.	$C_3H_4 + H$	$\rightarrow$ $C_3H_3 + H_2$	5.00E + 12	0.0	6.28E + 03
78.	$C_3H_4 + O$	$\rightarrow$ $CH_2O + C_2H_2$	1.00E + 12	0.0	0.0
79.	$C_3H_4 + O$	$\rightarrow$ $C_2H_3 + CHO$	1.00E + 12	0.0	0.0
80.	$C_3H_4 + OH$	$\rightarrow$ $C_2H_4 + CHO$	1.00E + 12	0.0	0.0
81.	$C_3H_4 + OH$	$\rightarrow$ $C_2H_3 + CH_2O$	1.00E + 12	0.0	0.0
82.	C_3H_4	$\rightarrow$ $C_3H_3 + H$	5.00E + 17	0.0	3.70E + 05
83.	$C_3H_3 + O$	$\rightarrow$ $C_3H_2 + OH$	3.20E + 12	0.0	0.0
84.	$C_3H_3 + O$	$\rightarrow$ $CO + C_2H_3$	3.80E + 13	0.0	0.0
85.	$C_3H_3 + O_2$	$\rightarrow$ $C_2HO + CH_2O$	6.00E + 12	0.0	0.0
86.	$C_3H_2 + H$	$\rightarrow$ C_3H_3	6.00E + 12	0.0	0.0
87.	$C_3H_2 + O$	$\rightarrow$ $C_2H + CHO$	6.80E + 13	0.0	0.0
88.	$C_3H_2 + OH$	$\rightarrow$ $C_2H_2 + CHO$	6.80E + 13	0.0	0.0
89.	$C_4H_2 + O$	$\rightarrow$ $C_3H_2 + CO$	2.70E + 13	0.0	7.20E + 03
90.	$C_4H_2 + OH$	$\rightarrow$ $C_3H_2 + CHO$	3.00E + 13	0.0	0.0
91.	$C_4H_2 + C_2H$	$\rightarrow$ $C_6H_2 + H$	3.50E + 13	0.0	0.0

steps 39 and 44 in Table 3.1). Additional reaction steps of ethanol (C_2H_5OH) are also included for completeness, which can be found in Table 3.2 (Warantz, 1984). The total reaction scheme thus comprises of 41 species and has 116 forward reaction steps. Normally, chemical reactions can proceed in both the forward direction (reactants forming products, rate constant k_f) and the reverse direction (reaction products reforming reactants, rate constant k_b); all the reverse reactions in Tables 3.1 and 3.2 are computed from relevant equilibrium constants, which will be described in section 3.4.1.3. The units in Tables 3.1 and 3.2 are in cm (centimeter), s (seconds), mol (mole), J (Joules), and K (Kelvin).

There are some circumstances in field modeling where the great amount of chemical information produced by a detailed reaction mechanism, represented in Tables 3.1 and 3.2, is not entirely necessary and a simpler reaction

Table 3.2 Extended reaction mechanism for C_2 hydrocarbon combustion for ethanol (C_2H_5OH). Rate coefficient in the form $k_f = AT^n \exp(-E_a/R_u T)$.

	Reaction			A	n	E
92.	C_2H_5OH		$\rightarrow$ $CH_3O + CH_3$	3.10E + 15	0.0	3.37E + 05
93.	$C_2H_5OH + OH$		$\rightarrow$ $s\text{-}C_2H_5O + H_2O$	8.00E + 06	1.78	− 6.45E + 03
94.	$C_2H_5OH + OH$		$\rightarrow$ $C_2H_5O + H_2O$	1.14E + 06	2.0	3.82E + 03
95.	$C_2H_5OH + OH$		$\rightarrow$ $p\text{-}C_2H_5O + H_2O$	2.56E + 06	2.06	3.60E + 03
96.	$C_2H_5OH + O$		$\rightarrow$ $s\text{-}C_2H_5O + OH$	6.00E + 05	2.46	7.74E + 03
97.	$C_2H_5OH + O$		$\rightarrow$ $C_2H_5O + OH$	4.82E + 13	0.0	2.87E + 04
98.	$C_2H_5OH + O$		$\rightarrow$ $p\text{-}C_2H_5O + OH$	5.00E + 12	0.0	1.85E + 04
99.	$C_2H_5OH + H$		$\rightarrow$ $s\text{-}C_2H_5O + H_2$	4.40E + 12	0.0	1.91E + 04
100.	$C_2H_5OH + H$		$\rightarrow$ $C_2H_5O + H_2$	1.76E + 12	0.0	1.91E + 04
101.	$C_2H_5OH + H$		$\rightarrow$ $p\text{-}C_2H_5O + H_2$	2.00E + 12	0.0	3.97E + 04
102.	$C_2H_5OH + O_2$		$\rightarrow$ $s\text{-}C_2H_5O + HO_2$	4.00E + 13	0.0	2.14E + 05
103.	$C_2H_5OH + O_2$		$\rightarrow$ $C_2H_5O + HO_2$	2.00E + 13	0.0	2.34E + 05
104.	$C_2H_5OH + O_2$		$\rightarrow$ $p\text{-}C_2H_5O + HO_2$	4.00E + 13	0.0	2.13E + 05
105.	$p\text{-}C_2H_5O$		$\rightarrow$ $s\text{-}C_2H_5O$	1.00E + 11	0.0	1.13E + 05
106.	$s\text{-}C_2H_5O + M$		$\rightarrow$ $C_2H_4O + H + M$	5.00E + 13	0.0	9.15E + 04
107.	$s\text{-}C_2H_5O + H$		$\rightarrow$ $C_2H_4O + H_2$	2.00E + 13	0.0	0.0
108.	$s\text{-}C_2H_5O + OH$		$\rightarrow$ $C_2H_4O + H_2O$	1.50E + 13	0.0	0.0
109.	$s\text{-}C_2H_5O + O$		$\rightarrow$ $C_2H_4O + OH$	9.04E + 13	0.0	0.0
110.	$s\text{-}C_2H_5O + O_2$		$\rightarrow$ $C_2H_4O + HO_2$	8.40E + 15	− 1.2	0.0
111.	$s\text{-}C_2H_5O + O_2$		$\rightarrow$ $C_2H_4O + HO_2$	4.80E + 14	0.0	2.09E + 04
112.	C_2H_5O		$\rightarrow$ $CH_2O + CH_3$	1.00E + 15	0.0	9.04E + 04
113.	$C_2H_5O + O_2$		$\rightarrow$ $C_2H_4O + HO_2$	9.77E + 10	0.0	6.65E + 03
114.	$C_2H_5O + OH$		$\rightarrow$ $C_2H_4O + H_2O$	1.00E + 14	0.0	9.75E + 04
115.	$C_2H_5O + H$		$\rightarrow$ $C_2H_4O + H_2$	1.00E + 14	0.0	0.0
116.	$C_2H_5O + O$		$\rightarrow$ $C_2H_4O + OH$	1.21E + 14	0.0	0.0

mechanism should suffice. In many combustion investigations, detailed mechanisms have been developed and validated for simpler C_1/C_2 hydrocarbon combustion. Nevertheless, numerical models that consider large-scale flames in two- or three-dimensional geometries may not be able to incorporate detailed kinetic mechanism because of the enormous computational costs that would likely be involved. For a given reaction mechanism, the computational costs depend predominantly on the number of chemical species rather than on the total number of reactions. Conventional numerical solution techniques for partial differential equations indicate that the computer time requirements are roughly proportional to N^2, where N is the number of species. As such, the total number of species to be considered in combustion can dramatically influence the computer requirements to resolve reacting flows.

Returning to the detailed oxidation mechanism of methane, one possible simplification is to adopt a quasi-global reaction mechanism as suggested by Edelman and Fortune (1969), which combined a single reaction of fuel and oxygen to form carbon monoxide CO and hydrogen H_2 together with a detailed reaction mechanism for CO and H_2 oxidation. For example, the fuel consumption reaction of methane to produce CO and H_2 can be modeled by the global reaction step

$$CH_4 + \frac{1}{2}O_2 \rightarrow CO + 2H_2 \qquad \text{(R3)}$$

According to Westbrook and Dryer (1981), the reaction rate constant k for methane is

$$k = 2.3 \times 10^7 \exp\left(-\frac{30}{R_u T}\right)$$

There are also other reaction rate constants for important gaseous phase fuels besides methane that have been established in Westbrook and Dryer (1981). Equation (R3) is further combined with 21 elementary reactions and species in the CO-H_2-O_2 system, of which the reactions and rate parameters are tabulated in Table 3.3 after Westbrook and Dryer (1981). The units are cm (centimeter), s (seconds), mol (mole), kcal (kilocalorie), and K (Kelvin).

With reference to the single-step mechanism of equation R2, computational solutions are required for five species including nitrogen. The CO-H_2-O_2 mechanism includes 10–12 species (H, O, H_2, O_2, OH, H_2O, N_2, CO, CO_2, HO_2, H_2O_2, and CH_4), while the detailed reaction mechanism for methane involves 41 species. In terms of computer time requirements, the quasi-global model is roughly between the simplest model and the most detailed model. The weakness of this approach is that the flame structure and species concentrations in the combustion zone cannot be accurately predicted due to the neglect of the detailed chemistry. Nevertheless, the strength of this approach allows accurate values to be predicted for the adiabatic

TABLE 3.3 Reaction mechanism used in quasi-global mechanism for CO-H_2-O_2 system. Rate coefficient in the form $k_f = AT^n \exp(-E_a/R_u T)$

	Reaction			A	n	E
1.	$H + O_2$	$\rightarrow$	$O + OH$	2.2E + 14	0.0	16.8
2.	$H_2 + O$	$\rightarrow$	$H + OH$	1.8E + 10	1.0	8.9
3.	$O + H_2O$	$\rightarrow$	$OH + OH$	6.8E + 13	0.0	18.4
4.	$OH + H_2$	$\rightarrow$	$H + H_2O$	2.2E + 13	0.0	5.1
5.	$H + O_2 + M$	$\rightarrow$	$HO_2 + M$	1.5E + 15	0.0	− 1.0
6.	$O + HO_2$	$\rightarrow$	$OH + O_2$	5.0E + 13	0.0	1.0
7	$H + HO_2$	$\rightarrow$	$OH + OH$	2.5E + 14	0.0	1.9
8.	$H + HO_2$	$\rightarrow$	$H_2 + O_2$	2.5E + 13	0.0	0.7
9.	$OH + HO_2$	$\rightarrow$	$H_2O + O_2$	5.0E + 13	0.0	1.0
10.	$HO_2 + HO_2$	$\rightarrow$	$H_2O_2 + O_2$	1.0E + 13	0.0	1.0
11.	$H_2O_2 + M$	$\rightarrow$	$OH + OH + M$	1.2E + 17	0.0	45.5
12.	$HO_2 + H_2$	$\rightarrow$	$H_2O_2 + H$	7.3E + 11	0.0	18.7
13.	$H_2O_2 + OH$	$\rightarrow$	$H_2O + HO_2$	1.0E + 13	0.0	1.8
14.	$CO + OH$	$\rightarrow$	$CO_2 + H$	1.5E + 07	1.3	− 0.8
15.	$CO + O_2$	$\rightarrow$	$CO_2 + O$	3.1E + 11	0.0	37.6
16	$CO + O + M$	$\rightarrow$	$CO_2 + M$	5.9E + 15	0.0	4.1
17.	$CO + HO_2$	$\rightarrow$	$CO_2 + OH$	1.5E + 14	0.0	23.7
18.	$OH + M$	$\rightarrow$	$O + H + M$	8.0E + 19	0.0	103.7
19.	$O_2 + M$	$\rightarrow$	$O + O + M$	5.1E + 15	0.0	115.0
20.	$H_2 + M$	$\rightarrow$	$H + H + M$	2.2E + 14	0.0	96.0
21.	$H_2O + M$	$\rightarrow$	$H + OH + M$	2.2E + 16	0.0	105.0

flame temperature and the equilibrium post-flame chemical composition, since all of the important elementary reactions and species in the CO-H_2-O_2 system are considered. Note that the adiabatic flame temperature, by definition, is the temperature of the combustion products, where the combustion process takes place adiabatically and with no work or changes in kinetic or potential energy. This represents the maximum temperature that can be achieved for the given reactants, as any heat transfer from reacting substances and any incomplete combustion would have a tendency to lower the temperature of the products. The complete combustion dictated by equation (R1) in pure oxygen stream or equation (R2) in air will yield the adiabatic flame temperature. A sample calculation of the adiabatic flame temperature and a table tabulating typical adiabatic flame temperatures for a range of fuels are presented in Appendix B.

Alternatively, an even simpler global reaction scheme for hydrocarbon fuels has been proposed by Jones and Lindstedt (1988). In essence, these reaction schemes comprise of two competing fuel breakdown reactions and equilibrium assumptions that have been employed to derive the initial estimates of the rate expressions. For example, the four-step reaction mechanism for methane is deduced as

$$CH_4 + \frac{1}{2}O_2 \rightarrow CO + 2\,H_2 \qquad\qquad (R4)$$

$$CH_4 + H_2O \rightarrow CO + 3\,H_2 \qquad\qquad (R5)$$

$$H_2 + \frac{1}{2}O_2 \rightleftharpoons H_2O \qquad\qquad (R6)$$

$$CO + H_2O \rightleftharpoons CO_2 + H_2 \qquad\qquad (R7)$$

Appropriate values of the rate constants for the four-step reaction steps between equations (R4)–(R7) can be estimated according to

$$k_{R4} = 4.4 \times 10^{11} \exp\left(-\frac{30}{R_u T}\right)$$

$$k_{R5} = 3.0 \times 10^{8} \exp\left(-\frac{20}{R_u T}\right)$$

$$k_{R6} = 2.5 \times 10^{16} T^{-1} \exp\left(-\frac{40}{R_u T}\right)$$

$$k_{R7} = 6.8 \times 10^{16} T^{-1} \exp\left(-\frac{40}{R_u T}\right)$$

The units for the preceding reaction rate constants are in kg (kilogram), m (meter), s (seconds), kmol (kilomole), kcal (kilocalorie), and K (Kelvin). From the preceding, only 7 species (H_2, O_2, H_2O, N_2, CO, CO_2, and CH_4) are considered, which further reduces the computer time requirements in comparison to the quasi-global reaction mechanism. This proposed mechanism has demonstrated excellent agreement with measured major chemical species for a wide range of premixed and diffusion flames computed. In Jones and Lindstedt (1988), reactions rate constants for gaseous alkane hydrocarbons have been formulated up to butane. One of the many attractions of this particular global reaction mechanism is the inherent simplicity it represents and can therefore be viewed as realistic alternatives to fundamental reaction schemes, where such schemes are either unknown or precluded on computational grounds. The extension of the schemes to include other fuels of practical interest, such as gaseous phase kerosene and octane, appears to be comparatively straightforward. We also note other similar global reaction schemes, which have been developed in the similar genre as aforementioned. In the study by Liu et al. (2003), a different four-step chemical kinetic mechanism is employed for the simulation of methane combustion of a spatially developing transitional free jet flame at moderate Reynolds number. The mechanism entails the following four reaction steps:

$$CH_4 + 2H + H_2O \rightarrow CO + 4H_2 \qquad (R8)$$

$$CO + H_2O \rightleftharpoons CO_2 + H_2 \qquad (R9)$$

$$H + H + M \rightarrow H_2 + M \qquad (R10)$$

$$O_2 + 3H_2O \rightleftharpoons 2H + 2H_2O \qquad (R11)$$

Reaction rate constants expressed in the Arrhenius form of $k_j = A_j T^{n_j} \exp(-E_{a,j}/R_u T)$ and suitable equilibrium constants required to evaluate the reverse reactions just indicated have been obtained from Seshadri and Peters (1988).

Admittedly, in spite of the established science in aptly determining the detailed chemical reactions for a range of simpler alkane hydrocarbon fuels such as methane, ethane, propane, and to some extent heptane, comprehensive combustion chemistry for complex fuels is still yet to be further determined. Specifically for fires, the detailed combustion chemistry of solid fuels such as cellulose or polymethylmethacrylate (PMMA) remains elusive. It is not entirely surprising that many field modeling investigations are still much inclined in the use of a simple single-step irreversible reaction in the form of equation (3.2.2), because of the absence of any *a priori* knowledge in handling the combustion of these combustible materials.

In considering turbulent combustion, the problem is nevertheless further compounded by the presence of turbulent fluid flow being intimately coupled with the chemical kinetics. Based on either the Reynolds or Favre decomposition and averaging, there is a great difference between the time-averaged or Favre-averaged reaction rate and the rate of reaction based on averaged quantities. Returning to the simple one-step irreversible reaction described in equation (3.2.3), noting that $C_{fu} = (\rho Y_{fu})/M_{fu}$ and $C_{ox} = (\rho Y_{ox})/M_{ox}$, where Y_{fu} and Y_{ox} are the mass fractions of fuel and oxygen, the instantaneous rate of reaction using the ideal-gas law of equation (2.3.2) and employing the Reynolds decomposition and averaging takes the form:

$$R_{fu} = \frac{Ap^2}{R_u} \left(\overline{T} + T'\right)^{n-2} \exp\left(-\frac{E_a}{R_u} \frac{1}{\overline{T} + T'}\right) \left(\overline{Y}_{fu} + Y'_{fu}\right) \left(\overline{Y}_{ox} + Y'_{ox}\right) \qquad (3.2.5)$$

The exponential term in the preceding equation may be written as

$$\exp\left(-\frac{E_a}{R_u}\frac{1}{\overline{T}+T'}\right) = \exp\left(-\frac{E_a}{R_u\overline{T}}\right)\exp\left(\frac{E_a}{R_u\overline{T}}\frac{T'/\overline{T}}{1+T'/\overline{T}}\right)$$

Expressing in terms of series expansions, the time-averaged reaction rate may be expanded to yield

$$\overline{R}_{fu} = \frac{Ap^2}{R_u}\overline{T}^{n-2}\exp\left(-\frac{E_a}{R_u\overline{T}}\right)\overline{Y}_{fu}\overline{Y}_{ox}(1+F) \tag{3.2.6}$$

with F given by

$$F \equiv \frac{\overline{Y'_{fu}Y'_{ox}}}{\overline{Y}_{fu}\overline{Y}_{ox}} + (P_2 + Q_2 + P_1Q_1)\frac{\overline{T'^2}}{\overline{T}^2} + (P_1 + Q_1)\left(\frac{\overline{T'Y'_{fu}}}{\overline{T}\,\overline{Y}_{fu}} + \frac{\overline{T'Y'_{ox}}}{\overline{T}\,\overline{Y}_{ox}}\right)$$

$$+ P_1\frac{\overline{T'Y'_{fu}Y'_{ox}}}{\overline{T}\,\overline{Y}_{fu}\overline{Y}_{ox}} + P_2\left(\frac{\overline{T'^2Y'_{fu}}}{\overline{T}^2\,\overline{Y}_{fu}} + \frac{\overline{T'^2Y'_{ox}}}{\overline{T}^2\,\overline{Y}_{ox}}\right) + (P_3 + Q_3)\frac{\overline{T'^3}}{\overline{T}^3} + \cdots$$

$$\tag{3.2.7}$$

where

$$P_n \equiv \sum_{k=1}^{n}(-1)^{n-k}\frac{(n-1)!}{(n-k)![(n-1)!]^2}\frac{1}{n}\left(\frac{E_a}{R_u\overline{T}}\right)^n \tag{3.2.8}$$

$$Q_n \equiv \frac{(\alpha-2)(\alpha-1)\ldots(\alpha+1+n)}{n!} \tag{3.2.9}$$

The term F in equation (3.2.6) includes the influence of turbulence on the time-averaged reaction rate. It should be noted that a development similar to the Reynolds decomposition can be followed for Favre decomposition and averaging. The increase of complexity of handling the time-averaged reaction rate of equation (3.2.6) results primarily from the need to close the system of equations by appropriately treating the second-, third-, or even higher-order correlation terms. By assuming that the series of F is convergent, the procedure can be followed by considering suitable transport equations for the second-order correlation terms: $\overline{Y'_{fu}Y'_{ox}}, \overline{T'Y'_{fu}}, \overline{T'Y'_{ox}}, \overline{T'^2}$. In most cases, the common practice is to neglect the third- and higher-order correlation terms by equating them to zero. This, of course, can lead to erroneous results, since these zero values may be inconsistent with the inequalities that exist between moments of various orders (Bilger, 1980). Even by considering the second-order correlation terms, this method is only applicable for relatively simple chemistry such as the single-step irreversible reaction. The extension of the method to the reduced four-step kinetic mechanisms remains plausible but challenging. More importantly, the series of F is convergent only if (E_a/R_u) is not much larger than the mean temperature $\overline{T}$, and the fluctuation levels are small. In practice, $(E_a/R_u\overline{T})$ is usually much greater than unity for many combustion problems. This presents a severe restriction on the applicability of equation (3.2.6) to practical problems. In order to overcome the difficulty in modeling these non-linear terms, various combustion models, capable of better handling practical flames, are explained in detail in the next section. The many possible alternative approaches to modeling turbulent combustion will be discussed in light of the assumptions imposed, the model formulation, their physical implications, and their validity to different flame flow configurations.

3.3 Overview of Combustion Modeling Approaches

Most chemical reactions have high rates. A limiting case nonetheless exists when the chemical reaction time is negligibly short in comparison to the mixing time due to turbulence, which is represented by the fast time scales at the extreme left-hand side of the chemical time scale in Figure 3.5. In this limit, the *fast chemistry assumption* for the turbulent flame drastically simplifies the problems of chemistry-turbulence interactions, since the molecular-species concentrations are directly related to a conserved scalar, and the statistics of all thermodynamics variables are obtainable from the knowledge of statistics of that scalar (Bilger, 1980). This so-called *conserved-scalar approach* removes the need to evaluate the mean reaction rates; all equilibrium chemical composition, temperature, and density of a gaseous mixture can be adequately determined if the elemental mass fractions of all the chemical species, the pressure,

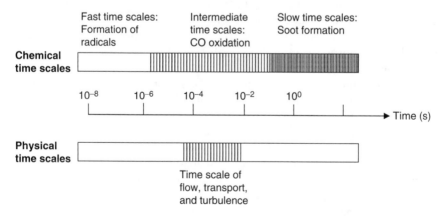

Figure 3.5 Chemical and physical time scales (adapted after Maas and Pope, 1992).

and the enthalpy are known. A typical choice for strictly conserved (i.e., zero source) scalar variable is the mixture fraction. The use of probability density function (PDF) links the means and higher moments of chemical species and temperature to those of the conserved scalar. This approach is commonly known as the *flame sheet approximation or mixed-is-burnt model*.

Early attempts to analyze turbulent diffusion flames using the *mixed-is-burnt model* have entailed the consideration whereby the chemical reaction possessed the features of being single-step, irreversible, and infinitely fast such as in the form of equation (3.2.1) or (3.2.2). It should be noted that such an approach, despite its simplicity, is still prevalently used in field modeling and will continue to be adopted so long as the detailed chemistry for the combustion of most solid fuels is absent. As a further extension to the mixed-is-burnt model, the *chemical equilibrium model* could be adopted to express all the intermediate species within a hydrocarbon fuel as a function of the conserved scalar (or mixture fraction). This mixed-is-burnt model assumes the chemistry is rapid enough for chemical equilibrium to always exist at the molecular level.

Another strategy to better describe the combustion process of turbulent non-premixed combustion is the assumption that where the microscopic element in the model, describing the local mixture state and burning, can be taken to have the structure of an undisturbed laminar diffusion flame. This flame sheet according to Williams (1975) and Peters (1984) can be treated as an ensemble of counterflow diffusion flames called "flamelets". Essentially, the laminar diffusion flame through these flamelets provides a unique relationship for all thermochemical variables in terms of the conserved scalar (or mixture fraction). Similar to the mixed-is-burnt model, these relationships are then averaged through the use of PDF for the conserved scalar for the turbulent flame. Also known as the *laminar flamelet model*, this approach differs from that of mixed-is-burnt model in the use of laminar flamelet relationships in place of equilibrium. In laminar hydrocarbon-air diffusion flames, the mixture state close to the origin is not predicted simply by chemical equilibrium, but involves a balance between chemical reaction and transport processes. Such notions are extended in this approach to permit not only the incorporation of more realistic chemistry but also to accommodate non-equilibrium effects.

In some circumstances, such as fast chemistry for localized high-temperature regions or fast mixing for high turbulence intensity, the overall features of turbulent flames can be taken to be independent of the detailed chemistry. Owing to the *eddy mixing* process, the rate of combustion in this limit is assumed to be determined by the rate of intermixing on a molecular scale of fuel and oxygen eddies, which is presented by the rate of dissipation of eddies. Since the reactants comprising of fuel and oxygen appear as fluctuating intermittent quantities, a relationship can be derived between the fluctuations and the mean concentration of the species. The rate of dissipation can thus be expressed in terms of either by their mean concentration or concentration fluctuation of the reacting species.

These so-called *eddy break-up models* and *eddy-dissipation models*, which are applicable for high Reynolds numbers, represent alternative approaches to treat turbulent non-premixed flames in addition to the mixed-is-burnt model, chemical equilibrium model, and the laminar flamelet model. Appropriate source terms representing the preceding consumption rates of combustion of fuel and oxygen are incorporated into their respective transport equations to solve for the distribution of the species in the flow field. It should be noted that this particular approach is still mainly constricted to single-step, irreversible, and infinitely fast chemical reactions, although the model could possibly be extended to accommodate a simple reduced reaction mechanism.

Modeling approaches for turbulent assisted combustion based on the mixed-is-burnt model, chemical equilibrium model, laminar flamelet model, eddy break-up model, and eddy-dissipation model are applicable for most practical fires in a well-ventilated environment. More details on their conceptual formulations and their applications in the context of field modeling are further described in the next section.

3.4 Combustion Models

3.4.1 Generalized Finite-Rate Formulation

3.4.1.1 Background Theory

There are a number of basic sets of assumptions of various limiting cases where the combustion process of buoyant diffusion flames could be treated. The knowledge of whether the chemical reaction is dominated either by chemical kinetics or mixing in the flaming zone of fuel and oxidant represents an important consideration in the development of suitable combustion models. One useful dimensionless parameter to characterize their relative contributions in a turbulent flame is the *Damköhler* number *Da*. By definition, *Da* can be expressed as

$$Da \equiv \frac{\tau_s}{\tau_k} = \frac{\text{time scale of tubulent mixing}}{\text{time scale of combustion chemistry}} \qquad (3.4.1)$$

In the mixing zone of fuel and oxidant, the single-step chemical reaction for the consumption of fuel in laminar combustion takes the following Arrhenius form:

$$R_{fu} = A_o \rho^2 \exp\left(-\frac{E_a}{R_u T}\right) Y_{fu} Y_{ox} \qquad (3.4.2)$$

Employing Favre averaging to the above expression and neglecting density and turbulent fluctuations yields

$$\overline{R}_{fu} = A_o \overline{\rho}^2 \exp\left(-\frac{E_a}{R_u \tilde{T}}\right) \tilde{Y}_{fu} \tilde{Y}_{ox} \tag{3.4.3}$$

Equation (3.4.3) reverts exactly for laminar flame combustion when the random fluctuations of the fluid flow are absent. According to Borghi (1973), the ratio between the kinetic energy of turbulence k and the dissipation ε—that is, k/ε—yields the diffusion time scale. Following Spalding (1976), the turbulent time scale τ_s that accounts for the mixing time in connection with the mean motion of the fluid and the random turbulence (defined as the stretching of the flame eddies) can be expressed as

$$\tau_s = \left[\left|\frac{\partial \tilde{u}_i}{\partial x_i}\right| + \frac{\varepsilon}{k}\right]^{-1} \tag{3.4.4}$$

In order to determine the flame characteristics under different fluid flow conditions, the preceding diffusion time scale should be compared against the chemical time scale τ_k, which is given by

$$\tau_k = \left[A_o \overline{\rho}\left(\tilde{Y}_{fu} + r\tilde{Y}_{ox}\right)\exp\left(-\frac{E_a}{R_u \tilde{T}}\right)\right]^{-1} \tag{3.4.5}$$

During the combustion process, the smaller of the two reaction rates is generally taken to represent the effective controlling rate. Evidently, the smaller time-scale rate of reaction corresponds to the larger of the time scales; it is natural that the reaction is controlled by the process that takes the larger portion of time. In regions where τ_s is larger than τ_k, $Da \gg 1$, the mixing of reactants is slow and the reactants are at suitable concentrations and temperature to react as soon as they are intimately mixed. The reaction is diffusion controlled; it is therefore governed by eddy mixing. However, when τ_s is very small, $Da \ll 1$, this corresponds to a large dissipation rate of eddies and rapid mixing. The reaction is kinetically influenced, which implies that the reactants are in intimate contact with each other but the reaction will only proceed at appropriate concentrations and temperature.

3.4.1.2 Species Transport Equations

The generalized finite-rate formulation, which can be applied to model the mixing and transport of chemical species through solving conservation equations describing convection, diffusion, and reaction sources for each chemical species, is suitable for a wide range of applications including laminar or turbulent reaction systems, and combustion with premixed, non-premixed, or partially premixed flames. Multiple simultaneous chemical reactions can be incorporated to characterize the combustion process. The key success to employ the generalized finite-rate formulation for turbulent reacting flames is mainly entrenched in the appropriate modeling of the reaction sources in turbulent flows.

In field modeling, it is more convenient to work in terms of mass fractions rather than mole fractions. The primary reason is that mass is always perfectly conserved during the combustion process, while moles are not necessarily conserved. By definition, the mass fraction of any ith species is given by

$$Y_i \equiv \frac{m_i}{\sum\limits_{i=1}^{N} m_i} \quad \text{where} \quad \sum\limits_{i=1}^{N} Y_i = 1 \tag{3.4.6}$$

for N different fluid phase chemical species in a given system. The weight m_i of the ith species can be taken to be equivalent to the product of the number of moles n_i and molecular weight M of ith gas, which is $m_i = n_i M_i$. The transport equation under consideration for the local mass fraction of each species Y_i follows the same form of the transport equation for the scalar property derived in Chapter 2. According to equation (2.4.45), the conservation equation in laminar combustion takes the following general form

$$\frac{\partial}{\partial t}(\rho Y_i) + \frac{\partial}{\partial x_j}(\rho u_j Y_i) = \frac{\partial}{\partial x_j}\left[\rho D_i \frac{\partial Y_i}{\partial x_j}\right] + R_i \tag{3.4.7}$$

where R_i is the net rate of production of ith species by chemical reaction. An equation of this form will be solved for $N - 1$ species. The Nth mass fraction is usually evaluated via $Y_N = 1 - \sum_{i=1}^{N} Y_i$. This Nth species is usually chosen as that species having the overall largest mass fraction, which in air is nitrogen (inert gas). Employing Favre averaging to equation (3.4.7), the transport equation for ith species is

$$\frac{\partial}{\partial t}(\bar{\rho} \tilde{Y}_i) + \frac{\partial}{\partial x_j}(\bar{\rho} \tilde{u}_j \tilde{Y}_i) = \frac{\partial}{\partial x_j}\left[\left(\rho D_i + \frac{\mu_T}{Sc_T}\right)\frac{\partial \tilde{Y}_i}{\partial x_j}\right] + \overline{R}_i \tag{3.4.8}$$

The diffusion coefficient of the diffusion term that appears in equations (3.4.7) and (3.4.8)—that is, ρD_i—can usually be approximated by the following approaches. In practice, it is rather convenient to express the laminar diffusion coefficient as

$$\rho D_i = \frac{\mu}{Sc_i} = \frac{\mu}{Sc} \tag{3.4.9}$$

where Sc_i is the laminar Schimdt number for ith species. Different Schmidt numbers could be prescribed for the reacting chemical species that are involved in the combustion. In most practical cases, a global value is normally chosen for computational simplicity and efficiency. The default value for the global laminar Schmidt number is usually prescribed as $Sc = 0.7$. Alternatively, a more complicated approach based on the direct evaluation of the coefficient D_i could also be carried out by considering the binary diffusion of ith species-diffusing to nitrogen (chemically inert gas). The binary diffusion coefficient according to the Chapman-Enskog theory (Strehlow, 1984) can be expressed as

$$D_i = D_{l,inert} = 5.943 \times 10^{-6} \frac{T^{3/2}}{p(\sigma_o)^2_{l,inert} \, \Omega_{D_{l,inert}}} \sqrt{\frac{1}{M_l} + \frac{1}{M_{inert}}} \qquad (3.4.10)$$

where $(\sigma_o)_{l,inert}$ is the average of the two gases: $(\sigma_{o,l} + \sigma_{o,inert})/2$. The diffusion collision integral $\Omega_{D_{l,inert}}$ as a function of temperature T is provided in Appendix D. It is worthwhile to note that turbulent diffusion generally dominates over laminar diffusion, and the specification of detailed laminar diffusion properties such as described by equation (3.4.10) is usually not required; the simpler form of equation (3.4.9) should suffice. For direct numerical simulation of the turbulent flows (see Chapter 5)—that is, without time-averaging or Favre-averaging the governing equations—the detailed evaluation of the binary diffusion coefficient in equation (3.4.10) should be adopted.

Concerning the conservation of energy equation, the inter-diffusion process as previously described in section 2.4.3 for multi-component reacting flows, may have a significant effect on the total enthalpy field. If the average enthalpy associated with the lth component is designated as h_l, and since there is a mass flux $\rho_l V_l$ of the lth component across a surface moving with the mass-average velocity of the gas-mixture, its molecules will carry across the surface an extra enthalpy according to $h_l \rho_l V_l$ or $\rho h_l Y_l V_l$. The total enthalpy flowing relative to the mass average motion of the mixture is thus $\rho \sum_{l=1}^{N} h_l Y_l V_l$. The transport of enthalpy due to species diffusion is given by

$$-\frac{\partial}{\partial x_i}\left(\rho \sum_{l=1}^{N} h_l Y_l V_l\right) \qquad (3.4.11)$$

where $V_l = -\dfrac{D_{l,inert}}{Y_l}\dfrac{\partial Y_l}{\partial x_i}$. For reacting flows, the total enthalpy of the mixture, including heats of formation of the species, is

$$h = \underbrace{\int_{T_{ref}}^{T} C_p d\,T + \frac{1}{2}\left(u^2 + v^2 + w^2\right)}_{h_s} + \sum_{l=1}^{N} Y_l \Delta H^o_{f,l} \qquad (3.4.12)$$

Incorporating the expression of equation (3.4.11) as an additional source term into the energy equation, and in the absence of radiation, the transport equation becomes

$$\frac{\partial(\rho h)}{\partial t} + \frac{\partial(\rho u_i h)}{\partial x_i} = \frac{\partial p}{\partial t} + \Phi + \frac{\partial}{\partial x_j}\left[k\frac{\partial T}{\partial x_j}\right] + \frac{\partial}{\partial x_j}\left(\sum_{l=1}^{N} h_l \rho D_{l,inert} \frac{\partial Y_l}{\partial x_j}\right)$$

(3.4.13)

where the total enthalpy h_l for lth species is approximated as

$$h_l = \int_{T_{ref}}^{T} C_{p,l} dT + \Delta H_{f,l}^o + \frac{Y_l}{2}\left(u^2 + v^2 + w^2\right)$$

(3.4.14)

Note that the first two terms in equation (3.4.12) without the inclusion of heats of formation of the species is usually defined as the sensible enthalpy h_s. Substituting equation (3.4.12) into equation (3.4.13), then applying equation (3.4.7), and after some mathematical manipulation, a transport equation for the sensible enthalpy is obtained according to

$$\frac{\partial(\rho h_s)}{\partial t} + \frac{\partial(\rho u_j h_s)}{\partial x_j} = \frac{\partial p}{\partial t} + \Phi + \frac{\partial}{\partial x_j}\left[k\frac{\partial T}{\partial x_j}\right] + \frac{\partial}{\partial x_j}\left(\sum_{l=1}^{N} h_l \rho D_{l,inert} \frac{\partial Y_l}{\partial x_j}\right) + Q_R$$

(3.4.15)

Here, the heat release source term Q_R is given by

$$Q_R = -\sum_{l=1}^{N} h_l R_l$$

(3.4.16)

where R_l is the reaction rate for the lth species and h_l is defined as:

$$h_l = \sum_{T_{ref}}^{T} C_{p,l} dT + \frac{Y_l}{2}\left(u^2 + v^2 + w^2\right)$$

In equations (3.4.13) and (3.4.15), the thermal conductivity k may also be similarly approximated according to the evaluation of the binary diffusion coefficient from kinetic theory or by the simpler Sutherland's formula. The thermal conductivity of the lth species is given by

$$k_l = \frac{\mu_l C_{p,l}}{Pr}$$

(3.4.17)

where in air, the Prandtl number Pr is 0.7. Based on Chapman-Enskog theory, the viscosity of the lth species is given as

$$\mu_l = 26.69 \times 10^{-7} \frac{\sqrt{M_l T}}{\sigma_{o,l}^2 \Omega_\mu} \tag{3.4.18}$$

with the diffusion integral Ω_μ approximated as a function of temperature similar to the form of the diffusion collision integral $\Omega_{D_{l,inert}}$. The diffusion integral Ω_μ as a function of temperature T is also provided in Appendix B. According to Sutherland's formula, the coefficient of molecular viscosity μ for a perfect gas with fixed composition as a function of temperature takes the following form:

$$\mu_l = \mu_{l,ref} \frac{T_{l,ref} + C}{T + C} \left(\frac{T}{T_{l,ref}}\right)^{3/2} \tag{3.4.19}$$

where C is the Sutherland constant. Some typical values for C are 120 K at reference viscosity $\mu_{l,ref}$ and temperature $T_{l,ref}$ of 18.27×10^{-6} Pa s and 291.15 K for air and 111 K at reference viscosity $\mu_{l,ref}$ and temperature $T_{l,ref}$ of 17.81×10^{-6} Pa s and 300.55 K for nitrogen, respectively. The specific heat of the mixture defined in equation (3.4.12) can be determined by

$$C_p = \sum_{l=1}^{N} Y_l C_{p,l} \tag{3.4.20}$$

where $C_{p,l}$ represents the specific heat of the lth species at constant pressure. This constant-pressure specific heat $C_{p,l}$ is usually a function of T only, and can be represented by the polynomial function

$$C_{p,l} = \sum_{k=0}^{K} a_{k,l} T^k \tag{3.4.21}$$

Typical low-order polynomials in the preceding equation for some gases are listed in Table B.4. More accurate expressions resulting in higher-order polynomials can be obtained from *NASA Thermochemical Polynomials*. On the basis of equation (3.4.20), the mixture viscosity μ and subsequently the thermal conductivity k could also be sensibly evaluated in most practical cases as

$$\mu = \sum_{l=1}^{N} Y_l \mu_l$$
$$k = \sum_{l=1}^{N} Y_l k_l$$

(3.4.22)

For more accurate evaluation to equation (3.4.22), the mixture viscosity μ and thermal conductivity k can be computed by more complicated formulae according to *Wilke's Law* as

$$\mu = \sum_{l=1}^{N} \frac{X_l \mu_l}{\sum_{n}^{N} X_n \mu_n \phi_{l,n}}$$

$$k = \sum_{l=1}^{N} \frac{X_l k_l}{\sum_{n}^{N} X_n k_n \phi_{l,n}}$$

(3.4.23)

where X_l is the mole fraction for lth component in terms of mass fraction defined as

$$X_l = \frac{Y_l / M_l}{\sum_{m}^{N} (Y_m / M_m)}$$

(3.4.24)

Note that equation (3.4.24) equally applies for the evaluation of the mole fraction X_n in equation (3.4.23). The inter-collisional parameter $\phi_{l,n}$ is determined according to

$$\phi_{l,n} = \frac{1}{\sqrt{8}} \left(1 + \frac{M_l}{M_n} \right)^{-1/2} \left[1 + \left(\frac{\mu_l}{\mu_n} \right)^{1/2} \left(\frac{M_l}{M_n} \right)^{1/4} \right]^2$$

(3.4.25)

As stipulated in Chapter 2, the weakly compressible assumption neglects the kinetic energy in the definition of enthalpy, the pressure work term, and the dissipation function that represents the source of energy due to work done deforming the fluid element in the conservation equation. In many reacting systems, the *Lewis* (or Lewis-Seminov) number represents another useful dimensionless parameter in characterizing the diffusion processes associated with heat and mass transfer. It is defined as

$$Le \equiv \frac{k}{\rho C_p D} = \frac{\text{rate of energy transport}}{\text{rate of mass transport}} \qquad (3.4.26)$$

On the basis of the definitions of the *Prandtl* number and *Schmidt* number as $Pr \equiv \mu C_p/k$ and $Sc \equiv \mu/\rho D$, equation (3.4.26) can be alternatively expressed in terms of these dimensionless parameters as

$$Le \equiv \frac{Sc}{Pr} \qquad (3.4.27)$$

For most non-reacting flows, Le is usually very nearly unity. It is often slightly less than unity in combustible gas mixtures. The approximation $Le \approx 1$ is frequently adopted in many field modeling investigations. Based on this, the turbulent Schmidt number Sc_T is taken to have a value of 0.7, since $Pr_T = 0.7$. Also, $Le \approx 1$ allows the inter-diffusion process to be safely neglected in most cases. Hence, the Favre averaged transport equations for the total enthalpy and sensible enthalpy in the absence of inter-diffusion process and radiation with an alternative expression for the laminar diffusion component term based on the weakly compressible assumption become:

$$\frac{\partial \left(\overline{\rho}\tilde{h} \right)}{\partial t} + \frac{\partial \left(\overline{\rho}\tilde{u}_i\tilde{h} \right)}{\partial x_i} = \frac{\partial}{\partial x_i}\left[\left(\frac{\mu}{Pr} + \frac{\mu_T}{Pr_T} \right) \frac{\partial \tilde{h}}{\partial x_i} \right] \qquad (3.4.28)$$

$$\frac{\partial \left(\overline{\rho}\tilde{h}_s \right)}{\partial t} + \frac{\partial \left(\overline{\rho}\tilde{u}_i\tilde{h}_s \right)}{\partial x_i} = \frac{\partial}{\partial x_i}\left[\left(\frac{\mu}{Pr} + \frac{\mu_T}{Pr_T} \right) \frac{\partial \tilde{h}_s}{\partial x_i} \right] + \overline{Q}_R \qquad (3.4.29)$$

where $\tilde{h} = \int_{T_{ref}}^{\tilde{T}} C_p dT + \sum_{l=1}^{N} \tilde{Y}_l \Delta H_{f,l}^{o}$ and $\tilde{h}_s = \int_{T_{ref}}^{\tilde{T}} C_p dT$, respectively. It should be noted that for flow cases where Le is far from unity, it is imperative that the species diffusion term is included in the transport equations above.

The local density of the mixture is dependent on the pressure, reactant, and product concentrations, and on the mixture temperature. Recalling the equation of state defined in Chapter 2, assuming weakly incompressible, it can be calculated according to

$$\overline{\rho} = \frac{p_0}{R_u \tilde{T} \sum\limits_{l=1}^{N} \frac{\tilde{Y}_l}{M_l}} \qquad (3.4.30)$$

The evaluation of the mixture temperature can nevertheless be determined from the definition of the total enthalpy or sensible enthalpy.

3.4.1.3 Laminar Finite-Rate Chemistry

Some basic understanding on how detailed chemical kinetics can be treated through the reaction rate source terms that appear in the species transport equations is presented in this section. The laminar finite-rate chemistry model is first described.

Consider the list of stoichiometric chemical reactions as tabulated in Tables 3.1 and 3.2 for methane. In general, any jth reaction from a set of M chemical reactions between species of concentration C_i, where $i = 1, 2, \ldots, N$, can be written in the form of

$$v'_{1,j}C_{1,j} + v'_{2,j}C_{2,j} + \ldots \underset{k_{b,j}}{\overset{k_{f,j}}{\rightleftharpoons}} v''_{1,j}C_{1,j} + v''_{2,j}C_{2,j} + \ldots \tag{3.4.31}$$

where $v'_{i,j}$ and $v''_{i,j}$ are the stoichiometric coefficients for the reactants and products and $k_{f,j}$ and $k_{b,j}$ represent the forward and backward rate constants for jth reaction, respectively. For non-reversible reactions, the backward rate constant $k_{b,j}$ is simply omitted. The summation of the stoichiometric reactant and product concentrations in equation (3.4.31) are for all chemical species in the reacting system. However, only species that appear as reactants or products will have non-zero stoichiometric coefficients. Species that are not involved in the reaction will drop out automatically of the equation. Consider the chemical reaction described by the CO oxidation:

$$CO + OH \rightleftharpoons CO_2 + H \text{ (Step 12, Table 3.1)}$$

The preceding reversible reaction rate has CO and OH as the reactants and CO_2 and H as the products. The stoichiometric coefficients: $v'_{CO, 12}$, $v'_{OH,12}$, $v''_{CO_2,12}$, and $v''_{H,12}$ are equivalent to unity for all reactants as well as for all products.

The reaction rate of the creation/destruction of ith species for the jth reaction can be defined as

$$R_j = \left(\sum_{i=1}^{N} \gamma_{i,j} C_i \right) \left(k_{f,j} \prod_{i=1}^{N} [C_i]^{n'_{i,j}} - k_{b,j} \prod_{i=1}^{N} [C_i]^{n''_{i,j}} \right) \tag{3.4.32}$$

where $\gamma_{i,j}$ represents the efficiency of ith species as a third body $\left(0 \leq \gamma_{i,j} \leq 1\right)$ for jth reaction, and $n'_{i,j}$ and $n''_{i,j}$ are the forward rate and backward rate exponents for each reactant and product of ith species, which correspond to the stoichiometric coefficients of the reactants and products in the jth reaction. From the preceding reaction rate of CO oxidation, the forward rate and backward rate exponents: $n'_{CO,12}$, $n'_{OH,12}$, $n''_{CO_2,12}$, and $n''_{H,12}$ are all respectively unity. No third body efficiency is present; the reaction rate is thus given by

$$R_{12} = k_{f,12}C_{CO}C_{OH} - k_{b,12}C_{CO_2}C_H \tag{3.4.33}$$

The forward rate constant $k_{f,j}$ for the jth reaction can be computed using the Arrhenius expression as

$$k_{f,j} = A_{o,j}T^{n,j}\exp\left(-\frac{E_{a,j}}{R_uT}\right) = A_{o,j}T^{n,j}\exp\left(-\frac{T_{f,j}}{T}\right) \tag{3.4.34}$$

where $T_{f,j}(= E_{a,j}/R_u)$ is the activation temperature for the jth reaction. If the reaction is reversible, the backward reaction rate constant $k_{b,j}$ can be obtained from the forward reaction constant via the equilibrium constant K_j as

$$k_{b,j} = \frac{k_{f,j}}{K_j} \tag{3.4.35}$$

This equilibrium constant can be ascertained from the change on Gibb's free energy of the reaction at the actual temperature T, and atmospheric pressure p_{atm}, by

$$K_j = \exp\left(-\frac{\Delta G^o_j}{R_uT}\right)\left(\frac{p_{atm}}{R_uT}\right)^{\sum\limits_{i=1}^{N}\left(n''_{i,j}-n'_{i,j}\right)} \tag{3.4.36}$$

where

$$\Delta G^o_j = \sum_{i=1}^{N}v''_i g^o_i - \sum_{i=1}^{N}v'_i g^o_i$$

$$g^o_i = H_i - TS_i = \text{Gibbs free energy}$$

$$S_i = \int\limits_{T_{ref}}^{T}\frac{C_{p,i}}{T}dT + S^o_i = \text{species entropy} \tag{3.4.37}$$

$$H_i = \int\limits_{T_{ref}}^{T}C_{p,i}dT + H^o_i = \text{species enthalpy}$$

where S^o_i and H^o_i are the standard-state entropy and standard-state enthalpy (heat of formation) of ith species, which could be readily obtained from *JANAF Thermochemical Tables* or *NASA Thermochemical Polynomials*. On the basis of equation (3.4.36), the equilibrium constant for the CO oxidation, for the purpose of illustration can be obtained from *NASA Thermo-chemical Polynomials* as follows. For a temperature of 1800 K, the values of the species enthalpy are $H_{CO} = -61035.3$ kJ kmol^{-1}, $H_{OH} = 85893.1$ kJ kmol^{-1}, $H_{CO_2} = -314135.5$ kJ kmol^{-1}, and $H_H = 249184.1$ kJ kmol^{-1}. For the species entropy, they are $S_{CO} = 254.7$ kJ kmol^{-1} K^{-1}, $S_{OH} = 238.6$ kJ kmol^{-1} K^{-1}, $S_{CO_2} = 302.8$ kJ kmol^{-1} K^{-1}, and $S_H = 152.0$ kJ kmol^{-1} K^{-1}. The Gibbs free energies for the respective species are $g^o_{CO} = -519582.5$ kJ kmol^{-1}, $g^o_{OH} = -343596.3$ kJ kmol^{-1}, $g^o_{CO_2} = -859171.7$ kJ kmol^{-1}, and $g^o_H = -24360.9$ kJ kmol^{-1} of which ΔG^o_j yields a value of -20353.8 kJ kmol^{-1}. With the universal constant R_u given as 8.31431 kJ kmol^{-1} K^{-1}, the equilibrium constant is evaluated as

$$K_{12} = \exp\left(-\frac{\Delta G^o_j}{R_u T}\right)\left(\frac{p_{atm}}{R_u T}\right)^{(2-2)} = \exp\left(\frac{20353.8}{8.31431 \times 1800}\right) = 3.9$$

a more detailed explanation in the evaluation of the equilibrium constant can also be further ascertained in Kuo (1986).

The net source of ith chemical species due to reaction R_j can be computed as the sum of the Arrhenius reaction sources over M reactions that the species participate in the form

$$R_i = M_i \sum_{j=1}^{M} R_j \qquad (3.4.38)$$

Equation (3.4.38) is substituted into equation (3.4.7) in order to determine the transport of all participating ith chemical species within the reacting flow field. The model is exact for laminar flames. In order to extend directly to turbulent flames, the model is unfortunately inaccurate due to the recourse of setting $F = 0$. This entails the omission of the highly non-linear correlation terms as shown in equation (3.2.6). This laminar model may be adequately considered to specifically treat reacting flows where the combustion possesses are strictly for relatively slow chemistry and small turbulent fluctuations. Most fuels in practical fires are nonetheless *fast* burning; the overall rate of reaction is strongly governed by the turbulent characteristics. In other words, the combustion is said to be controlled by turbulent mixing. In the next section, an alternative approach to turbulence-chemistry interaction based on the mixing-limited concept that has been widely adopted in field modeling, is described.

3.4.1.4 Eddy Break-up and Eddy Dissipation

Spalding (1971) has shown the instantaneous-reaction assumption can be used for turbulent flames of sufficiently high Reynolds number flows. The reaction rate can therefore be sufficiently determined from the prevailing conditions at each point in space, and the consequent profiles of concentration, temperature, density, and velocity can be allowed to develop as dictated by the partial differential equations. A different reaction-kinetic model based on the eddy break-up concept is essentially proposed, which allows the local state of turbulence to have a strong influence on the rate of local reaction where in most cases is the dominant influencing factor in turbulent flames.

As aforementioned in section 3.2, the difficulty while using the time-averaged reaction rate of equation (3.2.6) for a highly turbulent reacting system is the evaluation of the additional equations for the conservation of the higher- order nonlinear correlation terms. In order to circumvent the limitations that the laminar finite-rate chemistry model imposes, the eddy break-up model formulates the reaction rate that aims to enlarge the influence of the local turbulence level as well as at the same time diminishes that of the chemical-kinetic constants, at least over a wide range of conditions. According to Spalding (1971), the eddy break-up reaction rate can be taken to be based simply on the species concentration fluctuations and the rate of break-up of eddies. For the single-step irreversible reaction for the consumption of fuel, the eddy break-up reaction can be expressed as

$$\overline{R}_{fu} = C_R \overline{\rho} \frac{\varepsilon}{k} \left(\widetilde{Y''^2_{fu}} \right)^{1/2} \tag{3.4.39}$$

The preceding reaction rate requires the evaluation of the entity $\widetilde{Y''^2_{fu}}$. As reported by Bilger (1975), Hutchinson et al. (1977), and Bray (1984), the Favre averaged conservation equation for the transport of $\widetilde{Y''^2_{fu}}$ follows the similar form of the convection-diffusion equation for its mean counterpart

$$\frac{\partial \left(\overline{\rho} \widetilde{Y''^2_{fu}} \right)}{\partial t} + \frac{\partial \left(\overline{\rho} \tilde{u}_i \widetilde{Y''^2_{fu}} \right)}{\partial x_i} = \frac{\partial}{\partial x_i} \left[\left(\frac{\mu}{Sc} + \frac{\mu_T}{Sc_T} \right) \frac{\partial \widetilde{Y''^2_{fu}}}{\partial x_i} \right] + \overline{R}_{\widetilde{Y''^2_{fu}}} \tag{3.4.40}$$

where the mean source rate $\overline{R}_{\widetilde{Y''^2_{fu}}}$ can be modeled according to

$$\underbrace{C_{g_1} G_{fu}}_{Term\ I} - \underbrace{C_{g_2} \overline{\rho} \frac{\varepsilon}{k} \widetilde{Y''^2_{fu}}}_{Term\ II} - \underbrace{2A_o \frac{\overline{\rho}^2}{M_{fu}M_{ox}} \tilde{Y}_{fu}\tilde{Y}_{ox} \exp\left(-\frac{E_a}{R_u \tilde{T}} \right) \left(\frac{\widetilde{Y''_{fu} Y''_{ox}}}{\tilde{Y}_{ox}} + \frac{\widetilde{Y''^2_{fu}}}{\tilde{Y}_{fu}} \right)}_{Term\ III}$$

$$\tag{3.4.41}$$

and the default constants of $C_{g_1} = 2.0$ and $C_{g_2} = 2.0$ are commonly adopted. Term I denotes the production of concentration fluctuations due to the non-uniformity of the fuel mass fraction of which G_{fu} can be formulated as

$$G_{fu} = \left(\frac{\mu}{Sc} + \frac{\mu_T}{Sc_T} \right) \frac{\partial \tilde{Y}_{fu}}{\partial x_i} \frac{\partial \tilde{Y}_{fu}}{\partial x_i} \tag{3.4.42}$$

Term II describes the dissipation of the fluctuations due to molecular diffusion; the term ε/k that is the reciprocal of the turbulent time scale can be regarded as the rate of decay multiplier for the turbulence kinetic energy k. Term III appears due to the presence of the finite rate effect on the fluctuations of $Y_{fu}^{\prime 2}$. Based on Khalil (1977), the entity $\widetilde{Y_{fu}^{\prime\prime} Y_{ox}^{\prime\prime}}$ that appears in the above equation can be obtained from the corresponding convection-diffusion conservation equation alongside with the mean source rate following equation (3.4.41) as

$$
\frac{\partial(\bar{\rho}\widetilde{Y_{fu}^{\prime\prime}Y_{ox}^{\prime\prime}})}{\partial t} + \frac{\partial(\bar{\rho}\tilde{u}_i\widetilde{Y_{fu}^{\prime\prime}Y_{ox}^{\prime\prime}})}{\partial x_i} = \frac{\partial}{\partial x_i}\left[\left(\frac{\mu}{Sc}+\frac{\mu_T}{Sc_T}\right)\frac{\partial\widetilde{Y_{fu}^{\prime\prime}Y_{ox}^{\prime\prime}}}{\partial x_i}\right] + \bar{R}_{\widetilde{Y_{fu}^{\prime\prime}Y_{ox}^{\prime\prime}}} =
$$

$$
\left[\underbrace{C_{g_1}\left(\frac{\mu}{Sc}+\frac{\mu_T}{Sc_T}\right)\frac{\partial\tilde{Y}_{fu}}{\partial x_i}\frac{\partial\tilde{Y}_{ox}}{\partial x_i}}_{\text{Term I}} \underbrace{- C_{g_2}\bar{\rho}\frac{\varepsilon}{k}\widetilde{Y_{fu}^{\prime\prime}Y_{ox}^{\prime\prime}}}_{\text{Term II}}\right. \tag{3.4.43}
$$

$$
\left.\underbrace{-A_o\frac{\bar{\rho}^2}{M_{fu}M_{ox}}\tilde{Y}_{fu}\tilde{Y}_{ox}\exp\left(-\frac{E_a}{R_u\tilde{T}}\right)\left((\tilde{Y}_{fu}+r\tilde{Y}_{ox})\frac{\widetilde{Y_{fu}^{\prime\prime}Y_{ox}^{\prime\prime}}}{\tilde{Y}_{fu}\tilde{Y}_{ox}}+\frac{\widetilde{Y_{ox}^{\prime\prime 2}}}{\tilde{Y}_{ox}}+r\frac{\widetilde{Y_{fu}^{\prime\prime 2}}}{\tilde{Y}_{fu}}\right)}_{\text{Term III}}\right]
$$

The estimation of the relative contribution of the chemical kinetics and turbulent mixing to the mean source rates in the conservation equations of $\widetilde{Y_{fu}^{\prime\prime 2}}$ and $\widetilde{Y_{fu}^{\prime\prime}Y_{ox}^{\prime\prime}}$ can be best explained through the *Damköhler* number *Da*. In flame situations where the reactants are in intimate contact due to rapid mixing, characterized by large values of ε/k, the reaction is kinetically influenced—that is, $Da \ll 1$. The influence of chemical kinetics on the formation of the entities $\widetilde{Y_{fu}^{\prime\prime 2}}$ and $\widetilde{Y_{fu}^{\prime\prime}Y_{ox}^{\prime\prime}}$ is hence negligible and only Terms I and II contribute to the generation/destruction of $\widetilde{Y_{fu}^{\prime\prime}2}$ and $\widetilde{Y_{fu}^{\prime\prime}Y_{ox}^{\prime\prime}}$. Nevertheless, in the flame situations where the temperature is high enough for the reaction to proceed but the reactants are not in intimate contact, the reaction is considered to be controlled by mixing; the chemical time τ_k is very small relative to τ_s—that is, $Da \gg 1$. Here again, if the combustion process is suitably represented by a single-step reaction such as in the form of equation (3.2.2), the respective relationships for the mean reaction rates can be obtained

$$
\tilde{R}_{fu} = \frac{1}{r}\tilde{R}_{ox} = -\frac{1}{1+r}\tilde{R}_{pr} \tag{3.4.44}
$$

On the basis of equation (3.4.40), the eddy break-up model requires the solution to the mass fractions of the fuel, oxidant and products, in addition to three conservation equations for the transport of second order correlations: $\widetilde{Y''^2_{fu}}$, $\widetilde{Y''_{fu} Y''_{ox}}$, and $\widetilde{Y''^2_{ox}}$. The transport equation for the entity $\widetilde{Y''^2_{ox}}$ is formulated similar to equations (3.4.40) and (3.4.41).

Magnussen and Hjertager (1976) propose an alternative combustion model of which they have considered the reaction rate to be governed by the mean species concentrations rather than the species concentration fluctuations. Known as the eddy dissipation model, the consumption rate of the fuel in a single-step irreversible reaction in terms of mass fractions is given by the smaller (i.e., limiting value) of these two expressions:

$$\overline{R}_{fu} = C_R \overline{\rho} \frac{\varepsilon}{k} \min \left[\tilde{Y}_{fu}, \frac{\tilde{Y}_{ox}}{r} \right] \tag{3.4.45}$$

$$\overline{R}_{fu} = C'_R \overline{\rho} \frac{\varepsilon}{k} \left(\frac{\tilde{Y}_{pr}}{1 + r} \right) \tag{3.4.46}$$

Similar to the preceding eddy break-up model, the reaction rate will proceed whenever turbulence is present (small values of ε/k) and an ignition source is not required to initiate combustion. Equation (3.4.45) is applicable for non-premixed flames, but the inclusion of equation (3.4.46) according to Magnussen and Hjertager (1976), allows the present model to handle premixed flames. They postulated that in premixed flames the fuel and oxygen eddies will be separated by eddies containing hot combustion products. The rate of combustion will, in this case, be determined by the dissipation of the hot eddies where the concentration of hot combustion products is low. It is commonplace to introduce the Arrhenius reaction rate as described in equation (3.4.2) (the term F set to zero) to act as a kinetic "switch" for flaming regions where the combustion may be governed by chemical kinetics. This is usually useful to characterize the ignition/extinction of the flames. Once the flame is ignited, the eddy dissipation rate is the governing rate, since it is usually smaller than the Arrhenius rate and the reaction is mixing-limited. The net rate is thus taken as the minimum of the rates given in equations (3.4.45) and (3.4.46) and the Arrhenius reaction rate (see equation (3.4.2)). Another possible approach is to adopt the *Damköhler* number of which the eddy dissipation rates can be modified according to

$$\overline{R}_{fu} = C_R C_A \overline{\rho} \frac{\varepsilon}{k} \min \left[\tilde{Y}_{fu}, \frac{\tilde{Y}_{ox}}{r} \right] \tag{3.4.47}$$

$$\overline{R}_{fu} = C'_R C_A \overline{\rho} \frac{\varepsilon}{k} \left(\frac{\tilde{Y}_{pr}}{1 + r} \right) \tag{3.4.48}$$

where

$$C_A = \begin{cases} 1.0 & Da \geq Da_{ie} \\ 0.0 & Da < Da_{ie} \end{cases}$$

The limiting *Damköhler* number Da_{ie} can be set to a small value of 10^{-3}. In most cases, the constants C_R and C'_R are usually taken to have default values of 4.0 and 2.0. This model can be classified as the *collision mixing model*. In a later study by Magnussen et al. (1979), the reactions rates described by equations (3.4.47) and (3.4.48) were improved to include the capacity of predicting slow reactions. The approach accounts for the consideration that the dissipation is not homogeneously distributed in the turbulent fluid but occurs mainly in concentrated, highly strained regions that occupy only fractions of the total volume. These regions are occupied by fine structures, with characteristic dimensions of the same magnitude as the Kolmogorov microscale. These fine structures are specifically responsible for the dissipation of turbulence into heat. Within these structures, the reactants are assumed to be mixed at a molecular scale. With the knowledge of the volume fraction occupied by fine structures, and performing similarity considerations of the transfer of energy from the macroscale to the fine structures, the following reaction rate as proposed by Magnussen et al. (1979), which occurs in all the fine structures can be expressed as

$$\overline{R}_{fu} = 23.6 \left(\frac{\mu \varepsilon}{\overline{\rho} k^2} \right)^{1/4} \overline{\rho} \frac{\varepsilon}{k} \min \left[\tilde{Y}_{fu}, \frac{\tilde{Y}_{ox}}{r} \right] \tag{3.4.49}$$

For premixed flames, it should be noted that not all the fine structures will be sufficiently heated to react. This is obviously the case where both fuel and oxygen are present in the fine structures. The fraction of the fine structures that reacts can be assumed to be proportional to the ratio of the local concentration of reacted fuel and the total fuel concentration. Defining,

$$\chi = \frac{\tilde{Y}_{pr}/(1+r)}{\tilde{Y}_{pr}/(1+r) + \tilde{Y}_{fu}}$$

and substituting the preceding into equation (3.4.49), the rate of combustion at infinite rate between fuel and oxygen for premixed flames is

$$\overline{R}_{fu} = 23.6 \left(\frac{\mu \varepsilon}{\overline{\rho} k^2} \right)^{1/4} \overline{\rho} \frac{\varepsilon}{k} \chi \min \left[\tilde{Y}_{fu}, \frac{\tilde{Y}_{ox}}{r} \right] \tag{3.4.50}$$

This model is generally classified as the *viscous mixing model*. For nonpremixed flames, equation (3.4.50) is applied without the consideration of χ.

As formulated from the preceding the eddy dissipation model based on the collision mixing model and viscous mixing model is closely related to the eddy break-up model. It is nonetheless evidently clear that the eddy dissipation model differs in especially relating the dissipation of eddies to the mean concentration of intermittent quantities instead of the concentration fluctuations. Significantly, this model does not call for solution of equations for the entities: $\widetilde{Y''^2_{fu}}$, $\widetilde{Y''_{fu}Y''_{ox}}$, and $\widetilde{Y''^2_{ox}}$. Nevertheless, the simplifications of the eddy dissipation model should be viewed from the proposition whereby the mean quantities appear intermittent within the turbulent fluid.

Strictly speaking, the eddy dissipation and finite-rate/eddy dissipation models should only be used for one-step (reactant → product) or two-step (reactant → intermediate, intermediate → product) global reactions. To extend the eddy dissipation model to include detailed reaction mechanisms, it is more preferable to adopt the generalized eddy dissipation concept model developed in Magnussen (1981). Herein, the combustion is assumed to proceed with species reacting in fine structures over a time scale

$$\tau^* = 0.4082 \left(\frac{\mu}{\bar{\rho}\varepsilon} \right)^{1/2} \tag{3.4.51}$$

which is governed by the Arrhenius rates. The reaction rate of any mean ith species can be modeled as

$$\overline{R}_i = \frac{\bar{\rho}\xi^{*2}}{\tau^* \left[1 - \xi^{*3} \right]} \left(Y_i^* - \tilde{Y}_i \right) \tag{3.4.52}$$

where Y_i^* is the fine-scale species mass fraction after reacting over the time scale τ^* of which can be determined through the laminar finite-rate model and ξ^* is the volume fraction of the fine scales according to Gran and Magnussen (1996) as

$$\xi^* = 2.1377 \left(\frac{\mu\varepsilon}{\bar{\rho}k^2} \right)^{3/4} \tag{3.4.53}$$

3.4.2 Combustion Based on Conserved Scalar

3.4.2.1 Description of Approach

Based upon the pioneering work by Burke and Schumann (1928), who considered the oxygen as a negative fuel, and so obtained a scalar quantity that is conserved under the chemical reaction, they discovered that the reaction possessed features of being single-step, irreversible, and infinitely fast.

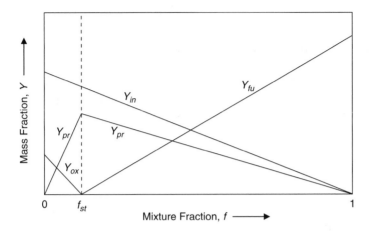

Figure 3.6   Single Chemical Reacting System (SCRS) relationships.

Their physical insights led to the belief that non-premixed fuel and oxidant concentrations could be derived in the limit of *fast chemistry*, where the reaction was taken to be so rapid that the fuel and oxidant could not co-exist everywhere except within an infinitely thin flame sheet. On the basis of this model approximation, it was inherently assumed that the chemical reaction occurs within a very thin reaction zone (flame sheet), as illustrated in Figure 3.6.

Consider the elemental mass fraction Z_i for some ith element, the Favre-averaged conservation equation of elemental mass fraction can be written in the similar form to the scalar property equation derived in Table 2.5 of Chapter 2 without any chemical source term as

$$\frac{\partial}{\partial t}\left(\bar{\rho}\tilde{Z}_i\right) + \frac{\partial}{\partial x_j}\left(\bar{\rho}\tilde{u}_j\tilde{Z}_i\right) = \frac{\partial}{\partial x_j}\left[\left(\bar{\rho}D + \frac{\mu_T}{Sc_T}\right)\frac{\partial \tilde{Z}_i}{\partial x_j}\right] \tag{3.4.54}$$

The chemical source term vanishes in equation (3.4.54), because in chemical reactions elements are conserved. Since there are essentially $L - 1$ variables for a system involving L elements, note that $\sum_{i=1}^{L}\tilde{Z}_i = 1$, solution of the $L - 1$ equations for $\tilde{Z}_i$ from equation (3.4.54) yields the mean elemental composition throughout the reacting turbulent flow field. By assuming *fast chemistry*, these molecular-species composition can be determined from this elemental composition. If the combustion process can be suitably represented by a single-step reaction such as in the form similar to equation (3.2.2), then simpler conserved scalars may be employed

Shvab (1948), Zel'dovich (1949), and William (1965) concluded that the linear combination of the species conservation equations for the non-premixed reactants, fuel, and oxidant yields an equation whose form is identical to the one governing the conservation of chemically inert species—the same form of equation (3.4.54). Assuming that the turbulent transport coefficient for the reactants and the products at each point in the flow field are equal, and the fuel and oxidant always combine in a stoichiometric ratio r to form $(1 + r)$ kg of products, the so-called Shvab-Zel'dovich coupling parameters expressed in terms of the Favre-averaged mass fractions of fuel $\tilde{Y}_{fu}$, oxidant $\tilde{Y}_{ox}$, and products $\tilde{Y}_{pr}$, that can be formulated according to

$$\tilde{\xi}_{FO} = \tilde{Y}_{fu} - \frac{\tilde{Y}_{ox}}{r}$$

$$\tilde{\xi}_{FP} = \tilde{Y}_{fu} + \frac{\tilde{Y}_{pr}}{1+r} \qquad (3.4.55)$$

$$\tilde{\xi}_{OP} = \tilde{Y}_{ox} + \frac{r\tilde{Y}_{pr}}{1+r}$$

which are essentially the conserved scalars describing the mixing in a non-premixed reacting flame. If all the conserved scalars are linearly related, then the solution for one scalar yields solutions for all others; the choice of the conserved scalar is thus arbitrary. Equation (3.4.55) indicates that linear relationships among all conserved scalars exist only when there are two uniform feeding streams. It should be noted that the Shvab-Zel'dovich coupling parameters are useful only when a single-step reaction is involved. For more complex chemical systems involving many multi-step and competitive reactions with high release rates, it is more convenient to employ conserved scalars based on elements, which will be further explained in section 3.4.2.6.

3.4.2.2   Definition of Mixture Fraction

Under a set of simplifying assumptions, the basis of the non-premixed modeling approach is that the instantaneous thermochemical state of the fluid is related to a conserved scalar quantity known as the mixture fraction f. The concept of mixture fraction is essentially a numerical construct used in the analysis of non-premixed combustion to describe the degree of scalar mixing between the fuel and oxidant. It is a local quantity within the flow field that varies both spatially and temporally.

The mixture fraction is best understood by visualizing the mixing process in a generic combustion chamber, as illustrated in Figure 3.7. Supposing that a resultant mixture (M) leaves at one end after the fuel (F) and air (A) enter the system via two feeding inlets and thereafter thoroughly mix in the chamber, the resulting mixture from this two-stream mixing process can be written as

$$f\beta_F + (1 - f)\beta_A = \beta_M \qquad (3.4.56)$$

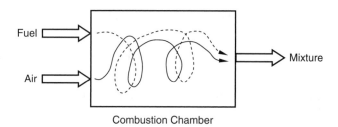

Combustion Chamber

Figure 3.7   Mixing of steady fuel and air streams in a combustion chamber.

The preceding equation subsequently leads to

$$f \equiv \frac{\beta_M - \beta_A}{\beta_F - \beta_A} \tag{3.4.57}$$

This property β of the mixture that is free from sources and sinks obeys the relation of a conserved property. In a non-reacting mixture of pure fuel and air, the mass fraction of fuel is equivalent to the value of f, and the mass fractions of oxygen and nitrogen are equivalent to $1 - f$. If $Y_{fu,F}$ is taken to be the mass fraction of fuel in the fuel stream (unity for pure fuel) and $Y_{ox,A}$ and $Y_{in,A}$ be the mass fractions of oxygen and nitrogen in the air stream, the mass fractions of fuel, oxygen, and nitrogen in a completely *unburnt* mixture are respectively:

$$Y_{fu} \equiv f \, Y_{fu.F} \tag{3.4.58}$$

$$Y_{ox} \equiv (1 - f) Y_{ox,A} \tag{3.4.59}$$

$$Y_{in} \equiv (1 - f) Y_{in,A} \tag{3.4.60}$$

Equations (3.4.58), (3.4.59), and (3.4.60) provide the necessary relationships between the mixture fraction and the mass fractions of fuel, oxygen, and nitrogen during the mixing of fuel and air in the absence of combustion.

In the event where combustion occurs, either the fuel or oxidant (oxygen in air) will have zero concentration in the mixture (M) state. Assuming that the chemical reaction is complete within the mixing chamber, the mixture fraction may be expressed in terms of any of the Shvab-Zel'dovich coupling parameters, since they also obey the relation of conserved properties. The mixture fraction for the two-feed system with equal species mass diffusivities can be defined as

$$f \equiv \frac{\beta_M - \beta_A}{\beta_F - \beta_A} = \frac{\xi - \xi_A}{\xi_F - \xi_A} \tag{3.4.61}$$

where ξ is the instantaneous conserved scalar represented in the similar form by

$$\xi = Y_{fu} - \frac{Y_{ox}}{r} \text{ or } \xi = Y_{fu} + \frac{Y_{pr}}{1+r} \text{ or } \xi = Y_{ox} + \frac{rY_{pr}}{1+r}$$

Within the combustion chamber, a special value of the mixture fraction, f_{st}, exists, which divides the flame into two distinct regions. For a single-step chemical reaction, with no oxidant in the fuel feeding stream and no fuel in the oxidant stream, f_{st} corresponds to the stoichiometric condition

$$f_{st} \equiv \frac{Y_{ox,A}}{rY_{fu,F} + Y_{ox,A}} \tag{3.4.62}$$

For fuel lean cases with infinitely fast chemistry, the first region of the flame is one where the oxidant and products co-exist. The following relationships for the burnt mixture $(0 < f < f_{st})$ are

$$Y_{fu} = 0$$
$$Y_{ox} = Y_{ox,A}\left(1 - \frac{f}{f_{st}}\right) \tag{3.4.63}$$

For fuel rich cases with infinitely fast chemistry, the second region is one where the fuel and products co-exist. The relationships $(f_{st} < f < 1)$ are

$$Y_{ox} = 0$$
$$Y_{fu} = Y_{fu,F}\left(\frac{f - f_{st}}{1 - f_{st}}\right) \tag{3.4.64}$$

Nitrogen in air does not participate in the chemical reaction, since it is chemically inert. The mass fraction follows the same relationship as derived in equation (3.4.5). The mass fraction of products is obtained according to the following conservation:

$$Y_{fu} + Y_{ox} + Y_{pr} + Y_{inert} = 1 \tag{3.4.65}$$

3.4.2.3   Flame Sheet Approximation

The *Flame Sheet Approximation*, also known as the *Mixed-Is-Burnt*, is a combustion model developed specifically for the simplest reaction scheme. This approach assumes that the chemistry is infinitely fast and irreversible, with fuel and oxidant species never co-exist in space and proceeds into complete combustion resulting in a one-step conversion to final products. Such a chemical reaction allows the species mass fractions to be determined directly from the given reaction stoichiometry without any required knowledge for the reaction rate or chemical equilibrium information.

Consider the fast chemical reaction for a stoichiometric combustion of one mole of fuel in an oxidant stream consisting only of nitrogen and oxygen:

$$F + v_{O_2}\left(O_2 + \frac{0.79}{0.21}N_2\right) \rightarrow v_{CO_2}CO_2 + v_{H_2O}H_2O + v_{O_2}\frac{0.79}{0.21}N_2 \qquad (R12)$$

For the combustion methane, the stoichiometric coefficients of v_{O_2}, v_{CO_2}, and v_{H_2O} as described in equation R2 are 2 moles, 1 mole, and 2 moles, respectively. Equations (3.4.63), and (3.4.64) require values at the free streams of air and fuel—namely, the mass fractions $Y_{ox,A}$ and $Y_{fu,F}$. In the fuel stream, if pure fuel exists, the mass fraction of fuel $Y_{fu,F}$ is equivalent to unity at $f = 1$. Since oxygen and nitrogen co-exist together in the air stream, the evaluation of the mass fraction of oxygen $Y_{ox,A}$ can be ascertained by the number of moles in the reactants' side (left-hand side) of the single-step reaction equation (R12). In other words,

$$Y_{ox,A} = \frac{2 \text{ mole} \times 32 \text{ g}}{(2 \text{ mole} \times 32 \text{ g} + 7.52 \text{ mole} \times 28 \text{ g})} = 0.233$$

at $f = 0$. Invoking mass balance, the mass fraction of nitrogen $Y_{in,A}$ in the air stream yields a value of 0.767. In the air and fuel streams, combustion products such as CO_2 and H_2O are not present. They are thus zero at $f = 0$ and $f = 1$. At the special value of $f = f_{st}$, the stoichiometric combustion process of reaction (R12) results in the one-step conversion to final products. The maximum mass fractions of products, for example, due to combustion of methane, can be determined based on the number of moles in the products' side (right-hand side) according to

$$Y_{CO_2,st} = \frac{1 \text{ mole} \times 44 \text{ g}}{(1 \text{ mole} \times 44 \text{ g} + 2 \text{ mole} \times 18 \text{ g} + 7.52 \text{ mole} \times 28 \text{ g})} = 0.151$$

$$Y_{H_2O,st} = \frac{2 \text{ mole} \times 18 \text{ g}}{(1 \text{ mole} \times 44 \text{ g} + 2 \text{ mole} \times 18 \text{ g} + 7.52 \text{ mole} \times 28 \text{ g})} = 0.124$$

Employing equation (3.4.61), the stoichiometric mixture fraction f_{st} for methane based on $Y_{ox,A} = 0.233$ and $Y_{fu,F} = 1$ with $r = (2 \text{ mole} \times 32 \text{ g}) / (1 \text{ mole} \times 16 \text{ g}) = 4$ yields a value of 0.055.

The linear relationship between the instantaneous mass fractions of fuel, oxygen, carbon dioxide, water vapor, and nitrogen and the mixture fraction f for a non-premixed combustion system in the limit of fast chemistry, is shown

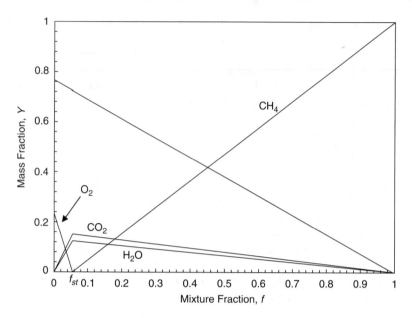

Figure 3.8   Single step combustion state relationships for methane.

in Figure 3.8. Relationships for other fuels besides methane could also be similarly derived following the procedure just described.

Equation (R12) could also be further generalized by introducing the equivalence ratio ϕ, which is expressed (including excess fuel or oxygen) as

$$F + \frac{1}{\phi}v_{O_2}\left(O_2 + \frac{0.79}{0.21}N_2\right) \rightarrow$$

$$\left(1 - \frac{1}{\phi}\right)F + \left(\frac{1}{\phi} - 1\right)v_{O_2}O_2 + \frac{1}{\phi}v_{CO_2}CO_2 + \frac{1}{\phi}v_{H_2O}H_2O + \frac{1}{\phi}v_{O_2}\frac{0.79}{0.21}N_2$$

$$\text{(R13)}$$

The equivalence ratio may be defined as the ratio of actual fuel-oxidant ratio (= mass of fuel / mass of oxidant) to the ratio of fuel-oxidant for a stoichiometric process. It can be related to the mixture fraction as

$$\phi = \frac{f}{1-f}\frac{1-f_{st}}{f_{st}} \tag{3.4.66}$$

For fuel-lean condition, the system lies in between $0 < \phi < 1$. At stoichiometric, $\phi = 1$. For fuel-rich condition, we have $1 < \phi < \infty$.

In fire science and combustion, the equivalence ratio is a useful parameter to interpret the results of experimental studies of the composition of the smoke layer under the ceiling of a compartment fire (Pitts, 1994). Careful attention was paid to the dependence of the yields of CO on the equivalence ratio. If the burning of fires is starved from sufficient entrainment of air such as with restricted ventilation, the yield of incompletely burnt products will increase. Generally speaking, high yields of CO are associated with high equivalence ratios. With reference to the detailed chemistry in Table 3.1, as for example the CO oxidation determined from Step 12, significant competition for the hydroxyl radicals between CO and other partially burnt products can cause the effective reduction of the likelihood for the complete conversion of CO to CO_2. Experimental evidence also suggests this conversion is also suppressed in the presence of soot particles, which are known to react with hydroxyl radicals (Puri and Santoro, 1991).

On the basis where equal diffusivities are assumed, the species transport equations can be reduced to a single equation for the mixture fraction f. The reaction source terms in these equations cancel and thus f is a conserved quantity similar to the elemental mass fraction Z_i as considered in section 3.4.2.1. The Favre-averaged transport equation for the mixture fraction with equal diffusivities is

$$\frac{\partial}{\partial t}\left(\bar{\rho}\tilde{f}\right) + \frac{\partial}{\partial x_j}\left(\bar{\rho}\tilde{u}_j\tilde{f}\right) = \frac{\partial}{\partial x_j}\left[\left(\frac{\mu}{Sc} + \frac{\mu_T}{Sc_T}\right)\frac{\partial \tilde{f}}{\partial x_j}\right] \tag{3.4.67}$$

Although the assumption of equal diffusivities appears to be problematic for laminar flows, it is generally acceptable for turbulent flows to adopt a global mass diffusion coefficient, since turbulent convection generally overwhelms the molecular diffusion; the specification of detailed laminar diffusion properties is therefore unwarranted.

3.4.2.4 State Relationships

The power of the mixture fraction modeling approach allows the chemistry to be reduced by the description of state relationships for the mass fraction of each species as a function of the mixture fraction f. For the combustion of methane in air as described in the previous section, the instantaneous mass fractions of fuel, oxygen, carbon dioxide, water vapor, and nitrogen in the limit of fast chemistry are just simply linear relationships, as shown in Figure 3.8. Since no reaction rates or equilibrium calculations are required, the model that is based on the flame sheet approximation is easily computed and yields a fast rate of convergence.

Specific assumptions or simplifications made to this particular combustion model are nonetheless noted: (1) no co-existence of fuel and oxidant in space and complete single-step conversion of the reactants to final products, and (2) mass diffusion coefficients of all species are equal. On the basis of the latter, the conservation of Favre-averaged mixture fraction as described by equation (3.4.66) is required to calculate the reacting flows rather than individual transport equations for the mass fraction of each species. For the special case where the fluctuation of the mixture fraction is neglected, this model is appropriate for turbulent diffusion flames where excessive air is employed. However, this particular model is an over-simplification of the features of real flame situation. It cannot predict the intermediate species formation or dissociation effects, which often results in an over-prediction of the peak flame temperatures. However, the model's inherent simplicity and effectiveness allow the possibility of handling combustion chemistry for a number of complex fuels of interest.

By assuming that the chemistry is sufficiently rapid for chemical equilibrium to always exist the molecular level, it is possible to utilize the equilibrium chemistry assumption to compute the intermediate species occurring in the combustion of methane in air. An algorithm based on the minimization of the Gibbs free energy can be used to compute these species mass fractions from the mixture fraction f. The NASA CEA (Chemical Equilibrium with Applications) program by Gordon and McBride (1994) or other available routines such as GASEQ—Chemical Equilibrium Program for Windows, which are freely downloadable from the World Wide Web, could be used to determine the equilibrium compositions. In the commercial code of ANSYS Inc., Fluent, the supplemental routine PrePDF Version 4 allows the possibility of carrying out chemical equilibrium calculations in order to pre-determine the mass species relationships prior to combustion investigations. The equilibrium model is rather powerful, since it does not require *a priori* knowledge of detailed chemical kinetic data. Instead of defining a specific multi-step reaction mechanism, the equilibrium model only requires the identification of important chemical species that will be present in the reacting system. Figure 3.9 illustrates the resulting mass fractions for a reacting system that includes 11 species for the combustion of methane in air. The equilibrium model is nevertheless without its deficiencies. As observed in Figure 3.9a, the major problem of this model is the excessive prediction of the mass fractions of CO near the stoichiometry limit. In addition, it is also observed that the significant level of decomposition of the fuel into graphitic carbon in the fuel-rich region. Generally speaking, the formation of CO is strongly influenced by the flow, transport, and turbulence as indicated by the comparable time scales in Figure 3.5. Needless to say, the formation of soot as also seen from Figure 3.5, when compared to the fluid mechanical mixing processes, is even slower and requires dedicated models that will be further described in the next chapter.

As an alternative to the equilibrium model, generalized state relationships established by Sivathanu and Faeth (1990) could be applied of which reasonable correlations have been found for major gas species for non-premixed combustion of diffusion flames, which included the molar fuel H/C ratios in the range 1–4 and equivalence ratios (or mixture fraction) in the range of 0.01–100.

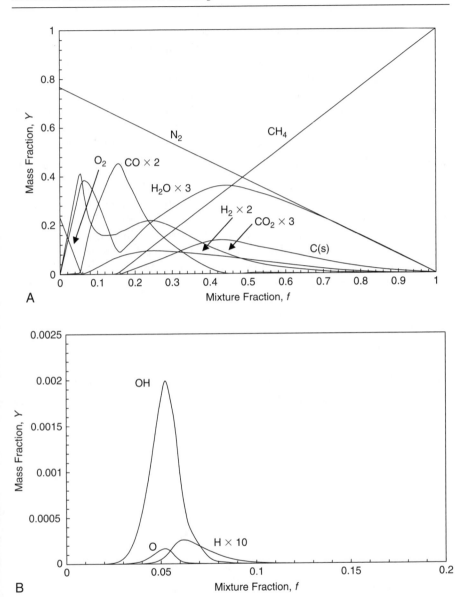

Figure 3.9 Chemical equilibrium state relationships for methane: (a) Major gas species and (b) Minor gas species.

These generalized state relationships have been found to approximate the thermodynamic equilibrium for fuel-lean conditions and departure from equilibrium in a relatively universal manner for near-stoichiometric and fuel-rich conditions. The generalized state relationships for the concentration of major gas species involve the prescription of species-specific values of a parameter ψ for each species as a function of the equivalence ratio ϕ. The species considered include N_2, O_2, CO_2, H_2O, CO, H_2, and fuel. For convenience, the state relations functions, $\psi = \psi(Y_i)$, as well as the form of $Y_i(\psi)$, are summarized in Table D.6 in Appendix D. The normalizing factors $\left(Y_{N_2}^{st}, Y_{CO_2}^{st}, \text{ and } Y_{H_2O}^{st}\right)$ can be determined experimentally or theoretically. For methane, $Y_{N_2}^{st} = 0.725$, $Y_{CO_2}^{st} = 0.151$, and $Y_{H_2O}^{st} = 0.124$. Appropriate values of other normalizing factors for propane, heptane, acetylene, and ethylene are also given in Sivathanu and Faeth (1990). The specifies-specific values of ψ for each value of ϕ are provided in Table D.7 in Appendix D. Note that Table D.7 yields the parameter ψ in terms of the equivalence ratio. From equation (3.4.66), the mixture fraction is related to the equivalence ratio as

$$f = \frac{\phi}{\left(\frac{1-f_{st}}{f_{st}} + \phi\right)} \tag{3.4.68}$$

Figure 3.10 illustrates the state relationships of the major gas species corresponding to the mixture fraction. It is worthwhile noting the significant

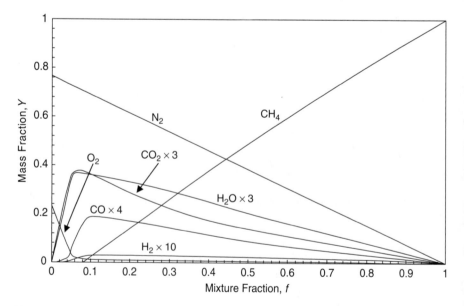

Figure 3.10   Generalized state relationships for major gas species of methane.

lower prediction of the mass fractions of CO as well as H_2 via the current empirical approach, when compared to the chemical equilibrium model as shown in Figure 3.9a.

Mass fractions of major gas species have also been measured experimentally for a handful of hydrocarbon fuels. For example, Norton et al. (1993) have undertaken a rather detailed experimental investigation of a laminar, methane/air diffusion flame, in which measurements have been made on major stable species as well as minor species such as OH, H, O CH, and CH_3. Such experiments provided not only considerable insights into the flame structure, but also aimed to improve our understanding of the chemical heat release of the oxidation process. Combustion experiments are usually very challenging and tedious to perform; hence, limited data are only available for a handful of fuels. These species profiles are usually expressed as functions of the local mixture fraction and in some circumstances on the scalar dissipation rate, which are of significant interest for describing the chemical composition and the strain field. The latter, which concerns the laminar flamelet approach, will be further expounded in section 3.4.2.6. Experimentally measured profiles could be readily applied if the corresponding fuels in the experiments as well as in the field modeling investigations are the same. In the next section, the extension of the conserved scalar approach to characterize the strong coupling between chemical heat release and turbulent mixing—that is, chemistry-turbulence interactions—is described.

3.4.2.5 Probability Density Function (PDF) of Turbulence-Chemistry

In turbulent reacting flow, the fuel and oxygen concentration fluctuations generally have a measurable value that varies from one location to another in the flame zone. Owing to these fluctuations as observed by Bilger and Kent (1972) and El Ghobashi (1974), the fuel and the oxidant exist at the same location but at different times. A number of closure methods are available to account for the strong interactions between the combustion chemistry and turbulent mixing. In this section, the probability density function (PDF) approach is described to treat non-premixed combustion of diffusion flames. In essence, this particular method employs a statistical description of the turbulent field together with the conservation equations governing the fluid flow.

The concept of the probability density function is illustrated in Figure 3.11. From a physical viewpoint, the fluctuating value of the mixture fraction f spends some fraction of time in the range denoted as Δf (see right-hand side of figure). Denoting the probability density function as $P(f)$, it should take on values such that the area under its curve in the band denoted by Δf is equal to the fraction of time that spends in this range. From a mathematical viewpoint, it takes the form

$$P(f)\Delta f = \lim_{t \to \infty} \frac{1}{t} \sum_i \tau_i \qquad (3.4.69)$$

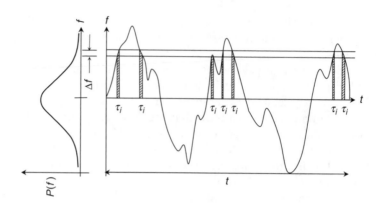

Figure 3.11   Probability density function of the mixture fraction.

where t is the time scale and T_i is the amount of time that f spends in the Δf band. From equation (3.4.69), it also implies that the sum of the values of $P(f)\Delta f$ over a long period of time must be equal to unity—that is,

$$\int_{-\infty}^{\infty} P(f)df = 1 \tag{3.4.70}$$

Hence, the Favre-averaged value of $\tilde{f}$ can be expressed as

$$\tilde{f} = \int_{-\infty}^{\infty} fP(f)df \tag{3.4.71}$$

which represents the first moment of $P(f)$ about the origin $f = 0$. The effect of turbulent fluctuations on the local flow properties can be introduced by including second and higher-order correlations of the concentration. The concentration fluctuation g can be defined as

$$g = \widetilde{f''^2} = \int_{-\infty}^{\infty} \left(f - \tilde{f}\right)^2 P(f)df \tag{3.4.72}$$

which the preceding equation constitutes a second-order closure. In addition to solving the Favre-averaged mixture fraction, which is described by equation (3.4.67), the local values of the concentration fluctuation g can also be obtained from the corresponding conservation equation expressed in the convection-diffusion form as

$$\frac{\partial}{\partial t}(\bar{\rho}g) + \frac{\partial}{\partial x_j}(\bar{\rho}\tilde{u}_j g) = \frac{\partial}{\partial x_j}\left[\left(\frac{\mu}{Sc} + \frac{\mu_T}{Sc_T}\right)\frac{\partial g}{\partial x_j}\right] + C_{g_1}\left(\frac{\mu}{Sc} + \frac{\mu_T}{Sc_T}\right)\frac{\partial \tilde{f}}{\partial x_j}\frac{\partial \tilde{f}}{\partial x_j}$$

$$- C_{g_2}\bar{\rho}\frac{\varepsilon}{k}g$$

(3.4.73)

The last two terms of equation (3.4.73), which represents the source terms of the governing equation, denote the production of concentration fluctuation due to non-uniformity of mixture fraction and dissipation of the fluctuations due to the molecular diffusion. Note the similarity of the preceding governing equation with the conservation equations of the scalar entities $\widetilde{Y''^2_{fu}}$ and $\widetilde{Y''_{fu}Y''_{ox}}$ without the finite rate terms in equations (3.4.40) and (3.4.43), respectively. Values determined through $\tilde{f}$ and g are used to determine the appropriate probability distribution.

The mean properties based upon Favre averaging can be obtained from the knowledge of the distribution of $P(f)$, which in turn can be ascertained via different PDF shapes of varying complexity. The shape of the function $P(f)$ is strongly influenced by the nature of the turbulent fluctuations in f. In practice, $P(f)$ is expressed as a mathematical function that approximates the PDF shapes that have been observed experimentally. The prescriptive approach based on assumed shapes of the probability density functions is described.

Three mathematical functions are examined:

- Double delta functions at $f = 0$ and $f = 1$—that is, square wave distribution of f with time
- A Gaussian distribution between $f = 0$ and $f = 1$ together with two delta functions at $f = 0$ and $f = 1$
- A beta function that automatically bounds between $f = 0$ and $f = 1$

Each of the preceding functions is explained in more detail following. In turbulent combustion, the probability density function approach has the beneficial property of determining the Favre-averaged values of variables that depend on f. At any location in space, the density-weighted average of any scalar property ϕ, which can be expressed as a function of the instantaneous mixture fraction f, may be obtained from

$$\tilde{\phi} = \int_0^1 \phi(f)P(f)df \tag{3.4.74}$$

For the mean mass fraction of ith species, equation (3.4.73) yields

$$\tilde{Y}_i = \int_0^1 Y_i(f)P(f)df \tag{3.4.75}$$

Through the conserved scalar approach, the relevant relationships between Y_i and f are usually obtained through state relationships as described in the previous section. Depending on whether the models of mixed-is-burnt, chemical equilibrium, generalized empirical relationships, or experimentally determined profiles are employed, different levels of complexity for the chemistry of the non-premixed combustion can be resolved.

Double delta function PDF: The double delta function is the simplest form of the probability density function that can be employed and is most easily computed. Applications of the square wave PDF shape can be found in El Ghobashi (1974) and Khalil (1977). The shape produced by these functions depends on the mean mixture fraction $\tilde{f}$ and concentration fluctuation g. Figure 3.12 illustrates the probability distributions for a square wave variation of mixture fraction.

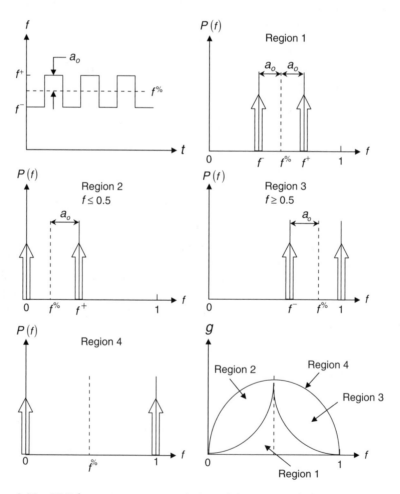

Figure 3.12   PDF for a square wave variation of the mixture fraction.

Four separate regions are indicated for the assumed probability distribution and the corresponding temporal distribution.

In the first region, $0 < \tilde{f} < 1$, the probability density function is

$$P(f) = 0.5\delta(f^-) + 0.5\delta(f^+) \quad \text{where}$$
$$f^- = \tilde{f} - \sqrt{g} \quad \text{and} \quad f^+ = \tilde{f} + \sqrt{g} \tag{3.4.76}$$

In the second region, $\tilde{f} < 0.5$ and $f^- = \tilde{f} - \sqrt{g} < 0$, the probability density function is

$$P(f) = \left(\frac{g}{\left(\tilde{f}^2 + g\right)}\right)\delta(0) + \left(\frac{\tilde{f}}{\tilde{f} + g/\tilde{f}}\right)\delta(f^+) \quad \text{where}$$
$$f^- = 0 \text{ and } f^+ = \tilde{f} + \sqrt{g} \tag{3.4.77}$$

In the third region, $\tilde{f} > 0.5$ and $f^+ = \tilde{f} + \sqrt{g} > 1$, the probability density function is

$$P(f) = \left(\frac{1 - \tilde{f}}{1 - \tilde{f} + g/\left(1 - \tilde{f}\right)}\right)\delta(f^-) + \left(\frac{g}{\left(\left(1 - \tilde{f}\right)^2 + g\right)}\right)\delta(1) \quad \text{where}$$
$$f^- = \tilde{f} - \sqrt{g} \text{ and } f^+ = 1$$
$$\tag{3.4.78}$$

In the fourth region where large oscillations are imposed such that either the fuel or oxidant is present, the probability density function is

$$P(f) = \left(1 - \tilde{f}\right)\delta(0) + \tilde{f}\delta(1) \tag{3.4.79}$$

and

$$g_{max} = \left(1 - \tilde{f}\right)\tilde{f} \tag{3.4.80}$$

Clipped Gaussian PDF: The conventional Gaussian distribution that extends beyond the limits of $f = 0$ and $f = 1$ is physically unrealistic, since such an assumption does not exist for scalars to attain negative or excessively large values. Lockwood and Naguib (1975) and Kent and Bilger (1977) proposed the use of a clipped Gaussian distribution of which the difficulty stems from the presence of the unwanted tails of the conventional Gaussian distribution is taken care by the imposition of two Dirac delta functions at $f = 0$ and $f = 1$.

The resultant clipped Gaussian probability distribution is given by

$$P(f) = \frac{1}{\sigma\sqrt{2\pi}}\exp\left[-\frac{1}{2}\left(\frac{f-\mu}{\sigma}\right)^2\right][D(f) - D(f-1)]$$

$$+ \int_{-\infty}^{0} \frac{1}{\sigma\sqrt{2\pi}}\exp\left[-\frac{1}{2}\left(\frac{f-\mu}{\sigma}\right)^2\right]df\delta(0)$$

$$+ \int_{1}^{\infty} \frac{1}{\sigma\sqrt{2\pi}}\exp\left[-\frac{1}{2}\left(\frac{f-\mu}{\sigma}\right)^2\right]df\delta(1) \tag{3.4.81}$$

where μ is the value of f giving the maximum probability, σ is the variance, $D(f)$ is the Heaviside step function where $D(\xi) = 0$ when $\xi < 0$ and $D(\xi) = 1$ when $\xi > 0$, and $\delta(f)$ is the Dirac delta function. Probable values of μ and σ^2 between $f = 0$ and $f = 1$ are determined from the values of $\tilde{f}$ and g as defined in equations (3.4.71) and (3.4.72):

$$\tilde{f} = \int_0^1 \frac{f}{\sigma\sqrt{2\pi}}\exp\left[-\frac{1}{2}\left(\frac{f-\mu}{\sigma}\right)^2\right]df \tag{3.4.82}$$

$$g = \int_0^1 \frac{(f-\tilde{f})^2}{\sigma\sqrt{2\pi}}\exp\left[-\frac{1}{2}\left(\frac{f-\mu}{\sigma}\right)^2\right]df \tag{3.4.83}$$

Given the predicted values of $\tilde{f}$ and g, the preceding equations can be inverted to obtain μ and σ. Equations (3.4.82) and (3.4.83) can either be approximated by numerical integration (Simpson's rule or higher order approximation methods) or be reduced to comparatively simple analytical forms (Liew et al., 1981). The inversion of the non-linear algebraic relationships in the preceding equations nonetheless necessitates an iterative computational procedure. Corresponding values of μ and σ that are subsequently ascertained to aptly determine the appropriate probability distribution of the clipped function described in equation (3.4.81). Figure 3.13 illustrates the probability distributions for a clipped Gaussian variation of mixture fraction.

Beta function PDF: The use of the beta function PDF as utilized by Jones (1980) represents another alternative strategy, which permits an approximation to the probability density distribution of the conserved scalar besides the clipped Gaussian function. For $0 < \tilde{f} < 1$, the beta function is defined as

$$P(f) = \frac{f^{\alpha-1}(1-f)^{\beta-1}}{\int_0^1 f^{\alpha-1}(1-f)^{\beta-1}df} \tag{3.4.84}$$

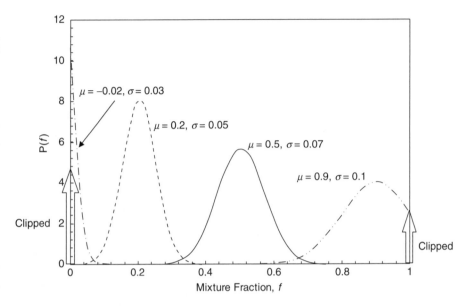

Figure 3.13 Different values of μ and σ depicting the characteristic behavior of the clipped Gaussian PDF.

It can be shown that α and β can be determined explicitly from the mean mixture fraction $\tilde{f}$ and concentration fluctuation g as

$$\alpha = \tilde{f} \left[\frac{\tilde{f}\left(1 - \tilde{f}\right)}{g} - 1 \right]$$

$$\beta = \left(1 - \tilde{f}\right)\frac{\alpha}{\tilde{f}}$$

(3.4.85)

The integral in the denominator in equation (3.4.84) is called the beta function, which can usually be expressed in terms of several Gamma functions as $\Gamma(\alpha)\Gamma(\beta)/\Gamma(\alpha + \beta)$, where $\Gamma(z) = (z - 1)!$. Figure 3.14 illustrates the probability distributions for a beta function variation of mixture fraction.

$$\alpha = 1.0, \quad \beta = 10.0 \quad \alpha = 12.0, \quad \beta = 1.0$$

At the limits of $f = 0$ and $f = 1$, it is noted that the integrand of the beta function PDF becomes singular when the parameters α and β tend to be appreciably small. In order to circumvent such a problem, Abou-Ellail and Salem (1990) proposed a blended distribution function consisting of the beta function

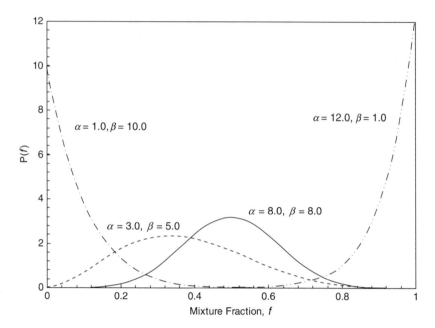

Figure 3.14 Different values of α and β depicting the characteristic behavior of the beta function PDF.

PDF and two Dirac delta functions at $f = 0$ and $f = 1$ to represent the spikes of air-rich and fuel-rich mixtures in the mixing region. Similar to the clipped Gaussian distribution, the alternate beta function PDF is given by $\alpha = 8.0$, $\beta = 8.0$

$$\alpha = 3.0, \quad \beta = 5.0$$

$$P(f) = \gamma_A \delta(0) + \gamma_M \frac{f^{\alpha-1}(1-f)^{\beta-1}}{\int\limits_0^1 f^{\alpha-1}(1-f)^{\beta-1}\,df} + \gamma_F \delta(1) \qquad (3.4.86)$$

It is further assumed that α is limited to a minimum value of α_{min} in the mixing region. If at any location in the flame where α is greater than α_{min}, then $\gamma_A = \gamma_F = 0$ and $\gamma_M = 1$ and equation (3.4.86) recovers the original form of equation (3.4.84). Based on Abou-Ellail and Salem (1990) model, the contributions of $\gamma_A, \gamma_M,$ and γ_F have been formulated according to known quantities given as

$$\gamma_A = \frac{\left(1 - \tilde{f}\right)\left[g\left(\alpha_{\min} + \tilde{f}\right) - \tilde{f}^2\left(1 - \tilde{f}\right)\right]}{\tilde{f}\left[\left(\alpha_{\min} + \tilde{f}\right)\left(1 - \tilde{f}\right) - \tilde{f}\left(1 - \tilde{f}\right)\right]}$$

$$\gamma_M = \frac{\left(\alpha_{\min} + \tilde{f}\right)\left[\tilde{f}\left(1 - \tilde{f}\right) - g\right]}{\tilde{f}\left[\left(\alpha_{\min} + \tilde{f}\right)\left(1 - \tilde{f}\right) - \tilde{f}\left(1 - \tilde{f}\right)\right]} \qquad (3.4.87)$$

$$\gamma_F = \frac{g\left(\alpha_{\min} + \tilde{f}\right) - \tilde{f}^2\left(1 - \tilde{f}\right)}{\left[\left(\alpha_{\min} + \tilde{f}\right)\left(1 - \tilde{f}\right) - \tilde{f}\left(1 - \tilde{f}\right)\right]}$$

where the parameter α_{min} is set approximately to a value of 6.5.

3.4.2.6 Laminar Flamelet Approach

The laminar flamelet approach can be best understood by the schematic description of a free standing fire, as illustrated in Figure 3.15. In hindsight, the laminar flamelet concept views a turbulent flame as an ensemble of laminar diffusion flamelets. Similar to the fast chemistry assumption where the notion is taken whereby the chemical time scale is very much smaller than the convective and diffusive time scales, the fuel and oxidant are considered to react in narrow regions in the vicinity of the stoichiometric flame surfaces. Combustion

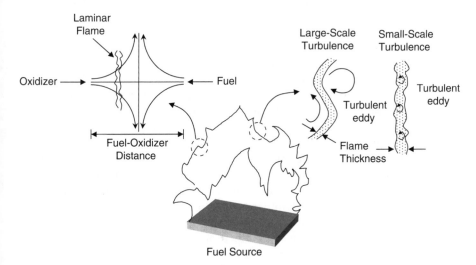

Figure 3.15 Schematic representation of the laminar flamelet concept and effect of the scale of turbulence on the structure of the flame front.

occurs rapidly within the local confines of one-dimensional structures normal to the stoichiometric contour. These structures can thus be assumed to resemble the thin flame sheets of so-called flamelets that are responsible for combustion in a laminar flame. In a turbulent reacting flow, these flamelets will be stretched and strained by the fluid flow and turbulence as demonstrated by the small- and large-scale turbulence affecting the flame front in Figure 3.15.

On the basis of the fast chemistry assumption, the local chemical equilibrium introduces an important simplification, since it eliminates the need to account for the chemical kinetics in the analysis. In turbulent flows, where the local diffusion time scales can vary considerably, the fast chemistry assumption may not be locally valid. If the average diffusion time scales approach the order of magnitude of the chemical time scales, local quenching of the turbulent flame will occur. Further reduction of the diffusion time scales then leads to lift-off and even blow-off of the entire turbulent flame. Even in globally stable flames, the variation of diffusion time scales may plausibly or selectively interact with different chemical processes within the reacting system. For such flames, the non-equilibrium effects are indispensable and should be incorporated in modeling by including at least one additional "progress" variable, which will be described following.

All the preceding models, except the chemical equilibrium and generalized eddy dissipation concept models, either adopt single-step combustion or accommodate limited degrees of combustion chemistry, and in most cases, they do not predict intermediate and minor species. The laminar flamelet model that will be described in this section is essentially an extension of the conserved scalar formulation to include non-equilibrium effects. Also, a key advantage of the model is the feasibility of incorporating detailed chemistry with relative ease and simplicity. Numerical calculations can be performed economically to evaluate important aspects of the combustion process and to aptly approximate the energy release due to combustion. The most important aspect of the laminar flamelet model is the decoupling of the combustion chemistry modeling from the calculations of the flow field.

There are two methods of generating the laminar flamelets. The first method involves solving governing equations for opposed flow diffusion flame situations in the physical space. Figure 3.16 illustrates the two typical configurations. Through the delivery of the fuel and oxidant at opposite locations, a diffusion flame is seen to be established close to the stagnation plane for the configurations of the planar opposed flow diffusion laminar flame and stagnation point for Tsuji's burner (Tsuiji and Yamaoka, 1967). For the latter, the flow is also assumed to be laminar, stagnation-point flow in cylindrical coordinates, and the configuration is considered to be infinitely wide and axisymmetric. The steady boundary layer equations with chemical reaction source terms, which are transformed into a system of one-dimensional ordinary differential equations, a series of prescribed values of the strain rate corresponding to the stretching conditions in practical turbulent flows (0.1–5000 s-1) are later imposed to generate the laminar profiles in the physical space. There are a

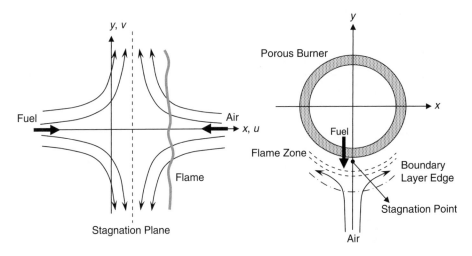

Figure 3.16 Schematic representations of the opposed flow flame and Tsuji burner configuration.

number of available computer programs for laminar flamelet calculations such as RUN1DL (Rogg, 1993, and Rogg and Wang, 1997), OPPDIF (Lutz et al., 1997), and more recently, COSILAB (http://www.softpredict.com) of which the reader may be interested in applying these codes in field modeling. The second method, which is concerned with transforming the governing equations of the opposed flow diffusion flames into mixture fraction space, will be further elaborated following.

According to *Bilger's mixture fraction formula* (Bilger, 1988), a conserved scalar most preferably employed for combustion purposes is the mass fractions Z of the chemical elements (C, H, O) that can be related to the mass fractions Y of species by

$$Z_j = \sum_{i=1}^{N} \frac{a_{ij} M_j}{M_i} Y_i \qquad (3.4.88)$$

where a_{ij} is the number of atoms of element j in a molecule of species i, M_i is the molecular weight of species, i and M_j is the molecular weight of element j. Using these element mass fractions, the conserved scalar for the variable β for a typical reaction $v_C C + v_O O + v_H H \rightarrow$ Products where v_j is the number of atoms of element j can be defined as

$$\beta = \frac{Z_C}{v_C M_C} + \frac{Z_H}{v_H M_H} - 2 \frac{Z_O}{v_O M_O} \qquad (3.4.89)$$

Using equation (3.4.57), the mixture fraction becomes

$$f = \frac{\dfrac{Z_C}{v_C M_C} + \dfrac{Z_H}{v_H M_H} - 2\dfrac{\left(Z_O - Z_{O,ox}\right)}{v_O M_O}}{\dfrac{Z_{C,fu}}{v_C M_C} + \dfrac{Z_{H,fu}}{v_H M_H} - 2\dfrac{Z_{O,ox}}{v_O M_O}} \tag{3.4.90}$$

As previously indicated, the subscripts fu and ox refer to the fuel and oxidant streams.

On the basis of the developments suggested by Peters (1984, 1986), the flamelet equations for the mass fraction of species and temperature can be derived by using the co-ordinate transformation of the Crocco-type. In mixture fraction space, they are:

Species

$$\rho\frac{\partial Y_i}{\partial t} - \rho D\left(\frac{\partial f}{\partial x_j}\right)^2\frac{\partial^2 Y_i}{\partial f^2} - \omega_i + \underbrace{R(Y_i)}_{\text{higher order terms}} = 0 \tag{3.4.91}$$

Temperature

$$\rho\frac{\partial T}{\partial t} - \rho D\left(\frac{\partial f}{\partial x_j}\right)^2\frac{\partial^2 T}{\partial f^2} - \frac{1}{C_p}\frac{\partial p}{\partial t} + \sum_{i=1}^{N}\frac{h_i}{C_p}\omega_i + \underbrace{R(T)}_{\text{higher order terms}} = 0 \tag{3.4.92}$$

The boundary conditions are

Oxidant stream, $f = 0$, $T = T_{ox,A}$ $Y_i = Y_{i,ox,A}$ $i = 1, \ldots, N$
Fuel stream, $f = 1$, $T = T_{fu,F}$ $Y_i = Y_{i,fu,F}$ $i = 1, \ldots, N$

The influence of the flow field is introduced into equations (3.4.91) and (3.4.92) by the instantaneous scalar dissipation (s^{-1}) defined by

$$\chi = 2D\left(\frac{\partial f}{\partial x_j}\right)^2 = 2D\left[\left(\frac{\partial f}{\partial x}\right)^2 + \left(\frac{\partial f}{\partial y}\right)^2 + \left(\frac{\partial f}{\partial z}\right)^2\right] \tag{3.4.93}$$

The scalar dissipation is a parameter that controls mixing and represents the non-uniformity of the mixture fraction, which is related to the strain. When the flame strain increases, the scalar dissipation rate also increases; it can thus be considered as the parameter of which describes the departure from equilibrium chemistry. At very low scalar dissipation rates, the combustion takes place in conditions that are very close to chemical equilibrium. Nevertheless, at very high scalar dissipation rate, the flame is highly strained; in this instance, the flame is close to extinction. To generate the laminar flamelet profiles in mixture fraction space, equations (3.4.91) and (3.4.92) are usually solved for given initial and boundary conditions for the fuel and oxidant as well as the temperature for a series of prescribed values of the scalar dissipation rate χ. For the purpose of illustration, we demonstrate the different flamelet relationships of the temperature and mass fractions of major and minor species in Figure 3.17, for

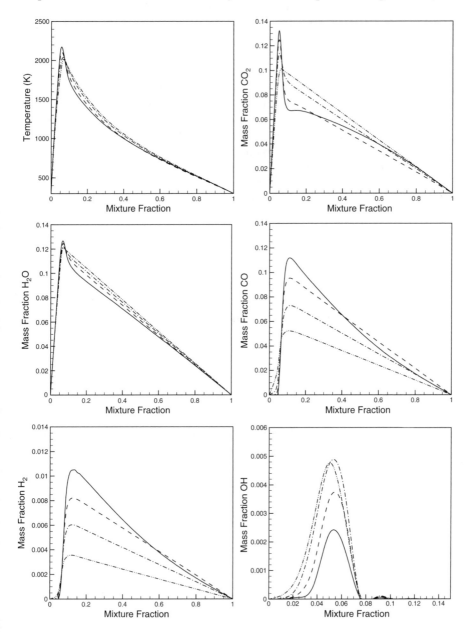

Figure 3.17 A sample of laminar flamelet profiles for methane combustion: $\chi = 0.01$ (—), $\chi = 0.1$ (----), $\chi = 1.0$ (—·—), and $\chi = 10.0$ (—··—).

different scalar dissipation rates of a C_1 skeletal reaction mechanism of methane combustion determined through an in-house computer program.

Different scalar dissipation levels yield different temperature and mass species profiles. With increasing scalar dissipation rates, the departure from chemical equilibrium is evident and at some critical scalar dissipation rate, which is just above $\chi = 10$, the flame is subsequently quenched. It should be noted that the preceding flamelet relationship have been obtained neglecting the higher-order terms involving convection and curvature along the mixture fraction surface—$R(Y_i)$ and $R(T)$. A more accurate representation of the flame structure could be obtained by the inclusion of these terms into the set of governing partial differential equations. The simpler forms of the one-dimensional equations have nevertheless been found to be sufficiently adequate for most practical purposes in fire investigations (Kang and Wen, 2004).

In turbulent flow fields, the mixture fraction and the scalar dissipation rate are statistically distributed. To evaluate the mean quantities, it is necessary to be aware of the statistical distribution these parameters in the form of a joint PDF $P(f, \chi)$. The mean scalar property are then evaluated from

$$\tilde{\phi} = \int_0^\infty \int_0^1 \phi(f, \chi) P(f, \chi) df d\chi \tag{3.4.94}$$

Assuming statistical independence for the mixture fraction and the scalar dissipation rate, the average value of the scalar property in equation (3.4.94) is now given by

$$\tilde{\phi} = \int_0^\infty \int_0^1 \phi(f, \chi) P(f) P(\chi) df d\chi \tag{3.4.95}$$

From the preceding, the PDF for the mixture fraction can be evaluated by the double Dirac delta, clipped Gaussian, or beta functions. It has been discovered in the literature that the distribution of χ conforms to the log-normality and the PDF for the scalar dissipation may therefore be written as

$$P(\chi) = \frac{1}{\chi \sigma \sqrt{2\pi}} \exp\left[-\frac{1}{2\sigma}(\ln \chi - \mu)^2\right] \tag{3.4.96}$$

where the parameters μ and σ related to the first and second moments of χ by

$$\tilde{\chi} = \exp\left[\mu + \frac{1}{2}\sigma^2\right] \tag{3.4.97}$$

$$\widetilde{\chi''^2} = \tilde{\chi}^2[\exp(\sigma^2) - 1] \qquad (3.4.98)$$

In principal, the two-equation models of turbulence provide the mean scalar dissipation rate by relating it to the concentration fluctuations and the turbulent time scale ε/k. The mean scalar dissipation rate can be formulated as

$$\tilde{\chi} = C_\chi \frac{\varepsilon}{k} g \qquad (3.4.99)$$

Normally, the constant C_χ in equation (3.4.99) is assigned a value of $C_\chi = 2.0$. The variance $\widetilde{\chi''^2}$ is usually not known *a priori*, and thus suitable values for σ must be deduced by experiments. Based on Seernivasan et al. (1977), the variance in the log-normal PDF is taken to be $\sigma^2 = 2.0$. It is noted that this value certainly presents only a first guess, which may be expected to be valid in the fully turbulent part of the fluid flow. In the intermittent parts, substantial changes of σ and of χ may be expected.

By determining $\tilde{\chi}$ from equation (3.4.99), and with the prescribed value of σ^2, the parameter μ can subsequently be determined via equation (3.4.97) to determine the distribution of $P(\chi)$. According to Bray and Peters (1994), it has nonetheless become common practice to ignore the scalar dissipation fluctuations. On the basis of this, the PDF $P(\chi)$ should suffice by the mere representation of double dirac delta functions for convenience with a constant variance to speed up computations. For accurate calculations, the log-normal distribution of $P(\chi)$ is still preferred.

In practice, the omission of the scalar dissipation fluctuations greatly simplifies the integration of equation (3.4.51). The evaluation of the mean scalar property thus reduces to

$$\tilde{\phi} = \int_0^1 \phi(f, \tilde{\chi}) P(f) df \qquad (3.4.100)$$

The preceding integration is usually not carried out during the flow field calculations. Rather, a flamelet library is generated for the mean scalar property as functions of the mean mixture fraction, concentration fluctuations, and mean scalar dissipation rates. On the basis of the range of values being prescribed for the mean mixture fraction, concentration fluctuations, and mean scalar dissipation rates, numerical integration is performed on equation (3.4.100) to yield the appropriate values of the mean scalar property. During the flow field calculations, the CFD model looks up pre-integrated values that have been tabulated from the created library.

3.5   Guidelines for Selecting Combustion Models in Field Modeling

A range of combustion modeling approaches has been illustrated. In the context of field modeling, which approach should be adopted as the preferred methodology to characterize the combustion process of practical fires? One important consideration in describing the combustion process is the principal knowledge of whether the combustion is governed by chemical kinetics or turbulent mixing. On the basis of this, pertinent guidelines via the *Damköhler* number are described in this section for determining the suitable combustion models in field modeling.

In the limit of *fast chemistry*—that is, $Da \gg 1$—the conserved scalar approach offers many benefits over the need to solve a large number of species transport equations with complex chemical kinetics. Depending on the different state relationships that can be applied, different levels of complexity of the combustion chemistry could be incorporated. Besides the mixed-is-burnt model, which is primarily catered for single-step chemistry, the chemical equilibrium model permits intermediate (radical) species prediction and dissociation effects, and rigorous turbulence-chemistry coupling could be realized when coupled with the statistical representation via the assumed shapes of the probability density function. To incorporate non-equilibrium effects, the laminar flamelet model extends the chemical equilibrium model to better resolve the combustion process affected by different levels of flame straining or stretching through the scalar dissipation rate, and it has the ability to predict the flame up to the point of flame quenching or extinction. Alternatively, the combustion process of naturally flaming fires could be assumed to be mainly governed by turbulent mixing. The eddy break-up model allows the recourse of actually dispensing with the need for expensive Arrhenius chemical kinetic calculations. On the other hand, the eddy dissipation model, which solves the intermittent mean concentrations instead of concentration fluctuations, further simplifies the turbulent flaming calculations.

Whereas conditions of $Da \approx 1$ and $Da \ll 1$ may persist, the influence of chemical kinetics becomes more prevalent and the suitability of the preceding models in the limit of fast chemistry is questionable—especially the need of modeling the ignition and flame spreading phenomena. For relatively slow chemistry and small turbulent fluctuations, the *laminar finite-rate chemistry* model, which is exact for combustion under laminar conditions, can be purposefully used to handle such combustion possesses. In order to better treat turbulent reacting flows, the *generalized eddy dissipation concept* model may be applied. Detailed reaction mechanisms that could be accounted for in the model, provide the feasibility of accurately describing a range of flaming regions and for a wider range of Da numbers. It should be noted that solving such a system of transport equations for the species mass fractions is usually rather stiff and involves enormous computational resources.

It should be emphasized that the assumed shapes of the probability distributions in the conserved scalar approach are essentially a mathematical representation of the problem in order to feasibly characterize the temporal distribution of the scalar properties and their corresponding PDFs. Along the centerline of the flame, numerous experimental evidence have shown that the probability distribution is single modal in the form of near Gaussian (Kennedy and Kent, 1978) or a beta function (Krambeck et al., 1972). Away from the centerline, the prevalence of an intermittent nature due to the growth of intermittency spikes may, however, result in a bi-modal characterization of the PDF of the burnt gas and unburnt mixture. In view of the different observed probability distributions, the *PDF method*, which entails a more sophisticated approach to combustion modeling by solving additional transport equation for the joint PDF of velocities and reactive scalars, can be applied (Dopanzo, 1994, Pope, 1985). The inclusion of finite-rate chemistry makes the *PDF method* applicable to not only non-premixed combustion but also to premixed and partially premixed combustion. By solving the PDF transport equation, values of the PDF are now dependant on the rate of reaction, time scales of turbulence, and reaction on the flow properties. The Monte Carlo method is normally employed to treat the turbulent reacting flow as an ensemble of particles so that each particle has its own position composition and travels in the flow with instantaneous velocity. Particle state is described by its position, velocity, and instantaneous reactive scalars (e.g., temperature, species mass fractions), and scalar properties are described by the stochastic Lagrangian models. No averaging is imposed, since all the flow and scalar variables are solved based on the instantaneous field.

As an alternative to the PDF method for non-premixed combustion, the *Conditional Moment Closure* (Bilger, 1993, Kilmenko and Bilger, 1999, Kim, and Huh, 2002) could also be adopted for infinitely fast and finite-rate chemistry, which overcomes the problem of limited range of validity that other closures have met. Better known by its acronym *CMC*, the main concept behind this approach is to ascertain how the reactive scalars depend on the mixture fraction. In essence, it aims to calculate conditional moments at a fixed location within the flow field, using modeled transport equations without any imposed assumptions on the small scale structure of reaction zones or on the relative timescale of chemistry and turbulence. The mean reaction rate based on the Reynolds or Farve averaging for turbulent flow (see equation (3.26)), usually contains higher order moments; variances and co-variances. Accommodation of the higher correlations is required in order to achieve accurate results. The whole purpose of the *CMC* is to bypass the problem of calculating the mean reaction rate by performing a conditional average, in the hope that the conditional fluctuations are small enough so that they can be considered to be negligible—that is, $F \approx 0$. CMC has been shown to be capable of adequately predicting the phenomena associated with local flame extinction and re-ignition. More detailed descriptions of the PDF method and CMC are provided in Appendix C.

Principally, the development of the PDF method and CMC focuses on the specific need of the extensive usage of the finite-rate chemistry to resolve a wide range of flaming conditions. It has been argued by Bilger (2000) that the laminar flamelet model neglects the influence of spatial terms in the equations and the effects of variations in the scalar dissipation rate. Concerning the questionable validity of the laminar flamelet model in aptly accounting the presence of local extinction and re-ignition, such flaming conditions are thoroughly accommodated through the PDF method and CMC. Nevertheless, the main drawback of employing the PDF method and CMC for turbulent combustion, is that such systems with complex combustion chemistry involve large numerical integration calculations and they are generally very computational-intensive. The computational cost can be expected to vary as

PDF method > Conditional Moment Closure > Laminar Flamelet Model > Mixed-is-Burnt, Eddy Break-up, or Eddy Dissipation Model

From the aspect of computational efficiency, the authors strongly recommend, similar to the selection of turbulence models, an incremental approach of systematically introducing the level of model sophistication into the numerical calculations. It is therefore advisable to begin by applying the simplest models that reside at the bottom level of the computational cost. This will bring significant benefits to over-prescribing the problem with unnecessary complexities. If required, more complex models could be applied to enhance the computational predictions.

3.6 Worked Examples on the Application of Combustion Models in Field Modeling

3.6.1 Single-Room Compartment Fire

In this worked example, field modeling that incorporates increasingly complex illustration of the chemical process to characterize the Steckler's single-compartment fire is described through the application of different combustion modeling approaches. The parametric study includes systematic representations of the fire source by different combustion models. Numerical simulations are performed through an in-house computer code FIRE3D. For the purpose of validating and verifying the different combustion models that could be applied in fire engineering, comparison of the computed results is made not only against measurements made by Steckler et al. (1984) but also the numerical results obtained from Lewis et al. (1997) for the same heat release rate of 62.9 kW. Special emphasis on the range of fire studies performed by Lewis et al. (1997) allows the feasibility of direct comparison of similarly applied combustion models in order to establish confidence in the usage of these models in fire safety investigations.

Numerical features: Numerical solutions to a system of three-dimensional Favre-averaged equations for the transport of mass, momentum, and enthalpy with the addition of mass fractions of gas species as well as mixture fraction accompanied by its fluctuation scalar are ascertained. Instead of the specific need of imposing a physical volume of the fire source within the computational domain, the fuel flow rate of 0.0013 kg/s corresponding to the heat release rate of 62.9 kW is now specified at the surface of the gas burner. All other boundary conditions remain the same as previously featured in the previous worked example, as illustrated in section 2.16.1.

For the comparative study against Lewis et al. (1997) numerical solutions, every attempt has been made to adopt similar or identical pressure-velocity linkage methods, numerical discretisation schemes, and turbulence models. Simulations for the single compartment fire are carried out using the SIMPLE pressure correction algorithm alongside the *hybrid differencing scheme* and standard k-ε turbulence model. The eddy dissipation combustion (EDM) model of Magnussen and Hjertager (1976) based on single step chemistry of methane and the conserved scalar approach employing the Sivathanu and Faeth (1990) state relationships, are assessed against similar combustion models employed in Lewis et al. (1997). Since this room configuration is symmetrical about the vertical plane bisecting the doorway and burner, mesh generation for the compartment fire is only carried out on half of the room, thus improving the resolution of the flow and thermal fields. The mesh density of 83160 grid nodes comparable to the finest mesh used by Lewis et al. (1997) of 70432 grid nodes is employed for the comparison of results.

The inclusion of an extended region away from the doorway is an important modeling consideration to correctly predict the migration of combustion products and entrainment of ambient air through the doorway. In retrospect, the size of the extended region attached to the compartment fire is usually not known *a priori*. To determine the extent of the open boundaries of the extended boundaries affecting the flow and thermal characteristics at the doorway, two regions having a size of 3 m × 2.8 m in plan and 6.8 m in height and with the same plan area and a lower height of 3.8 m, are investigated. Numerical experiments have revealed that no appreciable differences could be found for the predicted temperature profiles in the two simulation cases. It is nevertheless demonstrated in Figure 3.18 that the predicted velocity profiles just above the floor level for the large extended region, appear to be closer to the experimental data than those of the small extended region. The entrainment characteristic at the doorway can thus be inferred to be better accommodated through the requirement of a large extended region. Note the specific use of the large extended region for the simulations carried out for the worked example in section 2.16.1.

Numerical results: Predicted line graphs for the doorway temperature and velocity profiles utilizing different combustion models based on the eddy dissipation and conserved scalar by the in-house computer code are compared against those of Lewis et al. (1997) numerical results employing the eddy dissipation model and Steckler et al. (1984) experimental data. The numerical

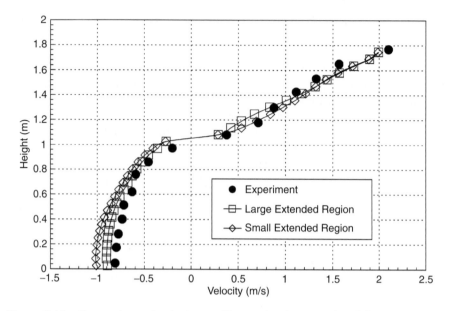

Figure 3.18 Comparison of velocity profiles at the doorway for different extended regions.

results obtained for the volumetric heat source approach from the previous worked example are also re-plotted in the same figures in order to assess the relative merits in characterizing the fire combustion as a result of modeling through the simpler or more complicated approach to field modeling. All relevant numerical results of the doorway temperature and velocity profiles and experimental data are shown in Figures 3.19 and 3.20, respectively.

Comparing the volumetric heat source and combustion modeling predictions in Figure 3.19, it is evident that the distinct separation of the hot and cold layers present within the compartment is better predicted by the latter than the former, which incidentally confirms the distinct two-layer structure observed during experiment. On the basis of the combustion simulations made by Lewis et al. (1997) and the in-house computer code for the temperature predictions near the top edge of the doorway, deviations observed in the numerical results could be attributed to a number of factors: (1) neglect of soot formation and its radiation and combustion efficiency to account for the incomplete combustion of the methane fuel, and (2) consideration of adiabaticity at the compartment walls of which the only probable path of the heat escaping from the compartment is through the doorway. Based on these, it is to be expected that the in-house predicted temperatures are likely to be higher than those predicted by Lewis et al. (1997).

In contrast to the predicted temperature profiles, the velocity profiles in Figure 3.20 does not appear to be strongly sensitive to the different approaches

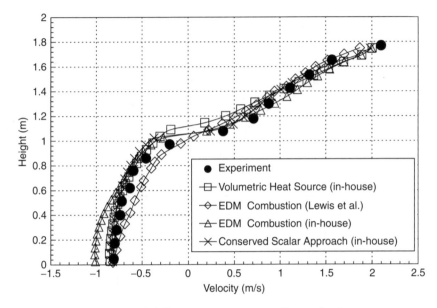

Figure 3.19   Comparison of different temperature profiles at the doorway.

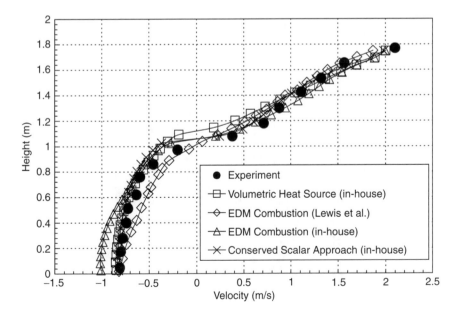

Figure 3.20   Comparison of different velocity profiles at the doorway.

adopted. Here again, the combustion modeling predictions appear to fair marginally better than the volumetric heat source approach in illustrating the distinct demarcation between the hot and cold layers that is evidently present within the compartment; cold air is entrained into the burn room at the bottom half, while combustion products are seen exhausting at the top half of the doorway.

Major species predictions of the concentration contours of H_2O and CO at the vertical symmetry plane dissecting the gas burner are reported in Figures 3.21 and 3.22, respectively. It is noted that actual measurements of

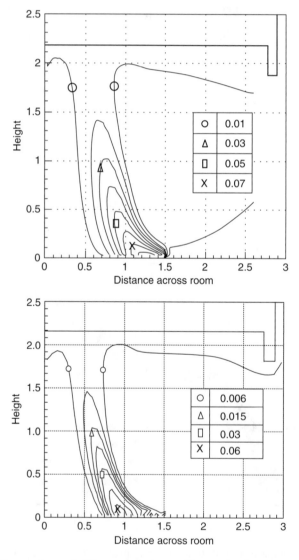

Figure 3.21 Comparison of different velocity profiles at the doorway.

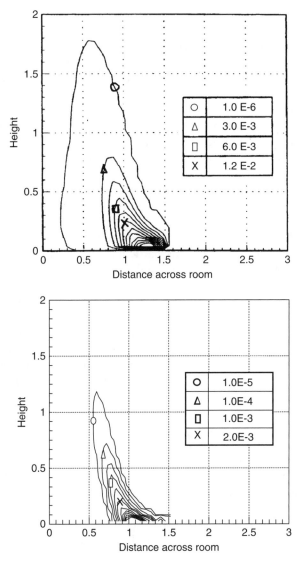

Figure 3.22 Comparison of different velocity profiles at the doorway.

the species in the Steckler et al. (1984) experiment are not available for comparison. Nevertheless, the capability of the conserved scalar approach using the Sivathanu and Faeth (1990) state relationships, could still be checked against Lewis et al. (1997) results to verify whether similar values or trends could be obtained. Since the compartment is well ventilated and the ambient levels of water vapor H_2O and carbon monoxide CO, remote from the fire source and plume regions, are, as expected, found to be very low. For H_2O, in-house predictions yield a peak level of 0.006, whereas a similar contour structure of 0.01 was obtained by Lewis et al. (1997). For CO, a peak level of 0.026 was predicted by Lewis et al. (1997), while a peak level of 0.01 is achieved through the in-house calculations. It is nevertheless noted that our computed CO levels are still within the same order of magnitude with those predicted by Lewis et al. (1997). Previously in Figure 3.20, doorway velocity profiles have indicated higher inflow and outflow mass fluxes than those predicted by Lewis et al. (1997). Since substantially more cold air is stipulated to be entering into the compartment, leaner fire combustion in turn is encouraged and promoted.

Conclusions: Coupled with a suitable two-equation turbulence model, application of the eddy dissipation combustion model of Magnussen and Hjertager (1976) and the conserved scalar approach employing the Sivathanu and Faeth (1990) state relationships to a full-scale compartment fire is demonstrated through this worked example. This particular approach does not depend on the *a priori* knowledge of the shape and size of the fire source, such as that required by the volumetric heat source approach, but instead depends only on the specification of the fuel flow rate at the burner surface corresponding to the required heat release of the fire to be simulated. More importantly, this worked example aptly shows the prevalence of a two-layer structure typical of compartment fires, through the added consideration of turbulent combustion models. The feasibility of the laminar flamelet representation for combustion in a turbulent fire is viable if the detailed reaction mechanisms allowing the prediction of intermediate chemical species are known for the fuel.

3.6.2 Two-Room Compartment Fire

It is imperative that the use and validation of the field model extends beyond the single-room compartment geometry, to more complex configurations that are usually encountered in practice. This worked example places emphasis on the assessment of field modeling and its feasibility in application against a series of two-room compartment fire experiments of turbulent buoyant diffusion flames, performed by Nielsen and Fleischmann (2000) at the University of Canterbury, New Zealand.

Figure 3.23 illustrates the schematic drawing of the particular geometry being investigated. The non-spreading buoyant fire was fueled by a square sand-box LPG burner of 0.3 m wide and elevated 0.3 m above the floor, centrally located in the burn room. Air was drawn into the two-room compartment through the open end of the adjacent room and entered the burn room through a doorway opening of 2.0 m high and 0.8 m wide, as depicted in Figure 3.23. Compartment walls and ceiling were insulated with Gib® Fyreline and Intermediate Service Board to minimize heat transfer and damage to the structure and instrumentation. Aspirated thermocouples were placed evenly throughout the compartment to enable the validation of vertical temperature

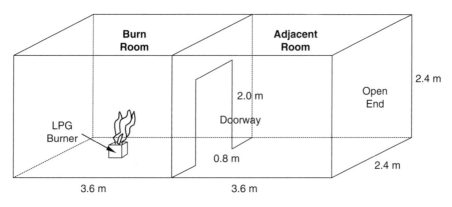

Figure 3.23   Schematic drawing of the two-room compartment structure.

	Burn Room					Adjacent Room		
Tree 1	Tree 2	Tree 3	Tree 4	Tree 5	Tree 6	Tree 7	Tree 8	Tree 9
+	+	+	+	+	+	+	+	+
0.15 m	0.9 m	1.8 m	2.7 m	3.6 m	4.5 m	5.4 m	6.3 m	7.2 m

Figure 3.24   Distribution of thermocouple tree positions (*Crosses indicate field trees*).

profiles. Figure 3.24 shows the spatially distributed thermocouple tress within the burn and adjacent rooms. More detailed information regarding the setup can be referred to in Nielsen and Fleischmann (2000). An in-house computer code FIRE3D is employed to generate the required numerical results. A heat release rate $Q^{\cdot}$ of 110 kW has been selected for the validation exercise.

Numerical features: The three-dimensional Favre-averaged equations for the transport of mass, momentum, and enthalpy are solved. A *hybrid differencing scheme* is employed for the convection terms. The velocity and pressure linkage is achieved through the SIMPLE algorithm. The eddy-viscosity concept is employed for the representation of the turbulent diffusivities in the governing equations due to turbulence. This is expressed by the solution of the standard k-ε turbulent model, with additional source terms to account for buoyancy effects (see work examples in section 2.16.1). An explicit equation for the mass fraction of fuel is solved for the eddy dissipation combustion model of Magnussen and Hjertager (1976). LPG (Liquefied Petroleum Gas) as described in Nielsen and Fleischmann (2000), comprises of approximately 80% of propane

(C_3H_8) and 20% of butane (C_4H_{10}). The global, one-step description of LPG combustion is assumed and can be written as

$$0.8C_3H_8 + 0.2C_4H_{10} + 4.5O_2 \rightarrow 3.2CO_2 + 4.2H_2O \qquad (3.6.1)$$

Figure 3.25 shows the grid distribution of the two-compartment geometry. Experimental observations by Nielsen and Fleischmann (2000) for the central fire in the burn room revealed that this configuration was symmetrical about the vertical plane bisecting the burner, doorway, and open end. Similar to the single-room compartment fire, mesh generation is thus carried out on half of the room, improving the resolution of the flow and thermal fields. Also, numerical experiments for the single-room compartment fire revealed that the inclusion of an extended region away from the open end was important to correctly model the flow through the open end. An extended region of 3.6 m × 4.8 m in plan and 6.5 m in height is attached to the two-compartment structure to isolate the end effects of the extended boundaries from the open end of the geometry. The computational grid is 85 × 44 × 25 (i.e., a total of 93,500 control volumes).

At the burner surface, the fuel flow rate is prescribed. The turbulence level is assumed to be weak; laminar prescription is enforced at this boundary.

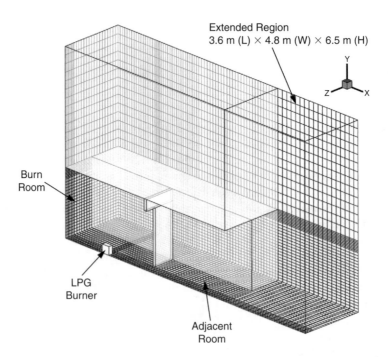

Figure 3.25 Mesh distribution of the two-room compartment geometry.

Temperature is set to be constant, based on the temperature of the unburnt fuel flowing through the burner. The mass fraction of fuel is set at unity. At the solid surfaces, the condition of no-slip is imposed by setting all velocities to zero. The normal gradients of the mass fraction of participating species are set to zero at these boundaries, due to impermeability of the walls. In order to resolve the momentum and heat fluxes near the wall region, conventional logarithmic wall function was applied. Adiabatic condition is imposed for the calculation of the wall temperatures. The enthalpy equation is correspondingly determined from the given wall temperature when solving the energy conservation equation. At the extended boundaries, the solution domain is treated as an entraining surface on which the ambient pressure is set to be constant. The normal gradients of all dependent variables are set to zero for in-flow or out-flow conditions, except for the temperature and mass fraction of participating species where ambient variables are specified at this plane when the flow enters the compartment.

Numerical results: Model predictions of the vertical temperature distribution above the fire source (Tree 3) and at the doorway (Tree 5) are shown in Figure 3.26. Focusing on the temperature distribution above the fire source, possible causes for the large discrepancy between the measured and predicted temperatures are provided following.

Firstly, temperatures that were measured through the uncovered bare-wire thermocouples did not actually reflect the real fluid temperatures. The occurrence of the heat transfer processes by convection and radiation within the

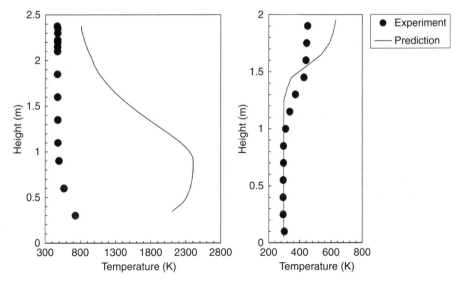

Figure 3.26 Comparison of predicted and measured temperature profiles above the fire source (Tree 3) and at the doorway (Tree 5).

sensor, surrounding surfaces, and fluid balanced each other to register temperatures that could possibly be between the surface and surrounding fluid temperatures. Wen et al. (2001) clearly showed that there were inherent errors in the raw thermocouple readings due to radiation for unshielded thermocouples. The differences between the uncorrected and corrected data subject to flame radiation could amount to a temperature difference greater than 100 K. Typical experimental observed temperatures within the flaming zone have been shown to be at least of the order 800°C or 1073 K, as exemplified in Drysdale (1986).

Secondly, a peak temperature of about 2300 K predicted by the field model is not entirely unexpected, since the combustion of LPG has been assumed to be merely represented by single step chemistry. In reality, the combustion process is never perfect or complete, and the gas-phase oxidation mechanism for LPG generally results in a number of intermediate species. Incomplete or inefficient combustion has a tendency of lowering the flame temperatures. More realistic flame temperatures could have been obtained by the consideration of detailed reaction mechanism for the fuel. From a practical viewpoint, the simple approach is retained. In addition, the absorption-emission radiative heat exchange phenomenon of the combustion products, of which such mechanism is absent in current field modeling consideration, generally contributes dramatically toward lowering the flame temperatures. This will be demonstrated in the subsequent worked example in section 3.13.2.

As seen in Figure 3.26, the unduly predicted high temperatures are confined to the region just above the fire source. At the doorway connecting the burn room and adjacent room, the predicted temperature distribution is nonetheless shown to compare rather well against the measured profile.

A distinct feature of compartment fires, as inferred earlier in Chapter 2 as well as in the previous worked example, is the existence of a two-layer structure; the hot layer comprises of hot combustion products at the top, while the cold layer contains the distribution of cold air entraining into the compartment at the bottom. In spite of the temperatures being rather substantially over-predicted when compared against the measured temperature profiles at the upper part of the burn room, the prevalence of the two-layer structure is still adequately represented as is evident in the predicted vertical temperature profiles in Figure 3.27 for thermocouple tress 1, 2, and 4 (away from the fire source). Predicted vertical post-flame temperature distributions in Figure 3.28 against the measured profiles, are nevertheless observed to be only marginally over-predicted for all the thermocouple trees represented in the adjoining room. The over-spilling effect of the high temperatures experienced in the burn room is clearly seen to be felt in the adjoining room and registering as far in the venting outlet or open end.

Conclusions: Field modeling investigations were carried out on a two-room compartment fire utilizing the three-dimensional Favre-averaged equations governing the conservation of mass, momentum, and energy coupled with a suitable two-equation turbulence model and the eddy dissipation combustion model of Magnussen and Hjertager (1976). The application of the combustion

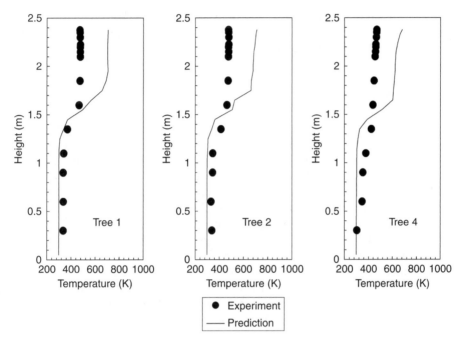

Figure 3.27 Comparison of predicted and measured temperature profiles in the burn room.

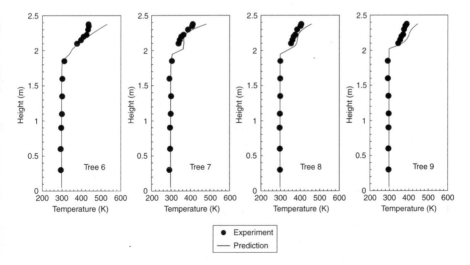

Figure 3.28 Comparison of predicted and measured temperature profiles in the adjoining room.

model to feasibly characterize the two-layer structure within the entire occupied space in both the burn room and adjacent room, is appropriately demonstrated through this worked example.

3.7 Summary

The generalized finite-rate formulation approach to combustion consists of primarily solving the conservation equations that describe the mixing, transport,
and chemical reactions for each chemical species in the fluid medium has been
explained. It can be applied to laminar or turbulent combustion covering a
wide range of premixed, non-premixed, or partially premixed flames. The laminar finite-rate chemistry model through neglecting the non-linear terms due to
Reynolds or Favre averaging and decomposition is exact for laminar flames. Its
immediate and direct application to turbulent reacting flow systems due to the
omission of turbulence-chemistry interaction is, however, not as straightforward, since, as previously discussed in section 3.2, erroneous results were
obtained when the model is applied for turbulent flames. The need to handle
such flames has led to developments of specific combustion models such as
the eddy break-up, eddy dissipation, and generalized eddy dissipation concept
in order to better resolve the turbulent combustion process.

Since most practical fires are naturally diffusion flames, and the reaction is so
rapid that the fuel and oxidant only co-exist within an infinitely thin flame sheet
in the limit of fast chemistry, the concept of conserved scalar can be purposefully exploited to directly relate the molecular-species concentrations within
it. The statistics of all thermodynamics variables are therefore ascertained based
on the knowledge of the statistics of this particular scalar. Computationally, the
model is efficient and requires only a single transport equation for the mixture
fraction (conserved scalar) to be solved. In turbulent combustion, an additional
transport equation for the concentration fluctuation of the mixture fraction is
required to characterize the random fluctuations of the turbulent flow. The
probability density function based on the square wave probability distribution
is by far simpler to apply and very easy to compute. However, it is invariably
less accurate in comparison to the more complex forms of clipped Gaussian
and beta function. Nevertheless, it is worth noting that the assumed shapes of
the probability distributions are essentially mathematical formulations of the
problem, and they are but convenient ways of characterizing the statistical
information of the random fluctuations in the turbulent flow. In spite of the
many complexities that exist within turbulent combustion, the clipped Gaussian or beta function is generally applicable for most practical purposes in field
modeling applications of reacting flows. The consideration of the laminar flamelet model offers the feasibility of incorporating detailed chemical kinetics
with minimal computational costs for computing turbulent flames.

As demonstrated through the worked examples, combustion modeling of
fires requires no *a priori* knowledge of the shape and size of the fire source

to be prescribed, as is evident in the volumetric heat source approach. More importantly, the additional consideration of turbulent combustion models aptly captures the commonly observed two-layer structure comprising of the hot layer of combustion products at the top and cold layer of ambient air at the bottom, in both the single-room and two-room compartment fires.

PART IV   RADIATION

3.8   Radiation in Fires

In practical fires, two modes of heat transfer exist: radiation and convection. As the body of the flame is heated up due to the energy release caused by chemical reactions, it will lose some part of the heat by convection (in a fluid such as air) and another part by radiation. The former usually persists at low temperatures from about 150°C to 200°C, while the latter becomes increasingly dominant for temperatures above 400°C. Specifically, thermal radiation involves the transfer of heat by electromagnetic waves that is confined to a relatively narrow "window" in the electromagnetic spectrum, and is that electromagnetic radiation emitted by the body as a result of its temperature. Figure 3.29 illustrates the electromagnetic spectrum of different wavelengths. Thermal radiation lies in the range from about 0.1 μm to 100 μm, which incorporates visible-light and extends toward the far infrared regions. The visible-light portion of the spectrum is nevertheless very narrow, extending only from 0.35 μm to 0.75 μm.

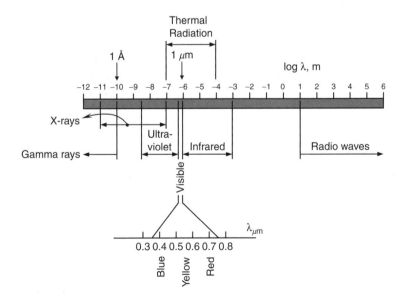

Figure 3.29   Electromagentic spectrum.

The emitted heat flux due to radiation is a strong function of the body temperature. As the temperature is increased, more radiation will be emitted of which the *emissive power E* defines the rate of the body's radiating surface of heat flow per unit time and per unit surface area. For an ideal emitter—that is, *black body*—the total energy emitted is proportional to the absolute temperature T to the fourth power:

$$E_b = \sigma T^4 \tag{3.8.1}$$

The preceding equation is generally referred to as the Stefan-Boltzmann law and σ is the Stefan-Boltzmann constant, which has the value

$$\sigma = 5.669 \times 10^{-8} \text{Wm}^{-2}\text{K}^{-4}$$

where E_b is in Wm^{-2} and T is in Kelvin. Equation (3.8.1) is called the *black body radiation*, because the body that obeys this law appears black to the eye; it is black because it does not reflect any radiation. Hence, a *black body* is also considered as one that absorbs all incident radiation. The *black body intensity*, which is used to determine the emitted intensity from surfaces to fluid, can also be expressed as

$$I_{black} = \frac{E_b}{\pi} = \frac{\sigma T^4}{\pi} \tag{3.8.2}$$

Nevertheless, it has been convenient to introduce the concept of a *gray body* (or an *ideal non-black body*) for which the emissivity ε is independent of wavelength to characterize a real body. While this is an approximation, it permits the simple use of the Stefan-Boltzmann equation. Following Kirchhoff's law where the absorptivity is equivalent to the emissivity of the body, as dictated by the first law of thermodynamics, the emissive power for a fictitious real body is given as

$$\varepsilon = \frac{E}{E_b} \tag{3.8.3}$$

Equation (3.8.3) can be employed to characterize the surface radiation properties of a material. Real surfaces usually emit less radiation than ideal black surfaces. Hence, the emissive power of a real surface is given by

$$E_w = \varepsilon \sigma T_w^4 \tag{3.8.4}$$

where T_w is the temperature of the surface material. The emitted wall intensity as a result of equation (3.8.3) is: $I_w = \varepsilon \sigma T_w^4 / \pi$. When radiant energy strikes a material surface, part of the radiation is reflected, part is absorbed, and part

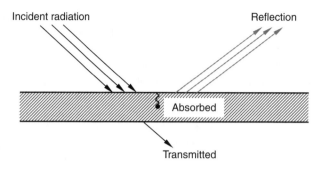

Figure 3.30   Schematic drawing showing effects of incident radiation.

is transmitted, as described in Figure 3.30. Defining the reflectivity ρ as the fraction reflected, the absorptivity α as the fraction absorbed, and the transmissivity T as the fraction transmitted, the sum of these fractions at the surface material will be equivalent to unity: $\rho + \alpha + T = 1$. Two types of reflection phenomena may be observed when radiation strikes a surface. Figure 3.31 illustrates the reflection caused by *specularly reflecting* and *diffusely reflecting* rays. A specular reflection occurs when the angle of reflection is equal to the angle of incidence of an incident beam. Note that it represents a mirror image of the source to the observer. In the case of diffusely reflecting, the rays leave the surface in all directions irrespective of the incident angle of the incoming beam. It should be noted that no real surface is either specular or diffuse, although most practical problems may be inclined to adopt the diffuse reflection behavior. Also, most solid bodies do not transmit thermal radiation, so for many applied problems they are considered to be opaque; the transmissivity may be taken as zero, $T = 0$. Consequently, $\rho + \alpha = 1$. In reality, all surface properties are highly dependent on the type of materials, surface finishing, surface roughness, and the presence of surface contaminants.

Air which contains traces of N_2 and O_2 and other gases of non-planar symmetrical molecules, are essentially transparent to radiation at low temperatures. As a result of the oxidation process in fires, combustion products such as carbon dioxide CO_2 and water vapor H_2O, and various hydrocarbon gases radiate to an appreciable extent. These molecules can interact with the

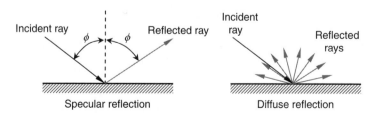

Figure 3.31   Schematic drawing illustrating specular and diffuse reflection.

electromagnetic radiation in the "thermal" region of the spectrum (0.1 μm – 100 μm). Such species generally do not exhibit the continuous absorption (and emission) throughout the volume of the gas. They tend to absorb (and emit) at several wavelength bands. In general, the presence of carbon dioxide and water vapor provides the mechanism of heat loss by radiation of the flame. Moreover, most practical fires will burn with luminous diffusion flames. Finely dispersed carbonaceous particles (soot) that can be found mainly on the fuel side of the reaction zone at high temperatures, act accordingly as individual minute black (or gray) body and emit continuously over a wide range of wavelengths. The presence of soot also significantly augments the radiant heat loss; the sootier the flame, the lower the flame temperature. The calculation of gas-radiation and soot-radiation properties is generally rather complicated. Suitable methods for evaluating the radiant exchange of these gases from practical engineering considerations will be provided in later sections of this chapter.

In any point in space of the physical system, incident radiation can be absorbed, transmitted, and/or scattered and radiation emitted for a volume of gas containing significant amounts of carbon dioxide, water vapor, and soot particles. In reacting flows, such a fluid is categorized as a *participating medium*. The strength of the interactions between a participating fluid medium and radiation can be measured in terms of its absorption coefficient K_a and its scattering coefficient σ_s. For the participating fluid medium at a fluid temperature T, the emitted intensity is simply the product of the absorption coefficient and the black body intensity: $I = K_a I_{black} = K_a \sigma T^4 / \pi$. The distribution of emitted intensity by a point radiation source in a participating fluid medium is taken to be uniform in all directions but not for the scattered intensity. In field modeling, according to Luo et al. (1997), the scattering coefficient for the dispersed soot particles in the order of diameters about 0.1 μm can nonetheless be considered to be negligible in comparison to the respective absorption coefficient. It is therefore appropriate to dispense with the specific consideration in modeling the scattering source in the equation of radiant energy transfer. The global radiation calculations are thus simplified for the radiant exchange in the fluid participating medium. In the next section, the expression for the radiative transfer equation ignoring scattering effects is described.

3.9   Radiative Transfer Equation

It is important that the physics governing the transport of radiation is properly understood in order to appreciate the difficulties involved in the analysis of radiative heat transfer. Consider a pencil of radiation traveling across a participating medium in Figure 3.32. From the Lagrangian viewpoint, the total variation of radiation over time Dt for monochromatic radiant energy can be written as

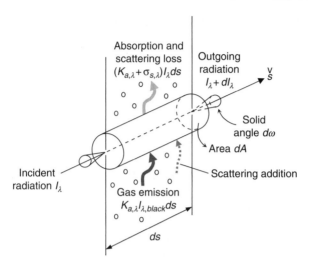

Figure 3.32 Change of intensity I_λ in direction $\vec{s}$ across an elemental volume.

$$\frac{DI_\lambda}{Dt} = \frac{\partial I_\lambda}{\partial t} + c\frac{\partial I_\lambda}{\partial s} \tag{3.9.1}$$

where c is the speed of light at which radiation travels and I_λ is the monochromatic radiation intensity, which is a function of wavelength, position, and direction. According to equation (3.9.1), the change in radiation DI_λ in a given direction is a function of time and space

$$DI_\lambda = (Dt)\frac{\partial I_\lambda}{\partial t} + c(Dt)\frac{\partial I_\lambda}{\partial s} \tag{3.9.2}$$

Since radiation propagates at the speed of light, at 3×10^8 m s^{-1}—approximately 10^5 times as fast as the speed of sound of common fluid encountered in engineering applications—to tranverse across a path length ds, the time required $(Dt = ds/c)$ is appreciably small. On the right-hand side of equation (3.9.2), the first term $(Dt)\partial I_\lambda/\partial t$ is comparatively smaller than the second term $c(Dt)\partial I_\lambda/\partial s$. This implies that for most engineering applications, radiation occurs in finite-size domain; it is reasonable to assume that

$$DI_\lambda \approx c(Dt)\frac{\partial I_\lambda}{\partial s} = ds\frac{\partial I_\lambda}{\partial s} \tag{3.9.3}$$

because the change of radiation intensity propagates so rapidly that the time dependence of I_λ can be safely disregarded without significant loss of accuracy.

According to Ozisik (1973) and Siegel and Howell (1981), the radiative transfer equation describing the variation of a monochromatic pencil of radiation across an elemental volume taken along the path $\vec{r}$ for steady state conditions and coherent isotropic scattering, is given by

$$\frac{dI_\lambda(\vec{r},\vec{s})}{ds} = -(K_{a,\lambda} + \sigma_{s,\lambda})I_\lambda(\vec{r},\vec{s}) + K_{a,\lambda}I_{black,\lambda}(\vec{r}) + \frac{\sigma_{s,\lambda}}{4\pi}\int_{4\pi} \Phi(\vec{s}',\vec{s})I_\lambda^-(\vec{s}')d\Omega'$$

$$(3.9.4)$$

where $I_{black,\lambda}$ is the Planck's intensity of black body radiation per unit wavelength at some temperature in the participating medium, $K_{a,\lambda}$ and $\sigma_{s,\lambda}$ are respectively the local spectral absorption and scattering coefficient for a wavelength λ, and Φ is the scattering phase function that specifies the fraction of incident intensity $I_\lambda^-(\vec{s}')$ from all possible $\vec{s}'$ that is scattered into direction $\vec{s}$ of which involves the integration over a unit sphere (i.e., a solid angle of 4π steradians) surrounding the point in the medium. The derivation of equation (3.9.4) is based on the radiant energy balance over an elemental solid angle $d\Omega$ in the direction $\vec{s}$ across an elemental volume like the one illustrated in Figure 3.32. As previously indicated in the previous section, scattering effects are normally neglected in field modeling; equation (3.9.4) can be reduced to

$$\frac{dI_\lambda(\vec{r},\vec{s})}{ds} = -K_{a,\lambda}I_\lambda(\vec{r},\vec{s}) + K_{a,\lambda}I_{black,\lambda}(\vec{r}) \qquad (3.9.5)$$

In words,

$$\begin{array}{lll}
\text{The rate of change} & \text{The net gain in} & \\
\text{of intensity per unit} & = \text{radiant energy} & (3.9.6) \\
\text{path length} & \text{about direction } \vec{s} &
\end{array}$$

The radiative transfer equation in the form of equation (3.9.4) is an *integrodifferential* equation while equation (3.9.5) is an ordinary partial differential equation, which they are extremely difficult to solve exactly for multidimensional geometries. These equations require solution along the relevant ray paths, since the transport of heat by radiation is usually three-dimensional. The calculation of radiation is thus numerically challenging, computationally intensive, and imposes high demands on computational resources. In attempting to solve the radiative transfer equation, algorithms need not only to compute the radiation intensity as a function of the Cartesian position (x, y, z) and angular direction (θ, ϕ) but also on radiation wavelength. Radiation properties of the participating medium are closely linked to the wavelength of radiation. For numerical simulations, it is necessary that either the full spectrally resolved radiation calculations are performed or the effect of this wavelength dependence is modeled by some reasonable approximations.

In essence, radiation modeling concerns two key issues. The first is the availability of appropriate solution methods to best handle the radiation heat transfer, which may entail different ways of treating the angular dependence and spatial variation of intensity within the physical domain. A range of popular radiation algorithms will be discussed in section 3.11. Secondly, appropriate evaluation of the absorption coefficient is required to account for the complex radiative properties of the participating medium, consisting of a gaseous mixture of combustion products. Possible approaches with different levels of sophistication are presented in the next section.

3.10 Radiation Properties of Combustion Products

Every combustion process produces combustion gases such as water vapor (H_2O) and carbon dioxide (CO_2). These gases do not scatter radiation significantly, but they are strong selective absorbers and emitters of radiant energy, such as illustrated from the radiation emitted from a flame in the infrared region in Figure 3.33. The radiation bands of water vapor are relatively broad and are spread over a wide wavelength spectrum of which the bands at 1.9 μm and 2.7 μm are important. Carbon dioxide, however, shows only two significant bands at 2.7 μm and 4.3 μm. In the latter region, carbon dioxide is almost opaque even at short path lengths. In contrast, the radiation spectrum of soot is continuous whereby the intensity varies strongly with the soot concentration and particle size. Consequentially, the variation of radiative properties especially for water vapor and carbon dioxide within the electromagnetic spectrum, needs to be aptly accounted for and spectral calculations can be performed by dividing the entire wavelength spectrum into several bands. The absorption/emission characteristics of each of the chemical species are assumed so that they remain either uniform or change smoothly in a given functional form over these bands. Obviously, the accuracy of the predictions increase as the width of these bands becomes narrower. Considering the diversity of products and the probability of having some or all of these gases in any volume element of the system, the prediction of radiative properties is certainly not an easy task. In order to present a systematic methodology for the prediction of the radiative properties of combustion products, this section will first focus on the discussion of suitable relations for the properties of the combustion gases and soot particles and simplifications that have been made in arriving to these relations. Note that the level of simplification for the properties to be determined, should be consistent with the level of sophistication of the radiative transfer and total heat transfer models. The relations for the radiative properties of individual constituents should also be compatible with each other, as well as with the radiative transfer models.

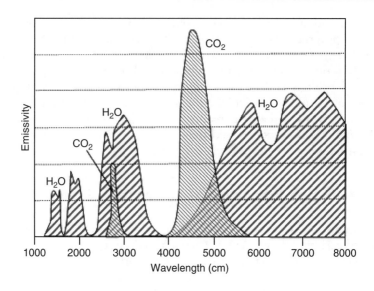

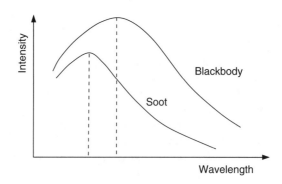

Figure 3.33 Radiating gases of a flame and infrared radiation spectrum of soot compared with spectrum of blackbody at the same temperature.

3.10.1 Gray Gas Assumption

For engineering calculations, it is always desirable to have some reliable yet simple models for predicting the radiative properties of the gases. In some circumstances, a detailed modeling of the radiative properties of combustion gases may not be warranted for the total accuracy of the total heat transfer predictions. If scattering is considered not to be important for the combustion gases, the gray absorption/emission coefficient can be obtained from Beer-Lambert's law. For a given mean path length L, the mean absorption coefficient for the combustion gases is given as

$$\overline{K}_{a,g} = -\frac{1}{L}\ln\left(1 - \varepsilon_{gray}\right) \tag{3.10.1}$$

It is possible to determine the so-called *gray* absorption and emission coefficients for each temperature, pressure, and path length to yield the same total absorptivity or emissivity of the CO_2-H_2O mixture. In equation (3.10.1), the corresponding mean path length must be properly evaluated. The characteristic cell size of the computational domain may be chosen as the mean path length or for a nearly homogeneous medium, the mean beam length based on Viskanta and Menguc (1987) for a volume (V) of a gas radiating to its entire surface area (A) can be evaluated according to

$$L = L_m = \frac{4CV}{A} \tag{3.10.2}$$

where C is the correction factor and for an arbitrary geometry its magnitude is 0.9 (Hottel and Sarofim, 1967). Considering the CO_2-H_2O mixture, the emissivity ε_{gray} in equation (3.10.1) may be ascertained as

$$\varepsilon_{gray} = C_{CO_2}\varepsilon_{CO_2} + C_{H_2O}\varepsilon_{H_2O} - \Delta\varepsilon_{CO_2-H_2O} \tag{3.10.3}$$

The gas emissivities of carbon dioxide ε_{CO_2} and water vapor ε_{H_2O} can be obtained from Hottel's charts of gas emittance as a function of gas temperature for different values of the product between the partial pressure and mean beam length as shown in Figure 3.34. While these diagrams apply to gas mixtures at a total pressure of 1 atmosphere, an effect of pressure broadening, which depends on the partial pressures of these species, influences the emission and must be accounted for. This refers to the $\Delta\varepsilon_{CO_2-H_2O}$ term in equation (3.10.3) of which an additional correction for the overlap of the wavelength about 4.4 μm of CO_2 and 4.8 μm of H_2O is necessary. Appropriate values of this term and the respective correction factors C_{CO_2} and C_{CO_2} can also be obtained by the Hottel's charts, as depicted in Figure 3.33. It is worth mentioning that this empirical method provides acceptable values of emissivity only up to about 1000 K. Above this temperature and at long path lengths, Hottel's method underestimates the emissivity, which could be due to the overestimation of the overlapping correction.

Alternatively, the emissivity of each individual constituent can be modeled according to Modak (1979) as

$$\ln(\varepsilon_{CO_2} \text{ or } \varepsilon_{H_2O}) = \sum_{i=0}^{2} T_i(a) \sum_{j=0}^{2} T_j(b) \sum_{k=0}^{3} c_{ijk} T_k(c) \tag{3.10.4}$$

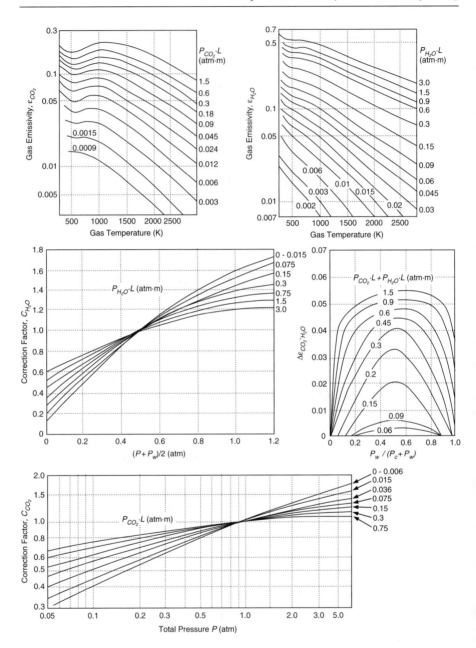

Figure 3.34   Hottel's Charts (after Hottel, 1954).

where

$$a = 1 + \frac{\ln(P_{CO_2} \text{ or } P_{H_2O})}{3.45}$$

$$b = \frac{2.555 + \ln(P_{CO_2}L \text{ or } P_{H_2O}L)}{4.345}$$

$$c = \frac{T_f - 1150}{850}$$

From the preceding P_{CO_2} and P_{H_2O} are the partial pressures of carbon dioxide and water vapor, $T_i(a), T_j(b)$, and $T_k(c)$ are Chebyshev polynomials of orders i, j, and k, respectively, and c_{ijk} are fitting coefficients for carbon dioxide and water vapor as formulated in Modak (1979). The term $\Delta\varepsilon_{CO_2-H_2O}$, which represents the overlapping of absorption bands for the gaseous combustion products, is given as

$$\Delta\varepsilon_{CO_2-H_2O} = F_1(\xi)F_2(P_{CO_2}, P_{H_2O}L)F_3(T) \qquad (3.10.5)$$

for $(P_{CO_2} + P_{H_2O})L \geq 0.1$ atm·m and is zero otherwise. The functions F_1, F_2, and F_3 are

$$F_1(\xi) = \frac{\xi}{10.7 + 101\xi} - \frac{\xi^{10.4}}{111,7}$$

$$F_2(P_{CO_2}, P_{H_2O}L) = [\log_{10}(101.3(P_{CO_2} + P_{H_2O})L)]^{2.76}$$

$$F_3(T) = -1.0204\left(\frac{T}{1000}\right)^2 + 2.2449\left(\frac{T}{1000}\right) - 0.23469$$

where ξ describes the local concentration of carbon dioxide and water vapor as a function of the partial pressures as

$$\xi = \frac{P_{H_2O}}{P_{H_2O} + P_{CO_2}} \qquad (3.10.6)$$

If the participating medium is in radiative equilibrium, the mean emission and absorption coefficients will be equal to each other. Otherwise, the absorption coefficient for the gas mixture $\overline{K}_{a,g}$ must be computed from the absorptivity (α_{gray}) (rather than to the emissivity) as

$$\overline{K}_{a,g} = -\frac{1}{L}\ln\left(1 - \alpha_{gray}\right) \qquad (3.10.7)$$

The absorptivity is related to the emissivity by

$$\alpha_{gray} = \varepsilon_{gray} \left(\frac{T}{T_s}\right)^{(0.6-0.2\xi)} \tag{3.10.8}$$

with T_s representing the black body source temperature and the gas emissivity is evaluated according to

$$\varepsilon_{gray} = \varepsilon_{CO_2} + \varepsilon_{H_2O} - \Delta\varepsilon_{CO_2-H_2O} \tag{3.10.9}$$

It is usually assumed that the soot particles are small when compared to the wavelength of radiation, and that the complex refractive index of soot is independent of wavelength. Under these assumptions, the spectral absorption coefficient of soot that is inversely proportional to wavelength and scattering effects are negligible. The spectrally integrated absorptivity of soot of a path length L according to Felske and Tien (1973) is given by

$$\alpha_{soot} = 1 - \frac{15}{\pi^4}\psi^{(3)}\left(1 + \frac{k_o\lambda_o T_s L}{C_2}\right) \tag{3.10.11}$$

where $k_o\lambda_o$ has been approximated by Hottel and Sarofim (1967) as a function of soot volume fraction f_v as $k_o\lambda_o \cong 7f_v$, C_2 is Planck's second constant with a value 1.4388 cm K, and $\psi^{(3)}$ is the pentagamma function (Abramowitz and Stegun, 1964). The total absorptivity of the CO_2-H_2O and soot mixture may now be approximated as

$$\alpha_T = \alpha_{soot} + (1 - \alpha_{soot})\alpha_{gray} \tag{3.10.12}$$

and the overall mean absorption coefficient is similarly evaluated according to equation (3.10.7) as

$$\overline{K}_a = -\frac{1}{L}\ln(1 - \alpha_T) \tag{3.10.13}$$

The method of Modak's is sufficiently accurate for maximum temperature that is below 2000 K . For a range of temperatures, pressures, and path lengths, the model has shown good agreement with spectral calculations, as well as measurements from a smoky ceiling layer formed in a room fire. More details on the model can be referred to in Modak (1979), as well as a sample computer program provided in the paper for immediate use.

The use of the absorption coefficients related to the mean beam length is a convenient way of scaling the radiation heat transfer in practical systems. Whereas difficulty may arise in finding an appropriate mean beam length, the mean absorption coefficient concept proposed by Hubbard and Tien (1978) can be used to calculate the mean absorption coefficient for the gas mixture. In this approach, the Planck's mean absorption coefficient, which is independent of the mean beam length, is determined. By definition, it is given by

$$\overline{K}_P = \frac{\int_0^\infty K_{a,\lambda} I_{black,\lambda} d\lambda}{\int_0^\infty I_{black,\lambda} d\lambda} = \frac{1}{\sigma T^4} \int_0^\infty K_{a,\lambda} E_{b,\lambda} d\lambda \qquad (3.10.14)$$

where $E_{b,\lambda}$ is the spectral black body emissive power. By adopting the Elsasser narrow-band model, where an array of equally intense, equally wide, equally spaced absorption lines is assumed to exist in a given spectral region, the Planck's mean absorption coefficient for the absorbing gases for the ith species can be expressed in discrete form as

$$\overline{K}_{P,i} = \sum_j a_j \frac{E_{b,\lambda,j}}{\sigma T^4} \qquad (3.10.15)$$

where a_j represents the integrated band intensity of each band. For the evaluation of the contribution of CO_2 to the Planck mean absorption coefficient, the vibration-rotation bands considered are at 15 μm, 10.4 μm, 9.4 μm, 4.3 μm, 2.7 μm, and 2.0 μm. For H_2O, the 20 μm pure rotation and the 6.3 μm, 2.7 μm, 1.9 μm, and 1.4 μm vibration-rotation bands are included. Figure 3.35

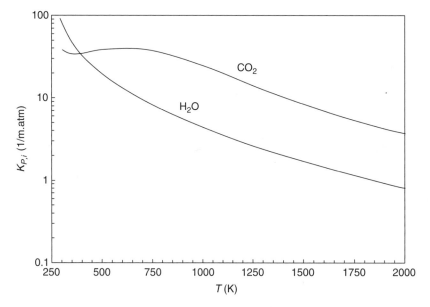

Figure 3.35 Planck's mean absorption coefficient for CO_2 and H_2O.

shows the respective mean absorption coefficient of the absorbing gases as functions of temperature. The rapid drop of the mean absorption coefficients of the absorbing gases with temperature results from the T^{-1} variation of a_j, and the shift of the black body curves away from the infrared spectrum where the absorption bands are located. For soot, the Rayleigh absorption limit is employed and the Planck's mean absorption coefficient is given by

$$\overline{K}_{P,s} = \frac{T}{C_2} \int_0^\infty C_0 \frac{15}{\pi^4} \frac{\zeta^4}{(e^\zeta - 1)} d\zeta \qquad (3.10.15)$$

where C_0 is a function of wavelength, which varies from 2 to 6 and $\zeta = C_2 \lambda / T$. The calculation of the coefficient $\overline{K}_{P,s}$ requires specification of the complex index of refraction at all wavelengths. A dispersion theory model is employed for this purpose. Figure 3.36 illustrates the mean absorption of soot as a function of temperature. According to Hubbard and Tien (1978), the overall absorption coefficient of the CO_2-H_2O and soot mixture is just the sum of the mean coefficients, which is

$$\overline{K}_a = \overline{K}_{P,s} f_v + \left(\overline{K}_{P,CO_2} P_{CO_2} + \overline{K}_{P,H_2O} P_{H_2O} \right) \qquad (3.10.16)$$

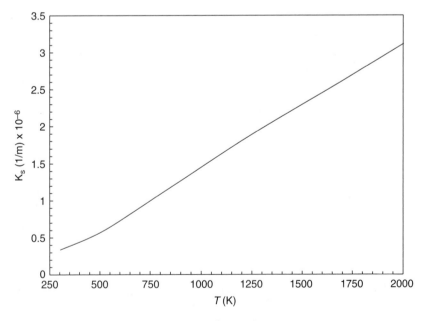

Figure 3.36   Planck's mean absorption coefficient for pure soot.

In field modeling, Wen and Huang (2000) have demonstrated the applicability of the Planck's mean absorption coefficient described in equation (3.10.16), to adequately represent the radiation properties of the CO_2-H_2O and soot mixture for confined jet fires in small and large compartments.

3.10.2 Weighted Sum of Gray Gases Model

The weighted sum of gray gases model (WSGGM) first introduced by Hottel and Sarofim (1967), represents another elegant radiative gas property model, which strikes a reasonable compromise between the oversimplified gray gas assumption and a complete model accounting for the entire spectral variations of radiation properties. In essence, the model postulates that the total emissivity and absorptivity may be represented by the sum of a gray gas emissivity weighted with a temperature dependent factor. The WSGGM entails the evaluation of the total emissivity over the distance L from the following expression:

$$\varepsilon_T = \sum_{i=0}^{I} a_{\varepsilon,i}\left(1 - e^{-k_i PL}\right) \tag{3.10.17}$$

where $a_{\varepsilon,i}$ denote the emissivity weighting factors for the ith fictitious gray gas as based on the gas temperature. The bracketed quantity in equation (3.10.17) is the ith gray gas emissivity with absorption coefficient k_i and partial pressure-path length product PL. For a gas mixture, P is the sum of the partial pressures of the absorbing gases—that is, $P = P_{CO_2} + P_{H_2O}$. Physically, the weighting factor $a_{\varepsilon,i}$ may be interpreted as the fractional amount of black body energy in the spectral regions where the gray gas coefficient k_i exists, as illustrated in Figures 3.37 and 3.38. The absorption coefficient k_0 is assigned a zero value in order to account for windows in the spectrum between spectral regions of high absorption $\left(\sum_{i=0}^{I} a_{\varepsilon,i} < 1\right)$ and the weighting factor for $i = 0$ is evaluated from

$$a_{\varepsilon,0} = 1 - \sum_{i=1}^{I} a_{\varepsilon,i} \tag{3.10.17}$$

From equation (3.10.17), only I values of the weighting factors need to be determined. A convenient representation of the temperature dependency of the weighting factors is of the polynomial form of order $J - 1$ given as

$$a_{\varepsilon,i} = \sum_{j=1}^{J} b_{\varepsilon,i,j} T^{j-1} \tag{3.10.18}$$

where $b_{\varepsilon,i,j}$ are the emissivity gas temperature polynomial coefficients. The absorption and polynomial coefficients have been determined through a

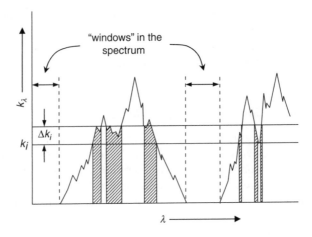

Figure 3.37   Interpretation of $a_{\varepsilon,i}$ in terms of the spectral energy distribution where $a_{\varepsilon,i}$ represents the fraction of black body energy in wave number region λ associated with Δk_i.

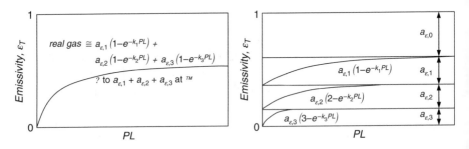

Figure 3.38   A one-clear and three-gray gas representation of a real gas.

number of different approaches. Employing a statistical *narrow-band* model and experimental spectral data, Taylor and Foster (1974) have fitted the parameters for CO_2-H_2O mixture by employing a one-clear 3-gray gases model for temperatures (see Figure 3.38) between 1200 K and 2400 K for the ratio of partial pressure of water vapor to partial pressure of carbon dioxide: $P_{H_2O}/P_{CO_2} = 1$ or 2. It is noted that in the case of natural gas combustion, the proportions of water vapor to carbon dioxide in the products of combustion can be shown that the ratio of their partial pressures P_{H_2O}/P_{CO_2} is approximately 2, corresponding to, for example, methane-air or methane-oxygen combustion. For oils and other fuels with the chemical formula $(CH_2)_x$, a ratio equaling to 1 exists. Most other hydrocarbon fuels have combustion products with a ratio of P_{H_2O}/P_{CO_2} in between 1 and 2. Smith et al. (1982) proposed a third-order polynomial fit of $b_{\varepsilon,i,j}$ for one-clear 3-gray gases model, to incorporate temperatures between 600 K and 2400 K for $P_{H_2O}/P_{CO_2} = 1$ or 2.

The initial emissivity data were generated from Edwards' exponential wide-band model (1976). For temperatures over 2000 K, Coppalle and Vervisch (1983) adjusted the total CO_2-H_2O emissivities calculated from the wide-band model of Edwards' for $P_{H_2O}/P_{CO_2} = 1$ or 2. The dependence of the weighting factors on temperature was linear, as in accordance with the proposal by Taylor and Foster (1974).

The values for the absorption and polynomial coefficients correlated from Smith et al. (1982) for gaseous fuels ($P_{H_2O}/P_{CO_2} = 1$) and oils ($P_{H_2O}/P_{CO_2} = 2$) are presented in Table D.8, Appendix D. The absorption coefficients are seen to be of the order of 0.1, 10, 100 (atm^{-1} m^{-1}), which may be interpreted as a real gas being represented by three gray gases corresponding to optically thin, intermediate optical thickness, and optically thick, respectively. As aforementioned, the polynomial coefficients $b_{\varepsilon,i,j}$ are presented by a third-order function of T:

$$a_{\varepsilon,i} = b_{\varepsilon,i,1} \times 10^{-1} + b_{\varepsilon,i,2} \times 10^{-4}T + b_{\varepsilon,i,3} \times 10^{-7}T^2 + b_{\varepsilon,i,3} \times 10^{-11}T^3$$

$$(3.10.19)$$

For $T > 2400$ K, the coefficients suggested by Coppalle and Vervisch (1983) are used.

In Taylor and Foster (1974), the polynomial coefficients $b_{\varepsilon,i,j}$ are nevertheless simpler and given by a linear function of T

$$a_{\varepsilon,i} = b_{\varepsilon,i,1} + b_{\varepsilon,i,2} \times 10^{-5}T$$

$$(3.10.20)$$

In addition to CO_2 and H_2O, the model developed by Beer, Foster, and Siddall (1971) further accounted the contribution of other gas species such as carbon monoxide CO and unburnt hydrocarbons (e.g. methane), which are also significant emitters of radiation. To incorporate these species, the term k_iP in equation (3.10.17) according to Beer, Foster, and Siddall (1971) can be generalized as

$$k_iP \rightarrow k_i(P_{CO_2} + P_{H_2O} + P_{CO}) + k_{HC_i}P_{HC}$$

$$(3.10.21)$$

Appropriate values of $b_{\varepsilon,i,1}, b_{\varepsilon,i,2}, k_i$, and k_{HC} for gaseous fuels and oils are provided in Table D.9, Appendix D. A one-clear 2-gray gases model similar to Truelove (1976) is presented, as well as the typical consideration of a one-clear 3-gray gases model. To extend the range of applicability for $T > 2400$ K, the coefficients formulated in Coppalle and Vervisch (1983) are also utilized as for the WSGGM of Smith et al. (1982).

For appreciably small distance L—that is, $L \leq 10^{-4}$ m—it can be shown that the change of radiation intensity in the WSSGM is identical to the gray gas assumption, independent of L, with the absorption coefficient:

$$\overline{K}_{a,g} = a_{\varepsilon,i}k_iP$$

$$(3.10.22)$$

For large distance where L is much greater than 10^{-4} m, the Beer-Lambert's law should be used instead, which is given by

$$\overline{K}_{a,g} = -\frac{1}{L}\ln(1 - \varepsilon_T) \tag{3.10.23}$$

For most practical purposes, equation (3.10.23) assumes that the absorptivity is taken to be equal to the emissivity. Such simplification is commonly adopted, and this assumption is fully justified if the medium is not optically thin and the temperature does not differ considerably from the gas temperature.

As exemplified in the previous section, the overall absorption coefficient of the presence of soot and CO_2-H_2O mixture can also be evaluated by summing the respective mean coefficients

$$\overline{K}_a = \overline{K}_s + \overline{K}_{a,g} \tag{3.10.24}$$

where the soot absorption coefficient can be calculated based on the expression from Kent and Honnery (1990) as

$$\overline{K}_s = 1862 f_v T \tag{3.10.25}$$

Alternatively, the overall absorption coefficient of soot and CO_2-H_2O mixture can also be expressed according to WSSGM, as suggested by Smith et al. (1987) by

$$\overline{K}_a = \sum_{j=1}^{J} \sum_{i=0}^{I} k_{i,j} a_{\varepsilon,i,j} \tag{3.10.26}$$

where the mixture absorption coefficient $k_{i,j}$ for the i, j gray gas assumption is evaluated from

$$k_{i,j} = k_{s,j} + k_{g,i} P \tag{3.10.27}$$

The mixture emissivity weighting factor is given by

$$a_{\varepsilon,i,j} = a_{s,j} a_{\varepsilon,i} \tag{3.10.28}$$

The gas emissivity factors $a_{s,j}$ can either be determined through equations (3.10.19) or (3.10.20), while the soot weighting factors $a_{s,j}$ are determined from third-order temperature polynomial function as

$$a_{s,j} = \sum_{k=1}^{K} b_{s,j,k} T^{k-1} \tag{3.10.29}$$

where the polynomial coefficients $b_{s,j,k}$ are given in Table D.8. The WSSGM considered for the soot and CO_2-H_2O mixture herein consists of a one-clear, 2-gray gases for the soot and 2-gray or 3-gray gases for CO_2-H_2O mixture.

3.10.3 Other Models

The simplifying assumption of adopting the gray gas assumption and a more sophisticated weighted sum of gray gases model is that the participating medium is usually taken to be homogeneous. For the inhomogeneous effects on radiative heat transfer in high temperature combustion gases, more complete models such as the statistical narrow-band or exponential wide-band models are necessary to accurately predict particularly the radiance emanating from non-isothermal, variable concentration carbon dioxide, and water vapor mixture.

A medium containing more than one component such as soot and CO_2-H_2O mixture, can be characterized by a spectral absorption, which is the sum of the spectral coefficients of each component. Similar to the mean absorption coefficient expression of equation (3.10.24), the spectral mixture coefficient can also be written as

$$K_{a,\lambda} = K_{s,\lambda} + K_{g,\lambda} \tag{3.10.30}$$

or in terms of the mixture transmissivity, which is simply the product of individual transmissivities

$$\tau_{a,\lambda} = \tau_{s,\lambda}\tau_{g,\lambda} \tag{3.10.31}$$

In line with band approximation, the mixture transmissivity is thus given as

$$\tau_{a,j} = \tau_{s,j}\tau_{g,j} \tag{3.10.32}$$

For absorbing gases, narrow-band models have been developed to approximate the average behavior of the spectral absorption coefficient in a small spectra interval, which generally lead to analytical expressions of the spectral transmissivity averaged over a spectral range $\Delta\lambda$ but small enough to assume the spectral black body intensity $I_{black,\lambda}$ remains constant inside it. Hundreds of individual absorption lines are contained within this interval; statistical assumptions are made for line locations, shapes, and intensities. Within the Goody statistical model (Goody, 1952), the individual spectral lines are randomly distributed within the interval $\Delta\lambda$. The *mean line-strength-to-spacing parameter S/d* can be found by summing up the contributions to line strength S from each rotational line divided by the distance d. Another important narrow-band parameter to this two-parameter model is the *mean strong-line parameter $1/d$*. For carbon dioxide, $\overline{S/d}$ and $\overline{1/d}$ can be obtained from the modified anharmonic oscillator/rotator model developed by Malkmus (1963a, 1963b). For water vapor, these two parameters can be obtained from tabulated data as a function of wave number and temperature by Ludwig et al. (1973). With the knowledge of the *mean line-strength-to-spacing parameter* and *mean strong-line parameter*, the Goody model yields only the transmissivity for an isothermal and homogeneous composition path and for Lorentz line shapes for each absorbing gas. The Curtis-Godson approximation (Goody, 1964) replaces $\overline{S/d}$ and $\overline{1/d}$ with suitable averages over the inhomogeneous composition path. It leads to accurate results for moderate line-width variations along the path (Soufiani et al., 1985, Young, 1977). For CO_2-H_2O mixture in the same spectral region, the mixture transmissivity $T_{g,\lambda}$ is given by the product of individual transmissivities, since spectra of different gases are not correlated. Accurate predictions require the narrow band width to have a width of about 25 cm^{-1}. The covered spectral range is between 150 cm^{-1} to 7000 cm^{-1} of which the total intensity calculation for CO_2-H_2O mixture requires scanning over about 300 narrow-band regions. It is not entirely surprising that the complexity of narrow-band models, and the large computational effort that they require, do not make them very attractive for engineering applications. Nevertheless, they are useful for model validation purposes. The reader may wish to obtain the computer code developed by Grosshandler (1993) from National Institute of Standards and Technology (NIST) called RADCAL, which exemplifies the essential features of the narrow-band model for radiation calculations in a combustion environment.

Wide-band models significantly reduce the inefficient spectral calculations from 300 down to a total of about 30 bands or less. Figure 3.39 shows a schematic illustration of a sample approximation made to the spectral absorptivities by the wide-band model, in comparison to the accurate narrow-band model. The radiative properties of gaseous species $T_{g,j}$ over the bandwidth $\Delta\lambda_j$ are determined using the exponential wide-band model of Edwards (1976). This model considers that the absorption and emission of infrared radiation by a particular species is generated in between one and six or eight wide bands that are associated with vibrational modes of energy storage by the species. A large number of spectral lines, which are associated with these rotational modes of energy storage, are deemed to exist within these vibrational bands. In contrast to those of narrow-band models, a detailed knowledge of the position and intensity of these rotational lines is considered to be unimportant in the wide-band model. The band shape is approximated by one of three simple exponential functions, depending upon whether the lower limit, upper limit, or band center wave number is used to prescribe the position of the band. The radiative properties of a given species are then determined by specifying three model parameters: (1) *integrated band intensity*, (2) *spectral line width parameter*, and (3) *bandwidth parameter*. Values of these parameters for carbon dioxide and water vapor can be found in Edwards and Balakrishnan (1973). This model is suitable for non-isothermal gases, and accounts for the effects of temperature and pressure on absorption and emission of radiation, and for the overlap of absorption bands for gaseous species such as the CO_2-H_2O mixture. More specific details on the method can be referred in Edwards (1976).

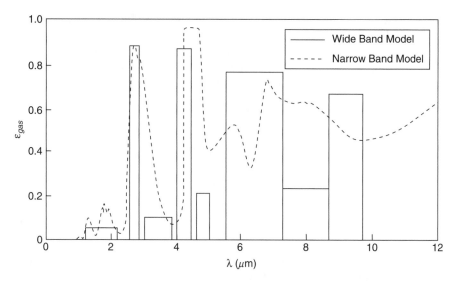

Figure 3.39   Two gas emissivity spectra, one with the wide-band model calculated from 15 bands of which 7 are clear and one with the narrow-band model consisting of 372 spectral regions, for gas temperature of 2000 K, a mean beam length of 2.83 m and partial pressures of CO_2 and H_2O at 0.08 atm and 0.15 atm, respectively (Wieringa et al., 1991).

The wide-band and narrow-band models are based on spectrally averaging an emissivity, absorptivity, or transmissivity, which introduces the need to specify a path length. For complex multi-dimensional fields, this introduces additional approximations of which an obvious choice of path length may not exist. The k-distribution method presents another means of performing spectral calculations more efficiently (Arking and Grossman, 1972, Goody and Yung, 1989). In essence, the method replaces a spectral integration over the wave number with the integration over the absorption coefficient. Here, the treatment of the Planck distribution function and scattering properties is considered constant over a spectral band of interest; a spectrally dependent parameter is ascertained for each specific value of the absorption coefficient. The spectral mean for the band is determined by integrating over the absorption coefficient weighted by the distribution function for the band (or integrating over the cumulative distribution function). Inhomogeneous media are appropriately handled using the related k-distribution (c-k) method (Goody and Yung, 1989, Goody et al., 1989, Lacis and Oinas, 1991).

As aforementioned, the radiation properties of absorbing gases are difficult to evaluate, since they emit and absorb electromagnetic radiation that only occur at wavelengths where the photon energies match the quantum changes in energy of the gas molecules. In comparison, soot tends to vary slowly with wavelength. It can therefore be assumed that the wavelength of radiation is greater than the soot particle diameter, and the complex refractive index of soot is independent of wavelength (Docherty and Fairweather, 1988). Under these assumptions, Mie theory, in the Rayleigh limit of small particles, predicts the scattering to be negligible. The spectral absorption coefficient can be given in the form

$$K_{s,\lambda} = C f_v \lambda \qquad (3.10.33)$$

where C is a constant with a value approximately equal to 7. The mean transmissivity of soot for band j may then be written according to Modak (1979) as

$$\tau_{s,j} = \frac{1}{2} \left[\exp\left(-C f_v \lambda_{j,lower} L\right) + \exp\left(-C f_v \lambda_{j,upper} L\right) \right] \qquad (3.10.34)$$

where $\lambda_{j,lower}$ and $\lambda_{j,upper}$ represent the lower and upper wavelength limits of band j which are related to the bandwidth by $\Delta\lambda_j = \lambda_{j,upper} - \lambda_{j,lower}$.

3.11   Radiation Methods for Field Modeling

Consideration of spectral variations of radiation properties tends to increase the complexity of an already extremely difficult problem. Only the exact *integrodifferential* equation (3.9.4) lends itself to an *a priori* spectral integration, using the narrow-band or wide-band models. Such a solution approach to evaluate the radiation properties quickly becomes intractable for all but the simplest geometries.

In practice, neither narrow-band nor wide-band models are actually required to accurately model the effects of radiative heat transfer in high temperature combustion gases. As a reasonable compromise, the concept of the weighted sum of gray gases approach (see section 3.10.2) as proposed by Modest (1991), can be applied for arbitrary solution methods in radiative transfer. In this method, the non-gray gas is replaced by a number of gray gases. The heat transfer rates are calculated independently, and the total flux is then determined by adding the fluxes of the gray gases after multiplication with certain weighting factors. It can be carried out to any desired accuracy, and since no spectral flux evaluations, followed by spectral integration, are required, computer time savings amount to factors of hundred and even thousands for comparable accuracy. The method may be applied to arbitrary geometries with varying absorption coefficients. Since non-scattering effects can be safely neglected in fires, the limitation of the method to cater for only non-scattering media within a blacked-wall or grayed-wall enclosure, is aptly applicable whenever radiation is considered for field modeling investigations.

According to Modest (1991), the exact integrodifferential equation (3.9.4) after some mathematical manipulation can be replaced by a system of the ith equations of transfer for the ith gray gases. Setting the total intensity as

$$I(\vec{r},\vec{s}) = \sum_{i=0}^{I} I_i(\vec{r},\vec{s}) \tag{3.11.1}$$

the gray intensity I_i satisfies the equation of transfer in the absence of scattering effects:

$$\frac{dI_i(\vec{r},\vec{s})}{ds} = -k_i I_i(\vec{r},\vec{s}) + k_i a_{\varepsilon,i} I_{black}(\vec{r}) \tag{3.11.2}$$

This is, of course, the equation of transfer for a gray gas with constant absorption k_i, with a black body intensity I_{black} (for medium as well as surfaces) replaced by a weighted intensity $a_{\varepsilon,i} I_{black}$. Hence, if the temperature field is known, the intensity field (or fluxes) can be determined for $i = 0, 1, \ldots, I$, using any standard solution methods. The results are then added to yield the total intensity (or radiative flux). The weighting factors $a_{\varepsilon,i}$ corresponding to particular absorption coefficient k_i, can be simply obtained from the various models proposed in section 3.10.2.

The integration of Equation (3.11.2) requires the ray paths originating at the boundaries of the physical domain, which are essentially the initial conditions at these surfaces. Consider, for instance, an interface separating a solid material from the fluid, as shown in Figure 3.40. An incoming radiation onto the wall generally causes an additional heat flux toward the interface and an extra outgoing heat flux associated with the emission of radiation. Denoting the incident radiative heat flux as q_w^- and the outgoing radiative heat flux as q_w^+, the overall energy balance for the interface between the fluid and the solid material requires that

$$q_{solid} + q_w^- = q_{fluid} + q_w^+ \tag{3.11.3}$$

Both the incident and outgoing radiative heat fluxes depend on the intensity of radiation. For most combustion-related radiative transfer problems, the diffuse-gray assumption is the most widely used boundary condition. The outgoing intensity for the ith gray gas can thus be expressed by

$$I_{w,i}^+ = \varepsilon_w a_{\varepsilon,i}(T_w) I_{black}(T_w) + (1 - \varepsilon_w) \int_0^{2\pi} \int_0^{\pi/2} I_{w,i}^-(\theta,\phi) \cos\theta \sin\theta d\theta d\phi$$

$$\tag{3.11.4}$$

where $I_{w,i}^-$ denotes the incident radiative intensity for the ith gray gas, which can be written as an integration over a unit of hemisphere (i.e., over a solid

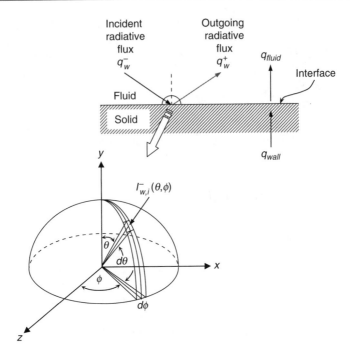

Figure 3.40 Boundary conditions with radiation.

angle of 2π steradians) surrounding the point, as indicated in Figure 3.38. The total outgoing radiative heat flux q_w^+ is simply given by

$$\frac{q_w^+}{\pi} = \sum_{i=1}^{I} I_{w,i} \qquad (3.11.5)$$

with the total incoming radiative heat flux q_w^- evaluated from

$$q_w^+ = (1 - \varepsilon_w)q_w^- + \varepsilon_w E_w = (1 - \varepsilon_w)q_w^- + \varepsilon_w \sigma T_w^4 \qquad (3.11.6)$$

There have been many solution methods that have been developed over the years to handle radiative heat transfer. These include various analytical approximation techniques and numerical methods. We will primarily focus on the latter algorithms that are commonly found in many general-purpose CFD applications, as well as in field modeling: Monte Carlo, P-1 radiation model, discrete transfer radiative model, discrete ordinates model, and finite volume method. Particular emphasis will be given in the proceeding sections to the formulation of non-gray radiative transfer methods incorporating the weighted sum of sum gray gases concept.

3.11.1 Monte Carlo

Monte Carlo methods for radiation heat transfer predictions are essentially purely statistical methods that yield solutions that are as accurate as exact methods. They can be used for any complex three-dimensional and non-Cartesian geometries, some known source of radiation incident on (or emitted within) the geometry, and complicated physical phenomena due to interaction of radiation with the participating medium. Monte Carlo is used in all branches of science and engineering that seeks solutions to complex problems, especially encountered in particle transport applications. For example, photon transport in atmospheric (Davis, 1978, McKee and Cox, 1974) or neutron transport studies (Lewis and Miller, 1984).

There are many different statistical approaches, which Monte Carlo method attains its name from. For the specific model for radiation calculations, the method assumes that the intensity is proportional to the differential angular flux of photons and the radiation field can be viewed as a photon gas. In its simplest form, the model consists of simulating a finite number of photon histories (energy bundles) through the use of a random number generator. For each photon history, random numbers between 0 and 1 are generated to determine the emission location and direction, and they are employed to sample appropriate probabilistic distributions for the distance traveled (path lengths) between collisions. In this section, the Monte Carlo method developed by Taniguchi et al. (1988) to specifically treat the radiative transfer of gray gas and particles in enclosures is presented. For a more in-depth understanding on the basic formulations of the Monte Carlo procedure for radiation calculations, the reader is strongly encouraged to refer to Siegel and Howell (2002) and Modest (2003).

The method generally begins by initially assigning each photon history a set of values, its initial position, energy, and angle. A photon history can be selected from either emission from boundary surface elements or volume elements within the media. For an example of a typical energy bundle emitted from surface element, as shown in Figure 3.41, the cone θ and circumferential ϕ angles for a diffuse emission (Siegel and Howell, 2002), assuming no dependence on the circumferential angle, are given as

$$\phi = 2\pi R_\phi \tag{3.11.7}$$

$$1 + \cos^2\theta = R_\theta \Rightarrow \theta = \cos^{-1}\sqrt{1 - R_\theta} \tag{3.11.8}$$

where R_θ and R_ϕ are chosen random. For the gray gas assumption, equations (3.11.7) and (3.11.8) are combined to yield Lambert's cosine law for the energy emitted from a small gray surface element dA in the direction of (θ, ϕ) within a solid angle $d\Omega$ as

$$d^2 Q_{w,emitted} = I_w^+ dA \cos\theta d\Omega \tag{3.11.9}$$

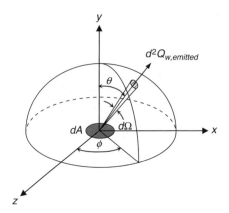

Figure 3.41   Radiative energy emitted from elemental area dA on a solid wall.

with the corresponding wall intensity I_w^+ given by

$$I_w^+ = \frac{q^+}{\pi} = \frac{\varepsilon_w \sigma T_w^4}{\pi} \tag{3.11.10}$$

Out of the radiative energy incident upon dA, only $\varepsilon_w dQ_{w,incident}$ is absorbed by the wall, while the remaining $(1 - \varepsilon_w)dQ_{w,incident}$ is reflected. The energy reflected in the direction of (θ, ϕ) within a solid angle $d\Omega$ can be expressed as

$$d^2Q_{w,reflected} = \frac{1}{\pi}(1 - \varepsilon_w)dQ_{wi}\cos\theta d\Omega \tag{3.11.11}$$

For a participating medium with soot particles, the temperature of the particles is generally assumed equal to that of the surrounding gas. This assumption is valid for small particles, which implies a large surface area-e ratio and smaller radiative transfer to the particles. The radiative energy emitted from a small volume element dV containing gas and particles in the direction of (θ, ϕ) within a solid angle $d\Omega$ as shown in Figure 3.42 is

$$d^2Q_{emitted} = \frac{1}{4\pi}\left[4\overline{K}_a\sigma T^4 dV - \frac{4}{3}\pi R^3 N_s \sigma T^4 dV + \varepsilon_{soot}4\pi R^2 N_s \sigma T^4 dV\right]d\Omega \tag{3.11.12}$$

where R is the particle radius and N_s is the particle number density.

The path of each photon history is tracked through the system by the attenuation of the intensity I of an incident radiation along the penetration distance s through a gas volume, which contains dispersed particles due to the absorption of the gas and particles, which can be expressed as

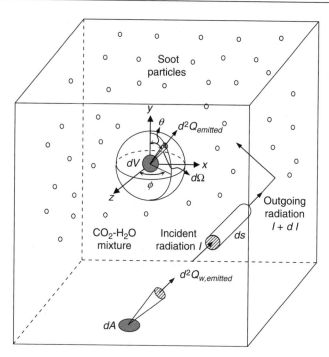

Figure 3.42 Radiative energy emitted from gas elemental volume dV within an enclosure containing a participating medium with soot particles.

$$dI = -\overline{K}_a I ds \tag{3.11.13}$$

The total absorption coefficient can be determined by

$$\overline{K}_a = \overline{K}_{a,g}\left(1 - \frac{4}{3}\pi R^3 N_s\right) + \varepsilon_{soot}\pi R^2 N_s \tag{3.11.14}$$

Integrating equation (3.11.13) yields the usual Beer's law

$$I(s) = I(0)e^{-\overline{K}_a s} \tag{3.11.15}$$

where $I(0)$ is the initial intensity. The penetration distance s is the length along which an energy particle travels before its extinction (absorption by gas and/or solid particle surface); it is determined by

$$s = -\ln\frac{(1 - R_s)}{\overline{K}_a} \tag{3.11.16}$$

which satisfies Beer's law of equation (3.11.15). The variable R_s denotes the random number for the penetration distance s. During its course of travel, a random number R_a is employed to determine whether it is absorbed before colliding with a wall. If $R_a > 0$, the photon history is absorbed by the gas and/or particle at the end of its penetration distance, and the pursuit of the photon history is terminated. However, if $R_a \leq 0$, the new penetration distance is determined by equation (3.11.16). Eventually, the photon history will strike another surface within the domain. A random number R_r is also prescribed to determine whether the photon history is absorbed or reflected. If $R_r > \varepsilon_w$, the photon history is taken to be absorbed by the wall and the pursuit of the photon history is thereby terminated. However, if $R_r \leq \varepsilon_w$, the photon history is reflected in the direction determined by equations (3.11.7) and (3.11.8). The new penetration distance from the reflected point is obtained by subtracting the length along which the photon history has already traveled from the original penetration distance. Following the termination of the photon history in whether it is absorbed in the gas volume or surface element, the procedure is repeated for successive collisions with the new set of assigned energy, position, and direction.

Monte Carlo calculations provide the necessary means of attaining appropriate radiation quantities such as the surface heat flux and volumetric sources or sinks, by tracking the energy absorbed or emitted within surface elements and volumes. The total energy absorbed and emitted by surfaces and volumes are usually recorded during the simulation. Additional information such as penetration distances could also be obtained and stored. The radiation source term for a control volume ΔV and surface heat flux for a surface element ΔA may be evaluated from

$$S_{rad} = \frac{1}{\Delta V} \left(\overset{\substack{absorbed \\ photon\ bundles}}{\sum} Q_{absorbed} - \overset{\substack{emitted \\ photon\ bundles}}{\sum} Q_{emitted} \right) \qquad (3.11.17)$$

$$q_s = \frac{1}{\Delta A} \left(\overset{\substack{absorbed \\ photon\ bundles}}{\sum} Q_{absorbed} - \overset{\substack{emitted \\ photon\ bundles}}{\sum} Q_{emitted} \right) \qquad (3.11.18)$$

where S_{rad} is the net radiative source or sink term per unit volume incorporated into the energy transport equation, and q_s is the net radiative heat flow of each surface element, which is required for the evaluation of the wall temperature. It should be noted that for the gas element

$$Q_{absorbed} = (1 - \alpha) 4 \overline{K}_a \sigma T^4 \Delta V \qquad (3.11.19)$$

while for the wall element

$$Q_{absorbed} = (1 - \alpha)\varepsilon_w \sigma T_w^4 \Delta A \qquad (3.11.20)$$

where α is the self-absorption ratio which represents the ratio of energy absorbed by the element itself to the total energy emitted by the element. Hence, $Q_{absorbed}$ in both equations (3.11.19) and (3.11.20) represents the total energy emitted from an element and absorbed by other elements.

In a majority of applications, numerous photon histories are generated to obtain estimates that closely reflect the physical quantities in the system. As the number of photons initiated from each surface or volume element increases, this method is expected to converge to the exact solution of the problem. It should nevertheless be noted that the directions of the photons are ascertained from a random number generator; hence the method is always subjected to statistical errors and lacks guaranteed convergence. Also, radiative heat transfer, by nature, is a three-dimensional phenomenon. Hence, the pursuit of photon histories must always be performed in three dimensions, even under one-dimensional or two-dimensional boundary conditions. It should be noted that the Monte Carlo method can be easily extended to handle non-gray gas problems. A Monte Carlo simulation is hardly affected by the number of gray gases in the case of applying the WSGGM, since the spectrum is just another independent parameter to be sampled. Hence, increasing the problem complexity leads to only gradual increase in the complexity of the Monte Carlo method and similar gradual increases in computer time.

3.11.2 P-1 Radiation Model

The spherical harmonics P-N differential approximation is one of the most tedious and cumbersome of the radiative transfer approximations. In spite of this, it is considered the most elegant approach to handle radiative energy transfer problems, because of its sound mathematical foundations. The method was first suggested by Jeans (1917) in connection with studies of certain problems in astrophysics.

In the spherical harmonics approximation, the radiation intensity is expressed by a series of spherical harmonics according to Case and Zweifel (1967) as

$$I(x, y, z, \theta, \phi) = \sum_{n-0}^{N} \sum_{m=-n}^{n} A_n^m(x, y, z) Y_n^m(\theta, \phi) \qquad (3.11.21)$$

where $Y_n^m(\theta, \phi)$ are the spherical harmonics given by

$$Y_n^m(\theta, \phi) = (-1)^{(m+|m|)/2} \left[\frac{2n+1}{4\pi} \frac{n-|m|!}{n+|m|!} \right]^{1/2} P_n^{|m|}(\cos\theta) e^{im\phi} \qquad (3.11.22)$$

and P_n^m are the associated Legendre polynomials, which are related to the Legrende polynomials. The upper limit N for the index n denotes the order of approximation of the method. Exact solution of the radiative transfer equation is obtained if N is taken as infinity. The approximation occurs when the series is truncated. When the distribution of intensity of equation (3.11.21) is substituted into the radiative transfer equation, an infinite set of differential equations involving A_n^m is obtained. For most practical calculations, a finite N is assigned. If $N = 1$, the first order P-1 spherical harmonic approximation, equivalent to the Eddington approximation (1988), is attained. If only four terms are used, this particular approximation comprises of a single diffusion-type partial differential equations for irradiance J (the spectral zero*th*-order moment of intensity I_0) for the mean gray gas absorption coefficient $\overline{K}_a$. In the absence of any scattering, it can be expressed as

$$\frac{\partial}{\partial x_j}\left(\frac{1}{3\overline{K}_a}\frac{\partial J}{\partial x_j}\right) = \overline{K}_a(4\pi I_{black} - J) \tag{3.11.23}$$

On the basis of Marshak's formulation (Ozisik, 1973), the boundary condition of the irradiance J for a diffuse gray surface without angular independence takes the following expression

$$\frac{1}{3\overline{K}_a}\frac{\partial J}{\partial n} = \pm\frac{\varepsilon_w}{2(2-\varepsilon_w)}(J - 4\pi I_{black}) \tag{3.11.24}$$

The radiation source term S_{rad} in the energy equation and the surface heat flux q_s can be directly evaluated from

$$S_{rad} = \frac{\partial}{\partial x_j}\left(\frac{1}{3\overline{K}_a}\frac{\partial J}{\partial x_j}\right) \tag{3.11.25}$$

$$q_s = \frac{1}{3\overline{K}_a}\frac{\partial J}{\partial n} \tag{3.11.26}$$

Fusegi and Farouk (1989) considered the spectrally dependent radiation of real gases by a non-gray P-1 approximation, by the incorporation of WSSGM. The original differential equation is replaced with a set of differential equations for the irradiance associated with each gray gas component. For the CO_2-H_2O mixture represented by a one-clear 3-gray gases model, this involves obtaining the solutions to additional three differential equations for the irradiance (only the gray gas irradiance components are solved). Equation (3.11.23) can be subsequently altered as

$$\frac{\partial}{\partial x_j}\left(\frac{1}{3k_i}\frac{\partial J_i}{\partial x_i}\right) = k_i\left(4a_{\varepsilon,i}\pi I_{black} - J_i\right) \quad i = 1,2,3 \tag{3.11.27}$$

The boundary condition for the irradiance for each gray gas is evaluated according to

$$\frac{1}{3k_i}\frac{\partial J_i}{\partial n} = \pm\frac{\varepsilon_w}{2(2-\varepsilon_w)}\left(J_i - 4a_{\varepsilon,i}\pi I_{black}\right) \quad i = 1, 2, 3 \tag{3.11.28}$$

with the radiation source term S_{rad} in the energy equation, and the surface heat flux q_s determined by

$$S_{rad} = \sum_{i-1}^{3}\frac{\partial}{\partial x_j}\left(\frac{1}{3k_i}\frac{\partial J_i}{\partial x_j}\right) \tag{3.11.29}$$

$$q_s = \sum_{i-1}^{3}\frac{1}{3k_i}\frac{\partial J_i}{\partial n} \tag{3.11.30}$$

For the soot and CO_2-H_2O mixture, the consideration of additional two gray gases for soot according to the expression in equation (3.10.26), now requires a total of six differential equations to be solved for the irradiance for the model comprising of one-clear, 2-gray gases for soot, and 3-gray gases for CO_2-H_2O mixture, which significantly increases the computational burden.

The P-1 approximation is very accurate if the optical dimension of the medium is large (i.e., greater than 2). However, it yields inaccurate results for thinner media particularly near the domain boundaries. If the radiation field is highly anisotropic where large temperature gradients exist in the medium, the P-1 approximation becomes less reliable in predicting the radiation transfer. Nonetheless, if the approximation of the spherical harmonics is expanded to the third order—that is, $N = 3$—the P-3 approximation can yield accurate results for an optical dimension as small as 0.5 (Ratzel and Howell, 1983) and for anisotropic fields. Naturally, the P-3 approximation results in additional equations, and they are usually more complicated than those of the P-1 approximation. For axisymmetric cylindrical geometry, there are four second order elliptic partial differential equations to be solved (Menguc and Viskanta, 1986), whereas for three-dimensional rectangular enclosures, six equations, are needed for the gray gas assumption. In spite of its superior accuracy, the P-3 approximation greatly suffers at the expense of the additional computational burden and becomes impractical if extended to WSSGM; hence, it has not enjoyed much success in practical applications when compared to the simpler P-1 approximation. More detailed descriptions and formulations regarding the P-3 approximation can be found in Menguc and Viskanta (1986).

3.11.3 Discrete Transfer Radiative Model

The discrete transfer radiative method (DTRM) proposed by Lockwood and Shah (1981) is principally built on the concept of solving representative rays in a radiating enclosure. To some extent, it closely resembles the Monte Carlo method, but the directions of the rays are now pre-specified in advance rather than being chosen at random. The rays are solved for only along paths between two boundary walls, rather than being partially reflected at walls and tracked to extinction.

Consider the hemisphere about the center point P on the surface element, as illustrated in Figure 3.43. Taking for an instance that the hemisphere can be divided into four equal segments, four representative rays may be visualized to impinge on point P from points Q_1, Q_2, Q_3, and Q_4 on the far walls of the enclosure. The intensities arriving at P are assumed constant over the

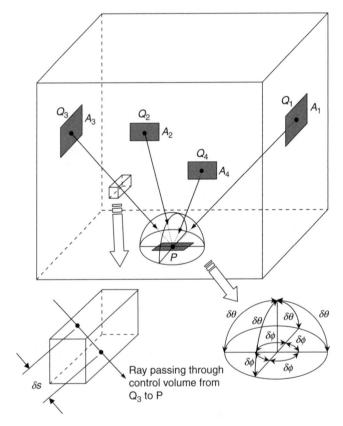

Figure 3.43 Illustration of the discrete transfer radiative method.

solid angles, and they are determined by the fundamental equation for the transfer of thermal radiation, assuming a homogeneous gas mixture, which is expressed by

$$\frac{dI}{ds} = -\overline{K}_a I + \overline{K}_a \frac{\sigma T^4}{\pi} \tag{3.11.31}$$

Integrating analytically, the preceding equation yields the recurrence relation:

$$I^{n+1} = I^n e^{-\overline{K}_a \delta s} + \frac{\sigma T^4}{\pi} (1 - e^{-\overline{K}_a \delta s}) \tag{3.11.32}$$

where δs is the distance traveled by the ray within the control volume, and I^n and I^{n+1} are respectively the radiation intensities entering and leaving the control volume as described in Figure 3.28. The recurrence relation as described in equation (3.11.32) is successively applied from a point Q_m to P_m to determine I^{n+1} from I^n for all the control volumes intercepted by the radiation rays.

The initial intensity value of each ray leaving a wall for a diffuse-gray surface, can be evaluated by equation (3.11.4) for a single gray gas as

$$I_w^+ = \varepsilon_w \frac{\sigma T_w^4}{\pi} (1 - \varepsilon_w) \underbrace{\int_0^{2\pi} \int_0^{\pi/2} I_w^-(\theta, \phi) \cos\theta \sin\theta d\theta d\phi}_{q^-} \tag{3.11.32}$$

In equation (3.11.32), the incident radiative flux q_w^- can be evaluated by the summation over all finite solid angles ($\delta\Omega$), which in discrete form is

$$q_w^- = \sum_{N_R} I_w^-(\theta, \phi) \cos\theta \delta\Omega = \sum_{N_R} I_w^-(\theta, \phi) \cos\theta \sin\theta \sin\delta\theta\delta\phi \tag{3.11.34}$$

where N_R denotes the number of rays arriving at the surface element. According to Shah (1979), the hemisphere can be discretised into N_θ equal polar angles and N_ϕ azimuthal angles according to:

$$\delta\theta = \frac{\pi}{2N_\theta} \quad \delta\phi = \frac{2\pi}{N_\phi} \tag{3.11.35}$$

The total 2π hemispherical solid angle on a surface element is the sum of all finite angles such that $N_\Omega = N_\theta \times N_\phi$. The values of $N_\theta = 1$ and $N_\phi = 4$ giving a total of 4 rays that have been arbitrarily chosen for the convenience of illustrating the DTRM in Figure 3.43, is obviously the coarsest angular discretisation (number of rays passing through the centriod of the spherical surface) that can be adopted in a three-dimensional domain. In practice, it is rather common to adopt a total of 8 rays or 16 rays spanning the hemisphere. It should be noted that the accuracy of the DTRM also depends on the surface discretisation (number of surface elements) in addition to the angular discretisation. With an adequately refined mesh and a sufficiently large number of rays, DTRM has shown to produce results that are rather comparable in accuracy with Monte Carlo solutions.

According to the boundary condition described in equation (3.11.6), the outgoing radiative flux q_w^+ depends on the value of the incident radiative flux q_w^-. Except for the case of black walls, the solution procedure is iterative. The procedure could begin by evaluating the initial intensity leaving the wall at the originating surface element by

$$I_w^+ = \frac{\varepsilon_w \sigma T_w^4}{\pi} \tag{3.11.36}$$

Estimates of the incident intensities will be thereafter made available after the first iteration to calculate the appropriate intensities leaving the boundary surfaces through equation (3.11.34). The process is repeated until successive values of the incident radiative flux converges within a specified limit. For each surface element, the net radiative heat flux can thus be ascertained by $q_s = q_w^+ - q_w^-$. The net gain or loss of radiation energy in a control volume constitutes the radiative source term, which is required for the energy conservation equation. With reference to Figure 3.43, this can be easily determined once all the PQ_1, PQ_2, PQ_3, and PQ_4 ray paths are known. The passage of the rays crossing the control volumes from all points of Q_m to P, in consideration of the definition of intensity, may be evaluated from

$$\begin{aligned}
S_{rad,m} &= (I^{n+1} - I^n) A_m \cos \theta_m \delta \Omega_m \\
&= (1^{n+1} - I^n) A_m \cos \theta_m \sin \theta_m \sin \delta \theta_m \delta \phi_m
\end{aligned} \tag{3.11.38}$$

where A_m is the area of the surface element from which the ray is emitted at Q_m. Overall energy gain or loss in the control volume is due to the intensity change for all rays that happened to traverse the specific finite control volume. For all N rays that traverse the control volume, the total radiation source is given by summing the individual source contributions of equation (3.11.38) divided by the finite size of the control volume as

$$S_{rad} = \frac{1}{\Delta V} \sum_{m-1}^{N} S_{rad,m} \tag{3.11.39}$$

DTRM extends readily to non-gray calculations, for example, based upon the WSSGM, the recurrence relation in equation (3.11.32) can be replaced by a system of i equations of transfer for i gray gases

$$I_1^{n+1} = I_i^n e^{-k_i, \delta s} + \frac{a_{\varepsilon,i}\sigma T^4}{\pi}(1 - e^{-k_i, \delta s}) \tag{3.11.40}$$

which is subject to the boundary condition for the outgoing wall intensity as

$$I_{w,i}^+ = \varepsilon_w a_{\varepsilon,i}(T_w)I_{black}(T_w) + (1 - \varepsilon_w)\sum_{N_R} I_{w,i}^-(\theta, \phi)\cos\theta\sin\theta\sin\delta\theta\delta\phi \tag{3.11.41}$$

The total radiation source attributed by the transverse rays is now determined according to

$$S_{rad} = \frac{1}{\Delta V}\sum_{i=1}^{M}\sum_{m=1}^{N} S_{rad,m,i} \tag{3.11.42}$$

where the individual source contribution for each gray gas is given by

$$\begin{aligned} S_{rad,m,i} &= (I_i^{n+1} - I_1^n)A_{m,i}\cos\theta_{m,i}\delta\Omega_{m,i} \\ &= (I_i^{n+1} - I_i^n)A_{m,i}\cos\theta_{m,i}\sin\theta_{m,i}\sin\delta\theta_{m,i}\delta\phi_{m,i} \end{aligned} \tag{3.11.43}$$

3.11.4 Discrete Ordinates Model

A discrete ordinate approximation to the radiative transfer equation, as the terminology suggests, is obtained by discretising the entire solid angle ($\Omega = 4\pi$) using a finite number of ordinate directions and corresponding weight factors. Originally proposed by Chandrasekhar (1960) for astrophysical problems, discrete ordinates method has enjoyed much success in applications to problems of neutron transport (Lathrop, 1976, Lewis and Miller, 1984).

Principally, the integrodifferential equation is solved for a set of directions spanning the solid angle around a point in space, and the integrals over solid angles are approximated using Gaussian or Lobatto quadratures. These are also called S_N-approximations to symbolize the discrete ordinates approximations in which there are N discrete values of positive and negative direction cosines ξ_n, μ_n, η_n, which always satisfy the identity $\xi_n^2 + \eta_n^2 + \mu_n^2 = 1$. Assuming a single gray assumption for the purpose of illustration, equation (3.11.2) may be written for each quadrature point n in Cartesian co-ordinates as:

$$\xi_n\frac{\partial I^n}{\partial x} + \mu_n\frac{\partial I^n}{\partial y} + \eta_n\frac{\partial I^n}{\partial z} = -\overline{K}_a I^n + \overline{K}_a\frac{\sigma T^4}{\pi} \tag{3.11.44}$$

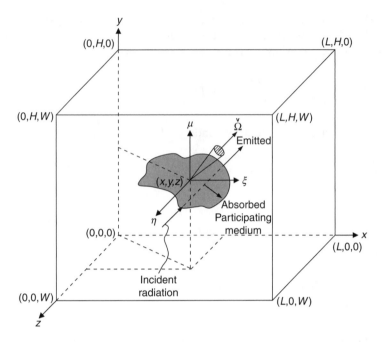

Figure 3.44 Illustration of the discrete ordinates method within a three-dimensional rectangular geometry.

Let us consider for the purpose of illustrating, this particular method via the three-dimensional rectangular enclosure containing a participating medium, as depicted in Figure 3.44. Solution of equation (3.11.44) requires boundary conditions at the wall as well as the temperature of the medium. By assuming the surroundings walls to be diffusely emitting-reflecting surfaces, the boundary conditions are

$$\text{at } x = 0: \quad I_w^n = \frac{\varepsilon_w \sigma T_w^4}{\pi} + \frac{(1 - \varepsilon_w)}{\pi} \sum_{\substack{n' \\ \xi_{n'} < 0}} w_{n'} |\xi_{n'}| I_w^{n'} \quad \text{for} \quad \xi_n > 0$$

$$\text{at } x = \text{L}: \quad I_w^n = \frac{\varepsilon_w \sigma T_w^4}{\pi} + \frac{(1 - \varepsilon_w)}{\pi} \sum_{\substack{n' \\ \xi_{n'} > 0}} w_{n'} |\xi_{n'}| I_w^{n'} \quad \text{for} \quad \xi_n < 0$$

$$\text{at } y = 0: \quad I_w^+ = \frac{\varepsilon_w \sigma T_w^4}{\pi} + \frac{(1 - \varepsilon_w)}{\pi} \sum_{\substack{n' \\ \mu_{n'} < 0}} w_{n'} |\mu_{n'}| I_w^- \quad \text{for} \quad \mu_n > 0$$

at $y = W$: $\quad I_w^+ = \dfrac{\varepsilon_w \sigma T_w^4}{\pi} + \dfrac{(1 - \varepsilon_w)}{\pi} \displaystyle\sum_{\substack{n' \\ \mu_{n'} > 0}} w_{n'} |\mu_{n'}| I_w^- \quad$ for $\quad \mu_n < 0$

at $z = 0$: $\quad I_w^+ = \dfrac{\varepsilon_w \sigma T_w^4}{\pi} + \dfrac{(1 - \varepsilon_w)}{\pi} \displaystyle\sum_{\substack{n' \\ \mu_{n'} < 0}} w_{n'} |\eta_{n'}| I_w^- \quad$ for $\quad \eta_n > 0$

at $z = H$: $\quad I_w^+ = \dfrac{\varepsilon_w \sigma T_w^4}{\pi} + \dfrac{(1 - \varepsilon_w)}{\pi} \displaystyle\sum_{\substack{n' \\ \eta_{n'} > 0}} w_{n'} |\eta_{n'}| I_w^- \quad$ for $\quad \eta_n < 0$

In the preceding equations, the values n and n' denote outgoing and incoming directions, respectively. Direction cosines ξ_n, μ_n, η_n, and associated weights w_n for basic discrete ordinates approximations for one octant according to Jamaluddin and Smith (1988), have been ascertained and they are tabulated in Table D.10, Appendix D. The different S_N-approximations considered which are S_2, S_4, S_6, and S_8 based on the relationship $n = N(N + 2)$ correspond to 8, 24, 48, and 80 permutations.

The most basic discrete ordinate approximation is S_2. Only one direction is represented in an eighth of sphere, as illustrated in Figure 3.45a. Usually, the symmetric representation is adopted for S_2. An improved approximation that is S_4 comprises of three principal directions in one-eighth of a sphere, as shown in Figure 3.45b. For higher-order approximations such as S_6 and S_8, they now contain six and ten directions spanning one-eighth of a sphere, which are schematically described in Figures 3.45c and 3.45d, respectively. In order to obtain the other directions in other octants of the sphere, appropriate negative and positive values can be obtained by reflection. The angular discretisation represents the solid angle subtended by a sphere. Hence, the sum of weights to the surface area of the unit sphere is given by

$$\sum_{n=1}^{N} w_n = 4\pi$$

The preceding expression can be easily verified for the respective S_2, S_4, S_6, and S_8 approximations given in Table D.10.

A discretised form of equation (3.11.44) can be obtained by integrating over the control volume shown in Figure 3.46, applying the usual finite volume approximations as follows

$$\xi_n(A_e I_e^n - A_w I_w^n) + \mu_n(A_n I_n^n - A_s I_s^n) + \eta_n(A_t I_t^n - A_b I_b^n)$$
$$= -\overline{K}_a I_p^n \Delta V + \overline{K}_a \frac{\sigma T_p^4}{\pi} \Delta V \tag{3.11.45}$$

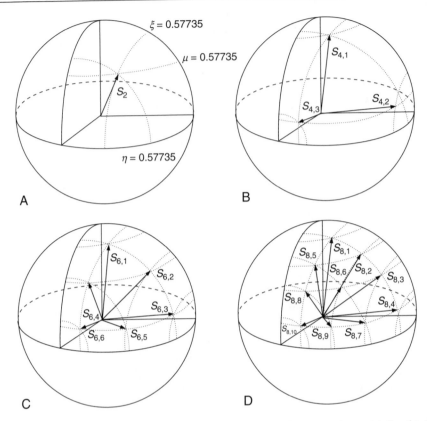

Figure 3.45 Discrete ordinates method in one octant of a unit sphere: (a) S_2, (b) S_4, (c) S_6, and (d) S_8 representations.

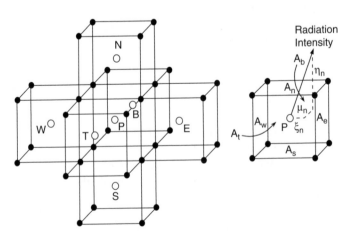

Figure 3.46 Three-dimensional representation of the discrete ordinates method.

The cell edge radiant fluxes can be related to the cell center radiant flux by a spatially weighted approximation that can be written as

$$I_p^n = (1 - \omega_x)I_w^n + \omega_x I_e^n = (1 - \omega_y)I_s^n + \omega_y I_n^n = (1 - \omega_z)I_b^n + \omega_z I_t^n$$

$$(3.11.46)$$

For all positive ordinate directions, equation (3.11.46) can be employed to eliminate the intensities I_e^n, I_n^n, and I_t^n, since I_w^n, I_s^n, and I_b^n are generally assumed known from boundary conditions prescribed at the boundary surfaces. Substituting equation (3.11.46) into equation (3.11.45) and solving for the intensity I_p^n yields

$$I_p^n = \frac{\dfrac{|\xi_n|}{\omega_x^n}A_x I_w^n + \dfrac{|\mu_n|}{\omega_y^n}A_y I_s^n + \dfrac{|\eta_n|}{\omega_z^n}A_z I_b^n + \overline{K}_a\left(\dfrac{\sigma T_p^4}{\pi}\Delta V\right)}{\dfrac{|\xi_n|}{\omega_x^n}A_x + \dfrac{|\mu_n|}{\omega_y^n}A_y + \dfrac{|\eta_n|}{\omega_z^n}A_z + \overline{K}_a\Delta V} \qquad (3.11.47)$$

where

$$A_x = \omega_x^n A_w + (1 - \omega_x^n)A_e$$
$$A_y = \omega_y^n A_s + (1 - \omega_y^n)A_n$$
$$A_z = \omega_z^n A_b + (1 - \omega_z^n)A_t$$

Equation (3.11.47) can be used to determine the intensity of the cell I_p^n starting from boundaries west, south, and bottom. The procedure is repeated for all ordinate directions of increasing x, y, and z. If negative direction cosines occur, the direction of the procedure reverses. For example, the process begins from the east, north, and top for all negative ordinate directions. It follows that the subscripts w, s, and b are replaced by e, n, and t, respectively. For non-black walls, iteration is required, since better estimates for the incoming intensities at the boundary surfaces can only be updated after the first sweep of the procedure.

Because equation (3.11.47) represents an extrapolation across a control volume, negative intensities may result when absorption cross sections are large or when inadequate spatial resolution is used, which is caused by the widely adopted diamond difference scheme where

$$\omega_x^n = \omega_y^n = \omega_z^n = 0.5 \qquad (3.11.48)$$

Nevertheless, if the spatial weights are equal to 1.0, the upwind difference scheme is invoked and it always ensures that the intensities are always non-negative. It provides, however, slightly less accurate predictions. An alternative approach to ensure non-physical intensities are the positive scheme as suggested by Kim and Lee (1988), and the exponential scheme by Chai et al. (1994). The former consists of calculating the weighing factors as

$$\omega_x^n = \max(0.5, \omega_x^{n'}), \quad \omega_y^n = \max(0.5, \omega_y^{n'}), \quad \omega_z^n = \max(0.5, \omega_z^{n'}) \quad (3.11.49)$$

The weighting factors of the latter are evaluated according to

$$\omega_x^n = \frac{1}{1 - \exp(-\tau_x)} - \frac{1}{\tau_x}, \quad \omega_y^n = \frac{1}{1 - \exp(-\tau_y)} - \frac{1}{\tau_y},$$

$$\omega_z^n = \frac{1}{1 - \exp(-\tau_z)} - \frac{1}{\tau_z} \tag{3.11.50}$$

where

$$\tau_x = \frac{\overline{K}_a \Delta x}{\xi^n}, \quad \tau_y = \frac{\overline{K}_a \Delta y}{\mu^n}, \quad \tau_z = \frac{\overline{K}_a \Delta z}{\eta^n}$$

It can be easily shown from the preceding that at the optically thin limit— that is, $\tau_x, \tau_y, \tau_z \to 0$, the exponential scheme approaches the diamond difference scheme, since $\omega_x^n, \omega_y^n, \omega_z^n \to 0.5$. At the optically thick limit, the exponential scheme becomes the upwind difference scheme, since $\omega_x^n, \omega_y^n, \omega_z^n \to 1.0$.

Once all the radiation intensities are determined, the wall radiative flux may be calculated as

$$q_w^- = \int_{2\pi} I_w^-(\vec{s} \cdot \vec{n}) d\Omega \tag{3.11.51}$$

For the three-dimensional rectangular enclosure illustrated in Figure 3.43, the surface fluxes are

$$\text{at } x = 0 : q_{w,x}^- = \sum_{\substack{n' \\ \xi_{n'} < 0}} w_{n'} |\xi_{n'}| I^{n'} \quad \text{for} \quad \xi_n > 0$$

$$\text{at } x = L : q_{w,x}^- = \sum_{\substack{n' \\ \xi_{n'} > 0}} w_{n'} |\xi_{n'}| I^{n'} \quad \text{for} \quad \xi_n < 0$$

$$\text{at } y = 0 : q_{w,y}^- = \sum_{\substack{n' \\ \mu_{n'} < 0}} w_{n'} |\mu_{n'}| I^{n'} \quad \text{for} \quad \mu_n > 0$$

at $y = W : q_{w,y}^- = \displaystyle\sum_{\substack{n' \\ \mu_{n'} > 0}} w_{n'}|\mu_{n'}|I^{n'}$ for $\mu_n < 0$

at $z = 0 : q_{w,z}^- = \displaystyle\sum_{\substack{n' \\ \eta_{n'} < 0}} w_{n'}|\eta_{n'}|I^{n'}$ for $\eta_n > 0$

at $z = H : q_{w,z}^- = \displaystyle\sum_{\substack{n' \\ \eta_{n'} > 0}} w_{n'}|\eta_{n'}|I^{n'}$ for $\eta_n < 0$

The radiative source term for the enthalpy equation may be evaluated from

$$S_{rad} = \overline{K}_a \sum_n w_n I^n - 4\overline{K}_a \sigma T^4 \tag{3.11.52}$$

Extending the discrete ordinates model to incorporate the WSGGM, can be easily accommodated by repeating the evaluation of the volume averaged intensity of equation (3.11.47) to additional gray gases for the absorption coefficient k_i via

$$I_{i,p}^n = \frac{\dfrac{\xi_n}{\omega_x^n} A_x I_{i,w}^n + \dfrac{\mu_n}{\omega_y^n} A_y I_{i,s}^n + \dfrac{\eta_n}{\omega_z^n} A_z I_{i,b}^n + a_{\varepsilon,i} K_i \left(\dfrac{\sigma T_p^4}{\pi}\right) \Delta V}{\dfrac{\xi_n}{\omega_x^n} A_x + \dfrac{\mu_n}{\omega_y^n} A_y + \dfrac{\eta_n}{\omega_z^n} A_z + k_i \Delta V} \tag{3.11.53}$$

The outgoing intensity for ith gray gas at the wall, for example at $x = 0$, is now calculated from

$$I_{w,i}^n = \frac{\varepsilon_w a_{\varepsilon,i}\sigma T_w^4}{\pi} + \frac{(1 - \varepsilon_w)}{\pi} \sum_{\substack{n' \\ \xi_{n'} < 0}} w_{n'}|\xi_{n'}|I_{w,i}^{n'}$$

Similar evaluations for the rest of boundary surface intensities are accordingly performed. The radiative source in conjunction with the WSSGM can be appended into the enthalpy equation by the summation of the individual contributions:

$$S_{rad} = \sum_{i=1}^M S_{rad,i} \tag{3.11.54}$$

where

$$S_{rad,i} = k_i \sum_n w_n I_i^n - 4k_i a_{\varepsilon,i} \sigma T^4 \qquad (3.11.55)$$

3.11.5 Finite Volume Method

In retrospect, the finite volume method bears many similarities with the discrete ordinates method. Here, the discretised radiative transfer equation is derived by integrating the ordinary differential equation over each control volume and each solid angle. For the gray gas assumption, it entails

$$\int\limits_{\Omega^n} \int\limits_{\Delta V} \frac{dI^n}{ds} dV d\Omega = \int\limits_{\Omega^n} \int\limits_{\Delta V} \left(-\overline{K}_a I^n + \overline{K}_a \frac{\sigma T^4}{\pi} \right) dV d\Omega \qquad (3.11.56)$$

By applying Gauss' divergence theorem to the volume integral on the left-hand side, the preceding equation can be rewritten as

$$\int\limits_{\Omega^n} \int\limits_{\Delta A} I^n (\vec{s} \cdot \vec{n}) dA d\Omega = \int\limits_{\Omega^n} \int\limits_{\Delta V} \left(-\overline{K}_a I^n + \overline{K}_a \frac{\sigma T^4}{\pi} \right) dV d\Omega \qquad (3.11.57)$$

where $\vec{n}$ is the unit normal vector of the control volume face. Assuming that the radiation intensity is constant on each of the control volume faces, the surface integral can be approximated by the sum over the control volume faces. It is also further assumed that the intensity is constant within the control volume and over the finite solid angle $\Delta \Omega^n$; equation (3.11.53) can thus be expressed in discrete form according to

$$\sum_{m=1}^{6} A_m I_m^n \int\limits_{\Omega^n} (\vec{s} \cdot \vec{n}) d\Omega = \left(-\overline{K}_a I_p^n + \overline{K}_a \frac{\sigma T^4}{\pi} \right) \Delta V \Delta \Omega^n \qquad (3.11.58)$$

In the discrete ordinates method, the direction $\vec{s}$ is taken as constant within the solid angle $\Delta \Omega^n$. However, this is not the case in the finite volume method where it changes following the variation of the polar angle θ and the azimuthal angle ϕ, as described by the coordinate system used to discretise the solid angle in Figure 3.42. In order to ensure positive intensities, the cell boundary intensities can be calculated by adopting the upwind difference scheme. For a set of discrete directions that span over the total solid angle of 4π, the equation for the volume averaged central point intensity of a three-dimensional rectangular control volume is given by

$$I_p^n \frac{A_w I_w^n |D_x^n| + A_s I_s^n |D_y^n| + A_b I_b^n |D_z^n| + \overline{K}_a \left(\frac{\sigma T_p^4}{\pi}\right) \Delta V \Delta \Omega^n}{A_x |D_x^n| + A_y |D_y^n| + A_z |D_z^n| + \overline{K}_a \Delta V \Delta \Omega^n} \quad (3.11.59)$$

where

$$\Delta \Omega^n = \int_{\Omega^n} d\Omega = \int_{\Delta\phi} \int_{\Delta\theta} \sin\theta \, d\theta d\phi \quad (3.11.60)$$

and

$$|D_x^n| = \int_{\Omega^n} (\vec{s}^n \cdot \vec{i}) d\Omega = \int_{\Delta\phi} \int_{\Delta\theta} (\vec{s}^n \cdot \vec{i}) \sin\theta d\theta d\phi = \int_{\Delta\phi} \int_{\Delta\theta} \cos\phi \sin\theta \sin\theta d\theta d\phi$$

$$(3.11.61)$$

$$|D_y^n| = \int_{\Omega^n} (\vec{s}^n \cdot \vec{j}) d\Omega = \int_{\Delta\phi} \int_{\Delta\theta} (\vec{s}^n \cdot \vec{j}) \sin\theta d\theta d\phi = \int_{\Delta\phi} \int_{\Delta\theta} \sin\phi \sin\theta \sin\theta d\theta d\phi$$

$$(3.11.62)$$

$$|D_z^n| = \int_{\Omega^n} (\vec{s}^n \cdot \vec{k}) d\Omega = \int_{\Delta\phi} \int_{\Delta\theta} (\vec{s}^n \cdot \vec{k}) \sin\theta d\theta d\phi = \int_{\Delta\phi} \int_{\Delta\theta} \cos\theta \sin\theta d\theta d\phi$$

$$(3.11.63)$$

In equations (3.11.57), (3.11.58), and (3.11.59), $\vec{i}, \vec{j}$, and $\vec{k}$ represent the base vectors of the Cartesian coordinate system. It is noted that the discrete ordinates equation employing the upwind difference scheme would be recovered from equation (3.11.59) by replacing $|D_x^n|, |D_y^n|$ and $|D_z^n|$ by $|\xi_n|, |\mu_n|$, and $|\eta_n|$, respectively, and removing the consideration of the finite solid angle $\Delta\Omega^n$ from the numerator and denominator.

The boundary condition on a solid wall is given as

$$I_w^n = \frac{\varepsilon_w \sigma T_w^4}{\pi} + \frac{(1 - \varepsilon_w)}{\pi} q_w^- \quad (3.11.64)$$

where the radiative heat flux on the wall is

$$q_w^- = \sum_{n'} I^{n'} \int_{\Delta\Omega^n} (\vec{s} \cdot \vec{n}_w) d\Omega = \sum_{n'} I^{n'} D_i^{n'} \quad (3.11.65)$$

In the preceding equation, the coefficients $D_l^{n'}$ are equivalent to $\pm D_x^{n'}$, $\pm D_y^{n'}$, and $\pm D_z^{n'}$ for a rectangular geometry in the Cartesian coordinate system, and they can usually be pre-determined for each wall element prior to calculating the volume averaged intensities of the domain. The radiative source term according to the finite volume method may be evaluated by

$$S_{rad} = \overline{K}_a \sum_n I^n \Delta\Omega^n - 4\overline{K}_a \sigma T^4 \qquad (3.11.66)$$

Similar to the discrete ordinates method, the procedure begins by marching across the domain in all three spatial directions by the known "upstream" intensities at the boundaries and in the "downward" direction upon reaching the other boundaries of the domain. Iterations are needed only for non-black walls. The extension of the procedure to non-gray participating medium is rather straightforward, which essentially entails solving additional components of the intensities for the number of gray gases considered in the WSGGM.

3.12   Guidelines for Selecting Radiation Models in Field Modeling

A number of radiation models that possess different levels of simplification or sophistication have been proposed and described. Unlike the rather straightforward guidelines as previously made available for the application of turbulence and combustion models, the selection of appropriate radiation models is complicated by not only the prospects of assessing feasible models to determine the local radiative properties of the combustion products but also the consistency between the level of simplification that is to be carried out for the radiation properties of absorbing gases and the level of sophistication of the radiative transfer and total heat transfer models. So what pertinent guidelines can be provided in the adoption of suitable radiation models in field modeling?

On the aspect of the local prediction of the radiative properties, the use of a reliable yet simple model offers many advantages. The gray gas assumption is probably the simplest approach that could be adopted in field modeling, to characterize the absorption/emission of the combustion products. A constant gas absorption/emission coefficient could be prescribed as a first step to include the radiation contribution of the burning fire, since consideration of spectral variations generally increases the complexity of an already extremely difficult problem. For the special case where the optical thickness lies in the thin limit—that is, $\ll 1$—the Planck's mean absorption coefficient, which does not require the specification of a path length, could be alternatively used to evaluate the local radiation properties in place of an obvious choice of a fixed value of the gas absorption/emission coefficient, which may not exist. For most practical fires, the absorbing/emission characteristics of the gases generally span a broad optical spectrum between the thin and thick limits. If the path length can be sufficiently ascertained, it would be more preferable to adopt, with ascending level of sophistication, the model proposed by Modak (1979), weighted sum of gray gases model, or even wide-band and narrow-band models. Such models, however, should be adjudicated against the increasing computational costs that may be incurred and the availability of computational resources to accommodate the extended calculations of the detailed spectral absorbing/emission characteristics of the gases.

On the next aspect regarding radiation methods, the expected computational requirements may be stipulated according to

Monte Carlo > Discrete Transfer Radiative Method > Discrete Ordinates Method or Finite Volume Method > P1-Radiation Model

In practice, the concept of weighted sum of gray gases approach as proposed by Modest (1991) for arbitrary solution methods in radiative transfer can be applied as a reasonable compromise instead of resorting to more complex non-gray models such as narrow-band or wide-band to account for the effects of radiative heat transfer in high-temperature combustion gases. For the consideration of only a single gray gas representation, which is effectively invoking the gray gas assumption, the P1-Radiation model should typically be employed, especially when the optical thickness is greater than unity. For optically thin limit, the P1-Radiation model suffers nevertheless a loss of accuracy. The discerete transfer radiative method, discrete ordinates method, and finite volume method, which spans the entire range of optical thicknesses, should be applied for optical thin problems. These models offer the flexibility of increasing the accuracy by increasing the number of rays and solve the surface-to-surface to participating radiation. Solving the problem with a large number of rays or fine angular discretisation, however, may be computationally intensive. For non-gray radiation using the gray-band approximation, P1-Radiation remains the most efficient model for optical thick problems. In covering a broader range of optical thicknesses, the demand on computational requirements will inadvertently increase as more complexities are accommodated into the models and amplify dramatically at the other extreme of the spectrum—that is, Monte Carlo. Acceptable turnaround of radiation calculations should still serve as a useful guideline in which radiation models should be preferred and best applied for the particular fire problems that are being solved.

3.13 Worked Examples on the Application of Radiation Models in Field Modeling

3.13.1 Single-Room Compartment Fire

Thermal radiation represents an important mode of heat transfer, since some proportion of the energy released by the combustion of fire is radiated to other parts of the flame and to external objects within the compartment. The influence of radiation heat transfer in the development of a fire in an enclosure is further demonstrated utilizing the benchmark case of the Steckler's single-room compartment fire in this worked example. Numerical results are obtained via the in-house computer code FIRE3D. For the purpose of assessing the model predictions, comparison of results is made for the same heat release rate of 62.9 kW against measurements made by Steckler et al. (1984) and the numerical results obtained from Lewis et al. (1997).

Numerical features: The system of governing equations is identical to those described in the previous worked example in section 3.6.1. All other boundary conditions remain the same as previously featured in the previous worked example in section 2.16.1. Numerical solutions have been attained by invoking the SIMPLE pressure correction algorithm to link the velocity and pressure, the *hybrid differencing scheme* to approximate the advective term in the governing equations, and the standard two-equation k-ε model for turbulence. The eddy dissipation combustion (EDM) model of Magnussen and Hjertager (1976), based on single-step chemistry of methane and the conserved scalar approach employing the Sivathanu and Faeth (1990) state relationships, is used to characterize the flaming fire. A mesh density consisting of 83160 grid nodes overlaying half of the burn room attached to an extended region having a size of 3 m $\times$ 2.8 m $\times$ 6.8 m is employed in this worked example.

The radiant exchange between fluid elements and the compartment boundaries are handled through the S_4 discrete ordinates model (DOM). In Lewis et al. (1997), the flame radiation has nonetheless been characterized by the application of the discrete transfer radiative method (DTRM). Their room fire simulations considered a total of 16 discrete rays emanated from the solid surface on each boundary cell. The number of rays that were chosen represented a compromise between computational economy and uniform coverage. For direct comparison of the model predictions against those of Lewis et al. (1997), a constant absorption coefficient of 0.2 is prescribed. This is in accordance with the approach adopted in their room fire simulations.

Numerical results: Predicted line graphs for the doorway temperature and velocity profiles utilizing the eddy dissipation combustion and conserved scalar model, with and without the consideration of radiation exchange, are depicted in Figures 3.47 and 3.48, along with the measured data of Steckler et al. (1984). When compared against predictions without the effects of radiation, the inclusion of a radiation heat transfer model is seen to significantly reduce and considerably sharpen the temperature profiles in the upper region at the doorway. In practical fires, radiation heat loss could account for as high as 30% of the total heat release rate of the fire. Subsequently, lower flame temperatures that are experienced above the fire source will result in lower advected temperatures away from the fire source. In spite of the significant discrepancies between the predicted and measured temperatures, the predicted velocity profiles are nevertheless not strongly sensitive to the effects of radiation, as shown by the results without radiation in Figure 3.47a and with radiation in Figure 3.48b.

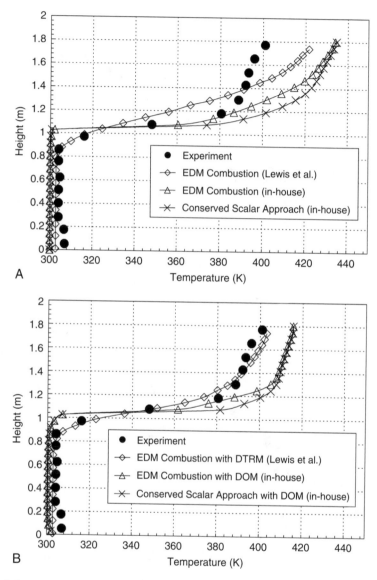

Figure 3.47 Comparison of doorway temperature profiles using the EBU combustion and conserved scalar models (a) without and (b) with radiation exchange.

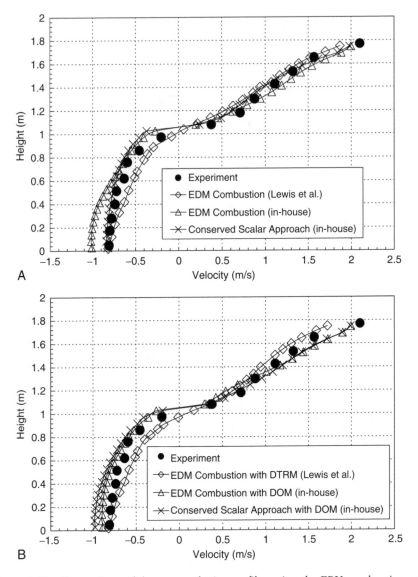

Figure 3.48 Comparison of doorway velocity profiles using the EBU combustion and conserved scalar models (a) without and (b) with radiation exchange.

Table 3.4 highlights the predictions of the height of the neutral layer, h_N, (relative to that of the doorway, h_O), the respective inflow and outflow mass fluxes, m_{in} and m_{out}, and the upper-layer temperature, $T_{upper\ layer}$ against Lewis et al. (1997) computational results and Steckler et al. (1984) experimental measurement. The definition of the neutral height and calculation of mass fluxes can be found in Lewis et al. (1997), and will not be repeated here. Discrepancies against experimental data for h_N / h_O, m_{in}, m_{out}, and $T_{upper\ layer}$ are respectively 3%, 6%, 9%, and 1%, based on the numerical results by Lewis et al. (1997). Present model predictions with radiative exchange yield discrepancies of 1%, 2%, 2%, and 1% for the solutions employing the EDM combustion model and 1%, 2%, 2%, and 4% for the conserved scalar model, respectively. Comparing the case where the EDM combustion model is employed in the numerical simulations, predictions of the inflow and outflow mass fluxes made by the present model fare better. One possible explanation for the improvement in our predictions could be the large, extended region to isolate the end effects of the extended boundaries on the doorway. Better predictions of the neutral height are obtained, as can be seen in Table 3.4. The conserved scalar model is found to yield similar results to those of the EDM combustion model.

Illustrative temperature profiles of the predicted room corner temperatures, are shown in Figure 3.49. The location of the thermocouple tree can be referred in the previous worked example in Figure 2.22 of Chapter 2. It is

TABLE 3.4 Comparison between model predictions against Lewis et al. (1997) and Steckler et al. (1984).

Grid/ No. of nodes	Combustion	Radiation	h_N / h_O	Inflow (kg/s)	Outflow (kg/s)	Temperature Upper Layer (°C)
Experiment			0.561	0.554	0.571	129
Lewis et al. (1997)						
70432	EDM	DTRM	0.546	0.521	0.523	128
Current						
83160	EDM	DOM	0.557	0.567	0.558	130
	Conserved Scalar	DOM	0.556	0.565	0.558	134

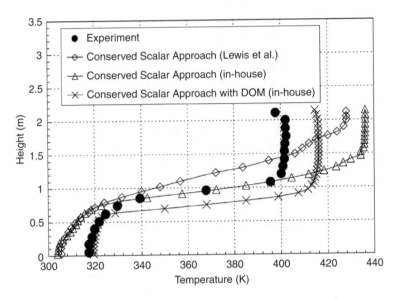

Figure 3.49 Comparison of predictions against measurements for corner rack temperature profiles using the conserved scalar approach with and without the consideration of radiation heat exchange.

clearly observed that predicted temperature profile by Lewis et al. (1997) using the conserved approach without radiation, fails to predict the distinct separation of the hot and cold layers as observed during the experiment. Present model predictions without the consideration of radiation behave otherwise. With the inclusion of radiation, the predicted temperature profile correctly replicates the behavior of the hot layer gas radiation heating the cold layer gas temperatures above the floor level.

A closer examination at the fire source is the apparent deflection of the fire plume toward the back wall, due to the incoming flow through the doorway at the bottom as evidently observed during the experiments. The temperature contours for the predicted fire plume at the vertical symmetry plane of the compartment are illustrated in Figure 3.50. Alternative to line contour plots (see Figure 2.26 in Chapter 2), *flooded* contour plots by a "gray-scaled" distribution, as shown in Figure 3.50, represent another effective graphical representation of the numerical data. On the basis of these two contour plots, the inclination of the fire for both of the two combustion models with radiative exchange produce plumes at an angle of approximately 45°. These encouraging results are in good qualitative agreement with the observation of Quintiere et al (1981) where they have observed a flame angle which lies between 33° and 43° in their experiments

Conclusions: It has been demonstrated through this worked example that flame radiation can play a significant role and should be properly accounted

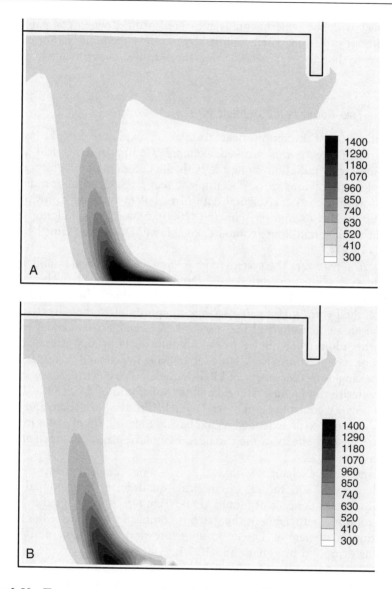

Figure 3.50 Temperature contours through the center of fire source and doorway: (a) Eddy dissipation model and (b) Conserved scalar mode.

in field modeling. Through the consideration of radiation heat exchange via the DOM, and when coupled with a two-equation turbulence model, eddy dissipation combustion model of Magnussen and Hjertager (1976) and the conserved scalar approach employing the Sivathanu and Faeth (1990) state relationships to characterize the turbulence and combustion of the turbulent buoyant fire, the case with radiation has been shown to yield much better agreement with the experimentally measured profiles when compared to the case without radiation, as exemplified by not only the temperature profiles at the doorway but also at the corner location of the compartment.

3.13.2 Two-Room Compartment Fire

The influence of radiation heat transfer is explored for the two-room compartment fire in this worked example. Comparison of field modeling predictions with radiation alongside with the computed temperature profiles obtained from previous worked example in section 3.6.2 without radiation are assessed against the experimental data of turbulent buoyant diffusion flames measured by Nielsen and Fleischmann (2000). Numerical simulations are performed through an in-house computer code FIRE3D with the same heat release rate of 110 kW.

Numerical features: The system of governing equations and boundary conditions are as described in section 3.6.2. Numerical solutions have been attained by invoking the SIMPLE pressure correction algorithm to link the velocity and pressure, the *hybrid differencing scheme* to approximate the advective term in the governing equations and the standard two-equation k-ε model for turbulence, with additional source terms to account for buoyancy effects. The eddy dissipation combustion (EDM) model of Magnussen and Hjertager (1976) based on single-step chemistry of LPG (Liquefied Petroleum Gas), is employed to characterize the flaming fire. A computational grid of $85 \times 44 \times 25$ (i.e., a total of 93,500) control volumes overlaying half of the burn room and adjacent room is attached to an extended region having a size of 3.6 m $\times$ 4.8 m $\times$ 6.5 m to isolate the end effects of the extended boundaries from the open end of the geometry.

Radiant heat exchange within the two-room compartment is handled through DOM via the S_4 numerical quadrature approximation. The mean absorption concept of Hubbard and Tien (1978) is adopted to calculate the absorption coefficients of the gaseous combustion products. For soot, the absorption coefficient is based on an expression from Kent and Honnery (1990) as expressed by equation (3.10.25).

Numerical results: In this worked example, results from the simpler strategy of employing the volumetric heat source approach to fire problems by specifying a fixed volume above the fire source to represent the flaming fire, are also obtained. On the basis of a prescribed volume size of 0.9 m $\times$ 0.3 m $\times$ 0.3 m is taken, the volumetric heat capacity of (110 W/ 0.081 m^3) is used as an input parameter for the source term of the energy equation. A constant value of 5.97 as recommended by Hubbard and Tien (1978) is used to account for the radiation heat loss due to combustion products.

Figure 3.51 illustrates the vertical temperature distribution above the fire source (Tree 3) and at the doorway (Tree 5), with the additional results provided by the inclusion of the effects of radiation in the field model through

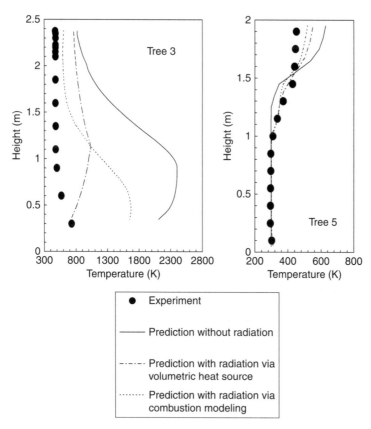

Figure 3.51 Comparison of predicted and measured temperature profiles above the fire source (Tree 3) and at the doorway (Tree 5) with and without gaseous radiation adopting the volumetric heat source and combustion modeling approaches.

the volumetric heat source and combustion modeling approaches. Substantially lower predicted temperatures above the fire source clearly identify the significance of radiation heat loss experienced by the fire. Nevertheless, it is observed that the use of the volumetric heat source approach grossly misrepresents the temperature behavior above the fire source, while the consideration of a combustion model is otherwise shown to aptly emulate the consistent burning behavior of the fire as evidenced by the predicted vertical temperature distribution. Away from the fire source, the post-combustion temperatures recorded at the doorway via the volumetric heat source approach are found to be of satisfactory agreement with the experimentally measured temperatures. This again confirms the unduly predicted high temperatures being confined to the region just above the fire source, and exerts only marginal influence on the temperature distributions away from the fire source. However, significant improvements to the temperature prediction are achieved by the consideration of combustion as well as radiation in the field model, not only above the fire source but also at the doorway joining the burn room and adjacent room (see results of Tree 5 in Figure 3.50).

Similar to the consideration of radiation in the single-room compartment fire as discussed in the previous worked example, the surrounding temperatures as demonstrated in Figure 3.52, are also better predicted within the burn room when radiation heat transfer is accommodated in the field model. Owing to the lower temperatures predicted in the burn room, the over-spilling of high temperatures are not as significantly felt in the adjacent room thereby resulting in the predicted temperatures being much closer to the experimental profiles, as seen in Figure 3.53. The predicted temperature profiles in the burn room and adjacent room correspond to the solutions attained from the field model adopting the combustion modeling approach. Figure 3.54 presents the line contour plots of the temperature at the symmetry plane for the cases with and without the consideration of radiation. Radiation contribution by the combustion products is seen to significantly reduce the thermal plume, confirming the lower-than-expected temperatures found in the adjacent room in contrast to higher predicted temperatures in the case where radiation is neglected.

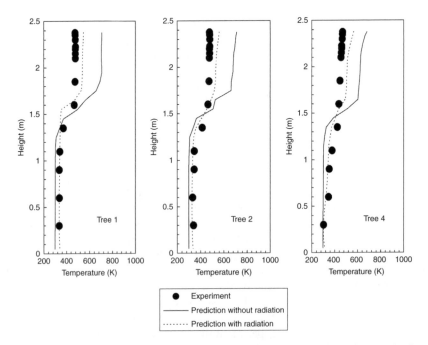

Figure 3.52 Comparison of predicted and measured temperature profiles in the burn room with and without gaseous radiation.

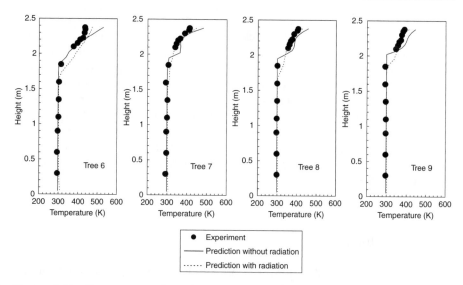

Figure 3.53 Comparison of predicted and measured temperature profiles in the adjoining room with and without gas

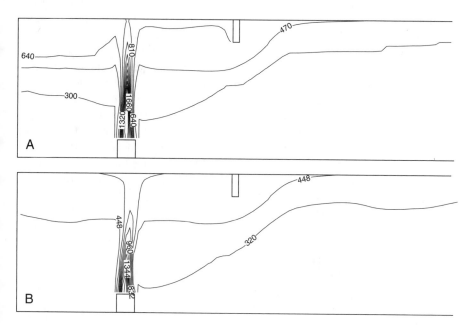

Figure 3.54 Predicted temperature distribution: (a) Without gaseous radiation and (b) With gaseous radiation.

Conclusions: Two important aspects are demonstrated through this worked example. Firstly, the simple approach based on representing the fire as a volumetric heat source is not recommended if knowledge of temperatures or other local burning characteristics within the flaming fire is required. Combustion modeling remains the only effective way of treating the burning fire. Consideration of combustion yields temperature distribution that is more consistent with measurements, as seen by the predicted temperature profiles in Figure 3.51. Secondly, the inclusion of radiation heat transfer appears to be an integral component in field modeling. Like in single-room compartment fire, flame radiation plays a significant role and should be incorporated in field modeling investigations.

3.14 Summary

It is well known that radiation from hot smoke is an important contributory factor to the development of fire within enclosed spaces. Radiation modeling, an essential component in field modeling, specifically involves the need to evaluate the complex radiative properties that are prevalent within the participating medium, and the knowledge in appropriately handling the radiant heat transfer through the treatment of the angular dependence and spatial variation of intensity within the physical domain. Depending on the different levels of simplification or sophistication, a variety of radiation models and methods have been developed.

Prediction of the complex spectral variation of radiative properties of the combustion products, which persists across the electromagnetic spectrum, is certainly an insurmountable task. Considering the diversity of products and the probability of having some or all of these combustion products in any volume element at any time, suitable relations for the properties of the combustion gases as well as soot particles and simplifications that have been made in arriving relations, have brought about a range of models of varying complexity. The simplest model that can be adopted is the gray gas assumption, with a constant gas absorption/emission coefficient. There are other more sophisticated models, such as the model proposed by Modak (1979), weighted sum of gray gases model, wide-band model, and narrow-band model, which attempt to better represent the spectral properties, provide a more accurate evaluation of the gas absorption/emission coefficient. Practical applications of these models are evident in their increasing usage in many field modeling investigations, and through generic commercial CFD codes.

The radiative transfer equation represents an equation in an integrodifferential form, which requires solution along the relevant ray paths within the three-dimensional space, and is generally very difficult to solve exactly for multi-dimensional geometries. This added complexity has brought about the development of dedicated numerical methods. In many practical CFD-cased applications, the commonly adopted methodologies with descending computational requirements are: Monte Carlo, Discrete Transfer Radiative Method (DTRM), Discrete Ordinates Method (DOM) or Finite Volume Method (FVM), and P1-Radiation Model. For most practical fires, the optical thicknesses that span a broad optical spectrum between the thin and thick limits, strongly govern the selection of suitable methods. The latter favors the application of the P1-Radiation Model. Radiation methods such as DTRM, DOM, and FVM are commonplace in field modeling investigations.

Consideration of the effect of gaseous radiation in the field model as demonstrated through the worked examples, contributes an integral part in dramatically lowering the surrounding temperatures within the burn room. Numerical results obtained with the consideration of radiation model and combustion model to characterize the fire chemistry are found to agree exceptionally well against measurements in the single-room and two-room compartment fire cases.

Review Questions

3.1. What are the four essential elements for a flaming fire?

3.2. Explain the difference between a diffusion flame and a premixed flame?

3.3. Is a naturally flaming fire more of a diffusion flame or a premixed flame? Why?

3.4. What is the difference between a laminar flame and a turbulent flame?

3.5. Why is a flame from a natural fire different from a jet flame?

3.6. What is combustion? Describe a state of complete combustion.

3.7. Write the complete reaction of propane (C_3H_8) and ethanol (C_2H_6O) occurring in air.

3.8. Define the heat of combustion.

3.9. What is the Arrhenius law?

3.10. Combustion processes are never complete in reality? Why?

3.11. Modeling turbulent combustion is difficult using the Arrhenius kinetic reactions. Why?

3.12. What are some of the practical combustion modeling approaches in the CFD-based fire model? For the particular model applied, explain the assumptions involved.

3.13. Define the Damköhler number. In which limit, the reaction is diffusion controlled? And in which limit, the reaction is kinetically influenced?

3.14. Define the Lewis number. What is the typical Lewis number adopted in field modeling?

3.15. Describe the eddy break-up and eddy dissipation models. How are they applied in field modeling?

3.16. What is the conserved scalar approach?

3.17. Explain the concept of mixture fraction and how is it useful in characterizing the combustion?

3.18. Describe the combustion model of Flame Sheet Approximation or Mixed-Is-Burnt. What does it entail?

3.19. Besides the state relationships defined in the Flame Sheet Approximation model, what other state relationships may also be feasibly applied?

3.20. What is the purpose of applying the probability density function to the conserved scalar approach?

3.21. Based on the prescriptive approach, what probability density functions are typically employed? How are they evaluated?

3.22. What is the laminar flamelet approach? Why is the approach useful in modeling the turbulent combustion process?

3.23. What are the two methods of generating the laminar flamelets? How are they normally used in the CFD-based fire model?

3.24. Two modes of heat transfer exist in practical fires. What are they?

3.25. Describe thermal radiation and in which range does it lie in the electromagnetic spectrum?

3.26. What is the difference between a black body and a gray body?

3.27. Explain the difference between a specular and a diffuse reflection.

3.28. What is a participating medium in radiation? Why is it important to be considered in field modeling?

3.29. Why is the radiative transfer equation significantly different from an ordinary partial differential equation?

3.30. Radiation modeling concerns two key issues. What are they?

3.31. What are the limitations in using the gray gas model to characterize the radiation properties of combustion products?

3.32. What is the weighted sum of gray gases (WSGGM) model?

3.33. In order to account for an inhomogeneous participating medium, what other sophisticated models can be applied in determining the radiation properties of combustion products?

3.34. In Monte Carlo, what is the dominant feature of this method?

3.35. What does the P-1 radiation model involve?

3.36. How is the radiation heat transfer treated in the Discrete Transfer Radiative Method?

3.37. How is the radiation heat transfer solved in the Discrete Ordinates Model? What is the difference between the Discrete Transfer Radiative Method and Discrete Ordinates Model?

3.38. How is the radiation heat transfer handled in the Finite Volume Method? What is the main difference between the Discrete Ordinates Model and Finite Volume Method?

4 Further Considerations in Field Modeling

Abstract

In most fires, the presence of fine carbonaceous particles (soot), which are invariably produced in the flame zone and dispersed in the smoke layer, significantly augments the global radiation process. Luminous soot radiation constitutes a portion of the radiative heat loss of the total heat release rate. The means of determining the distribution of soot particles require insights into the controlling physical and chemical mechanisms associated with the soot formation and soot oxidation. In an actual fire environment, the possible exposure of combustible condensed solids or wall linings housing the enclosure due to flame radiation may promote the prospect of flame spreading beginning from the items first ignited and subsequent fire development toward other nearby combustible materials. Underpinning all of this is the fundamental understanding of the pyrolysis occurring within these solid materials in fires.

Consideration of soot radiation and solid pyrolysis represents essential enhancements to the treatment of practical fires. Appropriate models are described in this chapter. The adoption of these supplemental models further increases the predictive capability and broadens the appeal of field modeling investigations to a wider range of fire problems.

PART V SOOT PRODUCTION

4.1 Importance of Soot Radiation

Particulate smoke (soot) is produced in almost all fires. As indicated by Rasbash and Drysdale (1982), most are formed in the gas phase as a result of *incomplete combustion* and *high temperature pyrolysis reactions at low oxygen concentrations*. Soot can be generated even if the original fuel is a gas or liquid in addition to those produced by the ablation of a condensed solid under high heat flux. Particularly in non-premixed flames, such as the buoyant laminar burning candle in air, as previously illustrated in Chapter 3 (see Figure 3.2), it is the presence of soot particles that provides its characteristic yellow luminosity. If by some means the flame height can be increased, the residence time will also increase, allowing more soot to be produced. The amount of soot that is formed nonetheless requires additional time to be oxidized in the upper part of the flame as it travels from the base to the tip of the flame. As soot travels through the flame, it radiates away energy and cools the combustion products—

Computational Fluid Dynamics in Fire Engineering
Copyright © 2009 by Academic Press. Inc. All rights of reproduction in any form reserved.

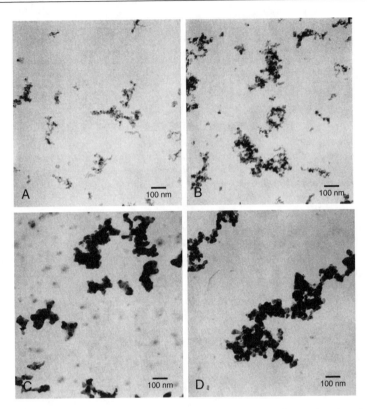

Figure 4.1 Typical TEM images of soot particles at four different heights above the burner along the flame centerline: (A) $z / D = 5$, (B) $z / D = 20$, (C) $z / D = 50$, and (D) $z / D = 80$, where z represents the axial height and D indicates the burner diameter. (After Hu and Koylu, 2004.)

the principal mechanism for *radiative heat loss*. At the *smoke point* flame height, radiative heat loss accounts for 30% of the total heat release rate (Markstein, 1986). The cooler combustion products prevent further oxidation of soot, and the soot formation/oxidation time is now equivalent to the diffusion (i.e., flow) time. Beyond the flame, soot particles are subsequently dispersed into the surrounding environment.

Smoke particles generally consist of agglomerations of minute soot particles that come together to form complex chains and clusters, which may have an overall size in excess of 1 μm (Drysdale, 1999). Figure 4.1 illustrates four typical Transmission Electron Microscope (TEM) photographs of the soot sampled from the centerline of a turbulent non-premixed acetylene flame at different heights above the burner, based on the experiment performed by Hu and Koylu (2004). Soot in the turbulent flame is clearly seen to be in the form of aggregates containing smaller primary particles. At a particular flame location, substantial variations in aggregate sizes and shapes indicate the poly-disperse characteristic accompanied by a complex soot structure. Nevertheless, the overwhelming observation through these images is that the primary particles remain rather spherical, and their nanometer diameters do not appear to vary significantly. At low-to-intermediate flame locations, young soot precursors are evident in the translucent particles captured by the TEM images. Their disappearance toward the flame tip is due to the completion of their carbonization before being released to the surroundings. In compartment fires, when their size becomes comparable to the size of the wavelength of light (0.3 μm–0.7 μm), these finely dispersed soot particles that are present in the smoke have a tendency to obscure the visibility by a combination of absorption and light scatter. Generally, the toxicity of soot does not play a pivotal role in fires. Rather, the poor visibility that the soot particles create significantly reduces the possibility of evacuation. Occupants within the confined enclosures may be subjected to other lethal combustion products such as carbon monoxide, which can arise due to the short exposure to high concentrations or long-duration exposure to low concentrations of such gases. The length of exposure increases if the visibility is extremely poor, making these occupants more unlikely to escape unaided.

More importantly, radiation heat transfer from the hot smoke layer is a significant contributory factor to the development of fire within enclosed spaces in compartment fires. In substantial amounts, soot is considered to be a much stronger absorber as well as emitter of radiation, in contrast to carbon dioxide and water vapor. During the growth or *pre-flashover* period of a compartment fire, the smoke layer under the ceiling, which may comprise mainly of finely dispersed soot particles, is a strong re-absorption of radiation to the boundaries encapsulating the enclosure. If the walls are combustible items, this smoke layer enhances the prospect of flames spreading, beginning from the item first ignited and spreading to nearby walls. Thereafter, a stage of the fully developed or *post-flashover* fire will be reached where flames will eventually engulf the entire volume. Transition from the pre-flashover to post-flashover fire usually involves a rapid spread from the area of localized burning to all combustible surfaces within the confinement. The burning behavior of these combustible materials in fires will be treated in more detail in Part VI within this chapter.

4.2 Overview and Limitations of Soot Modeling

In order to account for soot radiation in field modeling, the effective soot absorption coefficient requires the evaluation of the soot volume fraction, which is usually related to the concentration of soot particles. Essentially, the means of determining the concentration of soot particles require insights into the controlling physical and chemical mechanisms associated with the formation and oxidation of soot. A comprehensive model of the soot process must therefore include the consideration of both of these phenomena.

The main constituent of soot is mostly carbon; other elements such as hydrogen and oxygen are usually present in small amounts. Depending on the

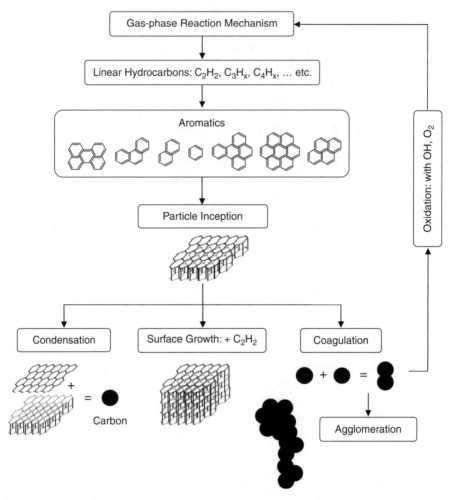

Figure 4.2   The process of soot formation and oxidation.

composition of the surrounding gas, other species may adsorb onto the surface of soot. In spite of the rigorous science to identify the many properties of soot, it is still not currently possible to uniquely define its chemical composition. As reviewed in Kennedy (1997), the various steps involved in the process of soot formation and oxidation are illustrated in Figure 4.2. The first step to the production of soot is the formation of the aromatic species such as cyclic benzene c-C_6H_6 and phenyl c-C_6H_5 in the gas phase. These aromatic species grow by the addition of other aromatic and smaller alkyl species into two-dimensional poly-aromatic hydrocarbons (PAH). The continued growth of the PAH leads to the smallest identifiable soot particles with diameters of the order 1 nm. Soot production is a chemically controlled phenomenon. During nearly all phases of soot production—inception, condensation, coagulation, surface growth, agglomeration, and oxidation—chemistry plays an important role. The inception of particles depicts the first occurrence, wherein two-dimensional PAHs merge into one three-dimensional particle. Once the soot particles are formed through the inception process, they can grow by three mechanisms: condensation, coagulation, and surface growth. The condensation and coagulation processes are typically physical in nature, where in the former, the particles grow via condensation of a two-dimensional PAH on a three-dimensional PAH, while the latter leads to the coalescence of particles, leading to the formation of a larger spheroid. Surface growth of particles proceeds in conjunction with coagulation. The particles grow via chemical reactions in the gas phase. It is generally agreed that acetylene C_2H_2 is mainly responsible for the growth of soot particles. Older particles will undergo agglomeration where large clusters of particles are subsequently formed. These clusters are now the primary soot particles in the system. Soot formation in diffusion flames usually occurs low in the flame and is followed by soot oxidation occurring at the tip of the flame. Oxidation occurs primarily as a result of attack by molecular O_2 and the OH radical and serves to contribute to the reduction of the particle sizes. During surface growth, oxidants may move to the particles and react with them to form some surface intermediates, which may then be desorbed and converted back into the gas phase.

   In real fires, soot is normally dependent upon the breakdown path of the parent fuel. To better understand the formation of soot, the *a priori* knowledge of the chemical composition and structure of the parent fuel is therefore of considerable importance. As reviewed in Drysdale (1999), fuels such as formaldehyde, formic acid, and methyl alcohol burn with non-luminous flames and are thus smokeless. In contrast, hydrocarbon fuels have a tendency to produce smoke by the introduction of *branching, unsaturation,* and *aromatic character,* while oxygenated fuels such as ethyl alcohol and acetone burn with considerable less smoke than the hydrocarbons. Similarly, hydrocarbon polymers such as polyethylene and polystyrene tend to yield far more substantial smoke than oxygenated fuels such as wood and polymethylmethacrylate, under the same free burning conditions. If the detailed chemistry of these parent fuels is known, models that describe the detailed elementary chemical reactions and physics of soot formation can be subsequently realized and applied. For hydrocarbon fuels, detailed kinetic models for the gas-phase phenomena (Frenklach and Wang, 1990) have been proposed to describe the successive chemical steps and the presence of pyrolysis products such as acetylene and PAH, which are generally considered as critical participating species in the nucleation process. For light hydrocarbons, reduced chemical mechanisms have been appropriately used to simplify the combustion and reduce the computational burden in modeling flames forming soot. Nonetheless, there is a need to formulate reduced chemical mechanisms to predict the consumption of heavy hydrocarbons. More so, there is a greater requirement in field modeling to gain further insights into the combustion chemistry of complex fuels (wood, polymethylmethacrylate, polyethylene, polystyrene, and so forth).

   Owing to the inadequate knowledge of the combustion chemistry of some parent fuels, and as a compromise between modeling simplicity and computational accuracy, practical models such as those based on single-step and semi-empirical approaches have been purposefully applied in field modeling to determine the concentration of soot particles. The degree of complexity of these models derives from the prospect in formulating the appropriate reactions rates that mimic the many important phenomenological descriptions of the physical processes of nucleation (inception), coagulation, surface growth, aggregation, and particle oxidation that occur in the soot process, as described in Figure 4.2. The main drawback of these models is the need to determine the necessary pre-exponential constants and activation energies that appear in the reaction rates, through some input from experimental data, which will become more apparent in the discussions of various models in the next section. For completeness, the description of detailed models that seek to solve the rate of equations for elementary reactions leading to soot is also described. These models, which attempt to directly characterize the soot process via an exhaustive analysis, should only be applied if the detailed chemistry of the parent fuel is fully realized.

4.3   Soot Models for Field Modeling

4.3.1   Single-Step Empirical Rate

This simple approach to soot modeling only requires a single transport equation for the mass fraction of soot Y_s. In the similar form to the scalar property equation, derived in Table 2.5 of Chapter 2, the Favre-averaged conservation equation can be written as

$$\frac{\partial}{\partial t}(\bar{\rho}\tilde{Y}_s) + \frac{\partial}{\partial x_j}(\bar{\rho}\tilde{u}_j\tilde{Y}_s) = D_{soot}^{th} + \frac{\partial}{\partial x_j}\left[\frac{\mu_T}{Sc_T}\frac{\partial\tilde{Y}_s}{\partial x_j}\right] + \bar{R}_{soot}^{+} + \bar{R}_{soot}^{-} \qquad (4.3.1)$$

where $\bar{R}_{soot}^{+}$ and $\bar{R}_{soot}^{-}$ are the mean reaction rates due to soot production and oxidation, respectively. In laminar flow, the diffusion of soot D_{soot}^{th} occurs exclusively by thermophoresis, which can be expressed in terms of mean quantities as

$$D_{soot}^{th} = -0.55\frac{\partial}{\partial x_j}\left[\tilde{Y}_s\frac{\mu}{\bar{T}}\frac{\partial\tilde{T}}{\partial x_j}\right] \qquad (4.3.2)$$

In turbulent flow, the turbulent diffusion term in equation (4.3.1) generally dominates over the diffusion term due to thermophoresis. The term D_{soot}^{th} can thus be safely neglected for most practical purposes.

For soot formation, the model proposed by Khan and Greeves (1974) that has been widely cited in literature for soot emissions from diesel engines, is described. As indicated in Figure 3.5 of Chapter 3, the soot time scale is generally slower than the time scale of flow, transport, and turbulence, which demonstrates a strong dependence on temperature. Khan and Greeves (1974) have assumed the production of soot particles in a flame is inherently a chemically controlled phenomenon and is governed entirely by the formation of soot particles—that is, by the soot inception rate. The global reaction rate expression $\bar{R}^+_{soot}$ chosen to characterize soot production in turbulent flow is

$$\bar{R}^+_{soot} = C_s \bar{P}_{fu} \phi^n exp\left(-\frac{E_a}{R_u \tilde{T}} \right) \tag{4.3.3}$$

where C_s is a constant, $\bar{P}_{fu}$ is the mean partial pressure of fuel, ϕ is the local unburnt equivalence ratio, E_a is the activation energy, and R_u is the universal gas constant usually taken to be equivalent to 1.9872 cal mol^{-1} K^{-1}. Modeling parameters such as C_s, n, and E_a have been ascertained through experiments performed in connection with diesel engines. For the exponent n and the activation energy E_a, values of 3 and 40200 cal mol^{-1} have been found to yield the best results against measurements, while the constant C_s can range from 0.01 (lightly sooty flame) – 1.5 (heavily sooty flame) kg N^{-1} m^{-1} s^{-1}. Soot production is essentially zero for the equivalence ratio ϕ less than that of the incipient soot limit and for ϕ in excess of a value corresponding roughly to the upper flammability limit. Following Khan and Greeves (1974), the upper and lower limits are set to 2 and 8, respectively. The applicability of the model to a wide range of fuels greatly suggests its suitability in field modeling.

For soot oxidation, Magnussen and Hjertager (1976) proposed a simple method similar to the eddy dissipation concept described in Chapter 3. Since the particle sizes are so small, it can be inferred that near-particle diffusion cannot possibly be a controlling process. Rather, the combustion is controlled by the rate of mixing of the particle-bearing vortices with adjacent oxygen bearing material. The consumption rate of soot is

$$\bar{R}^-_{soot} = -C_R \bar{\rho} \tilde{Y}_s \frac{\varepsilon}{k} \tag{4.3.4}$$

In regions where the oxygen concentration is low, the oxygen becomes the limiting species that controls the rate of consumption of soot. However, being a tracer element, soot must also compete for oxygen with the unburned fuel. This leads to

$$\bar{R}^-_{soot} = -C_R \bar{\rho} \left(\frac{\tilde{Y}_{ox}}{\tilde{Y}_s r_s + \tilde{Y}_{fu} r_{fu}} \right) \tilde{Y}_s \frac{\varepsilon}{k} \tag{4.3.5}$$

where C_R is a model constant assigned a value of 4, as previously indicated in Chapter 3, and r_s and r_{fu} are the soot and fuel stoichiometric ratios. The lower reaction rate of either equation (4.3.4) or (4.3.5) determines the local rate of soot consumption.

More recently, Lautenberger et al. (2005) proposed a simplified approach to modeling soot formation and oxidation in non-premixed hydrocarbon flames. Specifically, the model was formulated to make CFD calculations of fire radiation feasible in an engineering context. In order to retain simplicity and to minimize computational expense, the model considers the phenomena essential for obtaining sufficiently accurate predictions of the concentration of soot particles via the consideration of a single transport equation for soot. The basic form of the soot model is similar to the work by Kent and Honnery (1994), of which the soot formation rate could be estimated from only the local mixture fraction and temperature. The model postulated considered only *homogeneous* soot formation—moderately to heavily sooty flames. Nevertheless, both *heterogeneous* (surface area-dependent) and *homogeneous* (surface area-independent) soot formation processes are likely to contribute to the total soot formation rate in non-premixed hydrocarbon flames. The former are more important in lightly sooty flames such as lower alkanes, alcohol, and some cellulosics. Soot oxidation is treated by a global fuel-independent mechanism, which is also only a function of the mixture fraction and temperature. Here, the model assumes that the diffusion of molecular oxygen is the governing process rather than being controlled by the reaction of OH radicals impinging on the available soot surface area. For the formulation of the instantaneous soot formation and oxidation rates, they are simply determined from the product of an analytic function of mixture fraction (Z) and an analytic function of temperature (T) by

$$R_{soot}^+ = \dot{f}_{sf}'''(Z)g_{sf}(T) \tag{4.3.6}$$

$$R_{soot}^- = \dot{f}_{so}'''(Z)g_{so}(T) \tag{4.3.7}$$

Several analytic forms of these functions were considered by Lautenberger et al. (2005). The appropriate shapes of these functions are described following.

For laminar ethylene diffusion flame, a soot formation map analogous to $\dot{f}_{sf}'''(Z)g_{sf}(T) + \dot{f}_{so}'''(Z)g_{so}(T)$ envisaged by Kent and Honnery (1994) was shown to possess an approximately parabolic trend in mixture fraction. Peak soot formation rates occur at the mixture fraction values between 0.1 and 0.15, and over the temperature range between 1500 and 1600 K. Nevertheless, peak soot oxidation rates take place at mixture fraction values lower than the stoichiometric mixture fraction Z_{st}. Consistent with the observed trends, Lautenberger et al. (2005) proposed the set of analytic functions of $\dot{f}_{sf}'''(Z)$ and $\dot{f}_{so}'''(Z)$, which can be approximated by polynomial functions of third order. Typical shapes of the analytic mixture fraction functions for soot formation $\dot{f}_{sf}'''(Z)$ alongside with soot oxidation $\dot{f}_{sf}'''(Z)$ are illustrated in Figure 4.3. For soot formation, the function rises from a formation rate of zero at a mixture fraction of Z_L to a peak formation rate at a mixture fraction of Z_P and then falls back to zero at a mixture fraction of Z_H. The polynomial coefficients are determined

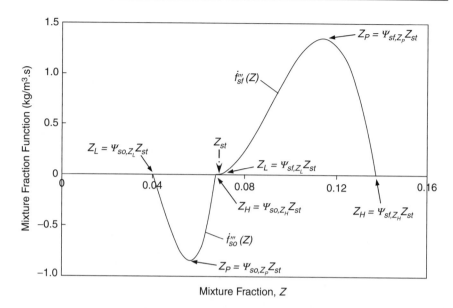

Figure 4.3 Polynomial functions of mixture fraction. (Adapted after Lautenberger et al., 2005.)

through specifying Z_L, Z_P, Z_H, and $\dot{f}_{sf}'''(Z_p)$, which are then solved through the resultant set of linear equations. In order to generalize the model to a range of fuels, the values of Z_L, Z_P, and Z_H for each polynomial are related to the fuel's stoichiometric mixture fraction Z_{st} by a parameter ψ of order unity. For soot formation, the soot oxidation mixture fraction function falls from a value of zero of Z_L to its peak negative value at Z_P, and then rises back to a value of zero at Z_H.

The analytic soot formation and oxidation temperature functions $g_{sf}(T)$ and $g_{so}(T)$ nevertheless show a less-discernible trend, as described in Figure 4.4. The soot formation function $g_{sf}(T)$ can be approximated as more or less parabolic, while the soot oxidation function $g_{so}(T)$ is assumed to be linear. For the former, a third order polynomial normalized between zero and unity takes on a value of zero at T_L, rises to a peak of unity at T_P, and falls back to zero at T_H. The endpoints at T_L and T_H can be interpreted as the minimum temperature at which soot forms and the temperature above which soot formation is absent. For the latter, soot oxidation diminishes at a critical temperature limit below 1400 K. It is noted that the maximum soot oxidation rate by the model may be stronger than the peak oxidation rate due to the function $g_{so}(T)$ exceeding unity.

To further extend the preceding model to turbulent combustion, statistical probability density function methods can be applied to account for the unresolved fluctuations in the instantaneous rates, as described by equations (4.3.6) and (4.3.7) to determine the appropriate mean rates $\bar{R}_{soot}^{+}$ and $\bar{R}_{soot}^{-}$ for equation

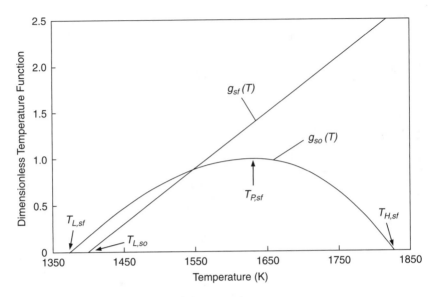

Figure 4.4 Polynomial functions of dimensionless temperature.
(Adapted after Lautenberger et al., 2005.)

(4.3.1). More details on the formulation of these rates as well as appropriate governing equations based on the conserved scalar approach, can be found in the dissertation of Lautenberger (2002).

4.3.2 Semi-Empirical Approach

This next level of soot modeling aims to incorporate some considerations of the physics and chemistry of the phenomenon, which usually leads to the development of rate of equations of soot precursors and particles with combustion chemistry description. In this approach, a transport equation for the particulate number density is introduced and solved in conjunction with the transport equation for the soot particles. The representation of soot properties is now characterized by two variables—the soot mass fraction or soot volume fraction and particulate number density. Three widely used models of soot formation for turbulent combustion developed by Tesner et al. (1971a, 1971b), Moss et al. (1988), and Leung et al. (1991), specifically based on the two-equation framework, are described in this section.

Tesner et al. (1971a, 1971b) have assumed that the soot formation occurs from a gaseous parent fuel in two stages. The first stage represents the formation of radical nuclei. Radical nuclei are defined to be the active sites on particles from which the soot deposits will eventually grow. According to Tesner et al. (1971a), the philosophy of the chain branching theory is that the increase of the rate of formation of particles is a result of a branched process, and the observed retardation is related to the acceleration of the destruction of active particles. This process is linked to the creation and rapid growth of the total surface of the soot particles on which the radical nuclei are being destroyed.

Similar to the scalar property equation derived in Table 2.5 of Chapter 2, the Favre-averaged form of the conservation equation for the concentration of radical nuclei $\tilde{n}_c$ can be expressed by

$$\frac{\partial}{\partial t}(\bar{\rho}\tilde{n}_c) + \frac{\partial}{\partial x_j}(\bar{\rho}\tilde{u}_j\tilde{n}_c) = D^{th}_{nuclei} + \frac{\partial}{\partial x_j}\left[\frac{\mu_T}{Sc_T}\frac{\partial \tilde{n}_c}{\partial x_j}\right] + \bar{R}^{+}_{nuclei} + \bar{R}^{-}_{nuclei} \qquad (4.3.8)$$

where D^{th}_{nuclei} is the diffusion occurs by thermophoresis, which in terms of mean quantities is given as

$$D^{th}_{nuclei} = -0.55\frac{\partial}{\partial x_j}\left[\tilde{n}_c\frac{\mu}{\bar{T}}\frac{\partial \tilde{T}}{\partial x_j}\right] \qquad (4.3.9)$$

Defining the particle number concentrations of nuclei according to

$$C_n = \bar{\rho}N_o\tilde{n}_c \qquad (4.3.10)$$

and soot particles as

$$C_s = \bar{\rho}\frac{\tilde{Y}_s}{m_p} \qquad (4.3.11)$$

where N_o is the Avogadro's number $(6.0223 \times 10^{26} \text{ mol}^{-1})$ and m_p is the mass of a soot particle, the rate of nuclei formation $\bar{R}^{+}_{nuclei}$ depends on a spontaneous generation and branching process described by

$$\bar{R}^{+}_{nuclei} = n_o + (f-g)C_n + g_oC_nC_s \qquad (4.3.12)$$

In the preceding equation (4.3.12), f and g are the linear branching and linear termination coefficients, and g_o is the coefficient of linear termination on soot particles. The spontaneous generation of radical nuclei n_o from the fuel is modeled according to the Arrhenius law as

$$n_o = a_o\bar{\rho}\tilde{Y}_{fu}exp\left(-\frac{E_a}{R_u\tilde{T}}\right) \qquad (4.3.13)$$

On the basis of a spherically shaped particle, the mass of a soot particle can be calculated from

$$m_p = \rho_{soot}\frac{\pi d_p^3}{6} \qquad (4.3.14)$$

The Favre-averaged conservation equation for soot particles also follows the same form of the scalar property equation derived in Table 2.5 of Chapter 2 as

$$\frac{\partial}{\partial t}(\bar{\rho}\tilde{Y}_s) + \frac{\partial}{\partial x_j}(\bar{\rho}\tilde{u}_j\tilde{Y}_s) = D^{th}_{soot} + \frac{\partial}{\partial x_j}\left[\frac{\mu_T}{Sc_T}\frac{\partial\tilde{Y}_s}{\partial x_j}\right] + \bar{R}^+_{soot} + \bar{R}^-_{soot} \qquad (4.3.15)$$

where D^{th}_{soot} is the thermophoresis diffusion given in equation (4.3.2). The rate of formation of soot particles depends on the interaction between the active particles and the original hydrocarbon molecules and on the termination process by the surface of the soot particles. The rate of soot formation $\bar{R}^+_{soot}$, which depends on the concentration of radical nuclei, is modeled as

$$\bar{R}^+_{soot} = m_p(a - bC_s)C_n \qquad (4.3.16)$$

The default values for all the soot parameters are those of acetylene fuel, which have been obtained from Tesner et al. (1971b). They are:

$$d_p = 1.785 \times 10^{-8} \text{ m}$$
$$a_o = 1.35 \times 10^{37} \text{ part kg}^{-1} \text{ s}^{-1}$$
$$E_a/R_u = 90000 \text{ K}$$
$$f - g = 100 \text{ s}^{-1}$$
$$g_o = 1.0 \times 10^{-15} \text{ m}^3 \text{ part}^{-1} \text{ s}^{-1}$$
$$a = 1.0 \times 10^5 \text{ s}^{-1}$$
$$b = 8.0 \times 10^{-14} \text{ m}^3 \text{ part}^{-1} \text{ s}^{-1}$$

Magnussen and Hjertager (1976) have employed the kinetic theory of soot formation by Tesner et al. (1971a, 1971b) to a turbulent acetylene flame. On the basis of the Eddy Dissipation Concept, they have proposed to conveniently calculate the mean combustion rates of the radical nuclei $\bar{R}_{nuclei}$ and soot particles $\bar{R}_{soot}$ from the rate of combustion of fuel. For the combustion of soot particles, the rate is given by

$$\bar{R}^-_{soot} = min\left[-C_R\bar{\rho}\tilde{Y}_s\frac{\varepsilon}{k}, -C_R\bar{\rho}\left(\frac{\tilde{Y}_{ox}}{\tilde{Y}_s r_s + \tilde{Y}_{fu}r_{fu}}\right)\tilde{Y}_s\frac{\varepsilon}{k}\right] \qquad (4.3.17)$$

The local radical nuclei can be assumed to be reduced by combustion according to

$$\bar{R}^-_{nuclei} = \frac{\tilde{n}}{\tilde{Y}_s}\bar{R}^-_{soot} \qquad (4.3.18)$$

This so-called Magnuseen soot model that combines the kinetic theory formation of Tesner et al. (1971a, 1971b) and oxidation rates in the respective equations (4.2.17) and (4.3.18) is widely applied in many field modeling investigations (for example, room fire simulations by Luo and Beck, 1996). It is also a standard feature in a majority of commercial CFD codes such as ANSYS Inc., Fluent, and ANSYS Inc., CFX. Note that the same model constants just tabulated have been successfully used to produce the flame data for other fuels, including methane.

Moss et al. (1988) have proposed a soot model that incorporates the essential physical processes of soot nucleation, coagulation, and surface growth influencing the soot volume fraction and particulate number density. While the kinetic theory of soot formation by Tesner et al. (1971a, 1971b) has been extensively employed, its focus solely on particle number density neglects the important role of surface growth in relation to soot mass addition. Particle size evolution could not be tracked, and without the knowledge of particle size, the aerosol surface area cannot be determined. Heterogeneous chemical process like oxidation would therefore be unsatisfactorily ascertained. This alternative model, which is represented by the soot volume fraction f_v and particulate number density n, is important to the description of the post-flame burnup, permit, but the soot aerosol surface area to be estimated from the average particle diameter according to

$$d_p = \left(\frac{6f_v}{n\pi}\right)^{1/3} \tag{4.3.19}$$

The Favre-averaged conservation equation for the soot particulate number density and soot volume fraction in terms of normalized variables can be written as

$$\frac{\partial}{\partial t}(\bar{\rho}\tilde{\zeta}_n) + \frac{\partial}{\partial x_j}(\bar{\rho}\tilde{u}_j\tilde{\zeta}_n) = D^{th}_{num\_dens} + \frac{\partial}{\partial x_j}\left[\frac{\mu_T}{Sc_T}\frac{\partial\tilde{\zeta}_s}{\partial x_j}\right] + \bar{R}^+_{num\_dens} + \bar{R}^-_{num\_dens} \tag{4.3.20}$$

$$\frac{\partial}{\partial t}(\bar{\rho}\tilde{\zeta}_s) + \frac{\partial}{\partial x_j}(\bar{\rho}\tilde{u}_j\tilde{\zeta}_s) = D^{th}_{vol\_frac} + \frac{\partial}{\partial x_j}\left[\frac{\mu_T}{Sc_T}\frac{\partial\tilde{\zeta}_s}{\partial x_j}\right] + \bar{R}^+_{vol\_frac} + \bar{R}^-_{vol\_frac} \tag{4.3.21}$$

where $\tilde{\zeta}_n = \tilde{n}/(\bar{\rho}N_o)$ and $\tilde{\zeta}_s = (\rho_s\tilde{f}_v)/\bar{\rho}$ with $D^{th}_{num\_dens}$ and $D^{th}_{num\_dens}$ given by

$$D^{th}_{num\_dens} = -0.55\frac{\partial}{\partial x_j}\left[\tilde{\zeta}_n\frac{\mu}{\tilde{T}}\frac{\partial\tilde{T}}{\partial x_j}\right] \tag{4.3.22}$$

$$D^{th}_{vol\_frac} = -0.55\frac{\partial}{\partial x_j}\left[\tilde{\zeta}_s\frac{\mu}{\tilde{T}}\frac{\partial\tilde{T}}{\partial x_j}\right] \tag{4.3.23}$$

In equations (4.3.20) and (4.3.21), the influence rate processes of nucleation, coagulation, and surface growth on the particulate number density and volume fraction are modeled as

$$\bar{R}^+_{num\_dens} = \underbrace{\bar{\alpha}}_{nucleation} - \underbrace{\bar{\rho}^2\bar{\beta}\tilde{\zeta}_n^2}_{coagulation} \tag{4.3.24}$$

$$\bar{R}^+_{vol\_frac} = \underbrace{\bar{\delta}}_{nucleation} + \underbrace{N_o^{1/3}\bar{\rho}\bar{\gamma}\tilde{\zeta}_s^{2/3}\tilde{\zeta}_n^{1/3}}_{surface\ growth} \tag{4.3.25}$$

The first term in equation (4.3.24) represents the increase in soot particle number density due to particle inception, which is given by

$$\bar{\alpha} = C_\alpha\bar{\rho}^2\tilde{T}^{1/2}\tilde{X}_{fu}\ exp\left(-\frac{T_\alpha}{\tilde{T}}\right) \tag{4.3.26}$$

The second term in equation (4.3.24) describes the loss of particles as a result of coagulation. According to Moss et al. (1988), it can be described by the Smoluchowski (Fuchs, 1964) expression as

$$\bar{\beta} = C_\beta\tilde{T}^{1/2} \tag{4.3.27}$$

The increase of the soot volume fraction in equation (4.3.25) is represented by the nucleation of new particles in the first term and the result of surface growth in the second term. On the basis of the particle nucleation expressed in equation (4.3.26), the accompanying mass growth of the former may be represented (for 12 carbon atoms initially) by

$$\bar{\delta} = 144\bar{\alpha} \tag{4.3.28}$$

The latter is modeled according to the surface growth of soot suggested by Syed et al. (1990), which contained a linear dependence on aerosol surface area, and is controlled by the rate relationship

$$\bar{\gamma} = C_\gamma\bar{\rho}\tilde{T}^{1/2}\ \tilde{X}_{fu}\ exp\left(-\frac{T_\gamma}{\tilde{T}}\right) \tag{4.3.29}$$

It should be noted that the surface growth rate based on Syed et al. (1990) in equation (4.3.25) depends on the soot surface area. The number density of an aerosol, in general, rapidly attains a nearly constant value as a result of a balance between the nucleation and coagulation. If the number density is approximately constant, the surface growth rate will lead to a predicted cubic time variation in soot volume fraction, since the terms with $\tilde{\zeta}_n$ and $\tilde{\zeta}_s$ are accounted for the soot surface area.

From a detailed chemical kinetic model of fuel pyrolysis, it is possible to identify the specific hydrocarbon species that are precursors to drive the nucleation and surface growth expressions through the mean fuel mole fraction $\tilde{X}_{fu}$. Depending on the parent fuel, the choice of acetylene and benzene as critical species for these processes is strongly supported by experimental evidence—for example, Harris and Weiner (1983a, 1983b) and Smyth et al. (1985). Generally, detailed reaction mechanisms, and thus distributions of minor species concentrations, are not accessible for more complex fuels. A more practical strategy that has been adopted by Syed et al. (1990) was to consider the mole fraction of the critical precursor to be that of the total C_xH_y concentration, summed over all the hydrocarbon species in the mixture locally. Methane combustion was studied in Syed et al. (1990). The pre-exponential constants C_α, C_α, and C_α and activation temperatures T_α and T_γ determined empirically for methane are:

$$C_\alpha = 65400 \text{ m}^3 \text{ kg}^{-2} \text{ K}^{-1/2} \text{ s}^{-1}$$
$$C_\beta = 1.3 \times 10^7 \text{ m}^3 \text{ K}^{-1/2} \text{ s}^{-1}$$
$$C_\gamma = 0.1 \text{ m}^3 \text{ kg}^{-2/3} \text{ K}^{-1/2} \text{ s}^{-1}$$
$$T_\alpha = 46100 \text{ K}$$
$$T_\gamma = 12600 \text{ K}$$

As will be demonstrated in the worked example later, the preceding soot parameters are applicable to other flames that are lightly sooting in nature, like methane. Moss and Stewart (1998) have shown that for quite heavily sooting flames such as propylene and methylmethacrylate (MMA), characteristic of polymeric materials encountered in room fires, modest changes to the model can be accommodated by simply altering the pre-exponential constants C_α, C_α, and C_α. On the basis of repeated numerical experiments and laser extinction measurements, the model coefficients for propylene (C_3H_6) are determined as:

$$C_\alpha = 1.3 \times 10^6 \text{ m}^3 \text{ kg}^{-2} \text{ K}^{-1/2} \text{ s}^{-1}$$
$$C_\beta = 2.0 \times 10^9 \text{ m}^3 \text{ K}^{-1/2} \text{ s}^{-1}$$
$$C_\gamma = 8.5 \times 10^{-13} \text{ kg}^{-2/3} \text{ K}^{-1/2} \text{ s}^{-1}$$

while for MMA (C_3H_6):

$$C_\alpha = 3.68 \times 10^5 \text{ m}^3 \text{ kg}^{-2} \text{ K}^{-1/2} \text{ s}^{-1}$$
$$C_\beta = 2.0 \times 10^9 \text{ m}^3 \text{ K}^{-1/2} \text{ s}^{-1}$$
$$C_\gamma = 8.5 \times 10^{-13} \text{ kg}^{-2/3} \text{ K}^{-1/2} \text{ s}^{-1}$$

From the preceding, it is observed that the changes to the coefficients (C_α) occur primarily in the expressions for soot particle nucleation alone.

In addition to the preceding soot formation model, Kaplan et al. (1996) have adopted the Nagle and Strickland-Constable (1962) rate for soot oxidation as the limiting mechanism for oxidation by O_2 in equations (4.3.20) and (4.3.21), which was also subsequently considered by Morvan et al. (2000) for the simulation of a buoyant methane/air radiating turbulent diffusion flame. In the model proposed by Nagle and Strickland-Constable (1962), they have assumed two types of active sites (A and B) that could be present for the oxidation of carbon. Type A sites react with oxygen to yield another A site and carbon monoxide:

$A + O_2 \rightarrow A + 2CO$ at a rate given by $\dfrac{k_A \bar{P}_{O_2}}{1 + k_Z \bar{P}_{O_2}} \chi$

where $\bar{P}_{O_2}$ is the mean partial pressure of oxygen. Type B sites react with oxygen to yield Type A sites and carbon monoxide:

$B + O_2 \rightarrow A + 2CO$ at a rate given by $k_B \bar{P}_{O_2}(1 - \chi)$

Finally, Type A sites thermally rearrange to give Type B sites as

$A \rightarrow B$ at a rate given by $k_T \chi$

The overall reaction rate (kg m^{-2} s^{-1}) is given by

$$\bar{R}_{ox} = 120 \left[\frac{k_A \bar{P}_{O_2}}{1 + k_Z \bar{P}_{O_2}} \chi + k_B \bar{P}_{O_2}(1 - \chi) \right] \tag{4.3.30}$$

where

$$\chi = \frac{1}{1 + \frac{k_T}{k_B \bar{P}_{O_2}}} \tag{4.3.31}$$

and

$$k_A = 2 \times 10^4 \, exp\left(-\frac{30000}{R_u \tilde{T}} \right)$$

$$k_B = 4.46 \, exp\left(-\frac{15200}{R_u \tilde{T}} \right)$$

$$k_T = 1.51 \times 10^8 \, exp\left(-\frac{97000}{R_u \tilde{T}} \right)$$

$$k_z = 2.13 \times 10^4 \, exp\left(\frac{4100}{R_u \tilde{T}} \right)$$

It is noted that the units for the preceding reaction rate constants are in kg (kilogram), m (meter), s (seconds), cal (calorie), K (Kelvin), and atm (atmospheres). The local particulate number density and soot volume fraction are reduced by combustion according to

$$\bar{R}_{num\_dens} = -N_o^{1/3} \left(\frac{36\pi}{\rho_{soot}^2} \right)^{1/3} \frac{\bar{\rho} \bar{R}_{ox} \tilde{\zeta}_n^{4/3}}{\tilde{\zeta}_s^{1/3}} \tag{4.3.32}$$

$$\bar{R}^-_{vol\_frac} = -N_o^{1/3}\left(\frac{36\pi}{\rho^2_{soot}}\right)^{1/3}\bar{\rho}\bar{R}_{ox}\zeta_n^{1/3}\zeta_s^{2/3} \qquad (4.3.33)$$

The soot formation model of Leung et al. (1991) differs from those of Tesner et al. (1971a, 1971b) and Moss et al. (1988), in the aspect of which assumed specifically acetylene as the precursor for soot nucleation and growth. An important characteristic of this model is that acetylene is *not* the actual fuel but assumed to be the product of the fuel breakdown process. Hence, the rates of soot nucleation and growth are directly proportional to the acetylene concentration rather than to the parent fuel concentration. This concept in retrospect is physically more plausible, although it complicates the modeling to some extent, since there is a concerted need to determine the acetylene mass fraction. If the detailed reaction mechanisms of the parent fuel are known, this does not represent a significant burden, especially when the laminar flamelet approach is used to ascertain the required state relationships for the major and minor chemical species of the buoyant diffusion flame. The model solves for the particulate number density n and soot mass fraction Y_s.

The model for soot formation in non-premixed flames is based on a simple kinetic mechanism, which entails the following four reaction steps:

nucleation $2C_2H_2 \rightarrow 2C(S) + H_2$
surface growth $nC(S) + C_2H_2 \rightarrow (n + 2)C(S) + H_2$
coagulation $nC(S) \rightarrow C_n(S)$
oxidation $C(s) + \frac{1}{2}O_2 \rightarrow CO$

From the reaction steps just described, the notation C(s) represents soot, and a mole of soot is taken as a mole of carbon atoms. The reaction step defined by the particle nucleation is similar to that outlined by Tesner et al. (1971a) for premixed acetylene-air flames. For simplicity, the rate of nucleation is assumed to be first order in acetylene concentration, which can be expressed in terms of mass fraction so that

$$\bar{R}_1 = 1 \times 10^4 \, exp\left(-\frac{12100}{\tilde{T}}\right)\left(\frac{\bar{\rho}\tilde{Y}_{C_2H_2}}{M_{C_2H_2}}\right)$$

where $\tilde{Y}_{C_2H_2}$ and $M_{C_2H_2}$ are the corresponding mean mass fraction and molecular weight of acetylene, and the units of the preceding rate are in kmol m^{-3} s^{-1}. Leung et al. (1991) also assumed the rate of surface growth to be first order similar to the rate of nucleation. An *ad hoc* assumption is made that the number of active sites is taken to be proportional to the square root of the total surface area available locally in the flame. The soot growth rate is given as

$$\bar{R}_2 = 6 \times 10^3 \, exp\left(-\frac{12100}{\tilde{T}}\right)\sqrt{\pi\left(\frac{6M_c}{\pi\rho_{soot}}\right)^{2/3}\left(\frac{\bar{\rho}\tilde{Y}_{C_2H_2}}{M_{C_2H_2}}\right)\left(\frac{\bar{\rho}\tilde{Y}_s}{M_C}\right)^{1/3}\tilde{n}^{1/6}}$$

Here, M_C refers to the molecular weight of carbon. The units are again in kmol m^{-3} s^{-1}. The decrease of the particle number density is accounted by the particle agglomeration of which this step is modeled using the normal square dependence—that is,

$$\bar{R}_3 = -2C_a \left(\frac{6M_C}{\pi\rho_{soot}}\right)^{1/6} \left(\frac{6\kappa\tilde{T}}{\rho_{soot}}\right)^{1/2} \left(\frac{\bar{\rho}\tilde{Y}_s}{M_C}\right)^{1/6} \tilde{n}^{11/6}$$

where C_a is the agglomeration rate constant taken to have a value of 9.0 and κ is the Boltzmann constant given as 1.381×10^{-23} J K^{-1}. The units are in m^3 s^{-1}. Instead of the Nagle and Strickland-Constable (1962) model, Leung et al. (1991) used the rate of soot oxidation proposed by Lee et al. (1962), which is due by the limiting mechanism of oxidation by O$_2$. The rate of soot oxidation where the dependence on local surface area has been retained is written in units of kmol m^{-3} s^{-1} as

$$\bar{R}_4 = -1 \times 10^4 T^{1/2} \, exp\left(-\frac{19680}{\tilde{T}}\right) \pi \left(\frac{\bar{\rho}6\tilde{Y}_s}{\pi\rho_{soot}\tilde{n}}\right)^{2/3} \left(\frac{\bar{\rho}\tilde{Y}_{O_2}}{M_{O_2}}\right) \tilde{n}$$

It is noted that the preceding oxidation rate has been adjusted by a factor of 14 to provide adequate agreement with the measurements of Garo et al. (1990). Again, such *ad hoc* adjustment is required in order to necessitate the neglect of OH radical as an oxidant of soot. Alternatively, an oxidation step involving OH could readily have been formulated in the context of the present model. Following investigations by Fenimore and Jones (1967), Puri et al. (1994), and Garo et al. (1990), the rate of soot by the OH radical, assuming a collision efficiency of 0.04 according to Brookes and Moss (1999), may be simply written as

$$\bar{R}_4 = -4.2325T^{1/2}\pi \left(\frac{\bar{\rho}6\tilde{Y}_s}{\pi\rho_{soot}\tilde{n}}\right)^{2/3} \left(\frac{\bar{\rho}\tilde{Y}_{OH}}{M_{OH}}\right) \tilde{n}$$

The Favre-averaged conservation equation for the soot particulate number density is

$$\frac{\partial}{\partial t}(\bar{\rho}\tilde{n}) + \frac{\partial}{\partial x_j}(\bar{\rho}\tilde{u}_j\tilde{n}) = D^{th}_{num\_dens} + \frac{\partial}{\partial x_j}\left[\frac{\mu_T}{Sc_T}\frac{\partial\tilde{n}}{\partial x_j}\right] + \bar{R}^+_{num\_dens} + \bar{R}^-_{num\_dens}$$

$$(4.3.34)$$

with $D^{th}_{num\_dens}$ given by

$$D^{th}_{num\_dens} = -0.55\frac{\partial}{\partial x_j}\left[\tilde{n}\frac{\mu}{\tilde{T}}\frac{\partial\tilde{T}}{\partial x_j}\right]$$

$$(4.3.35)$$

and the appropriate rates modeled as

$$\bar{R}^+_{num\_dens} = \frac{2N_o}{C_{min}} \bar{\rho}\bar{R}_1 \tag{4.3.36}$$

$$\bar{R}^-_{num\_dens} = \bar{\rho}\bar{R}_3 \tag{4.3.37}$$

where C_{min} is the number of carbon atoms in the incipient carbon particle. Fairweather et al. (1992) who adopted the soot reaction mechanism of Leung et al. (1991) for a methane air jet flame have assumed a value of 9×10^4 carbon atoms. For the soot mass fraction, the conservation equation is identical to the form derived in equation (4.3.15) with the source terms given by

$$\bar{R}^+_{soot} = 2M_C\bar{\rho}(\bar{R}_1 + \bar{R}_2) \tag{4.3.38}$$

$$\bar{R}^-_{soot} = M_C\bar{\rho}\bar{R}_4 \tag{4.3.39}$$

For all the soot formation and oxidation models described in this section, it should be noted that the chemical kinetic rates are evaluated based on the mean quantities. Unlike the concentrations of major species, which result from fast chemistry as compared with mixing, soot mechanisms are comparatively much slower and they tend to exhibit a degree of invariance, since the effect of turbulence will be filtered or smeared out due to the low chemical kinetic rates. Hence, the effect of turbulence on the soot chemistry particularly in buoyant fires is usually not accounted in order to enhance the computational efficiency. In the context of fire engineering, the prime consideration of these approximate models as supplemental models to field modeling is aimed to make fire radiation calculations practical in affecting the overall radiative heat transfer. The evaluation of mean soot concentrations should therefore suffice for determining the averaged optical property for soot radiation.

4.4 Population Balance Approach to Soot Formation

4.4.1 What Is Population Balance?

The viability of the population balance approach to handle a wide variety of particulate processes has certainly received unprecedented attention from both academic as well as industrial quarters. Particularly in modern chemical industry, CFD coupled with a micro-mixing models and the appropriate population balance models are increasingly being employed to predict the evolution of the particle size distribution (PSD) in order to improve the crystallization, precipitation, and polymerization processes of materials for a wide variety of applications including pharmaceutical, agriculture, and specialty chemical products, as well as the many aggregation-breakage processes in many other engineering applications associated with coagulation and rupture of flocs, addition and degradation of polymers, and coalescence and breakup of liquid drops. Mounting interest for population balances in the chemical, petrochemical, and mining industries has also resulted in significant design improvements to widely used bubble column reactors to promote increasing rates of mass transport between gases and liquids and to the more efficient mixing of competing gas-liquid reactions.

Nevertheless, what is *population balance*? According to Ramkrishna and Mahoney (2002), a population balance on any system is generally concerned with maintaining a record for the number of entities, which may be solid particles, liquid drops, gas bubbles, biological cells, or events whose presence or occurrence may dictate the behavior of the system under consideration. It can thus be regarded as a continuity statement for the evolution of entities. The equations that govern the PSD can be obtained by writing the relationships that describe the conservation of the mass of the entities. Such equations are usually referred to as the *population balance equations*.

The variables distinguishing an entity within the population balance equation may consist of *external* coordinates (its physical location) and *internal* coordinates (characteristic internal to the entity such as size, temperature, composition, and so forth). Mathematically, the collection of both of these coordinates is referred to as the state of the entity of which may be represented by a finite dimensional vector. This leads to the concept of the entities being distributed in a finite dimensional *state space*, which may change with time depending on its current realization and expected displacements because of random changes. In addition to the motion of these entities through the state space, it is usual to encounter *birth* processes that create new entities and *death* processes that destroy existing ones. These birth and death processes somewhat may depend on the states of the entities created or destroyed with an associated phenomenology. Nucleation of particles, breakup, and aggregation are some typical examples of such processes. By definition, a *population balance model* is formulated based on the collective phenomenology contained in the displacement of entities through the state space and the birth and death processes that terminate entities and produce new entities. It is imperative to note that the phenomenology is concerned with the behavior of the single entity in the *company* of its fellow entities for the population balance model to be a reasonable description of the system.

The structure of the population balance equations, generally expressed as an *integrodifferential* form of the particle size distribution, is generally very complex and their solution by analytical means remains elusive for all but the most idealized situations (Bove et al., 2005). Several numerical techniques have been proposed in the literature to best handle the population balance equations, not only achieving some considerable levels of accuracy but also obtaining the solutions in real time with moderate computational load. One of the well-known classical numerical approaches is the standard method of moments (SMM) developed by Frenklach (1985) in which the internal coordinate is integrated out and the PSD is determined through the respective moments.

Generally speaking, the mathematical difficulty of SMM lies in obtaining the appropriate closure. The simplest way of accomplishing this is to presume the functional form of the PSD function (Dobbins and Mulholland, 1984, Lee, 1983, Pratsinis, 1988). Whereas the PSD function is not *a priori* known, Hulburt and Katz (1964) proposed the consideration of only determining the lower-order of moments of the PSD that are sufficient to estimate the physical quantities, such as optical properties or volume fraction. Although the problem is condensed substantially by tracking this limited number of moments, it is severely limited by the ability to only handle the size-independent growth rate and the size-independent aggregation and breakage kernels. For a more general representation of the particle dynamics, the preferential treatment of higher-order moments becomes indispensable. To achieve closure, Frenklach (2002) has attested the validity of interpolation closure in obtaining these moments by first expressing the natural logarithmic of moments in terms of polynomial and later determining the required moments by separating the interpolation for positive-order and negative-order moments via the Lagrange interpolation among logarithms of the whole-order moments.

Another important numerical approach to population balance is the concept of classes method (CM). Here, the internal coordinate is discretised into a finite series of distinct bins or sizes. The PSD is now directly simulated instead of inferring to derivative variables (i.e., moments). In the zero-order CM, the PSD is considered to be constant within each class (or internal coordinate interval). Obviously, the larger the number of subdivisions across the size range, the more accurate the numerical solutions will be in restoring the autonomy of the discretised form of the population balance equations. A number of different zero-order CMs for aggregation-breakage of particles have been reviewed and compared by Vanni (2000). Yeoh and co-workers (Cheung et al., 2007, Yeoh and Tu, 2004, 2005, 2006) have successfully applied the zero-order CMs via the MUltiple-SIze-Group (MUSIG) model to explicitly track the size distribution of gas bubbles in bubbly flows with and without heat and mass transfer. Modeling the fundamental mechanisms of bubble coalescence and breakage as source terms has allowed the population changes to be realized for each class—the overall PSD is subsequently resolved. In the higher-order CMs, the PSD can be represented by a specific functional form, usually a low-order polynomial through cubic splines or orthogonal collocation, in each discrete size interval of the discretisation. They are usually more accurate but less robust and may suffer from dispersion effects when dealing with narrow initial PSD. Normally, zero-order and high-order CMs require a large number of classes to work with in order to attain good accuracy, especially where the range of the particle sizes is extremely wide.

As an attractive alternative to the SMM, the quadrature method of moment (QMOM) as proposed by McGraw (1997) could be employed to purposefully approximate the moment integrals. Here, nodes or abscissas and weights of the quadrature approximation can be determined from the moments of the distribution by using a very efficient algorithm (Gordon, 1968, McGraw, 1997,

McGraw and Wright, 2003). QMOM is a rather sound mathematical approach and represents an elegant tool in solving the population balance equations with limited computation burden. In essence, this method may be regarded as a presumed PSD method where the underlying distribution is assumed to be made of delta functions. It thus possesses many similarities with the conventional SMM where the PSD is assumed to be a monodisperse or lognormal distribution. Marchisio and Fox (2005) formulated a direct formulation (direct quadrature method of moment or DQMOM) for multi-dimensional problems, which this particular approach is able to cater for poly-disperse systems with two or more internal coordinates. Encouraging results attained by Barret and Webb (1998) and Marchisio et al. (2003a, 2003b) have certainly elevated DQMOM as a serious competing method that presents the main advantage of being extremely accurate in solving monovariate (i.e., one internal coordinate) problems and amenable for coupling with CFD calculations.

4.4.2 Formulation of Transport Equations and Rate Mechanisms

From the specific consideration of not overburdening the computational requirement, especially in the numerical treatment of fire combustion, two methods based on the population balance modeling of soot formation in turbulent flames are expounded in this section. Both of these methods are aptly applicable to field modeling investigations within the context of practical models for fire engineering. The basic framework of the first model is the moment method with a presumed PSD in describing the soot properties, which is principally based on the development carried out by Hong et al. (2005) for the predictions of trends in soot emissions for a wide range of operating conditions in diesel engines. Another novel and promising approach proposed by Zucca et al. (2006), the second model to be subsequently discussed, applies the DQMOM on the population balance equation in predicting the evolution of the size distribution of the soot particles generated by chemical reaction and/ or undergoing chemical and physical processes affecting their size.

Standard Method of Moments (SMM) with a Presumed PSD
In the method of moments, the evolution of the soot properties based on the model of Hong et al. (2005) is determined by only the first three moments, which represent the soot number density (M_0), soot volume fraction (M_1), and the average volume (M_2). With a presumed log-normal distribution as described by

$$n = \frac{1}{3\sqrt{2\pi}\log\sigma}\ exp\ \left(-\frac{\log^2(V/V_g)}{18\log^2\sigma} \right)\frac{1}{V} \qquad (4.4.1)$$

where n is the number density of spherical soot particles of volume V, V_g is the average particle volume, and σ is the standard deviation of the volume distribution. The first three moments that are considered herein in order to provide sufficient resolution of the soot properties can be expressed by the following relationships:

$$\tilde{M}_k = \tilde{M}_0 V_g^k \exp\left(\frac{9}{2}k^2\log^2\sigma\right) \tag{4.4.2}$$

$$V_g = \left(\frac{\tilde{M}_1^2}{M_0^{3/2}M_2^{1/2}}\right) \tag{4.4.3}$$

$$\log^2\sigma = \frac{1}{9}\log\left(\frac{\tilde{M}_0\tilde{M}_2}{\tilde{M}_1^2}\right) \tag{4.4.4}$$

It is noted that since a PSD is assumed, higher moments are not required to be determined, and significant computational savings can be realized. The assumption of a log-normal distribution for soot particles is fully supported by experimental observations (Annele et al., 2004, Harris and Maricq, 2002, and Haynes and Wagner, 1981). Analogous to the Favre-averaged transport equations describing the particle number density and soot mass fraction or volume fraction in the previous section, the transient features of the soot moments can be written as

$$\frac{\partial}{\partial t}(\bar{\rho}\tilde{M}_0) + \frac{\partial}{\partial x_j}(\bar{\rho}\bar{u}_j\tilde{M}_0) + 0.55\frac{\partial}{\partial x_j}\left[\tilde{M}_0\frac{\mu}{\tilde{T}}\frac{\partial\tilde{T}}{\partial x_j}\right]$$
$$= \frac{\partial}{\partial x_j}\left[\Gamma_T\frac{\partial\tilde{M}_0}{\partial x_j}\right] + \bar{R}_0^{pi} - \bar{R}_0^{coag} \tag{4.4.5}$$

$$\frac{\partial}{\partial t}(\bar{\rho}\tilde{M}_1) + \frac{\partial}{\partial x_j}(\bar{\rho}\bar{u}_j\tilde{M}_1) + 0.55\frac{\partial}{\partial x_j}\left[\tilde{M}_1\frac{\mu}{\tilde{T}}\frac{\partial\tilde{T}}{\partial x_j}\right]$$
$$= \frac{\partial}{\partial x_j}\left[\Gamma_T\frac{\partial\tilde{M}_1}{\partial x_j}\right] + \bar{R}_1^{pi} + \bar{R}_1^{sg} + \bar{R}_1^{ox} \tag{4.4.6}$$

$$\frac{\partial}{\partial t}(\bar{\rho}\tilde{M}_2) + \frac{\partial}{\partial x_j}(\bar{\rho}\bar{u}_j\tilde{M}_2) + 0.55\frac{\partial}{\partial x_j}\left[\tilde{M}_2\frac{\mu}{\tilde{T}}\frac{\partial\tilde{T}}{\partial x_j}\right]$$
$$= \frac{\partial}{\partial x_j}\left[\Gamma_T\frac{\partial\tilde{M}_2}{\partial x_j}\right] + \bar{R}_2^{pi} + \bar{R}_2^{sg} + \bar{R}_2^{ox} + \bar{R}_2^{coag} \tag{4.4.7}$$

where Γ_T is the turbulent diffusivity, $\bar{R}_k^{pi}$ is the particle inception rate, $\bar{R}_k^{sg+ox}$ is the combined surface growth and oxidation rates, and $\bar{R}_k^{coag}$ is the coagulation rate for $k = 0$, 1, and 2 moments. Particular attention of the source and sink terms in equations (4.45)–(4.47), to be further described following, have been formulated based on detailed physical and chemical sub-models for the soot processes, which include particle inception of soot primary particles coupled to the reaction kinetics, soot coagulation based on the collision theory, soot oxidation by hydroxyl radical OH and oxygen O_2, and soot surface growth using a modified Hydrogen-Abstraction-Carbon-Addition (HACA) mechanism.

As indicated in Frenklach (2002), the nucleation of soot primary particles is the least well-understood step in the soot formation. Frenklach and Wang (1994) have estimated the soot nucleation rates based on the consideration of higher PAH compounds of which the process involves the formation of two two-dimensional PAHs merging into the first three-dimensional structure, as well as the condensation of one two-dimensional PAH joining onto a three-dimensional PAH in the presence of acetylene (see Figure 4.1). The use of PAH as a soot precursor relies heavily on a relatively large reaction mechanism, which can result in increased computational costs and possibly engages in a larger degree of empiricism, if the supporting chemistry is unknown. Similar to Leung et al. (1991), Hong et al. (2005) simplified the nucleation of soot monomers by assuming acetylene as the primary soot precursor of which the particle inception rate for the kth moment is modeled according to

$$\bar{R}_k^{pi} = 1 \times 10^4 \ \exp\left(\frac{-2110}{\tilde{T}}\right) \left(\frac{\bar{\rho}\tilde{Y}_{C_2H_2}}{M_{C_2H_2}}\right) A_C V_s^k N_o \qquad (4.4.8)$$

where A_C is a correction factor with a value of 0.01, which has been calibrated and optimized under benchmark experiments, V_s^k is the specific volume of the soot primary particle raised to the power of the kth moment and N_o is the Avogrado's number.

The modified HACA mechanism (Markatou et al., 1993) with additional reaction paths as suggested by Colket and Hall (1994) is adopted for the combined surface growth and soot oxidation rates. A schematic illustration of the surface growth phenomenon alongside with a basic description of the soot cluster, can be envisaged in Figure 4.5. The kinetic mechanisms due to surface reaction and oxidation are:

$$\begin{array}{ll} C_{soot}H + H \rightleftharpoons C_{soot}^* + H_2 & \text{(R1)} \\ C_{soot}^* + H \rightleftharpoons C_{soot} + H & \text{(R2)} \\ C_{soot}^* + C_2H_2 \rightleftharpoons C^*C_2H_2 & \text{(R3)} \\ C_{soot}^* + C_2H_2 \rightarrow C_{soot}H + H & \text{(R4)} \\ C_{soot}^* + O_2 \rightarrow \text{Products} & \text{(R5)} \\ C_{soot}^*C_2H_2 + O_2 \rightarrow \text{Products} & \text{(R6)} \\ C_{soot}H + OH \rightarrow \text{Products} & \text{(R7)} \end{array}$$

Appropriate empirical rate constants associated with the preceding reaction rates are given in Table 4.1. The surface growth mechanism includes a chemical reaction path for the surface reactions (see reaction steps R1 and R2), which involves an active site on the surface of the soot particle bound to a hydrogen

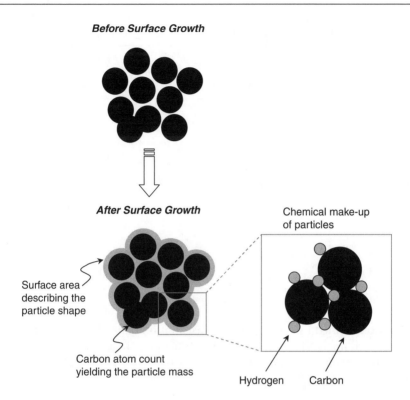

Figure 4.5 Obliteration of primary particles due to surface reaction and a basic description of an actual grown cluster.

radical $C_{soot}H$ and the radical active site C^*_{soot} due to the attack of H atoms, separates the acetylene addition process into a reversible formation of the radical adduct $C^*_{soot}C_2H_2$ (reaction R3) and a cyclization reaction (as depicted by reaction step R4). Soot oxidation proceeds as demonstrated by reaction steps (R5)–(R7), due to the attack of oxygen molecule O_2 and hydroxyl radical OH on the soot particle, radical adduct, and soot particle bound to a hydrogen radical. The rate constants for soot oxidation by O_2 and OH are obtained from Appel et al. (2000) and Neoh et al. (1981).

Contributions of the soot growth to the moment equations (4.46) and (4.4.7) are calculated using the empirical rate constants in Table 4.1, along with the mean rates given by

$$\bar{R}^{sg+ox}_k = \int\limits_0^\infty V_i^k k_X[X]\alpha S_i n_i dV_i \qquad (4.4.9)$$

where V_i^k is the volume of soot particle of ith size raised to the power of kth moment, S_j is the surface area of the jth soot particle, and n_j is the particle size

Table 4.1 Empirical rate constants for the modified HACA mechanism.

Reaction	$k = A \exp(- E / RT)$	
	A (cm^3 mol^{-1} s^{-1})	E (kcal mol^{-1})
R1f	2.5×10^{14}	12.0
R1b	4.0×10^{11}	7.0
R2f	2.2×10^{14}	–
R2b	2.0×10^{17}	109.0
R3f	2.0×10^{12}	4.0
R3b	5.0×10^{13}	38.0
R4	5.0×10^{10}	–
R5	2.2×10^{12}	7.5
R6	2.2×10^{10}	7.5
R7	Reaction probability $= 0.13$	

Note: f denotes the forward rate, while b denotes the backward rate.

distribution defined by the log-normal distribution defined in equation (4.4.1). The expression $k_X[X]$ in equations (4.4.9) denotes the product of the per-site rate coefficient and the concentrations of gas species involved in the surface reactions, which may be formulated as

$$k_X[X] = k_{f,4}[C_2H_2]\Psi^*_{Csoot} + k_5[O_2]\Psi^*_{Csoot} + k_7[OH] \qquad (4.4.10)$$

of which the surface area of the particle is multiplied by a factor Ψ^*_{Csoot} to account for the number of surface radicals per unit surface area. This factor is usually derived from a steady state assumption applied to the reactions steps (R1)–(R7), which is given by

$$\Psi^*_{Csoot} = \frac{k_{f,1}[H]}{k_{b,1}[H_2] + k_{f,2}[H] + k_{4,f}[C_2H_2] + k_6[O_2]} \times \Psi^*_{Csoot\_H} \qquad (4.4.11)$$

The number of $C_{soot}H$ sites per unit of soot particle ($\Psi^*_{Csoot\_H}$) can be estimated on the basis of dimensions of PAH rings and distance between PAH soot layers in soot yielding a value of about 2.3×10^{19} to 2.9×10^{19} m^{-2}. Subsequently, the steric factor α in equation (4.49), accounts for the probability of the gaseous species colliding with the reactive prismatic planes of a soot particle. According to Appel et al. (2000), it is generally found to exhibit a dependence on the temperature and mean particle size

$$\alpha = \tanh\left(\frac{12.65 - 15.3 \times 10^{-4}\tilde{T}}{\log(\tilde{M}_1/\tilde{M}_0)} + -1.38 + 6.8 \times 10^{-4}\tilde{T}\right) \qquad (4.4.12)$$

Concerning coagulation, the merging of two soot particles into one larger soot particle, the Smoluchowski's equation (Hinds, 1999) is used to describe the rate of coagulation of soot particle caused by collision

$$\bar{R}_k^{coag} = \int_0^\infty V_i^k \frac{1}{2} \int_0^{V_i} \beta(V_j, V_k) n_j n_k dV_j - \int_0^\infty V_i n_i \int_0^\infty \beta(V_j, V_i) n_j dV_j dV_i \qquad (4.4.13)$$

where $\beta(V_j, V_k)$ and $\beta(V_j, V_i)$ are the collision frequencies between soot particles, n_i and n_j are to number densities of soot particles of volumes i and j, V_i, and V_j are the volumes of soot particles of size i and j, and n_k is the number density for the kth moment corresponding to a soot particle volume V_k. When describing coagulation, three regimes that are free molecular, transition, and continuum can be appropriately described by the Knudsen number:

$$Kn \equiv \frac{\text{mean free path of gas molecules}}{\text{physical length scale}} = \frac{\lambda}{L} \qquad (4.4.14)$$

More specifically, the different regimes are:

- Free molecular, $Kn \gg 1$: The path between the particles is much *larger* than the particle diameter. It is anticipated that the particles in this regime are free to move around.
- Transition, $0.1 < Kn < 1$: This represents the state between the free molecular regime and the continuum regime
- Continuum, $Kn \ll 1$: The path between the particles is much *smaller* than the particle diameter. The particles are rather crowded and the movement in this regime is close to being a continuous flow.

By definition, the mean free path of gas molecules λ, which is the average distance the particle travels between collisions with other particles, may be calculated based on the particle diameter d from the following expression:

$$\lambda = \frac{k_B \tilde{T}}{\sqrt{2}\pi d^2 p_0} \qquad (4.4.15)$$

where k_B is the Boltzmann constant, which is essentially the ratio between the universal gas constant and the Avogadro's number ($\approx 1.38 \times 10^{-23}$ J / K), and p_0 is the fixed ambient pressure. In many combustion studies associated with the prediction of soot, it is common practice to adopt the particle radius as the physical length scale in which the Knudsen number is $Kn = 2\lambda/d$. On the basis of equation (4.4.15), the Knudsen number is directly dependent on only the particle diameter d. As dictated by the size of the soot particles, the collision frequency of the free molecular regime and the continuum regime appearing in equation (4.4.13) can be determined by

Free molecular regime

$$\beta(V_a, V_b) = K_F(V_a^{1/3} + V_b^{1/3})^2 \sqrt{\frac{1}{V_a} + \frac{1}{V_b}} \tag{4.4.16}$$

Continuum regime

$$\beta(V_a, V_b) = K_C(V_a^{1/3} + V_b^{1/3})\left(\frac{C(V_a)}{V_a^{1/3}} + \frac{C(V_b)}{V_b^{1/3}}\right) \tag{4.4.17}$$

In equation (4.4.16), K_F is obtained from

$$K_F = \left(\frac{3}{4\pi}\right)^{1/6} \sqrt{\frac{6k_B\tilde{T}}{\bar{\rho}}} \tag{4.4.18}$$

while in equation (4.417), K_C is given by

$$K_C = \frac{2k_B\tilde{T}}{3\mu} \tag{4.4.19}$$

where μ is the gas-phase viscosity and the Cunningham slip correction factor $C(V_l)$ can be obtained from

$$C(V_l) = 1 + 1.257Kn_l \tag{4.4.20}$$

Note that the Knudsen number changes accordingly with the particle diameter based on the specific size of the volume V_l in the preceding Cunningham slip correction factor. In the transition regime, the coagulation rate is typically determined by harmonic mean of the continuum and free molecular rate:

$$\bar{R}_k^{coag} = \frac{\bar{R}_k^{coag,C} \bar{R}_k^{coag,F}}{\bar{R}_k^{coag,C} + \bar{R}_k^{coag,F}} \tag{4.4.21}$$

Direct Quadrature Method of Moments (DQMOM) In describing the application of DQMOM in modeling soot formation in turbulent flames, the consideration of the population balance equation for the Favre-averaged number density $\tilde{n}$ is first described. It can be written similar to the transport equations for the moments as

$$\frac{\partial}{\partial t}(\bar{\rho}\tilde{n}) + \frac{\partial}{\partial x_j}(\bar{\rho}\tilde{u}_j\tilde{n}) + 0.55\frac{\partial}{\partial x_j}\left[\tilde{n}\cdot\frac{\mu}{\tilde{T}}\frac{\partial\tilde{T}}{\partial x_j}\right] = \frac{\partial}{\partial x_j}\left[\Gamma_T\frac{\partial\tilde{n}}{\partial x_j}\right] + \bar{S}_{\tilde{n}} \tag{4.4.22}$$

The main idea behind this approach as proposed by Zucca et al. (2006) is to obtain the solution of the closure problem by using a quadrature approximation of order N to predict the evolution of the moments of the PSD. This corresponds to the approximation of the number density as

$$\tilde{n} \approx \sum_{\alpha=1}^{N} w_\alpha \prod_{i=1}^{M} \delta[\xi_i - \xi_{i,\alpha}] \tag{4.4.23}$$

where w_α are the *weights* and $\xi_{i,\alpha}$ are the *abscissas* of the quadrature approximation, ξ_i is the discrete internal coordinate corresponding to the total number of internal coordinates M, and δ indicates the Dirac delta function. By considering only a single internal coordinate, for example the particle size as the internal coordinate ($\xi = L$), the quadrature approximation for this monovariate population balance yields

$$\tilde{n} \approx \sum_{\alpha=1}^{N} w_\alpha \delta[L - L_{i,\alpha}] \tag{4.4.24}$$

with the kth moment of the distribution subsequently expressed as

$$\tilde{M}_k = \int_0^{+\infty} \tilde{n} L^k \, dL \approx \sum_{\alpha=1}^{N} w_\alpha L_\alpha^k \tag{4.4.25}$$

On the basis of the development by Marchisio and Fox (2005), the approach of DQMOM consists of solving the transport equations of weights and abscissas. They are

$$\frac{\partial}{\partial t}(\bar{\rho} w_\alpha) + \frac{\partial}{\partial x_j}(\bar{\rho} \tilde{u}_j w_\alpha) + 0.55 \frac{\partial}{\partial x_j}\left[w_\alpha \frac{\mu}{\tilde{T}} \frac{\partial \tilde{T}}{\partial x_j}\right] = \frac{\partial}{\partial x_j}\left[\Gamma_T \frac{\partial w_\alpha}{\partial x_j}\right] + a_\alpha \tag{4.4.26}$$

$$\frac{\partial}{\partial t}(\bar{\rho} \mathcal{L}_\alpha) + \frac{\partial}{\partial x_j}(\bar{\rho} \tilde{u}_j \mathcal{L}_\alpha) + 0.55 \frac{\partial}{\partial x_j}\left[\mathcal{L}_\alpha \frac{\mu}{\tilde{T}} \frac{\partial \tilde{T}}{\partial x_j}\right] = \frac{\partial}{\partial x_j}\left[\Gamma_T \frac{\mathcal{L}_\alpha}{\partial x_j}\right] + b_\alpha \tag{4.4.27}$$

where $\mathcal{L} = w_\alpha L_\alpha$ represents the αth weighted abscissa and a_α and b_α are the respective source terms. According to Marchisio and Fox (2005), the source terms can be easily evaluated by solving a linear algebraic system, obtained from the population balance equation after application of the quadrature approximation and forcing the moments to be tracked with a higher level of accuracy. Setting $N = 2$ and $M = 1$, the source terms can be ascertained by solving the following linear system of the first four moments as

$$
\begin{aligned}
a_1 + a_2 &= \bar{S}_0 \\
b_1 + b_2 &= \bar{S}_1 \\
-L_1^2 a_1 - L_2^2 a_2 + 2L_1 b_1 + 2L_2 b_1 &= \bar{S}_2 + \bar{C}_2 \\
-2L_1^3 a_1 - 2L_2^3 a_2 + 3L_1^2 b_1 + 3L_2^2 b_1 &= \bar{S}_3 + \bar{C}_3
\end{aligned}
\tag{4.4.28}
$$

From the preceding, $\bar{S}_k$ represents the integrated source term of the kth moment:

$$
\bar{S}_k = \int\limits_{-\infty}^{+\infty} L^k \bar{S}_{\tilde{n}} \, dL
\tag{4.4.29}
$$

while $\bar{C}_k$ is a correction term which arises due to the quadrature approximation

$$
\bar{C}_k = k(k-1) \sum_{\alpha=1}^{N} L_\alpha^{k-2} w_\alpha \Gamma_T \frac{\partial D_\alpha}{\partial x_j} \frac{\partial D_\alpha}{\partial x_j}
\tag{4.4.30}
$$

In order to solve the linear system in equation (4.4.28), the appropriate source terms $\bar{S}_k$ are required to be evaluated as a result of the summation of several contributions where each of them corresponds to a specific process. These source terms are closely related to the nucleation, coagulation, surface growth, and oxidation of soot particles, which are further described following.

Considering the nucleation of soot particles, Zucca et al. (2006) have assumed that nucleation produces a uniform distribution of nuclei size $0 \leq L \leq L_\varepsilon$, where L_ε indicates the maximum possible size of the nuclei in order to avoid the abscissas of the quadrature approximation that may become null in the regions where there are no particles present. Based on this assumption and by means of the probability distribution algorithm (Gordon, 1968), the N abscissas corresponding to the production of the nuclei can be evaluated where w_α and L_α are null. With $N = 2$, the abscissas distribution of the nucleation are $D_1 = 0.2113 \cdot L_\varepsilon$ and $D_2 = 0.7887 \cdot L_\varepsilon$. The approximate expression for the source term of moments due to nucleation of a uniform distribution of nuclei is given by

$$
\bar{S}_k^{pi} \approx \frac{L_\varepsilon^k}{k+1} \bar{J}
\tag{4.4.31}
$$

Zucca et al. (2006) proposed the use of the kinetic rate proposed by Moss et al. (1995), which is based on acetylene as the soot precursor, for the nucleation rate $\bar{J}$. The rate expression is

$$
\bar{J} = 6 \times 10^6 \bar{\rho}^2 N_o \sqrt{\tilde{T}} \exp\left(\frac{-46100}{\tilde{T}}\right) \tilde{X}_{C_2 H_2}
\tag{4.4.32}
$$

where $\tilde{X}_{C_2H_2}$ is the mean mole fraction of acetylene.

As an alternative to the complex HACA mechanism, a simpler approach could suffice to account for the continuous size changes due to surface growth and oxidation of soot particles. Denoting the rate of continuous change of the particle size as G, the source terms of moments due to the combined surface growth and oxidation applying the quadrature approximation are

$$\bar{S}_k^{sg+ox} \approx k \sum_{\alpha=1}^{N} w_\alpha L_\alpha^{k-1} \bar{G}^{sg+ox} \tag{4.4.33}$$

Zucca et al. (2006) employed the models of Liu et al. (2003) and Said et al. (1997) for the surface growth and oxidation due to only the oxygen molecule O_2, respectively. The rate expressions for the combined surface growth and oxidation can be written as

$$\bar{G}^{sg+ox} = \frac{6}{D_f \rho_s} \left(\frac{R}{R_{c0}} \right)^{\frac{3-D_f}{3}} 2M_s \cdot 6 \, exp \left(\frac{-6038}{\tilde{T}} \right) \tilde{C}_{C_2H_2}$$

$$- \frac{\bar{p}}{D_f \rho_s} T^{-1/2} 6.5 \, exp \left(\frac{-26500}{\tilde{T}} \right) \tilde{Y}_{O_2} \tag{4.4.34}$$

where the subscript "0" indicates the primary particle $\tilde{C}_{C_2H_2}$ is the mean concentration of acetylene, and $\tilde{Y}_{O_2}$ is the mean mass fraction of oxygen. In the preceding equation (4.4.34), the fractal dimension D_f is usually defined through the collision radius (R_c) of the aggregates involved in the collision event by the following relationship:

$$R_c = \frac{D_0}{2} \left(\frac{V}{V_0} \right)^{1/D_f} \tag{4.4.35}$$

Spherical primary particles D_0 of 15 nm in size, supported by comparison with experimental data, is considered in Zucca et al. (2006). The radius of the primary particle (R_{c0}) is thus 7.5 nm according to equation (4.4.35). It is imperative to predict the fractal dimension of particles with sufficiently high accuracy. This is to gain not only information on particle morphology but also to calculate more realistic collision radii or diameters. Artelt et al. (2003) have modeled the evolution of the fractal dimension as

$$D_f = \begin{cases} D_{f,min} + (D_{f,0} - D_{f,min})^{1/\tau^s} \tau \leq 1 \\ D_{f,max} + (D_{f,max} - D_{f,0})^{\tau^s} \tau > 1 \end{cases} \tag{4.4.36}$$

The characteristic time τ in equation (4.4.36) essentially describes the ratio between the characteristic collision time t_c and characteristic restructuring time t_r— that is, $\tau = t_c/t_r$. If t_r is much smaller than t_c, adherent particles will nearly always be completely combined prior to the next collision event. Most of the isolated spherical particles will be formed with a fractal dimension $D_{f,max} = 3$. On the other hand, if t_c is much smaller than t_r, adherent particles will face their next collision event far before sintering has been terminated. A fractal dimension of around 1.7 ($= D_{f,min}$) will thus be generated. The characteristic collision time can be determined according to Rosner and Yu (2001) as

$$t_c = \frac{1}{\bar{\beta}\tilde{M}_0} \tag{4.4.37}$$

where $\bar{\beta}$ is the average aggregation kernel for two particles, and $\tilde{M}_0$ is the number concentration of particles (i.e., the zeroth moment order of the distribution). The characteristic restructuring time can be assumed to be equivalent to a turbulence micro-scale time so that the restructuring process is taken to be proportional to the shear rate inside the turbulent eddy

$$t_r = \sqrt{\frac{15\mu}{\bar{\rho}\varepsilon}} \tag{4.4.38}$$

In equation (4.4.36), the parameter s determines the slope of fractal dimension variation of which has been chosen to be equal to 1 as stipulated in Artelt et al. (2003). Accordingly, $D_{f,0}$ accounts for the fractal dimension at identical characteristic collision and restructuring and is assumed to be the arithmetic average value between the limiting cases; this equals 2.35.

For the coagulation process, the source term for the moments due to the aggregation of two particles of sizes D_1 and D_2 via applying the quadrature approximation can be written as

$$\bar{S}_k^{coag} \approx \frac{1}{2}\sum_{\alpha=1}^{N}\sum_{\gamma=1}^{N}(L_\alpha^3 + L_\gamma^3)^{k/3}\beta_{\alpha\gamma}w_\alpha w_\gamma - \sum_{\alpha=1}^{N}\sum_{\gamma=1}^{N}L_\alpha^k\beta_{\alpha\gamma}w_\alpha w_\gamma \tag{4.4.39}$$

The frequency (kernel) of aggregation of two particles with collision radius R_{c1} and R_{c2} is given by $\beta_{\alpha\gamma} = \beta(R_{c1}, R_{c2})$ in the preceding equation (4.4.39) which is evaluated based on Fuchs interpolation formula (Fuchs, 1964) as

$$\beta(R_{c1}, R_{c2}) = 4\pi(D_1 + D_2)\times$$

$$(R_{c1} + R_{c2})\left[\frac{(R_{c1} + R_{c2})}{R_{c1} + R_{c2} + \sqrt{g_1^2 + g_2^2}} + \frac{4(D_1 + D_2)}{\sqrt{c_1^2 + c_2^2}(R_{c1} + R_{c2})}\right]$$

$$\tag{4.4.40}$$

where c_i, D_i, and g_i are given by

$$c_i = \sqrt{\frac{8k_B \tilde{T}}{\pi m_i}},$$

$$D_i = \frac{k_B \tilde{T}}{6\pi\mu R_{ci}} \left(\frac{5 + 4Kn_i + 6Kn_i^2 + 18Kn_i^3}{5 - Kn_i + (8 + \pi)Kn_i^2} \right),$$

$$l_i = \frac{8D_i}{\pi c_i}$$

$$g_i = \frac{(2R_{ci} + l_i)^{1/3} - (4R_{ci}^2 + l_i^2)^{3/2}}{6R_{ci}l_i} - 2R_{ci}$$

(4.4.41)

On the basis of the soot density ρ_{soot}, the mass of a soot particle m_i can be calculated in equation (4.4.41) from

$$m_i = \rho_{soot} V_i \qquad (4.4.42)$$

Also in equation (4.4.41), the Knudsen number based on the ratio between mean free path of gas molecules and the particle radius is evaluated according to

$$Kn_i = \frac{2k_B \tilde{T}}{\sqrt{2}\pi d_i^3 p_o} \qquad (4.4.43)$$

4.5 Guidelines for Selecting Soot Models in Fire Modeling

During combustion, finely dispersed carbonaceous particles (soot) that emit at specific wavelength bands are present in many practical fires. In view of the complex and varied physical and chemical mechanisms governing the formation and oxidation of soot, what suitable guidelines can be provided in the selection of soot models in field modeling?

Categorically, fires can be considered to be of either lightly or moderately to heavily sooty flames. For moderately to heavily sooty flames of complex fuels, single-step empirical rate models offer the feasibility of ascertaining the soot concentration with minimal computational expense in order to augment the radiation contribution due to soot particles. These models are nevertheless only applicable for *homogeneous* soot formation. For lightly sooty flames, such as non-premixed combustion of most hydrocarbon fuels, *heterogeneous* soot formation process in addition to the *homogeneous* complement are both likely to contribute to the total soot formation rate. Semi-empirical models offer the flexibility of modeling the essential processes associated with nucleation (inception), coagulation, surface growth, aggregation, and particle oxidation in order to accommodate the heterogeneous soot formation while not significantly overburdening the computational load. Depending of whether the soot precursor is assumed to be of the actual fuel or acetylene, the explicit knowledge of the combustion chemistry pre-determines the application of these models. For soot nucleation and growth, the general approach developed by Tesner et al. (1971a, 1971b) and Moss et al. (1988) of using the parent fuel concentrations, offers great advantage especially in field modeling of fires due to the absence of comprehensive reaction mechanisms of practical flames. Nevertheless, the concept whereby soot nucleation and growth is activated by the presence of acetylene is physically more plausible, although it complicates the modeling by the need to incorporate detailed reaction mechanism in the combustion model to determine the acetylene mass fraction. Here, the model by Leung et al. (1991) provides computational tractability of resolving the soot process with minimal computational effort. In field modeling, such a model should provide sufficiently accurate soot concentrations for the consideration of radiation heat transfer in fires. Full-blown attempts to characterize the soot process via detailed models that seek to solve the rate of equations for elementary reactions leading to soot could nevertheless be adopted to predict the evolution of the size distribution of the soot particles generated by chemical reaction and/or undergoing chemical and physical processes. Such models are invariably more complicated and require substantial computational resources, but they tend to provide better prediction of the soot concentration levels. Note again that they should only be applied if the detailed chemistry of the parent fuel is fully realized.

4.6   Worked Examples on the Application of Soot Models in Field Modeling

4.6.1   Two-Room Compartment Fire

In this worked example, the two-room compartment fire case is used to further explore the importance of luminous soot radiation. A heat release rate of 110 kW for the fire is also adopted as the basis of comparison. Numerical simulations are performed through an in-house computer code, FIRE3D. Results incorporating soot and gaseous radiation are assessed against previous field modeling predictions with the consideration of only gaseous radiation from previous worked examples in section 3.13.2 and experimental data of turbulent buoyant diffusion flames measured by Nielsen and Fleischmann (2000).

Numerical features: This worked example solves the system of governing equations and boundary conditions as described in the worked example in section 3.6.2 along with the numerical models of turbulence, combustion, and radiation depicted in the worked example in section 3.13.2. For soot formation, three models are considered (i) the single-step empirical model of Khan and Greeves (1974), (ii) semi-empirical model of Tesner et al. (1971a,

1971b), and (ii) semi-empirical model of Moss et al. (1988). For convenience, these models are hereby referred as (i) Model 1, (ii) Model 2, and (iii) Model 3. These empirical soot models are attractive, since soot production is described simply in terms of concentration of the parent fuel and local temperature. The constants used for the soot formation equations in Model 1, Model 2, and Model 3 can be found in section 4.3.2. A constant $C_s = 0.01$ kg N^{-1} m^{-1} s^{-1} is assumed in Model 1, while a mean particle diameter of 22.5×10^{-9} m is prescribed in Model 2. In all the calculations, the soot density is taken to be 2000 kg·m^{-3}.

The importance of carrying out a grid sensitivity analysis is demonstrated by the numerical predictions obtained for the temperature profiles at the doorway connecting the burn room and adjacent room. Three grid meshes are tested: a coarse mesh of $47 \times 25 \times 15$ (a total of 17625 control volumes), a medium mesh of $65 \times 35 \times 20$ (a total of 45500 control volumes), and a fine mesh of $85 \times 44 \times 25$ (a total of 93500 control volumes). The case considering the soot model of Model 3 is adopted to investigate grid independency. Figure 4.6 illustrates the three predicted temperature profiles plotted against the experimentally measured profile. The coarse mesh grossly over-predicts the temperatures at the upper part of the doorway. With increasing mesh density, the predicted temperature distributions are more comparable with the measured data. On a closer examination, the numerical results do not show any appreciable differences between the medium and fine meshes. It can thus be concluded that the resolution of the fine mesh should suffice in adequately resolving the turbulent reacting flow within the two-compartment configuration.

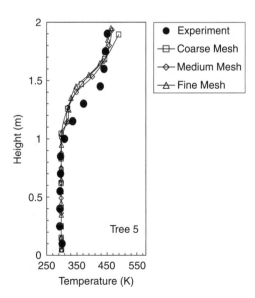

Figure 4.6 Grid sensitivity analysis on the predicted temperature profiles at the doorway.

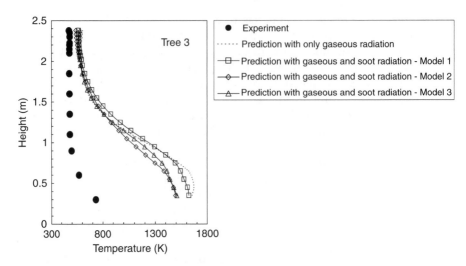

Figure 4.7 Comparison of predicted and measured temperature profiles above the fire source (Tree 3) with the consideration of only gaseous radiation and gaseous and soot radiation via Model 1, Model 2, and Model 3.

Numerical results: The predicted and measured vertical temperature distributions above the fire source (Tree 3) are shown in Figure 4.7. In spite of the additional consideration of soot radiation, the computed vertical temperature profiles are still appreciably higher than the measured data. As postulated in the worked example in section 3.6.2, Wen et al. (2001) have demonstrated that temperatures measured through bare-wire thermocouples do not actually reflect the real fluid temperatures. Heat transfer processes that generally occur by convection and radiation taking place within the sensor, surrounding surfaces, and fluid balance each other to record temperatures between the surface and surrounding fluid temperatures. This error is particularly amplified in high-temperature regions, especially temperatures above the fire source. On the basis of the analysis carried out by Wen et al. (2001), the thermocouple readings on this tree could be corrected by assuming the following heat transfer equilibrium equation:

$$\dot{q}_{convective\ to\ and\ from\ thermocouple} = \dot{q}_{radiative\ to\ and\ from\ thermocouple}$$

$$h(T_{gas} - T_{th}) = e_{th}(\sigma T_{th}^4 - \dot{q}_r) \tag{4.6.1}$$

where $\dot{q}_r$ is the local radiative flux, which may be obtained from the predicted values evaluated via DOM. The convective heat transfer coefficient, h, is estimated using correlation taken from Holman (1992) for a sphere in cross-flow, while the cross-flow velocity is set according to the predicted mean flow velocity across the sensor. The emissivity of the thermocouple e_{th} can be set to 0.2, according to the data in Perry and Chilton (1997). Applying equation (4.6.1),

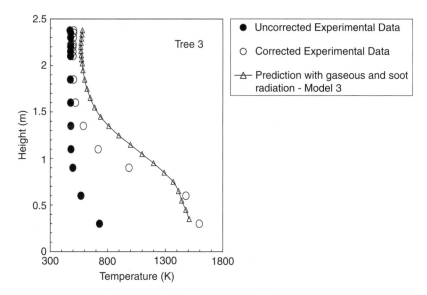

Figure 4.8 Comparison of uncorrected and corrected experimental data against predicted temperature profiles using the soot model of Model 3.

considering the local radiative flux evaluated from the soot model of Model 3, the corrected temperatures as shown in Figure 4.8 are seen to behave more consistently with typically observed temperatures (Drysdale, 1999).

Figure 4.9 compares the predictions of the vertical temperature profiles in the burn room against the spatial measurements carried out for thermocouple tress 1, 2, 4 (away from the fire source) and at the doorway by the thermocouple tree 5. Comparing against the results where only gaseous radiation is considered, the presence of soot radiation is seen to significantly augment the global radiation exchange by significantly improving the temperature predictions. Table 4.2 presents the measured floor temperatures against temperatures predicted by the various soot models. Floor temperatures in the burn room increase because of the backward radiation from the hot smoke layer below the ceiling of the compartment. Predicting floor temperatures by considering only gaseous radiation in the computational analysis does not yield satisfactory comparison to the experimental data measured at the locations of the thermocouple trees 1, 2, and 4 within the burn room. It is apparent that the radiation heat transfer is not only due to radiation contribution by the combustion products alone. By accounting further the luminous soot radiation, the model predicts temperatures that are inadvertently much closer to the measured temperatures. In Figure 4.9, temperatures at the upper part of the doorway are also found to be more comparable to the measured data due to the effect of soot radiation. From a modeling viewpoint, the presence of soot radiation improves the accuracy of predictions.

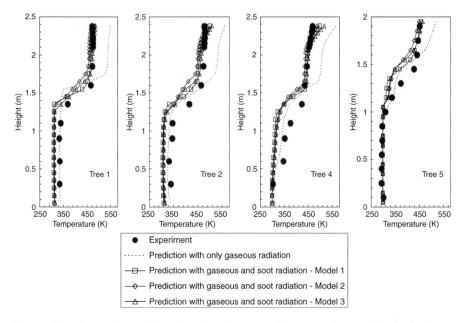

Figure 4.9 Comparison of predicted and measured temperature profiles in the burn room (Trees 1, 2, and 4) and at the doorway (Tree 5) with the consideration of only gaseous radiation and gaseous and soot radiation via Model 1, Model 2, and Model 3.

Table 4.2 Comparison of floor temperatures in the burn room.

	Tree 1 (K)	Tree 2 (K)	Tree 4 (K)
Experiment	398.95	413.65	423.75
Only gaseous radiation	495.76	515.91	499.04
Soot and gaseous radiation–Model 1	352.47	369.83	365.75
Soot and gaseous radiation–Model 2	353.74	368.95	364.00
Soot and gaseous radiation–Model 3	356.16	374.16	369.49

Numerical predictions of the vertical temperature profiles in the adjacent room and spatially measured temperatures carried out for thermocouple trees 6, 7, 8, and 9 are illustrated in Figure 4.10. All models considering soot radiation marginally under-predict the temperatures below the ceiling within the adjacent room. This could be attributed to the neglect of soot burnout or oxidation in the computational analysis. The measured and predicted floor temperatures in this adjacent room are tabulated in Table 4.3. As a consequence of a thinner smoke layer, the effect of backward radiation is not as pronounced as in the burn room. Predicted floor temperatures employing soot Model 3 are observed to be marginally closer to the measured data.

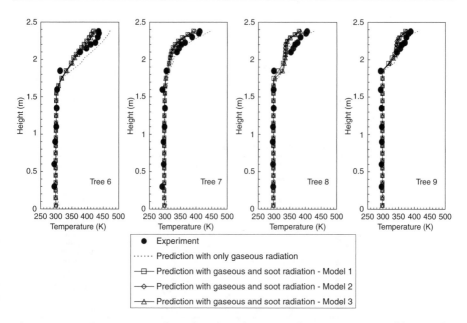

Figure 4.10 Comparison of predicted and measured temperature profiles in the adjacent room (Trees 6, 7, 8, and 9) with the consideration of only gaseous radiation and gaseous and soot radiation via Model 1, Model 2, and Model 3.

Table 4.3 Comparison of floor temperatures in the adjacent room.

	Tree 6 (K)	Tree 7 (K)	Tree 8 (K)	Tree 9 (K)
Experiment	326.05	318.45	317.55	295.75
Only gaseous radiation	403.47	381.67	367.34	343.37
Soot and gaseous radiation–Model 1	319.64	315.87	312.09	306.17
Soot and gaseous radiation–Model 2	316.26	315.28	311.84	306.15
Soot and gaseous radiation–Model 3	319.97	316.17	312.47	306.48

Line contour plots of the soot distribution at the symmetrical plane for the respective three soot models are shown in Figure 4.11. Among all the three models, Model 1 gives the lowest soot distribution, while higher soot yield is evidenced in Model 2. This is not entirely surprising, since Model 2 is based on pre-determined constants derived from experimental data fitted for acetylene flames. The consistency of low floor temperatures predicted in Tables 4.2 and 4.3 confirms such assertion. In hindsight, the pre-exponential constant in

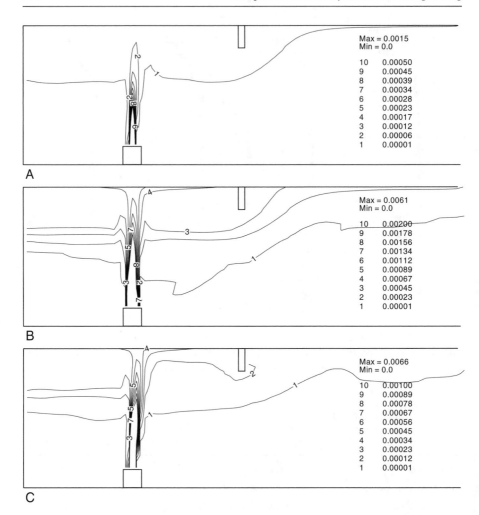

Figure 4.11   Predicted soot distribution: (a) Model 1, (b) Model 2, and (c) Model 3.

Model 1 could have been set higher to produce more soot. Increasing the soot loading, however, may significantly compromise the solutions of already lower floor temperatures thereby contradicting measurements. For such lightly sooty flame, Model 1, which only considers the soot inception rate as the dominant mechanism for the generation of soot, is clearly inadequate in predicting reasonable soot concentrations. Model 3 employs constants that were derived from methane combustion, a weakly sooting flame. Since LPG comprises of fuels of predominantly weakly sooting in nature, it is not surprising that reasonable soot levels are predicted when the same constants are applied in this worked example.

Conclusions: The consideration of soot for field modeling investigation of a two-room compartment fire is demonstrated in this worked example. On the basis of the preceding analysis, the inclusion of thermal radiation has shown to reduce the size of the fire plume where the maximum temperatures are located. More importantly, the presence of luminous soot radiation in conjunction with the radiation contribution by combustion products significantly improves the numerical predictions.

4.6.2 Multi-Room Compartment Fire

Amongst the increasing complexity that has been introduced into field modeling—turbulence, combustion, radiation, and soot sub-models, as demonstrated through the single-room and two-room compartment fires—it is imperative that the use of the fire model extends beyond these rather simple geometry configurations in order to ascertain the model's applicability in real building fires. In this worked example, the fire model is further assessed against fire experiments carried out by Luo and Beck (1994) on a full-scale multi-room compartment configuration. Numerical simulations are performed through an in-house computer code FIRE3D.

Figure 4.12 illustrates the plan layout of the first floor of the Experimental Building-Fire Facility. The area considered in this study, marked by the dotted lines, consisted of the Corridor and adjoining rooms designated by R101, R102, and R103. R102 during the experiment was chosen as the burn room.

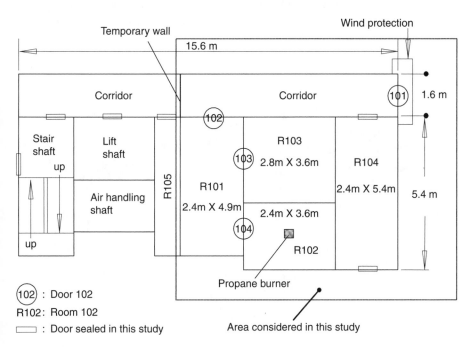

Figure 4.12 Schematic plan-view of the multi-room fire compartment building.

Door 104 that connected R102 and R101 by a doorway had dimensions of 0.8 m × 0.2 m. Other rooms in the compartment were connected to each other through a doorway having the same dimensions as Door 104. Except for Door 101, it had the full size of the corridor section—1.4 m × 2.5 m. A sand-box propane burner was centrally located in R102. Simulations carried out for comparisons against the numerical and experimental results were performed for a constant burn rate of 0.006 kg s^{-1} yielding a fire size of 300 kW. Thermocouple trees were spatially distributed in the Corridor, R101, R102, and R103 to measure the temperature distribution of the Building-Fire Facility. Chemical analyses for gas composition for CO_2 and CO were also measured.

Numerical features: The three-dimensional Favre-averaged equations for the transport of mass, momentum, and enthalpy are solved. A *hybrid differencing scheme* is adopted to approximate the convection terms in the transport equations and the SIMPLE algorithm is adopted. The eddy-viscosity concept is employed for the representation of the turbulent diffusivities, which is obtained through the solution of the standard k-ε turbulent model with additional source terms to account for buoyancy effects (more details are found in the worked example in section 2.16.1). The relevant sub-models include the conserved scalar approach employing the Sivathanu and Faeth (1990) state relationships to resolve the fire chemistry with DOM for radiative heat transfer, and a soot model accounting for soot formation process by Moss et al. (1988) and an oxidation process proposed by Lee et al. (1962).

Figure 4.13 shows the grid distribution of the multi-room compartment geometry. A mesh density of 135200 control volumes is generated for the entire geometry, with denser grids concentrated above the burner to resolve the fire

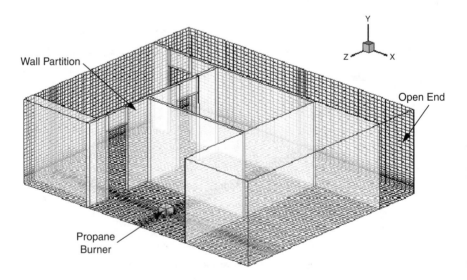

Figure 4.13   Mesh distribution of the multi-room compartment geometry.

chemistry. The normal velocity at the burner surface is evaluated from the fuel burn rate. The turbulence level is assumed to be weak; the laminar assumption is imposed at this boundary. Temperature is set to be constant based on the temperature of the fuel flowing through the burner. The mixture fraction and mass fraction of fuel is set at unity. The condition of no-slip is imposed at the inert solid surfaces by setting all velocities to zero. The normal gradients of the mixture fraction and its variance, participating species, and particulate normal density are set to zero at these boundaries due to impermeability of the walls. In order to resolve the momentum and heat fluxes near the wall region, conventional logarithmic wall function is applied. An adiabatic condition is imposed for the calculation of the wall temperatures. Correspondingly, the enthalpy equation is determined from the given wall temperature when solving the energy conservation equation. At external boundaries, the solution domain is treated as entraining surface on which the ambient pressure is set to be constant. The normal gradients of all dependent variables are set to zero for in-flow or out-flow conditions, except for the temperature, mixture fraction, and its variance, participating species and particulate normal density where ambient variables are specified at this plane when the flow enters the compartment.

Numerical results: Figure 4.14 shows the comparison between the measured and numerically ascertained temperatures by Luo and Beck (1994) above the fire source in the burn room and the predicted temperature distribution obtained via the in-house computer code FIRE3D. As seen by the result in Figure 4.14c, the spatial temperatures are in good agreement with the measured isotherms, as well as the numerical predictions obtained in Luo and Beck [17]. Predicted temperatures within the combustion zone, due to the presence of soot formation and burnout and the effect of soot radiation, are of reasonable flaming temperatures of at least 800°C, typically observed in practice.

Further assessment of the fire model is demonstrated through the comparison of the measured velocities with the predicted velocities at the centerline of Doors 102 and 104 in Figure 4.15. Here, good agreement between model predictions and measurements is also evidenced especially the ability of the model to accurately describe the transitional behavior between the inflow and outflow through the doorways. Fig. 4.16 illustrates the overall flow behavior within the entire multi-room compartment. Instead of sectional velocity vector plots, the means of using streaklines of massless particles represent another effective tool in CFD to describe the flow phenomena in a three-dimensional perspective view of the complex fluid flow. When the flow particle is released at point 1, ambient air flow is seen entraining into R103, raising up due to buoyancy and leaving the room at the top edge of Door 103 and entering R103 (the burn room) at the bottom of Door 104. At Point 2, the flow is observed entering into the plume area, while point 3 indicates the flow leaving R102 after passing through the burner, traveling along the ceiling of R101 and the corridor into the open end of the compartment. Flow beginning at point 4 indicates the ambient air entraining into compartment along the floor of the corridor and entering Door 104. At point 5, flow entering into the corridor

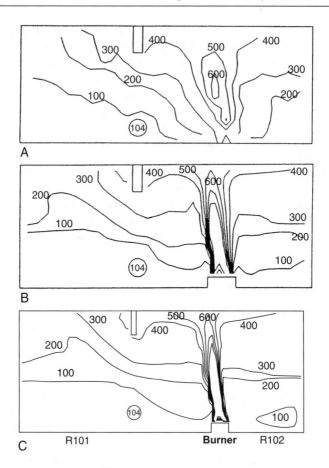

Figure 4.14   Temperature profiles (°C) in R101 and R102 on a vertical section across the centerline of Door 104: (a) Measured, (b) Luo and Beck (1994) prediction, and (c) Present prediction.

merges with the hot combustion products at the wall adjacent to Door 102, and travels along the ceiling of the corridor into the surroundings.

Predicted distributions of the combustion products CO_2 and CO and soot at a vertical section through R101 and R102 are represented in Figure 4.17. The measured mass fractions of CO_2 and CO are indicated in dark circles. The present model employing the conserved scalar combustion model with the state relationships of Sivathanu and Faeth (1990) yields predictions of the combustion species that are rather comparable to the measurements albeit of slightly over-predicted concentration levels. They are still nonetheless within the same order of magnitude. The significant amount of soot as observed in the hot layer and combustion zone is greatly seen to significantly enhance the absorption/emission of the global radiation, hence resulting in lower spatially distributed temperatures.

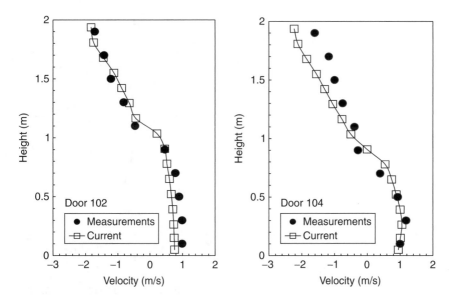

Figure 4.15 Comparison between predicted and measured velocities: (a) Door 102 and (b) Door 104.

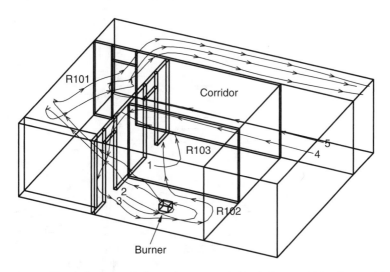

Figure 4.16 The behavior of three-dimensional fluid flow within the multi-room building via the illustration of streaklines beginning at various locations.

Conclusions: Field modeling on a multi-room compartment fire is investigated in this worked example for a 300 kW propane burner fire operating at steady-state conditions. Predicted temperatures of the present model have been found to be in good agreement with measured temperature data attained in Luo and Beck (1994), as well as the predicted temperature via a similar field

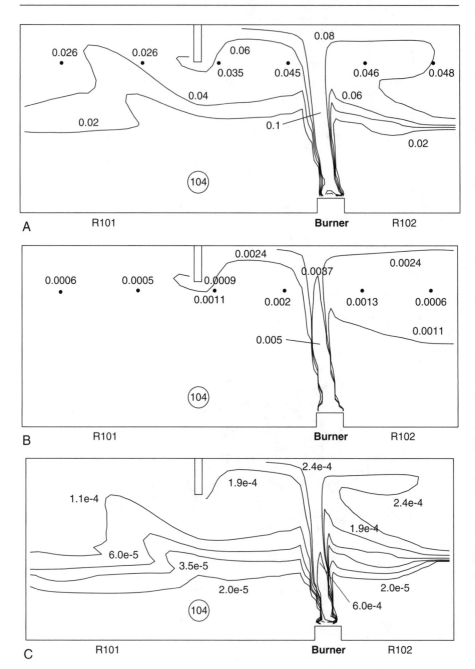

Figure 4.17 Concentration levels of (a) CO_2, (b) CO, and (c) soot on a vertical section across the centerline of Door 104. Measurements indicated by dark circles.

model employed in the same literature. The conserved scalar combustion model with the application of the state relationships of Sivathanu and Faeth (1990) is found to perform rather well in predicting the CO_2 and CO concentrations within the burn room. Like in the case of the two-room compartment fire in the previous worked example, the global radiative heat exchange due to soot also exerts a significant influence on the thermal behavior within the multi-room compartment building, consequently resulting in temperature predictions more comparable to the measurements.

4.7 Summary

In most practical fires, luminous soot radiation constitutes a substantial portion of the radiative heat loss of the total heat release rate. The demand of suitable soot models has been primarily driven by the prospect of feasibly obtaining the concentration of soot particles for the specific evaluation of the soot absorption coefficient in the radiation model. Particular emphasis has been placed in exploiting the possible use of practical models such as the single-step and semi-empirical approaches. In order to determine the concentration of soot particles, the former model retains simplicity and minimizes computational expense with the consideration of only a single transport equation for the soot mass fraction, while the latter model attempts to accommodate the consideration of many important physical processes associated with nucleation (inception), coagulation, surface growth, aggregation, and particle oxidation through two transport equations, one for the particulate number density, the other for the soot mass fraction or volume fraction. Full-blown attempts to characterize the soot process based on the population balance modeling of soot formation and oxidation in turbulent flames through the standard method of moments with a presumed particle size distribution and direct quadrature method of moments, represent the future generation of possible models to be applied in field modeling.

On the basis of the parametric study carried out for the different soot models for field modeling investigation on a two-room compartment fire in the first worked example, the addition of luminous soot radiation is shown to significantly improve the temperature predictions in contrast to the consideration of only accounting gaseous radiation in the model. The second worked example, which focuses on the application of a field model for a multi-room compartment fire, exemplifies the possible extension toward practical field modeling investigations of real.

PART VI   PYROLYSIS

4.8   Importance of Pyrolysis in Fires

Most frequently used materials in building fires are invariably carbon-based polymers. Typical examples are wood, polystyrene, polymethylmethacrylate, polyvinylchloride, and many others. These materials, either natural or synthetic derivatives, may exist in the solid phase or liquid phase. During the burning of these combustible solids or liquids, a process called *pyrolysis* is responsible for yielding products of sufficiently low molecular weight that are volatilized from the surfaces and enter the flaming region, thus maintaining the supply of combustible volatiles required for combustion.

In modern building design, *wood* is commonly utilized within many building structures. The ignition of this particular condensed solid and its subsequent development to sustainable combustion are significant considerations in building fires and their associated hazards. Underpinning the importance is the understanding of the behavior of this material in fires. Useful information, which includes the ignitability, rate of flame spread, rate of release of heat, and the release of smoke and its toxicity, have been experimentally attained. The cone calorimeter apparatus as depicted in Figure 4.18 represents one such industrial standard for testing of materials (ASTM E1354, 1990), which has provided the feasibility of measuring the mass-loss rates, heat-release rates, ignitability, and visible smoke production obtained from a small piece or

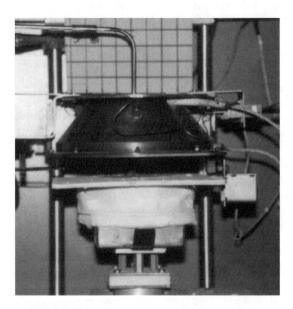

Figure 4.18   The cone calorimeter apparatus for the study of the ignition and combustion of wood.

sample material under controlled levels of radiant heating. Thermal radiation from the conical heater induces pyrolysis in the test specimen. An electric spark is applied at regular intervals above the surface of the material to promote ignition of the combustible volatiles, which subsequently leads to the development of a sustainable flame. An important investigation demonstrating the versatility of the cone calorimeter has been deduced from the experimental studies performed by Delichatsios et al. (2003), where they have, for example, conducted extensive experiments to determine the flammability properties of charring material based on Australian Radiata pine for the prediction of ignition and pyrolysis histories of wood.

Many empirical or semi-empirical relationships have been developed based on the ignition, heat, and smoke release data obtained through the cone calorimeter experiments. When applied alongside the field model in the gas phase, such simple models provided a viable strategy in handling the flame spread and room corner fire growth on combustible wall linings in actual geometry structures. Lockwood et al. (1988) developed a field model to predict the flashover phenomenon in a compartment fire. Three-dimensional equations governing mass, momentum, enthalpy, mixture fraction, turbulent kinetic energy, and its dissipation were solved. Thermal radiation was modeled by considering the fire as a grey body. A solid pyrolysis model was not included to predict the flame spread over the wall covering but rather experimentally measured heat release rates were treated as input requirements to obtain the numerical solution. Comparison of the computed temperature with the measured data showed favorable agreement. In the quest of establishing more effective predictions on ignition of combustible volatiles, flame spreading, and development of fires, it is nevertheless imperative that a more elaborate consideration on the pyrolysis, emission of volatile combustible gases, gas phase reactions, and feedback of radiant heat onto the wood for further generation of volatiles is undertaken to provide a more accurate representation of such processes.

Ever since the development of a one-dimensional (1-D) mathematical model for wood pyrolysis by Bamford et al. (1946), experiments using various sizes and shapes of heated wood samples inside furnaces with closely controlled temperatures and non-oxidizing environments have been performed to determine the kinetic data. Transient quantities such as the mass loss and temperature rise of the samples are usually measured. For dry wood, 1-D mathematical models to describe the wood pyrolysis by Tinney (1965), Kanury and Blackshear (1970a, 1970b), Kanury (1972a, 1972b), and Kung (1972, 1974) have been employed to estimate the kinetic data. Roberts (1970a, 1970b) reviewed the kinetic data and concluded that they were dependent on the size of the sample. Alves and Figueiredo (1989) further proposed a 1-D model for the pyrolysis of wet wood with six-reaction schemes, taking into account the different *major constituents* in wood such as hemicellulose, cellulose, and lignin; the corresponding kinetic data were obtained by method of multistage isothermal thermogravimetry. Fredlund (1988, 1993) and Di Blasi (1994a, 1996) developed two-dimensional (2-D) models with the consideration of pressure-driven internal convection of gases in the pyrolysing wood. The former considered moisture evaporation in wet wood pyrolysis, while the latter considered only dry wood pyrolysis with the primary and secondary reaction schemes formulated according to Broido and Nelson (1975) and Bradbury et al. (1979). The latter also recently applied the model to accommodate fast pyrolysis characteristics of cellulosic particles where extra-particle tar evolution has been described in addition to the primary char formation. Bonnefoy et al. (1993) further developed a three-dimensional (3-D) model for wood pyrolysis and obtained the kinetic data for beech wood pyrolysis using the experimental mass loss measurements. However, their model did not consider the moisture content, the anisotropic nature of wood, or the convective heat transfer as a result of the internal flows of volatile gases produced in the pyrolysis processes.

On the basis of a 2-D model, Di Blasi (1994b) predicted a vertical downward flame spread with an opposed laminar flow over a wood surface. The pyrolysis model consisted of a two-dimensional, unsteady, variable property mathematical model of the degradation of porous cellulosic fuels to volatiles and chars, including convective and conductive heat transfer, of which it was coupled to a quasi-steady, 2-D mathematical model, including the gas phase momentum, energy, and chemical species mass equations, to simulate downward flame spread. By assuming no accumulation of the generated volatiles in the solid, the mass flux at the surface was obtained by integration of a one-dimensional continuity equation. Only dry wood was considered, and the combustion kinetics were described by a one-step finite rate Arrhenius reaction.

Novozhilov et al. (1996) adopted a CFD model of wood combustion of which the model comprised of different sub-models for the gas phase and solid phase. In the gas phase, the fluid model consisted of three-dimensional Favre-averaged transport equations of mass, momentum, gas species concentrations, and enthalpy. A two-equation k-ε turbulence model was employed, and turbulent combustion was modeled via the eddy dissipation model of Magnussen and Hjertager (1976). The effect of soot radiation was treated by the discrete transfer method with the soot concentration determined via the conserved scalar approach. In the wood, a 1-D solid model based on Kung (1972) and Kung and Kalelkar (1973) was adopted with some simplifications. A single first-order Arrhenius equation was employed to describe the pyrolysis reaction. The model was validated with the experimental data for the thermal degradation of particle board subjected to a constant radiant flux in an inert atmosphere. Reasonable agreement was obtained in terms of the comparison of the surface temperature and mass loss rate from prediction and experiments. The model was also validated with experimental data for Pacific maple using a cone calorimeter, which was modeled as a rectangular prism. Sensitivity analysis for the back surface heat transfer coefficient, activation energy and pre-exponential factor and heat of pyrolysis for the Pacific maple, have been made by comparing the predicted and measured mass loss rate. It is nonetheless noted that the pressure gradients were neglected in their pyrolysis model for wood, and the volatiles were assumed to emerge from the wood surface immediately upon generation. The anisotropic properties such as conductivity and permeability of wood were not accounted.

In order to establish a basic understanding on wood pyrolysis, a phenomenological description of the mechanisms involved is reviewed in the next section. The physico-chemical processes of pyrolysis due to different *major constituents* such as hemicellulose, cellulose, and lignin are later expounded in order to provide the cornerstone for the development and formulation of a 3-D mathematical model for the pyrolysis of dry and wet wood. In brief, the pyrolysis reaction is modeled by six first-order Arrhenius reactions representing the competing thermal degradation reactions of various constituents such as the cellulose, hemicellulose, lignin, and other possible minor constituents. Evaporation of moisture inside the wood is handled through a consideration of the saturation vapor pressure. An energy equation incorporating heat conduction, internal convection due to movement of the water vapor, volatile and inert gases inside the pyrolysing wood, as well as the heats of pyrolysis and evaporation, is solved. The transport of gases and vapor through the charring solid is assumed to obey Darcy's Law. Mass conservation equations describing the vapor, volatile, and inert gases in wood are considered. The anisotropic properties of wood including conductivities and permeabilities due to the structure of the grains in wood are accommodated within the model. The thermophysical properties of the charring wood have been assigned an extent-of-reaction and porosity dependence. For computational efficiency, the present model adequately captures the essential chemical processes required for practical simulations when it is used in conjunction with the field model without substantially overburdening the computational resources in contrast to more comprehensive models developed by Di Blasi (2001, 2002).

4.9 Phenomenological Understanding of Pyrolysis Processes

Pyrolysis is essentially the process of *decomposition* or *degradation* of condensed solid such as wood by heat. It encompasses the processes by which gaseous fuel is liberated to support the fire by the breakdown of the fuel constituents under the influence of heat. The burning of wood can be considered as a combination of the complicated processes involving the pyrolysis of the constituents in wood, and the subsequent ignition and combustion of volatile gases produced. A sustained flaming combustion of wood also requires a continuous supply of heat to the unpyrolysed wood to maintain the pyrolysis and thus the production of combustible volatile gases. Such a supply of heat is usually brought about by the feedback mechanism of heat transfer from the flame itself due to *conduction, convection,* and *radiation,* such as illustrated in Figure 2.1 of Chapter 2.

A number of mechanisms involved in the thermal degradation of wood or its major constituents, which generally consist of cellulose, hemicellulose, and lignin, have been identified in order to better develop a phenomenological

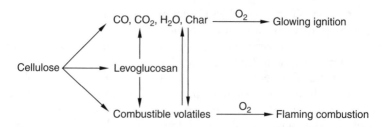

Figure 4.19   General reactions involved in pyrolysis and combustion of cellulose.

understanding of the pyrolysis processes. One excellent review by Shafizadeh (1968) has demonstrated the mechanisms of the thermal decomposition of wood and its major constituents, as illustrated in Figure 4.19 by the thermal degradation of cellulosic materials proceeding through a complex series of concurrent and consecutive chemical reactions. Heating at lower temperatures favored the dehydration and charring reactions forming CO, CO_2, H_2O, Char, and subsequently to glowing ignition, while heating at higher temperatures especially above 250°C led to the principal reaction involved in the pyrolysis of cellulose via de-polymerization to levoglucosan (i.e., 1,6-anhydro-β-D-glucopyranose), a principal intermediate compound. This compound further decomposed at elevated temperatures to yield volatile combustibles leading to a flaming combustion. Another possible route was the direct conversion by fragmentation from cellulose to combustible volatiles, which resulted in flaming combustion. These identifiable mechanisms for the pyrolysis have been confirmed subsequently by other authors including Lewellen et al. (1976) who studied the pyrolysis of cellulose by electrically heating it at various heating rates to achieve temperatures of 250°C–1000°C in helium. Moreover, the proposed pyrolysis mechanisms formed the foundation whereby they have been adopted in theoretical modeling of pyrolysis and ignition of wood. The mechanism for the transformation and carbonization of cellulose to form char was also discussed in detail in the review of Shafizadeh (1968). The yield and properties of the char depended on the rate of heating and flash pyrolysis by intense thermal radiation, leaving little char in contrast to heating at slowly rising temperatures, which resulted in the carbonization of cellulose.

By the comparison of results of differential thermal analysis of wood and its major components (i.e., cellulose, hemicellulose, and lignin), Shafizadeh and Chin (1977) concluded that the thermal degradation of wood reflects the sum of that of its three major components. The key mechanisms by which the major components decompose during pyrolysis are broadly outlined in Figure 4.20. In their study, they also confirmed that there was no significant interaction among the three major components during the thermal degradation of wood. Shafizadeh et al. (1979) reported that the pyrolysis of cellulose proceeded at a much faster rate at higher temperatures (300°C–500°C) to give a tar fraction containing mainly levoglucosan and glucose condensation products. At 400°C, the pyrolysis was essentially completed within 3 minutes,

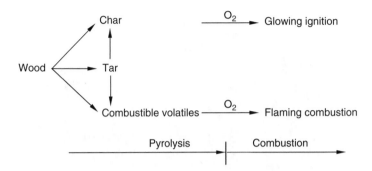

Figure 4.20 The pyrolysis and combustion of wood.

yielding a tar which contained 39% levoglucosan and, upon mild acid hydrolysis, giving 49% D-glucose. The yield could be further increased by washing or treatment of the cellulosic substrates with acids, which indicated the substrate dependency of the yield of the pyrolysis. This was not only important for improving industrial processes of wood but also provided invaluable insights into the pyrolysis of wood during the development of flame spread and fires.

Based on an experimental investigation performed by Lee et al. (1976), the pyrolysis process was shown to be strongly affected by the anisotropic properties of wood and char, relative to the internal flow of heat and gas. Lee and Diehl (1981) further ascertained that the pyrolysis process was delayed in wet wood due to vaporization, and that the volatile combustible gases were diluted by the water vapor based on the experimental measurements carried out for the combustion of dry and wet oak.

4.10 Physico-Chemical Description of Pyrolysis Processes

Wood is extremely inhomogeneous, and its structural and chemical variability is reflected by wide ranges in its physical properties such as permeability, capillary behavior, thermal conductivity, and the diffusion of bound water. The complexity of the three-dimensional structure of wood can be found, for example, in Meylan and Butterfield (1972). Figure 4.21 illustrates a three-dimensional view of a typical block of wood.

In order to better understand the behavior of woods in fires, it is essential to understand the topological, physical, and chemical properties of the wood of which such properties significantly influence its combustion and heat release characteristics. Among the many topological and physical characteristics are the structure of wood fibers and the pathways for moisture content, specific gravity, void volume, and thermal properties, while for chemical characteristics include the summative analysis and higher heating value. Wood, either softwood or hardwood, is anisotropic (i.e., properties are dependent on direction) and hygroscopic (i.e., loses and gains moisture). Its chemical components

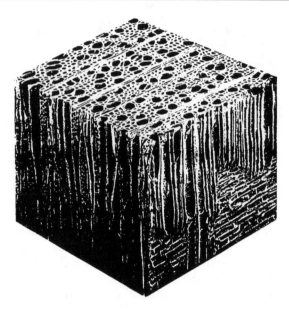

Figure 4.21 A three-dimensional view of a typical block of wood, showing the transverse plane at the top, the tangential longitudinal plane to the left, and the radial longitudinal plane to the right. Magnification: x 68. (After Meylan and Butterfield, 1972.)

include cellulose, hemicellulose, and lignin in varying amounts depending on species. In general, hardwoods contain more holocellulose (i.e., carbohydrates) and less lignin than softwoods.

The topological structure of wood, which includes the growth rings, wood rays, and grains, leads to the anisotropic nature of its thermophysical properties such as conductivity and permeability of wood. The outside layer of the cross section of a tree trunk is covered with bark, the outermost layer of which is made up of a corky material consisting of dead tissue, while the inner bark (phloem) is composed of soft living tissue. Between the bark layer and the stemwood (xylem) interface is a thin indistinguishable layer called the *cambium*, which is responsible for the production of new phloem and xylem tissue forming the growth rings. Wood rays originate from the cambium and extend to the pitch (i.e., center) and bark, running perpendicular to the growth rings. Wood grains that correspond to the fibers or tracheids are aligned in the longitudinal direction and are responsible for the relatively higher permeability and thermal conductivity of wood in this direction, compared with the transverse directions.

Burning of wood is a complicated phenomenon associated with the thermal decomposition of the different constituents of wood resulting in the production of combustible volatile gases, the migration of the gases to the surface of the wood, the emission and mixing of the volatile gases with air, and combustion in the gas phase. The major chemical constituents of wood are cellulose, hemicellulose, and lignin. Cellulose, in the form of microfibrils, is the structural framework. Hemicellulose is the matrix substance present between the micro-fibrils, while lignin is the encrusting substance binding the wood cells together and giving the strength to the cell wall. Extractives (i.e., low-molecular-weight organic compounds) are present as a minor constituent in most wood species. Softwoods contain 40%–50% cellulose, 11%–20% hemicellulose, and 27–30% lignin. Hardwoods contain approximately 45%–50% cellulose, 15%–20% hemicellulose, and 20%–25% lignin (Saka, 1993).

Cellulose is a linear polymer composed of β-D-glucopyranose units linked together by (1 → 4)-glycosidic bonds in a chair conformation, with 44.4% carbon, 6.2% hydrogen, and 49.4% oxygen. Every glucose unit is rotated over 180° with respect to its neighbors (see Figure 4.22), and each has one primary and two secondary hydroxyl groups. The degree of polymerization of wood cellulose can be as high as 10,000 (Goring and Timell, 1962). It is the main constituent in wood and presents predominantly in the secondary cell wall (Sjostrom, 1981) and has a strong tendency to form hydrogen bonds between adjacent glucose units as well as adjacent cellulose chains (Sjostrom, 1981). The hydroxyl groups can also easily form hydrogen bonds with water, which serves to explain its strong affinity to water and hence the hygroscopic behavior of wood. In wood, the cellulose is partly crystalline to 50%–60% and the remainder is amorphous. Bundles of cellulose molecules are aggregated together in the form of microfibrils, in which highly ordered (crystalline) regions alternate with less ordered (amorphous) regions. Microfibrils form fibrils and finally cellulose fibers (Sjostrom, 1981).

Hemicelluloses are polysaccharides formed through biosynthetic routes, which are different from the routes of cellulose (Sjostrom, 1981). They function as a binding material for the cellulose microfibrils in the cell walls. They are characterized by major monomeric units such as xylan, galactan, and mannan. In contrast with the linear structure of cellulose, the hemicelluloses exhibit branch structures. Functional groups associated with hemicelluloses include methyl, carboxyl, and hydroxyl units. Most hemicelluloses have a degree of polymerization of only 200 (Sjostrom, 1981). The much lower molecular weight distinguishes them from the cellulose. Hemicelluloses are the least stable and are readily degraded upon application of heat.

Lignins are usually isolated from extractive free wood as an insoluble residue. They are three-dimensional polymers composed mainly of phenylpropane units linked together by various means and often considered as the "glue" holding the wood structure together. Lignins encrust the intercellular space

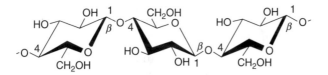

Figure 4.22   Structure of cellulose, in chair conformation.
(After Sjostrom, 1981.)

and any openings in the cell wall upon the deposition of the cellulose and hemi-celluloses. Their polymerization products are p-coumaryl, coniferyl, and sina-pyl alcohols (Siau, 1984), and they are found to be far less hygroscopic than cellulose and hemicelluloses and serve to reduce the hygroscopicity of wood.

The thermal decomposition of these major constituents is highly dependent on their respective chemical structures. Physico-chemical processes associated with the pyrolysis associated with the three major constituents of wood as well as overall thermal behavior of wood for the pyrolysis of these major constituents, are further expounded in subsequent sections.

4.10.1 Pyrolysis of Cellulose

In general, pyrolysis of cellulose produces char, tar, and fixed gases. Studies of the yields of pyrolysis products of cellulose have been reported by numerous investigators. Martin (1965), Bradbury et al. (1979), and Shafizadeh (1968) indicate that levoglucosan (i.e., 1,6-anhydro-β-D-glucopyranose) is found to be the major product of pyrolysis of cellulose. Ohlemiller et al. (1985) performed a more detailed analysis of gaseous products and found that the major portion of the gaseous products from the pyrolysis of cellulose contains carbon monoxide, carbon dioxide, and water vapor. The other gaseous pyrol-ysis products include hydrogen, methane, ethylene, ethane, propane, butane, and butene, with traces of the other compounds such as formaldehyde, acetal-dehyde, acetone, methanol, propanol, 2-methylfuran, acetic acid, furfural, 2,3-butanedione, methyl ethyl ketone, crotonaldehyde, and cyclopentane. The productions of some aromatic species such as furan, benzene, toluene, and phenol have also been reported. A comprehensive review of the pyrolysis of cellulosic materials by Shafizadeh (1968) ascertained that the cellulose pyroly-sis proceeded with a complex series of concurrent and consecutive chemical reactions. The general set of reactions involved in the pyrolysis of cellulose was outlined, which comprised of three reaction routes competing for the simultaneous consumption of cellulose, as described in Figure 4.7. It is noted that in Shafizadeh's model, gases and char can react to form combustible vola-tiles and vice versa through decomposition. In Shafizadeh (1968), Lipska and Parker (1966), Lee et al. (1976), Lee and Diehl (1981), and Ohlemiller et al. (1985), these reactions were shown to be highly influenced by physical condi-tions such as temperature, external radiant flux, external heating rate, total time of heating, sample type, ambient oxygen concentration, water content, pressure, and chemical impurities or additives in the pyrolysing substrate.

4.10.2 Pyrolysis of Hemicellulose

The pyrolysis of hemicellulose is similar to that of cellulose because it belongs to the family of polysaccharides. However, the hemicelluloses are found to be the least stable among the three major constituents of wood and decompose at 225°C–325°C. Shafizadeh and Lai (1972) analyzed the pyrolysis products of hemicelluloses. They found that the products consisted of tar in about 16% yield of which 17% was oligosaccharites. On the basis of further analysis of the products, the pyrolysis of the hemicelluloses bears many similarities to that of cellulose, which involves the cleavage of the glycosidic groups forming random condensation products. At higher temperatures, these products and the glycosyl units are further degraded to give a variety of volatile products. It has been observed that the pyrolysis products of hemicelluloses and cellulose are basically similar (Shafizadeh and Lai, 1972), and account for most of the volatile products in the thermal degradation of wood.

4.10.3 Pyrolysis of Lignins

Compared to the pyrolysis of cellulose and hemicellulose, the pyrolysis of lignin is relatively unexplored and not well understood. The thermal degradation of lignin generally proceeds gradually over a wide temperature range of 250°C–500°C, with the decomposition occurring most rapidly at 310°C–420°C (Shafizadeh and Chin, 1977). There are four main fractions within the pyrolysis products. The first fraction is char, a highly condensed carbonaceous residue accounting for 55% of the yield. The second fraction is an aqueous distillate which contains mainly water, methanol, acetone, and acetic acid. This fraction is produced in about 20% yield. The third fraction is tar, which is a mixture of phenolic compound produced in about 15% yield. The last fraction is fixed gases such as carbon monoxide, carbon dioxide, methane, and ethane produced in about 12% yield.

4.10.4 Pyrolysis of Wood

An approach which includes the primary activation reaction followed by secondary reactions, has been employed by Chan et al. (1985) and more recently by Di Blasi (1994a). The former adopted four competing first order Arrhenius reactions for the production of gases, water vapor, char, and tar. A secondary reaction was employed to describe further thermal degradation of tar to form secondary tars and gases. The latter adopted the mechanism of cellulose pyrolysis by Broido and Nelson (1975) and Bradbury et al. (1979), where an activation step was included before competing thermal degradation reactions begun. Alves and Figueiredo (1989) proposed a wood pyrolysis model with a six-reaction scheme, taking into account the competing reactions of different constituents such as hemicelluloses, cellulose, lignin, and other minor constituents, in which the kinetic data (i.e., the activation energy and pre-exponential factor for the first order Arrhenius expression) were obtained by the method of multistage isothermal thermogravimetry. Generally, the kinetic data vary with the species and the size of the wood and suitable kinetic data (i.e., 6 sets) for the six-reaction expressions require rather extensive experimental investigations, which remain elusive for most simulation purposes. Nevertheless, its inherent complexity presents an attractive model to be adopted for a general representation of the wood pyrolysis process.

The single step first order reaction has been found to be more popular due to its simplicity and relatively fewer kinetic parameters employed. A single step first order Arrhenius reaction has been used in the 1-D models of Kanury and Blackshear (1970a), Roberts (1970a, 1970b), Kanury (1972a), Kung (1972), Kung and Kalelkar (1973), and Tzeng and Atreya (1991); the 2-D model of Fredlund (1988, 1993); and the 3-D model of Bonnefoy et al. (1993). Reasonable agreement between the model predictions and experimental results demonstrated that the single step first order reaction can be viably applied for simulations of wood pyrolysis. A single step Arrhenius reaction is thus used to model the thermal decomposition of wood. The activation energy and pre-exponential factor are estimated by a method of best fit to the measured mass loss history reported by Bonnefoy et al. (1993).

4.11   Formulation of Governing Equations

In the burning of wood, the pyrolysis of a hygroscopic material such as wood, which involves the complex combination of the physical and chemical processes, is modeled via a 3-D model incorporating moisture evaporation, anisotropic properties of wood, and the internal convection of gases. A comprehensive pyrolysis model is developed under the following conditions: (i) wood is generally anisotropic and characterized by its different permeabilities and thermal conductivities for longitudinal and transverse directions to the grains, (ii) heat is transferred to the solid external boundary by convection and radiation, (iii) internal heat transfer includes conduction through the solid and convection of the volatile gases, (iv) internal mass transfer is driven by pressure gradient (Darcy's Law), (v) thermal decomposition reactions for the constituents in wood are modeled by Arrhenius expressions, (vi) evaporation of moisture is sufficiently rapid to attain thermodynamic equilibrium, (vii) escaping volatile gases and vapor are in thermal equilibrium with the solid matrix, (viii) migration of moisture through the solid matrix is assumed to occur only in the vapor phase, (ix) thermal properties are functions of solid and char densities and temperature, and (x) thermal swelling and shrinkage are negligible throughout the pyrolysis.

4.11.1   Conservation of Energy for Wood Pyrolysis

The rate of accumulation of energy per unit volume can be expressed in terms of the enthalpies of solid, moisture, volatile gas, vapor, and dry air as

$$\frac{\partial(\rho_s h_s + \rho_m h_m + \rho_g h_g + \rho_v h_v + \rho_i h_i)}{\partial t} \tag{4.11.1}$$

where h denotes the specific enthalpy and ρ is the mass per unit volume (or bulk density) with the subscripts s, m, g, v, and i indicating the solid, moisture in liquid phase, volatile gas, vapor, and dry air. For most practical purposes in the pyrolysis of wood in fire situations, equation (4.11.1) can be simplified to

$$\frac{\partial(\rho_s h_s + \rho_m h_m)}{\partial t} \approx \frac{\partial(\rho_s C_{ps} + \rho_m C_{pm})T_s}{\partial t} \tag{4.11.2}$$

where T_s is temperature of solid wood and C_{ps} and C_{pm} are the specific heat capacities of the pyrolysing solid and moisture (liquid phase).

Energy transfer in the solid occurs by convection due to the movement of vapor, volatile gases, and dry air through the porous wood and thermal conduction. The net rates of energy transfer due to convective movement of the volatile gas, vapor, and dry air in the Cartesian co-ordinate system are

$$\frac{\partial(m_g^x h_g + m_v^x h_v + m_i^x h_i)}{\partial x} + \frac{\partial(m_g^y h_g + m_v^y h_v + m_i^y h_i)}{\partial y}$$
$$+ \frac{\partial(m_g^z h_g + m_v^z h_v + m_i^z h_i)}{\partial z} \tag{4.11.3}$$

where m represents the mass flux with the subscripts and superscripts denoting the gaseous species and directions. Expressing the specific enthalpy of the gaseous species in terms of the solid wood temperature, equation (4.11.3) becomes

$$\frac{\partial(m_g^x C_{pg} + m_v^x C_{pv} + m_i^x C_{pi})T_s}{\partial x} + \frac{\partial(m_g^y C_{pg} + m_v^y C_{pv} + m_i^y C_{pi})T_s}{\partial y}$$
$$+ \frac{\partial(m_g^z C_{pg} + m_v^z C_{pv} + m_i^z C_{pi})T_s}{\partial z} \tag{4.11.4}$$

The net rates of energy transfer due to heat conduction can be expressed as

$$\frac{\partial}{\partial x}\left(k_s^x \frac{\partial T_s}{\partial x}\right) + \frac{\partial}{\partial y}\left(k_s^y \frac{\partial T_s}{\partial y}\right) + \frac{\partial}{\partial z}\left(k_s^z \frac{\partial T_s}{\partial z}\right) \tag{4.11.5}$$

where k_s^x, k_s^y and k_s^z represent the thermal conductivity in the x, y, and z directions, taking into account the anisotropic nature of wood.

The rate of energy production due to pyrolysis reactions, $\dot{q}_p$, is formulated according to

$$\dot{q}_p = -\left(\sum_{j=1}^{6} \Delta H_{pj} R_{pj}\right) \tag{4.11.6}$$

where ΔH_p is the heat of pyrolysis with a positive value being endothermic, and R_p is the rate of pyrolysis reaction. The subscript j represents competing pyrolysis reactions of up to six constituents that can be considered. Kung

(1972) discussed the importance of the reference datum in the formulation of a one-dimensional energy equation for pyrolysis of wood. The same reference datum (i.e., 0°C) has been taken for the heat of pyrolysis and evaporation of moisture in wood. The rate of energy production due to evaporation of moisture $\dot{q}_{ev}$ can be written as

$$\dot{q}_{ev} = -\Delta H_{ev} R_{ev} \tag{4.11.7}$$

where ΔH_{ev} is the heat of evaporation of moisture and R_{ev} is the rate of evaporation of moisture in wood.

Based on the consideration of the *first law of thermodynamics*, the energy equation can be derived for an elemental volume as the rate of change of energy equals to the net rate of heat plus the rate of heat added or removed by heat source.

$$
\begin{array}{ccccc}
\textit{The rate} & & \textit{The net rate} & & \textit{The rate of heat} \\
\textit{increase of} & = & \textit{of heat} & + & \textit{added or removed} \\
\textit{energy of the} & & \textit{added to the} & & \textit{by heat source on} \\
\textit{fluid element} & & \textit{fluid element} & & \textit{the fluid element}
\end{array}
\tag{4.11.8}
$$

The equation governing the conservation of energy becomes

$$
\begin{aligned}
& \frac{\partial(\rho_s C_{ps} + \rho_m C_{pm})T_s}{\partial t} + \frac{\partial(m_g^x h_g + m_v^x h_v + m_i^x h_i)}{\partial x} + \frac{\partial(m_g^y h_g + m_v^y h_v + m_i^y h_i)}{\partial y} \\
& + \frac{\partial(m_g^z h_g + m_v^z h_v + m_i^z h_i)}{\partial z} \\
& = \frac{\partial}{\partial x}\left(k_s^x \frac{\partial T_s}{\partial x}\right) + \frac{\partial}{\partial y}\left(k_s^y \frac{\partial T_s}{\partial y}\right) + \frac{\partial}{\partial z}\left(k_s^z \frac{\partial T_s}{\partial z}\right) - \left(\sum_{j=1}^{6} \Delta H_{pj} R_{pj}\right) - \Delta H_{ev} R
\end{aligned}
\tag{4.11.9}
$$

4.11.2 Conservation of Mass for Wood Pyrolysis

The flow of pyrolysis volatile gas and vapor is driven by pressure gradients and is assumed to conform to the Darcy's Law. The much slower moisture transfer in the liquid phase compared to that in the gas phase is, however, neglected. According to Darcy's law, the total mass fluxes of vapor, volatile, and dry air are

$$m^x = -\alpha_s^x \frac{\partial P_s}{\partial x} \tag{4.11.10}$$

$$m^y = -\alpha_s^y \frac{\partial P_s}{\partial y} \tag{4.11.11}$$

$$m^z = -\alpha_s^z \frac{\partial P_s}{\partial z} \tag{4.11.12}$$

where P_s is the sum of the partial pressures of the vapor, volatile, and dry air, and $\alpha_s^x = D_s^x/v_{mix}$, $\alpha_s^y = D_s^y/v_{mix}$ and $\alpha_s^z = D_s^z/v_{mix}$ are the mass transfer coefficients of the solid in the x, y, and z directions, taking into account the anisotropic nature of wood. The kinematic viscosity of the mixture is defined by $v_{mix} = \mu_{mix}/\rho_t$, where μ_{mix} is the dynamic viscosity of the mixture. The dynamic viscosity of the mixture is determined by the weighted sum of the individual dynamic viscosities of the gaseous species in wood, which will be further discussed later.

Apart from the vapor and volatile gases produced during pyrolysis, the dry air present initially inside wood prior to the pyrolysis can significantly affect the build up of the total gas pressure and hence the pressure gradient inside the wood. The amount of the dry air and its migration in the wood must be modeled. Also, its contribution to the internal convection heat transfer has been included in the energy balance equation, as discussed in the previous section. The mass fluxes of the vapor, volatile, and dry air are weighted with respect to their bulk densities. Mass fluxes of the volatile gas, m_g, vapor, m_v, and dry air, m_i in the Cartesian directions can be expressed as

$$m_g^x = \frac{\rho_g}{\rho_v + \rho_g + \rho_i} m^x, \quad m_g^y = \frac{\rho_g}{\rho_v + \rho_g + \rho_i} m^y, \quad m_g^z = \frac{\rho_g}{\rho_v + \rho_g + \rho_i} m^z$$

$$(4.11.13)$$

$$m_v^x = \frac{\rho_v}{\rho_v + \rho_g + \rho_i} m^x; \quad m_v^y = \frac{\rho_v}{\rho_v + \rho_g + \rho_i} m^y; \quad m_v^z = \frac{\rho_v}{\rho_v + \rho_g + \rho_i} m^z$$

$$(4.11.14)$$

$$m_i^x = \frac{\rho_i}{\rho_v + \rho_g + \rho_i} m^x; \quad m_i^y = \frac{\rho_i}{\rho_v + \rho_g + \rho_i} m^y; \quad m_i^z = \frac{\rho_i}{\rho_v + \rho_g + \rho_i} m^z$$

$$(4.11.15)$$

where ρ_g, ρ_v, and ρ_i are the volatile gas, vapor, and dry air densities, respectively.

The fundamental physical principle of mass conservation requires

The rate increase of mass of volatile gas, vapor, or dry air within the fluid element	$=$	*The net rate at which mass of volatile gas, vapor, or dry air enters the fluid element*	$+$	*The rate of mass added or removed on the fluid element*

$$(4.11.16)$$

The three-dimensional continuity equations for the vapor, volatile, and dry air can be expressed as

$$\frac{\partial \rho_v}{\partial t} = -\frac{\partial m_v^x}{\partial x} - \frac{\partial m_v^y}{\partial y} - \frac{\partial m_v^z}{\partial z} + R_{ev}$$

$$(4.11.17)$$

$$\frac{\partial \rho_g}{\partial t} = -\frac{\partial m_g^x}{\partial x} - \frac{\partial m_g^y}{\partial y} - \frac{\partial m_g^z}{\partial z} + \sum_{j=1}^{6} R_{pj} \tag{4.11.18}$$

$$\frac{\partial \rho_i}{\partial t} = -\frac{\partial m_i^x}{\partial x} - \frac{\partial m_i^y}{\partial y} - \frac{\partial m_i^z}{\partial z} \tag{4.11.19}$$

Note that there is no production of dry air in the pyrolysis of wood, so a source term is not found in equation (4.11.19). On the basis of equations (4.11.10)–(4.11.15), the governing equations for mass balance become

$$\frac{\partial \rho_v}{\partial t} = \frac{\partial}{\partial x}\left[\left(\frac{\alpha_s^x \rho}{\rho_t}\right)\frac{\partial P_s}{\partial x}\right] + \frac{\partial}{\partial y}\left[\left(\frac{\alpha_s^y \rho_v}{\rho_t}\right)\frac{\partial P_s}{\partial y}\right] + \frac{\partial}{\partial z}\left[\left(\frac{\alpha_s^z \rho_v}{\rho_t}\right)\frac{\partial P_s}{\partial z}\right] + R_{ev}$$
$$\tag{4.11.20}$$

$$\frac{\partial \rho_g}{\partial t} = \frac{\partial}{\partial x}\left[\left(\frac{\alpha_s^x \rho_g}{\rho_t}\right)\frac{\partial P_s}{\partial x}\right] + \frac{\partial}{\partial y}\left[\left(\frac{\alpha_s^y \rho_g}{\rho_t}\right)\frac{\partial P_s}{\partial y}\right] + \frac{\partial}{\partial z}\left[\left(\frac{\alpha_s^z \rho_g}{\rho_t}\right)\frac{\partial P_s}{\partial z}\right] + \sum_{j=1}^{6} R_{pj}$$
$$\tag{4.11.21}$$

$$\frac{\partial \rho_i}{\partial t} = \frac{\partial}{\partial x}\left[\left(\frac{\alpha_s^x \rho_i}{\rho_t}\right)\frac{\partial P_s}{\partial x}\right] + \frac{\partial}{\partial y}\left[\left(\frac{\alpha_s^y \rho_i}{\rho_t}\right)\frac{\partial P_s}{\partial y}\right] + \frac{\partial}{\partial z}\left[\left(\frac{\alpha_s^z \rho_i}{\rho_t}\right)\frac{\partial P_s}{\partial z}\right] \tag{4.11.22}$$

where ρ_t represents the total sum of the densities of volatile gases, vapor, and dry air—that is, $\rho_t = \rho_g + \rho_v + \rho_i$. By dividing equations (4.11.20)–(4.11.22) by their respective molecular weights and differentiating them with respect to time and taking the summation of the differentials, a three-dimensional "pressure" equation representing the conservation of mass for volatile gases, dry air, and vapor can be obtained as

$$\frac{\partial}{\partial t}\left[\left(\frac{\pi_g}{RT_s}\right)P_s\right] = \frac{\partial}{\partial x}\left[\left(\frac{\rho_g}{M_g} + \frac{\rho_v}{M_m} + \frac{\rho_i}{M_i}\right)\frac{\alpha_s^x}{\rho_t}\frac{\partial P_s}{\partial x}\right]$$
$$+ \frac{\partial}{\partial y}\left[\left(\frac{\rho_g}{M_g} + \frac{\rho_v}{M_m} + \frac{\rho_i}{M_i}\right)\frac{\alpha_s^y}{\rho_t}\frac{\partial P_s}{\partial y}\right]$$
$$+ \frac{\partial}{\partial z}\left[\left(\frac{\rho_g}{M_g} + \frac{\rho_v}{M_m} + \frac{\rho_i}{M_i}\right)\frac{\alpha_s^z}{\rho_t}\frac{\partial P_s}{\partial z}\right]$$
$$+ \frac{1}{M_g}\sum_{j=1}^{6} R_{pj} + \frac{R_{ev}}{M_m}$$
$$\tag{4.11.23}$$

Invoking the ideal gas law, the densities of the volatile gases, vapor, and dry air, ρ_g, ρ_v, and ρ_i, are assigned a dependence on porosity:

$$\rho_v = \frac{M_m \pi_g}{RT_s} P_v \tag{4.11.24}$$

$$\rho_g = \frac{M_g \pi_g}{RT_s} P_g \tag{4.11.25}$$

$$\rho_i = \frac{M_i \pi_g}{RT_s} P_i \tag{4.11.26}$$

where π_g is the porosity (i.e., pore volume occupied by the gas per unit volume) and that depends on the extent of pyrolysis and evaporation, P_g, P_v, and P_i are partial pressures of volatile gases, vapor, and dry air, and M_g, M_v, and M_i are the molecular weights of volatile gases, water, and dry air. The total pressure is thus given by $P_s = P_v + P_g + P_i$.

4.11.3 Modeling Wood Pyrolysis Source Terms

On the basis of the model of Alves and Figueiredo (1988), the thermal decomposition reaction of each of the six constituents can be represented by an Arrhenius equation with different kinetic parameters. The rate of pyrolysis of wood and the production of volatile gas are considered the summation of the effects of the competing thermal decomposition reactions of up to six constituents. Each of the six constituents can be assumed to consist of two phases: an active portion, which forms volatile gas upon thermal decomposition reaction, and a charcoal phase. The chemical formulae representing the thermal decomposition reaction of each of the six constituents are

$$S_{aj} \underset{E_{pj}, A_{pj}}{\overset{\Delta H_{pj}}{\rightarrow}} G \uparrow \quad j = 1, 2, \ldots, 6 \tag{4.11.27}$$

where the S_a represents the active portion of the constituent, $G\uparrow$ indicates the volatile gas produced and ΔH_p, E_p, and A_p denote the heat of pyrolysis, activation energy, and pre-exponential coefficient of the thermal decomposition reaction for each of the constituents (i.e., 1 to 6). The rate of mass loss (i.e., rate of the pyrolysis reaction) for each of the constituents, R_{pj}, is modeled by an Arrhenius reaction equation:

$$R_{pj} = -\frac{\partial \rho_{aj}}{\partial t} = \rho_{aj} A_{pj} \exp(-E_{pj}/R_u T_s) \tag{4.11.28}$$

where ρ_a denotes the density of the active portion of each of the constituents with the subscript $j = 1$ to 6 representing the six constituents and T_s is thermodynamic temperature of the solid. The density of dry solid (i.e., the active portion and char) is given by $\rho_s = \sum_{j=1}^{6} \rho_j + \rho_f$, where ρ_f is the final char density.

The conservation equations together with the Arrhenius type equations for pyrolysis kinetics are found to adequately describe the pyrolysis of dry wood. For the pyrolysis of wet wood, an additional equation is necessary for the representation of the rate of moisture evaporation. Several approaches for the calculation of the rate of moisture evaporation in wood have been proposed by Alves (1988), Atreya (1983), and Fredlund (1988). These different approaches are further expounded following.

It was proposed by Alves (1988) that moisture evaporation is governed by evaporation temperature. The evaporation temperature can be given by

$$T_{ev} = 1/\{2.13 \times 10^{-3} + 2.778 \times 10^{-4}\ln(X) + 9.997 \times 10^{-6}[\ln(X)]^2$$
$$-1.461 \times 10^{-5}[\ln(X)]^3\}$$

$$(4.11.29)$$

where T_{ev} is the evaporation temperature and X is the moisture content in percentage based on dry mass of wood for $1\% < X < 14.4\%$. When X falls below 1%, the evaporation temperature can be taken as $T_{ev} = 473\,K$. For $X \geq 14.4\%$, the evaporation temperature is assumed to be $T_{ev} = 373\,K$ with negligible discontinuity and error. When the solid temperature reaches a value T_{ev}, evaporation occurs at a rate as described by the following equation:

$$R_{ev} = \frac{1}{\Delta H_{ev}} \left\{ - \frac{\partial(\rho_s C_{ps} + \rho_m C_{pm})T_s}{\partial t} \right.$$

$$+ \frac{\partial}{\partial x}\left(\lambda_s^x \frac{\partial T_s}{\partial x}\right) + \frac{\partial}{\partial y}\left(\lambda_s^y \frac{\partial T_s}{\partial y}\right) + \frac{\partial}{\partial z}\left(\lambda_s^z \frac{\partial T_s}{\partial z}\right)$$

$$- \frac{\partial(m_g^x C_{pg} + m_v^x C_{pv} + m_i^x C_{pi})T_s}{\partial x}$$

$$- \frac{\partial(m_g^y C_{pg} + m_v^y C_{pv} + m_i^y C_{pi})T_s}{\partial y}$$

$$\left. - \frac{\partial(m_g^z C_{pg} + m_v^z C_{pv} + m_i^z C_{pi})T_s}{\partial z} - \left(\sum_{j=1}^{6} \Delta H_{pj} R_{pj}\right) \right\}$$

$$(4.11.30)$$

According to Atreya (1983) the phenomenon of evaporation of adsorbed moisture can be described by considering the breakage of the hydrogen bonds holding the water molecules to the cell walls. The rate of evaporation of moisture in wood (i.e., rate of desorption of moisture) is proportional to the instantaneous concentration of adsorbed moisture and to the probability that the water molecules possess the activation energy required to break the hydrogen bond. A first order Arrhenius expression is assumed

$$R_{ev} = -\frac{\partial \rho_m}{\partial t} = A_{ev}\rho_m \exp(-E_{ev}/R_u T_s) \tag{4.11.31}$$

where ρ_m is the density of the adsorbed moisture, A_{ev} is the pre-exponential coefficient, and E_{ev} is the activation energy required for the breakage of hydrogen bond. The values of $A_{ev} = 4.5 \times 10^3 \text{ s}^{-1}$, $E_{ev} = 10.5 \text{ kcal mol}^{-1}$, and $\Delta H_{ev} = 574 \text{ cal g}^{-1}$ had been determined by the method of best fit of the experiment performed by Atreya (1983) on an exposed slab of wood of ¾-inch thick to a low incident heat flux of 0.49 W cm^{-2} so that the surface temperature was always maintained below 175°C to avoid any thermal decomposition.

Fredlund (1988) assumed that the vaporization process of the moisture in wood is sufficiently rapid to achieve complete saturation of vapor in the pores. This phase equilibrium between water and vapor holds as long as there is water in liquid phase at the location in the solid. The saturation vapor pressure P_{vs}, was assumed to be given by the following function of the thermodynamic temperature of the solid T_s

$$P_{vs} = K_1 \exp(-K_2/T_s) \tag{4.11.32}$$

where K_1 and K_2 are constants, which depend on the temperature range. In the temperature range from 20°C to 1000°C, $K_1 = 4.143 \times 10^{10}$ Pa, and $K_2 = 4822$ K (Fredlund, 1988); $P_v = P_{vs}$ for $\rho_m > 0$. With this assumption of the thermodynamic phase equilibrium between the liquid and gas phases, and the known values of thermodynamic state variables of temperature and pressure, the rate of evaporation of the adsorbed moisture in wood can be determined by the continuity equation. The rate of evaporation of the adsorbed moisture R_{ev} can be determined by the following expression, which is obtained by rewriting Equation (4.11.23) as

$$R_{ev} = \frac{\partial \rho_v}{\partial t} - \frac{\partial}{\partial x}\left[\left(\frac{\alpha_s^x \rho_v}{\rho_t}\right)\frac{\partial P_s}{\partial x}\right] - \frac{\partial}{\partial y}\left[\left(\frac{\alpha_s^y \rho_v}{\rho_t}\right)\frac{\partial P_s}{\partial y}\right] - \frac{\partial}{\partial z}\left[\left(\frac{\alpha_s^z \rho_v}{\rho_t}\right)\frac{\partial P_s}{\partial z}\right]$$

$$\tag{4.11.33}$$

for $\rho_m > 0$. In contrast to the two approaches of Alves (1988) and Atreya (1983), the rate of evaporation requires knowledge of the distribution of not only the temperature but also the total pressure inside the wood material. This particular approach of Fredlund (1988) nonetheless corresponds well to the physical phenomena in which the moisture evaporation is determined by a thermodynamic equilibrium between vapor and its liquid state.

4.11.4 Thermophysical Properties of Wood Pyrolysis

Thermophysical properties including specific heat capacity, thermal conductivity, permeability, and dynamic viscosity of the pyrolysing wood, pyrolysis products, and vapor are crucial input parameters for the modeling of wood pyrolysis. The selected values and their variation due to changes in temperature, porosity, moisture content, and the degree of conversion from virgin wood to char are discussed.

Specific Heat Capacity of Partially Charred Wood

For partially charred wood, the specific heat capacity varies with the degree of conversion from virgin wood to char during the pyrolysis. According to Chan et al. (1985), the effect of charring is treated implicitly. In their model, the specific heat capacity of partially charred wood is related linearly to the decreasing solid density, which effectively reflects the degree of conversion from wood to char as

$$C_{ps} = A + B\rho_s \tag{4.11.34}$$

where $A = 1.339 \, \text{kJ kg}^{-1} \, \text{K}^{-1}$ and $B = 3.147 \times 10^{-3} \, \text{kJ m}^3 \, \text{kg}^{-2} \, \text{K}^{-1}$. However, a widely adopted formulation is used to assume the volumetric specific heat capacity varies linearly between the virgin wood and final char and can be expressed by

$$\rho_s C_{ps} = a\rho_w C_{pw} + (1 - a)\rho_f C_{pf} \tag{4.11.35}$$

where C_p is the specific heat capacity and ρ is the density with the subscripts s, w, and f denoting the pyrolysing solid, dry virgin wood, and final char. The mass fraction of virgin wood (i.e., unconverted wood) is expressed as

$$a = (\rho_s - \rho_f)/(\rho_w - \rho_f) \tag{4.11.36}$$

The preceding formulation has been adopted by previous workers such as Kung and Kalelkar (1973), Kung (1972), and Fredlund (1988). Similar expressions for the specific heat of partially charred wood have also been used by Alves and Figueiredo (1989), Bonnefoy et al. (1993), and Di Blasi (1994a).

Specific Heat Capacity of Dry Virgin Wood

The specific heat capacity of dry virgin wood as a function of temperature has been determined by Atreya (1983) for a temperature range from 0°C to 140°C. Fredlund (1988) has assumed that the relationship is valid even for temperatures above 140°C. The expression for the specific heat capacity of dry virgin wood as a linear function of temperature is given by

$$C_{pw} = C_{pw,o} + C_{pw,m}T_s \tag{4.11.37}$$

where $C_{pw,o} = 1.4\,\text{kJ kg}^{-1}\,\text{K}^{-1}$ and $C_{pw,m} = 3.0 \times 10^{-4}\,\text{kJ kg}^{-1}\,\text{K}^{-2}$. It is nonetheless noted that a constant specific heat capacity for dry virgin wood independent of temperature has also been assumed by a number of investigators such as Kanury and Blackshear (1970a), Kung (1972), Kung and Kalelkar (1973), Chan et al. (1985), Alves and Figueiredo (1989), Bonnefoy et al. (1993), and Di Blasi (1994a). Values ranging from $1.386\,\text{kJ kg}^{-1}\,\text{K}^{-1}$ to $2.52\,\text{kJ kg}^{-1}\,\text{K}^{-1}$ with most of them larger than $2.0\,\text{kJ kg}^{-1}\,\text{K}^{-1}$ have been typically employed.

Specific Heat Capacity of Char
A similar expression for the specific heat capacity of char as a linear function of temperature as proposed by Fredlund (1988), can be expressed as

$$C_{pf} = C_{pf,o} + C_{pf,m}T_s \qquad (4.11.38)$$

where $C_{pf,o} = 0.7\,\text{kJ kg}^{-1}\,\text{K}^{-1}$ and $C_{pf,m} = 6.0 \times 10^{-4}\,\text{kJ kg}^{-1}\,\text{K}^{-2}$. A value for the specific heat capacity of char of $0.672\,\text{kJ kg}^{-1}\,\text{K}^{-1}$ has also been given by Fredlund (1988) and by a number of other investigators such as Kanury and Blackshear (1970a), Kung (1972), Kung and Kalelkar (1973), Chan et al. (1985), Alves and Figueiredo (1989), Bonnefoy (1993), and Di Blasi (1994a). These ranged from $0.672\,\text{kJ kg}^{-1}\,\text{K}^{-1}$ to $2.52\,\text{kJ kg}^{-1}\,\text{K}^{-1}$. In some pyrolysis models (Kanury and Blackshear, 1970a, Kung, 1972), the specific heat capacity of char was not distinguished from that of the dry virgin wood material.

Specific Heat Capacity of Water and Vapor
In the wet wood pyrolysis model of Alves and Figueiredo (1989), the specific heat capacity of moisture (i.e., liquid water) is taken to be constant at $C_{pm} = 4.19\,\text{kJ kg}^{-1}\,\text{K}^{-1}$. For the range of temperature between 25°C and 100°C where moisture in wood is found to exist in the liquid phase, the specific heat capacity ranges from $4.181\,\text{kJ kg}^{-1}\,\text{K}^{-1}$ to $4.219\,\text{kJ kg}^{-1}\,\text{K}^{-1}$ (Rogers and Mayhew, 1980).

Alves and Figueiredo (1989) applied a constant value of specific heat capacity of vapor $C_{pv} = 1.88\,\text{kJ kg}^{-1}\,\text{K}^{-1}$. Fredlund (1988) presented graphically the variation of specific heat capacity of vapor with increasing temperature from 0°C to 1000°C. The reported values varied approximately linearly from $1.86\,\text{kJ kg}^{-1}\text{K}^{-1}$ to $1.90\,\text{kJ kg}^{-1}\text{K}^{-1}$. The constant value of Alves and Figueiredo (1989) is equivalent to the average value between the lower and upper values used by Fredlund (1988). In most practical cases, a linear variation of the specific heat capacity of vapor may be more preferentially adopted.

Specific Heat Capacity of Volatile Gases
Constant values of specific heat capacity for the mixture of volatile gases ranging from $1.008\,\text{kJ kg}^{-1}\,\text{K}^{-1}$–$1.2\,\text{kJ kg}^{-1}\,\text{K}^{-1}$ have been employed by Kanury and Blackshear (1970a), Kung (1972), Kung and Kalelkar (1973), Alves and

Figueiredo (1989), and Di Blasi (1994a). The specific heat capacity as a function of temperature has nonetheless been assumed by Atreya (1983).

Fredlund (1988) obtained the proportions of the chemical compounds in the pyrolysis products from the experimental result. The specific heat capacity for the volatile gases C_{pg} was evaluated based on the proportions of pyrolysis products. It is expressed as a function of temperature and is given by

$$C_{pg} = C_{pg,o} + C_{pg,m} T_s \qquad (4.11.39)$$

where $C_{pg,o} = 1.0 \text{ kJ kg}^{-1} \text{ K}^{-1}$ and $C_{pg,m} = 8.0 \times 10^{-5} \text{ kJ kg}^{-1} \text{ K}^{-2}$.

Specific Heat Capacity of Dry Air

Dry air (i.e., mixture of gases excluding the vapor portion, which is treated separately) is present initially in wood, and contributes to the total pressure inside the wood throughout the pyrolysis. This mixture of gases and their migration in wood must be modeled. In order to account for the internal convection heat transfer due to dry air migration, its specific heat capacity is needed. The specific heat capacity of the dry air can be expressed as

$$C_{pi} = C_{pi,o} + C_{pi,m} T_s \qquad (4.11.40)$$

where $C_{pi,o} = 0.994 \text{ kJ kg}^{-1} \text{ K}^{-1}$ and $C_{pi,m} = 2.0 \times 10^{-4} \text{ kJ kg}^{-1} \text{ K}^{-2}$ for the temperature range between $0°C$ and $1200°C$ (Rogers and Mayhew, 1980).

Thermal Conductivity

Fredlund (1988) revealed that the heat transfer during pyrolysis comprises of three components: conduction through the solid phase, radiant heat transfer in the pores, and convection heat transfer in the enclosed gases. The overall thermal conductivity is obtained as the sum of the conductivities of each of these components. Radiation heat transfer in the pore system is generally found to be negligible at room temperatures, even at high porosities. At elevated temperature and for porosities below 0.7, radiation heat transfer is still only marginally significant. In his subsequent pyrolysis modeling, Fredlund (1988) simply uses a thermal conductivity, which is a function of only the temperature of the material. On the basis of the pyrolysis models of Chan et al. (1985), the terms corresponding to the variations of porosity and radiant heat transfer in the pores are included. However, the significance of both terms has not been explored.

As applied in the one-dimensional pyrolysis models of Kung (1972), Kung and Kalelkar (1973), and Alves and Figueiredo (1989), the thermal conductivity of partially charred wood k_s has been taken to vary linearly between that of virgin wood and that of final char:

$$k_s = a k_w + (1 - a) k_f \qquad (4.11.41)$$

where k_w and k_f are the thermal conductivity of the virgin wood and final char; the mass fraction of virgin wood a is as defined in equation (4.11.36). In pyrolysis models by Kanury and Blackshear (1970a), Kung and Kalelkar (1973), and Kung (1972), constant values of thermal conductivity for virgin wood materials have been applied, which ranged from 0.1134 W m^{-1}K^{-1} to 0.21 W m^{-1}K^{-1}. Constant values of thermal conductivity ranged from 0.0412 W m^{-1}K^{-1} to 0.189 W m^{-1}K^{-1} for final char have nevertheless been used by Kung and Kalelkar (1973) and Kung (1972).

Di Blasi (1994a) has proposed the use of different thermal conductivities of wood and char in the longitudinal and transverse (i.e., tangential and radial) directions to the grains. In three dimensions, equation (4.11.41) can be extended to give the thermal conductivity of the partially charred wood in the Cartesian directions as

$$k_s^x = ak_w^x + (1-a)k_f^x \qquad (4.11.42)$$

$$k_s^y = ak_w^y + (1-a)k_f^y \qquad (4.11.43)$$

$$k_s^z = ak_w^z + (1-a)k_f^z \qquad (4.11.44)$$

where the subscripts s, w, and f denote the pyrolysing solid, dry virgin wood, and final char, respectively. According to Di Blasi (1994a), the thermal conductivities of virgin wood in the longitudinal and transverse direction of the grains have been assigned values of 0.255 W m^{-1}K^{-1} and 0.105 W m^{-1}K^{-1}, while the thermal conductivities of char in the longitudinal and transverse direction of the grains were set to 0.1046 W m^{-1}K^{-1} and 0.071 W m^{-1}K^{-1}. These values have been found to be within the ranges reported in Siau (1984).

Permeability

Owing to the anatomy of wood, the permeability of wood varies greatly with direction relative to the grain orientation. Skaar (1988) has indicated that the ratio of the measured longitudinal-to-tangential permeabilities for hardwoods can be as high as 80,000 to 1, and the longitudinal-to-radial ratio may be 50,000 to 1, while for the softwoods, the longitudinal-to-transverse ratio can be as high as 4×10^8 to 1. The permeabilities in the transverse directions relative to grain orientation are, however, almost the same.

When the virgin wood gradually pyrolyses to form carbonaceous char, the porosity of the partially charred wood increases. This results in an increase of the permeability until it reaches that of the final char. Fredlund (1988) described the theoretical estimation of the permeability of the partially charred wood as a function of the varying porosity as follows:

$$D_s^x = K_{D1}^x \exp\left[K_{D2}^x\left(1 - \frac{\rho_s - \rho_f}{\rho_w - \rho_f}\right)\right] \qquad (4.11.45)$$

$$D_s^y = K_{D1}^y exp\left[K_{D2}^y\left(1 - \frac{\rho_s - \rho_f}{\rho_w - \rho_f}\right)\right] \tag{4.11.46}$$

where K_{D1} and K_{D2} are constants depending on species and orientation of the wood grains with the superscripts x and y denoting the directions. The preceding expression, however, predicted only a 100-fold increase in the permeability when the porosity changes from 0.647 to 0.9. This is considered much smaller than the change expected in reality, where cracking in the carbon layer leads to considerable increase in the permeability. Di Blasi (1994a) assumed that the permeability varied linearly between the virgin wood and char in both the longitudinal and transverse direction, respectively, according to the degree of conversion from virgin wood to char. In other words,

$$D_s^x = aD_w^x + (1 - a)D_f^x \tag{4.11.47}$$

$$D_s^y = aD_w^y + (1 - a)D_f^y \tag{4.11.48}$$

In her model, an activation cellulose was produced as an intermediate product during pyrolysis, which was assumed to have the same thermal conductivity as the virgin wood.

In the mathematical modeling of wood pyrolysis and combustion, a good estimate of the permeability of the partially charred wood, which governs the migration of volatile gases and vapor through wood as well as their emission from the wood surface, is needed. Suitable expressions for the permeability of the partially charred (i.e., still pyrolysing) wood remain elusive. The current authors have adopted both experimental and modeling techniques for the determination of the expression for the permeability of partially charred wood and the proposed exponential expression similar to the exponential form of Fredlund (1988) is

$$D_s^x = K_{D1}^x exp\left\{K_{D2}^x\left[1 - \left(\frac{\exp(ab) - 1}{\exp(b) - 1}\right)\right]\right\} \tag{4.11.49}$$

$$D_s^y = K_{D1}^y exp\left\{K_{D2}^y\left[1 - \left(\frac{\exp(ab) - 1}{\exp(b) - 1}\right)\right]\right\} \tag{4.11.50}$$

In the experiment, the determined permeabilities show a similar trend of increase of permeability with percentage char for the partially charred wood. Although a theoretical derivation seems impossible, Shafizadeh (1968) gave an insight into an exponential formulation of permeability of the pyrolysis wood. The pyrolysis of wood comprises of competitive reaction paths of which the formation of levoglucosan (i.e., 1,6-anhydro-β-D-glucopyranose) at temperatures around 250°C is the major by-product, followed by competitive secondary reactions to form char, vapor, and gases at elevated temperatures. This greatly suggests that at the initial stage of pyrolysis the char formation is limited, thus a slower increase of permeability. An exponential expression, which gives a slow but increasing permeability, appears to be more plausible. The proposed expression consists of a second exponential term with a multiplying factor b, which controls the rate of increase of the permeability with respect to the mass fraction of virgin wood a. This constant b has been assigned different values to obtain the best-fit prediction of the average permeability to that determined experimentally. The predicted curve using $K_{D1} = 9.22 \times 10^{-18}$ m^2 and $K_{D2} = 10.88$ with a value of $b = 15$ is seen to provide a prediction of average permeability values, which fits extremely well to the experimentally determined values as illustrated in Figure 4.23. This expression may be employed to calculate the permeability of partially charred wood in modeling of pyrolysis and combustion of wood.

Dynamic Viscosity

The dynamic viscosity of the pyrolysis gas can be calculated by the proposal of Fredlund (1988), using the proportions of pyrolysis products according to the

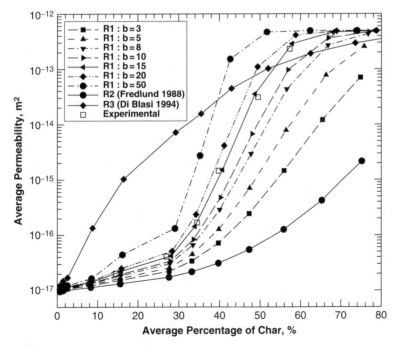

Figure 4.23 Average permeability against char content from experiment and computations.

experimental data. For the volatile gases, the dynamic viscosity can be expressed as a function of temperature in the form of

$$\mu_g = \mu_{g,o} + \mu_{g,m} T_s \tag{4.11.47}$$

where $\mu_{g,o} = 8.5 \times 10^{-6} \, \text{kg m}^{-1} \, \text{s}^{-1}$ and $\mu_{g,m} = 2.95 \times 10^{-8} \, \text{kg m}^{-1} \, \text{s}^{-1} \, \text{K}^{-1}$. The dynamic viscosity of the vapor, also by Fredlund (1988), is given by

$$\mu_v = \mu_{v,\rho} + \mu_{v,m} T_s \tag{4.11.48}$$

where $\mu_{v,o} = 8.5 \times 10^{-6} \, \text{kg m}^{-1} \, \text{s}^{-1}$ and $\mu_{v,m} = 3.75 \times 10^{-8} \, \text{kg m}^{-1} \, \text{s}^{-1} \, \text{K}^{-1}$. The dynamic viscosity of the dry air is assumed to be constant, $\mu_i = 3.178 \times 10^{-5} \, \text{kg m}^{-1} \, \text{s}^{-1}$, which is the median value for the temperature ranging between 300°C and 1000°C (Rogers and Mayhew, 1980).

4.12 Practical Guidelines to Pyrolysis Models in Field Modeling

There are a number of important aspects governing the effective use of a wood pyrolysis model in field modeling. The first aspect focuses on the need to ascertain suitable thermophysical properties, to provide closure to the transport equations in the model. This is because the parameters selected and their variation due to changes in temperature, porosity, moisture content, and the conversion of virgin wood are highly dependent on the specific wood specimen. For example, the properties of pine may be rather different from maple or even Douglas fir. An extensive but not entirely comprehensive review of the various thermophysical properties has been presented in section 4.11.4 for wood pyrolysis. A recommendation to applying the pyrolysis model in field modeling is that care should still be exercised in better quantifying the suitability of application of these critical input parameters on a case-by-case basis.

Since wood is highly anisotropic, the direction relative to the grain orientation greatly determines the heat and mass transfer behaviors inside the wood during pyrolysis. This represents the second aspect, particularly in choosing the appropriate pyrolysis model in field modeling investigation. Acute simplification to the consideration of the pyrolysis model can be made, which will lower the computational requirements of an already complex problem, if the migration of the volatile gases can be assumed to align with the direction of the grains. On the basis of this orientation, a 1-D model for the mass transfer can be imposed with sufficient accuracy to characterize the problem. Otherwise, the full 3-D model, as already described in section 4.11.2, should be conservatively applied.

Pyrolysis is mainly attributed by the surface activities occurring at the interface between the solid phase and gas phase. In field modeling, the third aspect revolves around the important requirement of mesh resolution immediately below and above the surface of the wood. Mesh sensitivity on the proper grid space should always be carried out in order to adequately capture the steep gradients that persist in the vicinity of the solid-gas interface. Appropriate boundary conditions governing the heat and mass transfer should also be imposed to characterize the mass flow of volatiles and surface temperature, due to adjacent convective flow and thermal radiation.

4.13 Worked Example on Ignition of Combustible of Charring Material in a Cone Calorimeter

In this worked example, the model for pyrolysis of wood is applied to the prediction of piloted ignition times, and the development of flaming combustion of wood in the cone calorimeter experiment as described in section 4.8 considering various irradiation levels. A numerical treatment for the simulation of the pilot ignition is included. During the numerical calculations, the irradiances from the conical heater on to the wood surface due to surface-to-surface radiation are evaluated analytically, which will be further demonstrated below. The effects of different initial moisture content of wood are investigated. Predicted ignition times are compared with experimentally determined values by Shields et al. (1993).

Numerical features: A schematic drawing of the computational model of a cone calorimeter is shown in Figure 4.22. The wood sample is circular and the side and lower faces of the sample are assumed to be insulated (adiabatic) and sealed (impermeable). An electrical pilot spark is simulated at regular time intervals of 1 second at an axial position 12.5 mm above the surface of the wood, as shown in Figure 4.24.

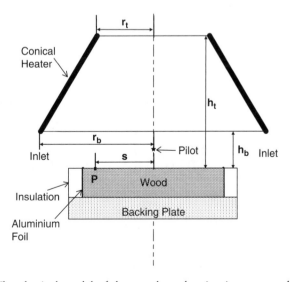

Figure 4.24   The physical model of the wood combustion in a cone calorimeter.

Governing Equations

The three-dimensional mathematical equations for the pyrolysis of wet wood, are expressed in non-orthogonal curvilinear system in order to conform to the circular shape wood sample. In compact form, the equation for the conservation of energy is given by

$$
\frac{\partial[(\rho_s C_{ps} + \rho_l \ell C_{pl}\ell)\theta_s]}{\partial t} + \frac{1}{\sqrt{g}} \frac{\partial[\beta_{ki}(\dot{m}_g^k C_{pg} + \dot{m}_v^k C_{pv} + \dot{m}_i^k C_{pi})\theta_s]}{\partial \xi^i}
$$

$$
= \frac{1}{\sqrt{g}} \frac{\partial}{\partial \xi^i}\left(\beta_{ki}\beta_{kj}\lambda_s^k \frac{\partial\theta_s}{\partial \xi^j}\right) + (\Delta H_p R_p + \Delta H_{ev}R_{ev})
$$

(4.13.1)

where C_p, ΔH_p and ΔH_{ev} are the specific heat capacities, heat of pyrolysis, and evaporation defined with reference to a datum T_{ref}, $\theta_s = T_s - T_{ref}$ is the modified solid temperature, and β_{ki}, β_{kj}, and $\sqrt{g}$ represent the geometric coefficients and the Jacobian of the non-orthogonal curvilinear transformation. The superscript $k = x$, y and z denotes the directions along (i.e., parallel) and across (i.e., perpendicular) grains to account for the anisotropic properties of wood.

The equations for the conservation of mass (i.e., continuity) of the vapor, volatile, and inert gases expressed in non-orthogonal curvilinear coordinate system are

$$
\frac{\partial\rho_v}{\partial t} = \frac{1}{\sqrt{g}} \frac{\partial}{\partial \xi^i}\left(\frac{\alpha_s^k \rho_v}{\rho_t}\beta_{ki}\beta_{kj}\frac{\partial P_s}{\partial \xi^j}\right) + R_{ev}
$$

(4.13.2)

$$
\frac{\partial\rho_g}{\partial t} = \frac{1}{\sqrt{g}} \frac{\partial}{\partial \xi^i}\left(\frac{\alpha_s^k \rho_g}{\rho_t}\beta_{ki}\beta_{kj}\frac{\partial P_s}{\partial \xi^j}\right) + R_p
$$

(4.13.3)

$$
\frac{\partial\rho_i}{\partial t} = \frac{1}{\sqrt{g}} \frac{\partial}{\partial \xi^i}\left(\frac{\alpha_s^k \rho_i}{\rho_t}\beta_{ki}\beta_{kj}\frac{\partial P_s}{\partial \xi^j}\right)
$$

(4.13.4)

A combined pressure equation representing the conservation of mass is obtained through

$$
\frac{\partial(\pi_g P_s/R_g T_s)}{\partial t} = \frac{1}{\sqrt{g}} \frac{\partial}{\partial \xi^i}\left[\frac{\alpha_s^k}{\rho_t}\left(\frac{\rho_v}{M_l\ell} + \frac{\rho_g}{M_g} + \frac{\rho_i}{M_i}\right)\beta_{ki}\beta_{kj}\frac{\partial P_s}{\partial \xi^i}\right] + \frac{R_{ev}}{M_l\ell} + \frac{R_p}{M_g}
$$

(4.13.5)

The rate of evaporation of moisture is given by

$$
R_{ev} = \frac{\partial\rho_v}{\partial t} - \frac{1}{\sqrt{g}} \frac{\partial}{\partial \xi^i}\left(\frac{\alpha_s^k \rho_v}{\rho_t}\beta_{ki}\beta_{kj}\frac{\partial P_s}{\partial \xi^j}\right)
$$

(4.13.6)

while the rate of pyrolysis is described by a first-order Arrhenius reaction

$$R_p - \frac{\partial \rho_a}{\partial t} = \rho_a A_p \exp[-E_p/(R_u T_s)]$$ (4.13.7)

The total mass flux is evaluated according to Darcy's law as

$$\dot{m}^k = -\frac{\alpha_s^k}{\sqrt{g}} \beta_{kj} \frac{\partial P_s}{\partial \xi^j}$$ (4.13.8)

Also in the gas phase, the equations of the conservation of mass, momentum, and energy are also transformed into non-orthogonal coordinates, consistent with the governing equations describing the pyrolysis of wet wood:

$$Continuity \quad \frac{\partial \rho}{\partial t} + \frac{1}{\sqrt{g}} \frac{\partial}{\partial \xi^l} \left(\rho \hat{U}^l \right) = 0$$ (4.13.9)

$$Momentum \quad \frac{\partial(\rho u_i)}{\partial t} + \frac{1}{\sqrt{g}} \frac{\partial(\rho \hat{U}^l u_i)}{\partial \xi^l} = \frac{1}{\sqrt{g}} \frac{\partial}{\partial \xi^l} \left(\mu \sqrt{g} g^{lm} \frac{\partial u_i}{\partial \xi^m} \right)$$
$$\frac{A_i^k}{\sqrt{g}} \frac{\partial}{\partial \xi^k} \left(\mu \frac{1}{\sqrt{g}} A_i^l \frac{\partial u_j}{\partial \xi^l} \right) + (\rho - \rho_{ref}) g_i$$ (4.13.10)

$$Energy \quad \frac{\partial(\rho h)}{\partial t} + \frac{1}{\sqrt{g}} \frac{\partial(\rho \hat{U}^l h)}{\partial \xi^l} = \frac{1}{\sqrt{g}} \frac{\partial}{\partial \xi^l} \left(\frac{\mu}{Pr} \sqrt{g} g^{lm} \frac{\partial h}{\partial \xi^m} \right)$$ (4.13.11)

where $\hat{U}^l = \sum\limits_{k=1}^{3} \beta_{kl} u_k$, ρ_{ref} is the reference density, Pr is the Prandtl number, and A_i^k, A_j^k, and A_i^j are the adjugate Jacobian metric elements, while g^{lm} is the contravariant metric element in the non-orthogonal coordinates transformation. In general, the complexity of the gas phase chemical reaction mechanism for the gas volatiles of wood precludes the knowledge of any detailed reaction mechanism. A one-step global reaction remains the common way to model the combustion. The generalized finite-rate model entails the solutions to the species equations for the mass fractions of fuel, oxygen, carbon dioxide, and water vapor:

$$\frac{\partial(\rho Y_{fu})}{\partial t} + \frac{1}{\sqrt{g}} \frac{\partial(\rho \hat{U}^l Y_{fu})}{\partial \xi^l} = \frac{1}{\sqrt{g}} \frac{\partial}{\partial \xi^l} \left(\frac{\mu}{Sc} \sqrt{g} g^{lm} \frac{\partial Y_{fu}}{\partial \xi^m} \right) - R_{fu}$$ (4.13.12)

$$\frac{\partial(\rho Y_{ox})}{\partial t} + \frac{1}{\sqrt{g}} \frac{\partial(\rho \hat{U}^l Y_{ox})}{\partial \xi^l} = \frac{1}{\sqrt{g}} \frac{\partial}{\partial \xi^l} \left(\frac{\mu}{Sc} \sqrt{g} g^{lm} \frac{\partial Y_{ox}}{\partial \xi^m} \right) - \frac{v_{ox} M_{ox}}{v_{fu} M_{fu}} R_{fu}$$ (4.13.13)

$$\frac{\partial(\rho Y_{co_2})}{\partial t} + \frac{1}{\sqrt{g}}\frac{\partial(\rho \hat{U}^l Y_{co_2})}{\partial \xi^l} = \frac{1}{\sqrt{g}}\frac{\partial}{\partial \xi^l}\left(\frac{\mu}{Sc}\sqrt{g}g^{lm}\frac{\partial Y_{co_2}}{\partial \xi^m}\right) + \frac{v_{co_2}M_{co_2}}{v_{fu}M_{fu}}R_{fu}$$

$$(4.13.14)$$

$$\frac{\partial(\rho Y_{h_2o})}{\partial t} + \frac{1}{\sqrt{g}}\frac{\partial(\rho \hat{U}^l Y_{h_2o})}{\partial \xi^l} = \frac{1}{\sqrt{g}}\frac{\partial}{\partial \xi^l}\left(\frac{\mu}{Sc}\sqrt{g}g^{lm}\frac{\partial Y_{h_2o}}{\partial \xi^m}\right) + + \frac{v_{h_2o}M_{h_2o}}{v_{fu}M_{fu}}R_{fu}$$

$$(4.13.15)$$

where Y and M are the mass fractions and molecular weights associated with the subscripts fu, ox, co_2, and h_2o representing the fuel, oxygen, carbon dioxide, and water vapor and Sc is the Schmidt number. The chemical reaction rate R_{fu} is evaluated based on the following Arrhenius expression:

$$R_{fu} = A\rho^2 Y_{fu}Y_{ox}\exp[-E/(R_u T)]$$

$$(4.13.16)$$

Coupling of Solid Phase and Gas Phase

The conservation equations in the solid phase and gas phase are coupled through energy and mass balances at the interface.

Considering the mass balance at the interface, the total mass of fuel species leaving the surface of the wood equals the mass flux of the fuel species due the bulk velocity of the mixture and mass flux of fuel species due to the mass diffusion at the interface. The mass balance equation of fuel can be written as

$$m_{fu}^n = m'' Y_{fu} - \rho D \frac{\partial Y_{fu}}{\partial n}$$

$$(4.13.17)$$

where ρ, and D are the density and mass diffusivity in the gas phase, m'' is the total mass flux normal to the interface, and n is the coordinate normal to the interface. Equation (4.13.17) can be rearranged to yield

$$\rho D \frac{\partial Y_{fu}}{\partial n} = m'' Y_{fu} - m_{fu}^n$$

$$(4.13.18)$$

On the basis of the above expression, the boundary conditions for the other species are similarly formulated as

$$\rho D \frac{\partial Y_{ox}}{\partial n} = m'' Y_{ox}$$

$$(4.13.19)$$

$$\rho D \frac{\partial Y_{co_2}}{\partial n} = m'' Y_{co_2}$$

$$(4.13.20)$$

$$\rho D \frac{\partial Y_{h_2o}}{\partial n} = m'' Y_{h_2o} - m''_{h_2o} \tag{4.13.21}$$

Invoking an energy balance at the interface, the conservation equation at the interface boundary becomes

$$-k \frac{\partial T}{\partial n} = -k_s \frac{\partial T_s}{\partial n} - F_{w-c} \varepsilon \sigma [T_{so}^4 - T_{cone}^4] - F_{w-o} \varepsilon \sigma [T_{so}^4 - T_{ref}^4] \tag{4.13.22}$$

where ε denotes the surface emissivity, σ is the Stefan-Boltzmann constant, T and T_s are respectively the gas and solid phase temperatures, k and k_s are the thermal conductivities in the gas and solid phases, T_{so} is the wood surface temperature, T_{cone} is the cone surface temperature, and T_{ref} is the ambient temperature. The shape factors F_{w-c} and F_{w-o} from any point P as seen in Figure 4.22 at the upper surface of the wood to the cone and to the surroundings can be derived from expressions for standard geometrical configurations (Siegel and Howell, 2002). They can be formulated as

$$F_{w-o} = 1 - F_{w-c} \tag{4.13.23}$$

$$F_{w-c} = \frac{1}{2} \left\{ \frac{s^2 + h_t^2 - r_t^2}{\sqrt{(s^2 + h_t^2 - r_t^2)^2 - 4(sr_t)^2}} - \frac{s^2 + h_b^2 - r_b^2}{\sqrt{(s^2 + h_b^2 - r_b^2)^2 - 4(sr_b)^2}} \right\} \tag{4.13.24}$$

where s represents the horizontal distance of P from the axis, h_t and h_b are the vertical distances of the top and bottom of the cone from the wood, and r_t and r_b are the radii of the top and bottom of the cone, as shown in Figure 4.22.

At the interface between the solid and gas phases, the velocity boundary conditions are given by the mass flux of the emerging stream of volatile gases from the solid: $\rho u = m''$. The pressure may fluctuate slightly due to the flow conditions. However, the fluctuations can be taken to be relatively insignificant when compared to the pressure variations as computed inside the wood, due to the evaporation of moisture and pyrolysis. The pressure at the interface is usually taken to be at atmospheric pressure ($P_o = 1.013$ bar).

Other Boundary Conditions

Figure 4.25 presents the boundary conditions imposed on the computational model of the cone calorimeter configuration in the gas phase. At the conical heater surface, the no-slip condition is applied for the velocity. *Neumann* condition is specified for the species mass fractions where their normal gradients at the boundary are zero, while for the temperature, constant heat flux values are imposed.

At the open boundaries, two types of boundary conditions, inflow and outflow boundaries, exist depending on the direction of the flow. For an inflow

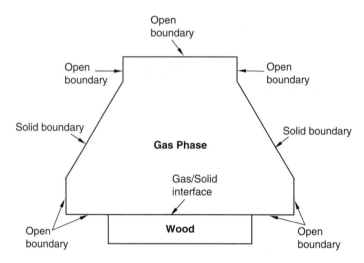

Figure 4.25   Boundary conditions in the gas phase.

boundary, the values of the velocity components u, v, and w are specified. Temperature and mass fraction of species are given by the upstream (ambient) values. Outflow boundaries are often located at positions where flows are unidirectional and stresses are known. The pressure is specified at atmospheric pressure. Adopting the *Neumann* condition, the gradients of the velocity, temperature, and mass fractions of species are assumed to be zero. In this worked example, pressure boundary conditions are applied at all the open boundaries.

Computational Mesh

A total of $13 \times 13 \times 15$ control volumes in the ξ^1 (axial), ξ^2 (circumferential), and ξ^3 (radial) directions are employed for the solid phase, while $38 \times 13 \times 24$ control volumes are used for the gas phase. Figure 4.26 represents the nonuniform mesh that has been used in the wood and the air regions, with a finely concentrated mesh immediately below and above the surface of the wood to sufficiently accommodate the steep gradients there. A sensitivity analysis of the mesh size is performed. Computed piloted ignition time remains unchanged, when the above mesh sizes in the gas and solid phase are increased to $18 \times 18 \times 21$ and $52 \times 18 \times 34$, respectively for a cone irradiance of $40\,\mathrm{kW}\;\mathrm{m}^{-2}$ and moisture content of 0%.

Numerical results: The three-dimensional model that comprises of both the wood pyrolysis and gas phase flaming combustion is applied to simulate the piloted ignition and burning of wood in a cone calorimeter configuration. A time step of 0.05 second is adopted. Physical properties and parameters that are required for the governing equations are tabulated in Figure 4.4. The initial temperature of the wood is set to 23°C, which corresponds to the standard conditions of the cone calorimeter tests. The specimen thickness is 20 mm and the emissivity of the surface of the wood is assumed to be 0.8 (Siegel

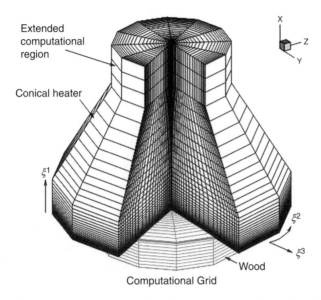

Figure 4.26 The computational mesh for the gas phase and wood with the x-axis along the direction of grains.

and Howell, 1992). Cone irradiances of 20, 40, 60, and 70 kWm^{-2}, corresponding to fires of different magnitudes, are imposed. In order to investigate the effects of moisture content of wood on piloted ignition and combustion, initial moisture contents of 0%, 5%, and 10% are considered. During the numerical simulations, the pilot, which is an electrical spark introduced at regular time intervals of 1 second in the cone calorimeter, has been introduced by raising the temperature of the gas at the control volume located at the pilot position to a temperature of 1000 K—a condition found to be sufficient for ignition of the fuel. This heat source is initiated intermittently at regular time intervals of 1 second until sustained flaming occurs. It was demonstrated that the energy contributed by the small pilot spark in this three-dimensional model causes no adverse effect on the piloted ignition time. Other temperature values of 1200 K and 1500 K have been tested. The time to ignition is found to vary within a range of 1 second at the irradiance 60 kWm^{-2}. In order to limit the undue heating effects on the computed ignition time, the value of 1000 K is therefore selected.

Figure 4.27 shows the transient temperature evolution at the pilot position of different irradiance levels and initial moisture contents. For dry wood (0% initial moisture content), the ignition time for the irradiance level at 40 kWm^{-2} occurs at 8 seconds of which burning is considered to be sustained after ignition. For wet wood with 5% and 10% initial moisture content, flashing (unsustained ignition) is seen to occur at 9 and 11 seconds. Here, the combustion could not be sustained and the flame is subsequently extinguished within 2 to 3 seconds due to an insufficient supply of fuel from the wooden specimen.

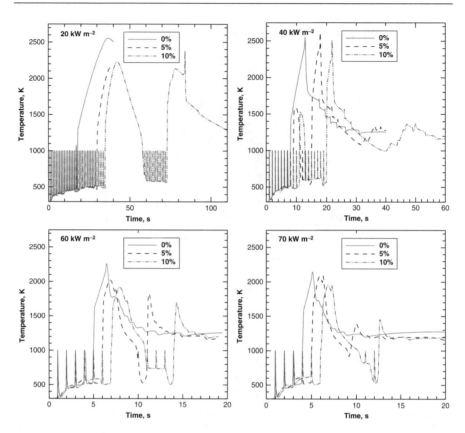

Figure 4.27 Transient temperature development at the pilot position with initial moisture contents of 0%, 5%, and 10% for irradiance: (a) 20 kW m^{-2}, (b) 40 kW m^{-2}, (c) 60 kW m^{-2}, and (d) 70 kW m^{-2}.

The pilot is turned off when flashing begins but is initiated again once it ceases. Sustained combustion is eventuated for both cases of moisture contents of 5% and 10%, at 15 and 20 seconds. Not surprisingly, the piloted ignition time increases with moisture content. Similar trends are also observed with the other investigations of wood with different irradiances from the conical heater and various initial moisture contents. An increase in the initial moisture content of wood causes a lengthening of the pilot ignition time at any irradiance level. Nevertheless, an increase in irradiance levels from the conical heater results in a reduction of the ignition time.

Table 4.5 shows the comparison between the predicted piloted ignition times and experimental results of Shields et al. (1993). During the experiments, the wood specimens were pre-conditioned carefully in a controlled environment of a temperature of 23°C and a relative humidity of 50% before pilot ignition experiments were performed. Although the initial moisture contents of the

Table 4.4 Physical properties of wood

Physical properties	Value
ρ_f (kg m^{-3})	111.3
C_{pw} (kJ kg^{-1} K^{-1})	$1.4 + 0.3\,T_s$
C_{pf} (kJ kg^{-1} K^{-1})	$0.70 + 0.6\,T_s$
C_{pg} (kJ kg^{-1} K^{-1})	$1.0 + 0.8\,T_s$
C_{pv} (kJ kg^{-1} K^{-1})	1.88
C_{pm} (kJ kg^{-1} K^{-1})	4.19
λ_w^{ℓ} (W m^{-1} K^{-1})	$0.166 + 0.396\ (\rho_m/\rho_w)$
E_p (kJ mole^{-1})	26.3
A_p (s^{-1})	0.54
ΔH_p (kJ kg^{-1})	0
ΔH_{ev} (kJ kg^{-1})	2260
E (kJ mole^{-1})	112.86
A (s^{-1})	3.13 x 10^9
ΔH (MJ kg^{-1})	-16.72

Note: H refers to the heat of combustion for the gas volatiles.

Table 4.5 Comparison of the computed pilot ignition times with the measured pilot ignition times*.

Irradiance	Computed pilot ignition time (s)			Measured pilot ignition time (s)*	
	% Initial moisture content				
(kW m^{-2})	0%	5%	10%	Spark[†]	Gas[‡]
20	18	30	73	306–405	145–437
40	8	15	20	19–34	20–29
60	5	11	14	6–12	6–11
70	4	9	12	4–11	6–12

[*] Experimental ignition times for piloted ignitions of wood of a moisture content of 10% with an electric spark or gas pilot flame, as reported by Shields et al. (1993).
[†] Electric spark pilot source by means of high voltage across electrodes.
[‡] Gas pilot source by means of a small naked flame.

sample were not reported, the average initial moisture contents of samples in this worked example are estimated to be 10%, according to the formula of Skaar (1988). The predicted pilot ignition time is 20 seconds for the irradiance 40 kWm^{-2}, which is well within the range of experimentally determined pilot ignition times of 19–34 seconds and 20–29 seconds for spark and gas pilot ignition sources, respectively. For the irradiance of 60 kWm^{-2}, the predicted pilot ignition time is 14 seconds, and it appears to be just marginally out of the range of the reported values of 6–12 seconds and 6–11 seconds for spark and gas pilot ignition source. At the irradiance 70 kWm^{-2}, the predicted pilot ignition time is ascertained to be 12 seconds, which is within the range of 6–12 seconds for the gas pilot ignition source from the experiments. However, the pilot ignition at irradiance of 20 kWm^{-2} is predicted to be 73 seconds. This discrepancy is most probably attributed to the inadequacy of a first order reaction to describe the gas combustion and the lack of considerations of soot formation and gas radiant heat transfer in the gas phase combustion model. Also, the volatile gases produced from the pyrolysis of wood are assumed to contain pure fuel (100% combustible gas) with an empirical chemical formula of CH_2O. In a similar study by Tzeng and Atreya (1991), the percentage of combustible gases was assumed to be as low as 20% of the total volatile gases produced in the pyrolysis, in order to improve their predictions. This was especially found to be important when the irradiance level was as low as 20 kWm^{-2}. Although such an adjustment could improve the predicted pilot ignition time, the transient values of the percentage of combustible gases cannot be verified through experiments, and is thus considered to be artificial.

Owing to the axisymmetric nature of the solution, it is sufficient to only illustrate the fluid flow and heat transfer on half of the cone calorimeter configuration. Figures 4.28 illustrates the computed temperature contours (on the left) and velocity vectors (on the right) for irradiance 40 kW m^{-2} with 10% initial moisture content, at 10 seconds time interval between 5 and 55 seconds. Similar observations are also found with the computed results for the pilot ignition and combustion of wood, with different initial moisture contents and irradiances of 20, 60, and 70 kWm^{-2}.

A gradually developing fuel rich region above the wood is found, as the temperature of wood increases during the period from 5 to 20 seconds. The temperature and velocity distributions remain steady right up to 20 seconds. Although flashing occurs between 11 and 13 seconds, the amount of heat generated within this short period is found to be unsustainable for combustion, because of the buoyant flow having sufficient drag to transport the heat away. Nevertheless, as the temperature increases at the wood surface, more fuel is produced to replenish the burnt fuel at the pilot ignition location, and ignition occurs at 20 seconds with sustainable combustion of the gas volatiles. The isotherms illustrate a central core of high temperatures above the wood, with the highest computed temperature predicted at around 1900 K. This temperature representing a zone of rapid combustion in the gas phase is much higher than those found in the actual combustion of wood, because of the absence of the gas radiant heat absorption and soot formation in the gas phase model. The strong buoyancy due to the fire, increases the velocities at the central axial region in the flow field, which are substantially higher values than those due to the conical heater at the sides.

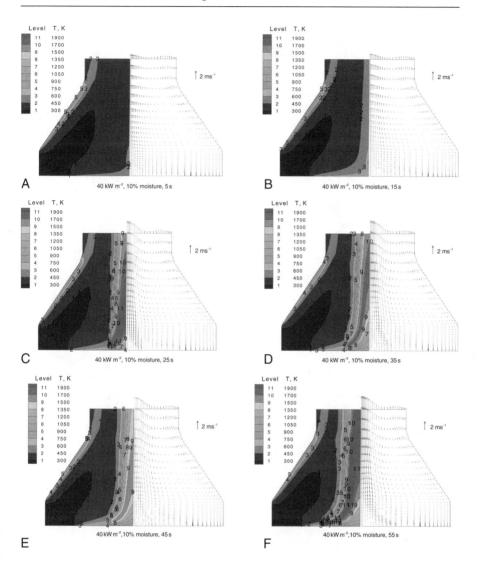

Figure 4.28 Transient isotherms (left) and velocity vectors (right) for irradiance 40 kW m^{-2} with an initial moisture content of 10% at (a) 5 seconds, (b) 15 seconds, (c) 25 seconds, (d) 35 seconds, (e) 45 seconds, and (f) 55 seconds.

Figure 4.29 demonstrates the distributions of temperature, char fraction ρ_c/ρ_w, and moisture fraction ρ_m/ρ_w at various depths from 0.35 to 5.51 mm below the surface at the center line of the wood at 40 kWm^{-2} irradiance and initial moisture content of 10%. Similar behaviors are also observed for initial

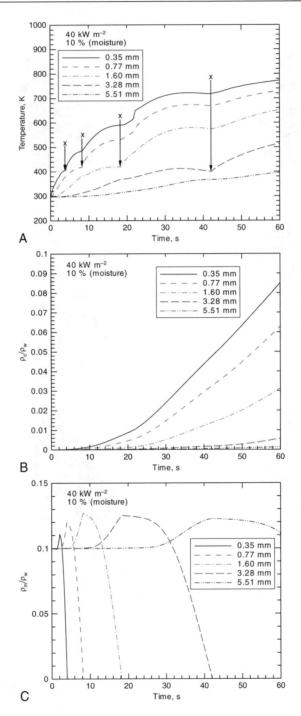

Figure 4.29 Transient distributions at various depth below the surface at the center of wood for irradiance 40 kW m^{-2} with initial moisture content of 10%: (a) Temperature, (b) Char fraction ρ_c/ρ_w, and (c) Moisture fraction ρ_m/ρ_w.

moisture content of 0% and 5%. At a depth of 0.35 mm (just below the exposed surface of wood), the temperature increases continuously and gradually to a value of around 610 K. Shortly after piloted ignition and the development of sustained combustion above the wood, a sudden jump of the temperature is observed at around 12 seconds, which corresponds to the heat being transferred downwards from the combustion zone to the wood. Similar jumps of the temperature curves for the depth of 0.35 mm below the wood surface are also observed after ignition when the temperature reaches around 610 K, which subsequently indicates sustained combustion developing above the wood in the gas phase. For wet wood, there appears to be a distinct change in the slopes of the curves corresponding to the depths of 0.35 to 3.28 mm, at the points marked "×." Before "×," absorption of latent heat during evaporation of moisture reduces the rate of rise of temperature. Unlike commonly found vaporization processes at ambient conditions where the pressure is normally constant (at atmospheric pressure), the pressure inside the wood tends to increase due to the accumulation of trapped vapor in wood, and the poor permeability of the porous wood substrate while undergoing vaporization. During the vaporization process, the temperature will still increase, albeit at a slower rate than that for dry wood. The vaporization process generally begins at 373 K and shortly terminates at around 420 K. After "×," the rate of rise of temperature is mainly increased due to the evaporation of the moisture. Some evaporation is noticeable at 5.51 mm, as demonstrated by the slight decrease of slope of the temperature curve. The computed temperatures at $40 \, \text{kWm}^{-2}$ irradiance with initial moisture contents of 5% and 10% agreed qualitatively with the measured transient temperatures of Blackshear and Kanury (1970), Lee and Diehl (1981), and Tran and White (1992). All the transient temperature curves demonstrate a plateau occurring around 100°C—the evaporation of moisture in wood.

Conclusion: A 3-D mathematical model based on the non-orthogonal curvilinear coordinate system to simulate the geometry and flow field within a cone calorimeter, is demonstrated in this worked example. Predicted results showing the transient thermal response and the combustion phenomenon of the wood are shown to correspond closely to typically observed cone calorimeter experiments. The effects of initial moisture content on ignition time show that higher initial moisture content delays ignition. Flashing ignitions are also observed for wood with initial moisture contents of 5% and 10%. The numerically predicted pilot ignition times over a range of external irradiance fluxes from $20 \, \text{kW m}^{-2}$ to $70 \, \text{kW m}^{-2}$ and initial moisture contents of 0%, 5%, and 10%, are found to be in good agreement with experimental measurements. However, the predicted pilot ignition results at irradiance $20 \, \text{kWm}^{-2}$ yield considerably lower than the experimentally reported values. This discrepancy is most probably attributed to the inadequacy of a first order reaction to describe the gas combustion and the lack of considerations of soot formation and gas radiant heat transfer in the gas phase combustion model.

4.14   Worked Example on Fire Growth and Flame Spread Over Combustible Wall Lining in a Single-Room Compartment

Flame spread and fire over cellulosic materials occur when the burning region supplies sufficient heat to the virgin solid to cause gasification. The application of combined models of the solid phase pyrolysis of wet wood and gas phase combustion, is demonstrated in this worked example to calculate the flame spread over a vertical timber wall in a full-scale compartment.

Figure 4.30 shows the room built according to the ISO/DIS 9705 (1990) standard, of which the flame spread experiment was performed. An untreated Radiate pine, 2.44 m wide by 2.44 m tall, with a thickness of 19 mm was placed at the back of the room. The sand-box propane burner served as the ignition source and was located 0.3 m above the floor at the center of the bottom edge of the timber wall. This was assumed to be flush with the floor level in the computations. The experiment was conducted with a heat release output from the burner of 40 kW for the first 5 minutes and later increased to 160 kW for a further 15 minutes. These times have been chosen based on standard experimental produces in the ISO/DIS 9705 (1990) standard. Development of the fire was captured by a video camera located just outside the room and near the doorway. The room was instrumented with Type K MIMS (Mineral Insulated Metal Sheathed) thermocouples, with a 310 stainless steel sheath, an outer diameter of 1.5 mm, and an unearthed junction.

Numerical features: for the solid phase, an energy conservation equation for cellulosic fuel, which describes the heat transfer and heat balance due to

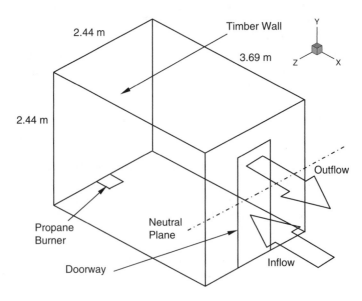

Figure 4.30   ISO/DIS 9705 (1990) standard room for fire tests.

thermal conduction, convection of internal volatiles, and heat of pyrolysis, is formulated. A single step first order reaction is assumed for the pyrolysis of wood in which the active constituent undergoes pyrolysis to become combustible volatiles. Thermal swelling, shrinkage, surface regression, and possible surface oxidation reactions of the virgin wood and char are neglected. The timber wall is assumed to be dry. Volatiles and water vapor flow only toward the heated surface, with no resistance to mass flow. The pressure is therefore assumed to be constant. Based on the stated assumptions, the model equations for the pyrolysis of wood are simplified to yield:

Virgin solid mass balance

$$R_p = \frac{\partial \rho_a}{\partial t} = -\rho_a A_p \exp\left(-\frac{E_p}{R_u T_s}\right) \tag{4.14.1}$$

Evaporation

$$R_v = \frac{\partial \rho_m}{\partial t} = -\rho_m A_m \exp\left(-\frac{E_m}{R_u T_s}\right) \tag{4.14.2}$$

Energy balance

$$\frac{\partial(\rho_m C_{pm} + \rho_s C_{ps})T_s}{\partial t} = \frac{\partial}{\partial x}\left(k_s \frac{\partial T_s}{\partial x}\right) + \frac{\partial}{\partial y}\left(k_s \frac{\partial T_s}{\partial y}\right) + \frac{\partial}{\partial z}\left(k_s \frac{\partial T_s}{\partial z}\right)$$

$$-\frac{\partial(\rho_v C_{pv} + \rho_g C_{pg})T_s}{\partial x} - R_p \Delta H_p - R_v \Delta H_v$$

Continuity $\hspace{7cm}$ (4.14.3)

$$\frac{\partial m_g}{\partial x} = -R_p \tag{4.14.4}$$

$$\frac{\partial m_v}{\partial x} = -R_v \tag{4.14.5}$$

The pyrolysis mass flux m_g and m_v are obtained by integrating equations (4.14.4) and (4.14.5), in other words,

$$m_g = \int_0^x -R_p dx \tag{4.14.6}$$

$$m_v = \int_0^x -R_v dx \tag{4.14.7}$$

Thermophysical properties such as the volumetric $\rho_s C_{ps}$ and thermal conductivity k_s of the pyrolyzing wood are assumed to vary linearly with char fraction from the initial value of the virgin wood to that of the char such as represented in equations (4.11.35) and (4.11.41), respectively. The final char density is assumed to be known *a priori*.

For the gas phase, the model for the turbulent diffusion flame consists of the three-dimensional, Favre-averaged equations of transport for mass, momentum, enthalpy, and scalar property, representing each chemical specie concentration is used. The governing equations in Cartesian coordinates are the same as those listed for the Favre-averaged Navier-Stokes equations in Table 2.5 of Chapter 2. Turbulence is handled via the standard k-ε model (refer to equations (2.11.8) and (2.11.9) derived in Chapter 2). The production of turbulence due to buoyancy and the effect of thermal stratification of the turbulence dissipation are also accounted as additional source terms in the turbulence model equations.

For the gas phase combustion, the rates of the combustion reactions in the gas phase are determined by the minimum of the two values: the rate given by the Arrhenius expression, and the eddy break-up rate due to the mixing of the turbulent eddies containing intermittent species concentrations. Since two gaseous fuels (i.e. the combustible volatiles G and propane fuel F) are involved in the combustion, two single-step reactions are assumed:

Volatile gas

$$\nu_G G + \nu_o^G O_2 \rightarrow \text{Products } (CO_2 \text{ and } H_2O) \tag{4.14.8}$$

Propane

$$\nu_F F + \nu_o^F O_2 \rightarrow \text{Products } (CO_2 \text{ and } H_2O) \tag{4.14.9}$$

where ν_G, ν_O^G, ν_F, and ν_O^F are the number of moles of the species involved during the combustion reactions, which are competing reactions. The reaction rates for the combustible volatiles can be written as

$$R_{G,chemical} = A' \exp\left(-\frac{E'}{R_u T}\right) (\rho \tilde{Y}_G)^{a'} (\rho \tilde{Y}_O)^{b'} \tag{4.14.10}$$

$$R_{G,diffusion} = C_R \rho \frac{\varepsilon}{k} \min\left\{ \tilde{Y}_F, \frac{\tilde{Y}_O}{(\nu_O^G M_O / V_G M_G)} \right\} \tag{4.14.11}$$

$$R_G = \min\{R_{G,chemical}, R_{G,diffusion}\} \tag{4.14.12}$$

while the reaction rates for the propane fuel are given by

$$R_{F,chemical} = A \exp\left(-\frac{E}{R_u T}\right) (\rho \tilde{Y}_F)^{a'} (\rho \tilde{Y}_O)^{b'} \tag{4.14.13}$$

$$R_{F,diffusion} = C_R \rho \frac{\varepsilon}{k} \min\left\{ \tilde{Y}_F, \frac{\tilde{Y}_O}{(v_O^F M_O / v_F M_F)} \right\} \tag{4.14.14}$$

$$R_F = \min\{R_{F,chemical}, R_{F,diffusion}\} \tag{4.14.15}$$

where $R_{G,chemical}$, $R_{G,diffusion}$, and R_G represent the Arrhenius, eddy break-up, and net reaction rates, while A', E', a', and b' are the pre-exponential constant, activation energy, and exponents for the combustible volatiles. $R_{F,chemical}$, $R_{F,diffusion}$, and R_F, A E, a, and b are the corresponding quantities or the propane fuel. Owing to the involvement of two gaseous fuels, the source terms of the governing equations for the conservation of species are modified, as shown in Table 4.6, conforms to the generic form of equation (2.5.1) given in Chapter 2.

The specific enthalpy of the gas phase is re-defined to reflect the contribution from the propane fuel as

$$\tilde{h} = \int_{T_{ref}}^{\tilde{T}} C_p dT + \gamma_G \tilde{Y}_G \Delta H + \gamma_F \tilde{Y}_F \Delta H' \tag{4.14.16}$$

where ΔH and $\Delta H'$ are the heating values for the combustible volatiles and propane and γ_G and γ_F are the combustion efficiencies. Considering the impurity of the fuels and incomplete combustion due to soot formation (the supplemental model of soot is not employed in this worked example in order to speed up the computations), the combustion efficiency γ_G for the combustible volatiles is taken to be 70% (Drysdale, 1986), while the combustion efficiency γ_F for the propane fuel is assumed to be 80% (Luo and Beck,

Table 4.6 Transport equations for various species in the flow field.

ϕ	Γ_ϕ	S_ϕ
$\tilde{h}$	$\dfrac{\mu}{Pr} + \dfrac{\mu_T}{Pr_T}$	S_{rad}
$\tilde{Y}_G$	$\dfrac{\mu}{Sc} + \dfrac{\mu_T}{Sc_T}$	$-R_G$
$\tilde{Y}_F$	$\dfrac{\mu}{Sc} + \dfrac{\mu_T}{Sc_T}$	$-R_F$
$\tilde{Y}_O$	$\dfrac{\mu}{Sc} + \dfrac{\mu_T}{Sc_T}$	$-\dfrac{v_O^G M_O}{v_G M_G} R_G - \dfrac{v_O^F M_O}{v_F M_F} R_F$
$\tilde{Y}_{CO_2}$	$\dfrac{\mu}{Sc} + \dfrac{\mu_T}{Sc_T}$	$\dfrac{v_{CO_2}^G M_{CO_2}}{v_G M_G} R_g + \dfrac{v_{CO_2}^F M_{CO_2}}{v_F M_F} R_F$
$\tilde{Y}_{H_2O}$	$\dfrac{\mu}{Sc} + \dfrac{\mu_T}{Sc_T}$	$\dfrac{v_{H_2O}^G M_{H_2O}}{v_G M_G} R_g + \dfrac{v_{H_2O}^F M_{H_2O}}{v_F M_F} R_F$

1994). The enthalpy equation is also modified by adding the contribution due to radiation heat transfer, which has been achieved through the discrete ordinates method (see Chapter 3).

Conservation equations in the solid phase and gas phase are coupled through energy and mass balance at the interface of the vertical timber wall according to

$$k_s \frac{\partial \tilde{T}_s}{\partial x} = k \frac{\partial \tilde{T}}{\partial x} + \varepsilon q_{rad} - \varepsilon \sigma \tilde{T}_s^4 \tag{4.14.12}$$

$$\bar{\rho} D \frac{\partial \tilde{Y}_G}{\partial x} = (m_g + m_v)(\tilde{Y}_G - \tilde{Y}_{G,solid}) \tag{4.14.13}$$

$$\bar{\rho} D \frac{\partial \tilde{Y}_F}{\partial x} = (m_g + m_v)\tilde{Y}_F \tag{4.14.14}$$

$$\bar{\rho} D \frac{\partial \tilde{Y}_O}{\partial x} = (m_g + m_v)\tilde{Y}_O \tag{4.14.15}$$

$$\bar{\rho} D \frac{\partial \tilde{Y}_{CO_2}}{\partial x} = (m_g + m_v)\tilde{Y}_{CO_2} \tag{4.14.16}$$

$$\bar{\rho} D \frac{\partial \tilde{Y}_{H_2O}}{\partial x} = (m_g + m_v)(\tilde{Y}_{H_2O} - \tilde{Y}_{H_2O,solid}) \tag{4.14.17}$$

$$T = \tilde{T}_s, \quad \bar{\rho}\tilde{u} = m_g, \quad \tilde{v} = \tilde{w} = 0 \tag{4.14.18}$$

where q_{rad} is the incoming radiative flux at the interface, ε is the wall emissivity, and $\tilde{Y}_{G,solid} + \tilde{Y}_{H_2O,solid} = 1$. At the burner interface, temperature is specified at the inlet fuel temperature recorded during the experiment (T_{fuel}), while the species concentrations are evaluated through the conservation of mass

$$\bar{\rho} D \frac{\partial \tilde{Y}_G}{\partial y} = m_b \tilde{Y}_G \tag{4.14.12}$$

$$\bar{\rho} D \frac{\partial \tilde{Y}_F}{\partial y} = m_b (\tilde{Y}_F - \tilde{Y}_{propane}) \tag{4.14.13}$$

$$\bar{\rho} D \frac{\partial \tilde{Y}_O}{\partial y} = m_b \tilde{Y}_O \tag{4.14.14}$$

$$\bar{\rho} D \frac{\partial \tilde{Y}_{CO_2}}{\partial y} = m_b \tilde{Y}_{CO_2} \tag{4.14.15}$$

$$\bar{\rho} D \frac{\partial \tilde{Y}_{H_2O}}{\partial y} = m_b \tilde{Y}_{H_2O} \tag{4.14.16}$$

$$T = T_{fuel}, \quad \bar{\rho}\tilde{v} = m_b, \quad \tilde{u} = \tilde{w} = 0 \tag{4.14.18}$$

where $Y_{propane}$ is equivalent to 0.95 (95% of pure propane). The burner mass flux m_b is pre-determined from experiments. At the walls, the no-slip condition is imposed for the velocities. The normal derivatives of species concentrations and the turbulent kinetic energy k are equated to zero, while the dissipation rate ϵ is evaluated from an empirical equation (see Chapter 2). The walls are assumed to be adiabatic and their temperature is calculated using an energy balance of the incoming and outgoing heat fluxes at the boundaries, including radiation as well as convection. For the momentum, heat and mass fraction fluxes to the wall, conventional logarithmic wall functions are used. At the extended computational region, the boundary conditions require that the normal derivatives of the velocities and turbulent quantities are set to be zero and the pressure is specified to be the atmospheric pressure. Temperature and species concentrations are extrapolated from their upstream values for flow exhausting to the surroundings, while for flow entering into the compartment, quantities at ambient conditions are specified at the boundaries. Pressure is specified at all times to be atmospheric pressure.

The non-uniform Cartesian grid distribution for the burn-room is shown in Figure 4.31. Note that fine grid distribution in the vicinity of the timber wall is required to resolve the active pyrolysis processes occurring at the surface. There are 10 cells across the thickness of the timber wall, 48 cells along the length, 28 cells along the width, and 25 cells along the height, giving a total

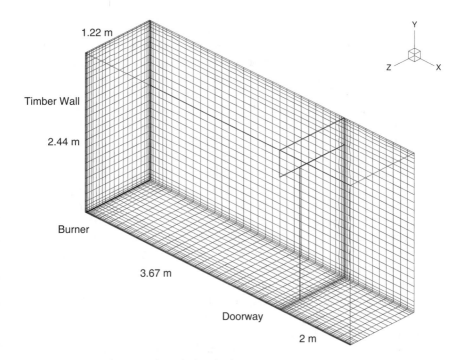

Figure 4.31 Computational mesh for the burn-room.

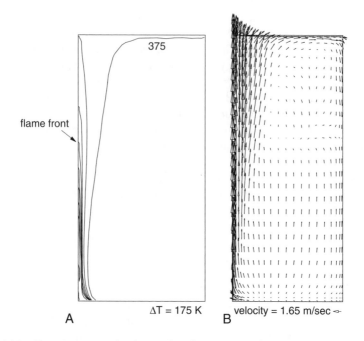

Figure 4.32 Temperature and velocity distribution immediately above the timber wall surface at 45 seconds for 0% moisture content.

of 40600 cells. Because of symmetry, only half the room is modeled. The solution domain is extended beyond the doorway where constant pressure boundary conditions are applied on the boundaries of the external region.

Numerical results: Physical properties and parameters for pine can be found in Tzeng and Atreya (1991), Di Blasi (1994a), and Yuen et al. (1995, 1997, 1998, and 2000). For propane, they are obtained from Westbrook and Dryer (1981). The initial temperature of the air and all walls is set to 27°C.

Figures 4.32, 4.33, and 4.34 demonstrate the computed temperature and velocity distributions for 0% moisture content (dry wood) over a y-z plane located at x = 0.029 m, at 45, 441, and 690 seconds. Owing to the symmetric nature of this particular solution, it is sufficient to show only half of the elevation in the figures. Upon the introduction of the propane flame at the initial stage, the timber wall subsequently burns due to the intense localized radiation from the burner flame promoting the pyrolysis of the virgin wood and causing the volatiles to emerge from the wood surface. Combustion occurs above the wood surface due to the sufficiently high temperatures in the vicinity when the volatiles and oxygen meet and react. Radiation feedback from the flame to the wood surface causes further pyrolysis of the virgin wood underneath the surface, and at neighboring locations at the timber surface. This results in a rapid upward spread of flame within 45 seconds, as indicated by the position of the flame front in Figure 4.32. The upward spread is found to be the dominant behavior of the flame spread phenomenon in

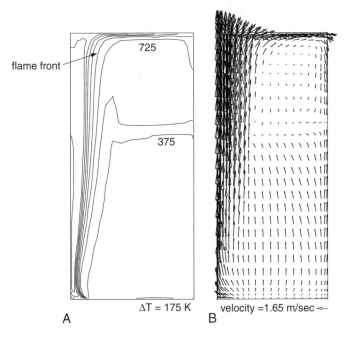

Figure 4.33 Temperature and velocity distribution immediately above the timber wall surface at 441 seconds for 0% moisture content.

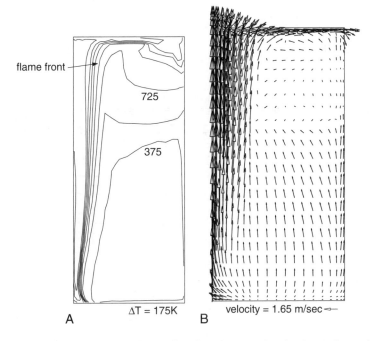

Figure 4.34 Temperature and velocity distribution immediately above the timber wall surface at 690 seconds for 0% moisture content.

comparison to the horizontal spread, because of the buoyancy force dragging the flames upward. However, the upward spread is impeded after 45 seconds as convection and radiation heat losses to the ambient surrounding increase.

At 300 seconds, the heat output of the propane burner increasing to 160 kW causes the fire to grow significantly. In Figure 4.33a, the upward spreading flame has reached the ceiling and is deflected horizontally, forming a ceiling jet. This jet then gives rise to a substantial increase of radiation and convection heat feedback to the unburnt timber underneath the ceiling causing the downward spread of the flame, as in Figure 4.34a. As rapid upward spreading is progressively gaining momentum, the horizontal spreads are also intensifying, as indicated by the broadening of the burnt areas. Figure 4.35 shows the temperature and velocity

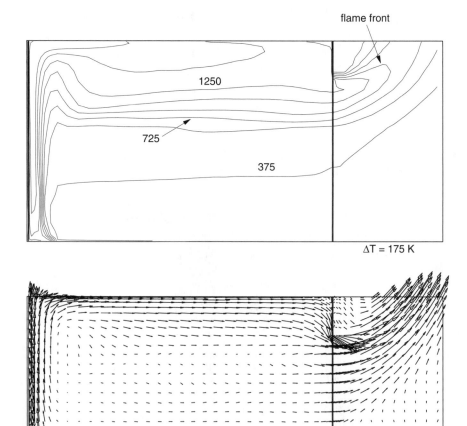

Figure 4.35 Temperature and velocity distribution at symmetry plane occurring at 742.5 seconds for 0% moisture content.

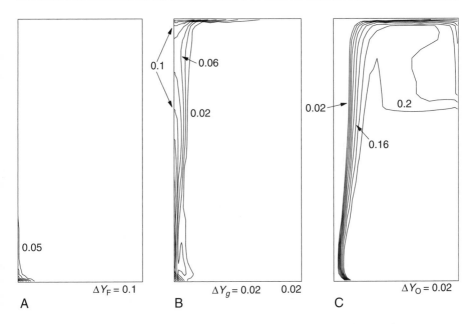

Figure 4.36 Mass fraction contours of (a) propane, (b) volatiles, and (c) oxygen immediately above the timber wall surface at 441 seconds for 0% moisture content.

distribution in the symmetry plane $z = 1.22$ m at 742.5 seconds. The flames are seen propagating away from the timber wall and covering the entire ceiling. In addition, a tongue of flame is observed escaping from the room through the doorway, which confirms the condition of flashover.

Figures 4.36 and 4.37 present the mass fractions of propane, volatiles, and oxygen at 441 and 690 seconds. Propane fuel is present only in the immediate vicinity of the propane burner, as shown by the mass fraction plots in Figures 4.36a and 4.37a. Although the propane burner has been introduced to ignite the timber wall, it has no significant influence on the flame spread behavior. The evolution of the combustible volatiles when the burner output is at 160 kW is clearly evident in contributing much of the burning, especially for the development of fire spreading at the upper layer below the ceiling. This can be best illustrated by the mass fraction contours of the volatiles in Figures 4.6 and 4.37b. At 441 seconds, a volatile region is developed and reaches the ceiling forming a thin layer of fuel just below the ceiling. More combustible volatiles are subsequently produced, and the enlargement of a pocket of fuel at the top corner between the ceiling and side wall at 690 seconds leads to a large fuel rich region and the occurrence of flashover at 742.5 seconds.

Figures 4.38 and 4.39 depict the video images taken during the experiment at 272, 441, 841, and 887 seconds. The flames stayed steady above the burner and had not reached the ceiling at 272 seconds. At 441 seconds, with the increased heat output of the propane burner from 40 kW to 160 kW, the

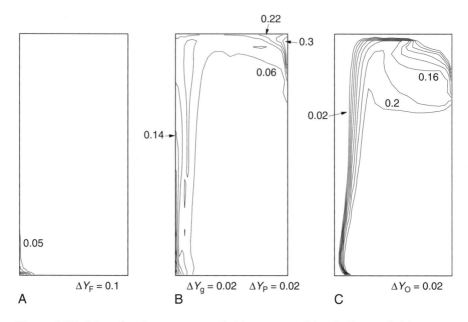

Figure 4.37 Mass fraction contours of (a) propane, (b) volatiles, and (c) oxygen immediately above the timber wall surface at 690 seconds for 0% moisture content.

flames propagated upward reaching the ceiling in a very short time, as shown in Figure 4.39b. The computed flame structure shown in Figure 4.33a corresponds extremely well to the experimentally observed shape. In Figure 4.39, the video images show the onset and occurrence of flashover at 841 and 887 seconds. The computations, however, predict an earlier time of 742.5 seconds in reaching this condition. This discrepancy could be possibly due to the neglect of moisture content in the wood pyrolysis model. In reality, there is always a small percentage of moisture that is present in the timber wall. Further computer simulation with an assumption of moisture content of 5%, results in flashover occurring at 930 seconds, which is closer to the experimental observation as shown in Figure 4.39. It should be noted that the consideration of moisture in the wood model only delays the flame spreading, but does not significantly alter the overall fire growth and development over the vertical timber wall in the room, such as already predicted during the dry wood computations. On another computer simulation with an assumption of 10% moisture content, an interesting situation is revealed whereby the flashover condition could not be attained. This actually corresponds to similar observations in experiments that have been conducted in humid conditions, and is believed to be due to the high level of absorbed moisture present in the timber.

Conclusion: A mathematical model coupling the three-dimensional wood pyrolysis model with the three-dimensional gas phase field model successfully computes the characteristics of the flame spread and combustion phenomenon over a vertical timber wall in a room. The predicted development of the flame

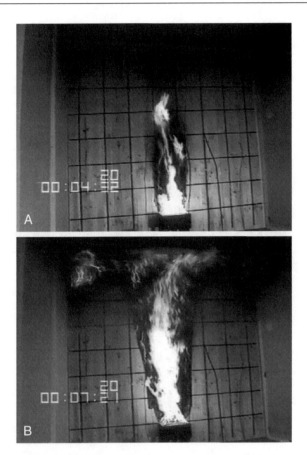

Figure 4.38 Video images of flame structures at (a) 272 and (b) 441 seconds.

spread and fire structure over the timber wall corresponds very well to the experimental observations. With the assumption of 5% moisture content in the wood, the time to flashover is closer to the particular experiment described in the worked example. The consideration of 10% moisture content in the timber does not result in the occurrence of flashover.

4.15 Summary

A three-dimensional mathematical model for the pyrolysis of wet wood is presented, which includes detailed considerations of the evaporation of moisture, anisotropic and variable properties, and pressure-driven internal convection of gases in wood. Although a single first-order Arrhenius reaction is generally used for computational simplicity, multiple competing reactions for up to six constituents have been formulated in the developed model. It has been

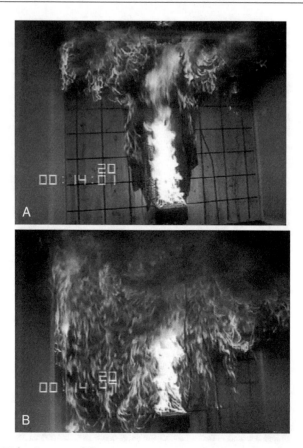

Figure 4.39   Video images of flame structures at (a) 841 and (b) 887 seconds.

demonstrated through the worked examples that the current three-dimensional model for pyrolysis of wet wood coupled with the gas phase combustion can be applied to adequately describe the sophisticated chemical and physical processes associated with the pyrolysis process in the cone calorimeter environment, and the flame spreading behavior over a combustible lining leading toward the occurrence of flashover in a full-scale enclosure. In general, the predicted results obtained from the coupled gas phase and solid phase models correspond very well to the experimental measurements and observations.

Review Questions

4.1. Soot is produced in most practical fires. Why?
4.2. How can soot be identified in non-premixed flames?
4.3. What is the principal mechanism for the radiative heat loss in fires?
4.4. What do smoke particles generally consist of?

4.5. In compartment fires, the consideration of particulate smoke is important. Why?

4.6. How is soot considered and incorporated in the field modeling approach?

4.7. What are the essential physical processes in soot formation and oxidation in formulating the soot model?

4.8. What is the typical soot precursor mainly responsible for the growth of soot particles?

4.9. What are the common approaches in the modeling of soot? Explain the advantages and disadvantages of each modeling approach.

4.10. The population balance concept represents another possible consideration in the predictive treatment of the soot process. What is population balance? How can it be feasibly used with the CFD-based fire model?

4.11. Explain the difference between *homogeneous* and *heterogeneous* soot formation.

4.12. Describe the pyrolysis process. Why is the advantage of considering pyrolysis in field modeling?

4.13. What standard experimental apparatus that can be used to measure the mass-loss rates, heat-release rates, and so forth for a solid combustible material?

4.14. The physico-chemical processes of pyrolysis of wood are due to different major constituents. What are they?

4.15. Describe the characteristics of the burning of wood in air.

4.16. On a phenomenological understanding of the pyrolysis process, what are the general reactions involved in the pyrolysis and combustion of wood? Compare with the pyrolysis and combustion of cellulose.

4.17. What are the three important properties of wood that significantly influence the combustion and heat release characteristics?

4.18. In the wood pyrolysis model, what are the governing equations? How is the flow of pyrolysis volatile gas, water vapor, and dry air considered?

4.19. Appropriate source terms are usually formulated in the governing equations to adequately describe the pyrolysis process. What are they and how are they treated in the wood pyrolysis model?

4.20. Thermophysical properties are required in the wood pyrolysis model. What are they and why are they important?

4.21. When can a one-dimensional model for the mass transfer be applied when considering pyrolysis in field modeling?

4.22. Pyrolysis is mainly attributed by the surface activities occurring between the solid and gas phases. In order to obtain accurate results, what is the most important consideration when accounting pyrolysis in field modeling?

5 Advance Technique in Field Modeling

Abstract

The prospect of faster and more powerful digital computers, as well as the ever-increasing development of efficient numerical algorithms, has made it possible to feasibly handle practical fires of technical relevance through more direct considerations. This chapter focuses on the description of the large eddy simulation of fires. Essentially, the concept centers on characterizing the turbulent reacting flow by solving the macroscopic large-scale motion through the governing equations of fluid mechanics, while the microscopic small-scale motion is approximated via appropriate models. Subgrid scale modeling of the microscopic flow processes is described, with special emphasis on a range of explicit models, the effects of subgrid fluctuations on the chemical heat release rate, and appropriate radiation and soot modeling for large eddy simulation of turbulent fires.

5.1 Next Stages of Development and Application

In field modeling, the difficulties associated with the analyses of fire-related phenomena stem from the wide range of length and time scales that exist within the turbulent reacting flow. Firstly, the combustion zone above the fuel source is a region where the local mixing of gasified fuel and air reacts to produce combustion products associated with the release of chemical energy and emission of radiant energy. These processes, considered to be *microscopic* in nature, can occur on length scales ranging from a fraction of a millimeter to a few centimeters. Secondly, the combustion zone represents a source of buoyancy, which induces large-scale mixing of the air and combustion products, forming a plume, which can prevail as an organized structure over length scales covering meters or tens of kilometers, depending on the fire scenario of interest. This refers to the *macroscopic* description of the fire dynamics. In compartment fires, this plume, in turn, acts as a giant pump that generates a flow pattern throughout the entire structure housing the indoor fire.

Subject to the availability of computing power and resources, much effort have been invested in the formulation of appropriate sub-models describing combustion, radiation, soot production, and pyrolysis, in order to dramatically increase the sophistication of modeling the fire dynamics. To reduce the computational effort to acceptable levels, the use of the so-called Favre-Averaged Navier-Stokes (FANS) approach is adopted to primarily dispense with the notion of resolving the temporal behavior due to the averaging process performed on the governing equations. As illustrated through the worked examples in previous

Computational Fluid Dynamics in Fire Engineering
Copyright © 2009 by Academic Press. Inc. All rights of reproduction in any form reserved.

chapters, good qualitative and quantitative agreement against experimental measured profiles as investigated in different compartment fire configurations, have been obtained. Within a limited class of turbulent reacting flows, the use of the conventional field models via k-ε representations of turbulence and appropriate sub-models, have clearly illustrated the feasibility of attaining useful numerical results. No doubt that the development of the FANS approach coupled with these complex sub-models have contributed immensely to the advancement of field modeling in fire engineering.

In anticipation of the ever-increasing power of digital computers and the development of quicker numerical algorithms, there is nonetheless a shift in focus and a greater emphasis being placed in fostering the next stages of development and application of fire models, in resolving the turbulent reacting flows of technical relevance by more *direct* means. Specifically, this involves the ability of calculating the fluid flow and combustion at sufficiently high enough spatial and temporal resolution subject to the availability of computational resources. In CFD, the most accurate approach to turbulence simulation is to solve directly the governing transport equations, without undertaking any averaging or approximation other than the consideration of appropriate numerical discretisations performed on them. Commonly known as the direct numerical simulation (DNS), this approach requires all significant turbulent structures to be adequately captured or fully resolved. This means that the domain of which the computation is to be carried out requires resolution of the largest as well as the smallest turbulent eddies. The computational resources required are obviously much larger that those associated with the FANS approach. Indeed, they tend to preclude simulations of many fires except for small-scale flames. Alternatively, consider another approach where the structure of turbulent flow is now viewed as distinct transport of large- and small-scale motions. On this basis, the large-scale motion that governs the mixing of gases is directly simulated on as fine a scale as the underlying computational grid will allow; the small-scale motion is modeled accordingly. Since the large-scale motion is generally much more energetic and by far the most effective transporters of the conserved properties than the small-scale ones, such an approach, recognized as the large eddy simulation (LES), which treats the large eddies exactly but approximates the small eddies, makes perfect sense. Computationally, LES is still regarded to be considerably more expensive than FANS, but much less costly than that of DNS. A schematic drawing highlighting the trade-off between the computational effort and the modeling complexity of different approaches, is shown in Figure 5.1. For FANS calculations, it is noted that all physical processes that occur at length scales larger than the integral length scale ($\Delta f > l$) are captured by the averaged transport equations and can thus be solved directly, whereas those occurring length scales at smaller than the threshold length scale ($\Delta f < l$) require modeling—additional Reynolds and scalar stress terms appearing within the averaged equations. For DNS, no modeling is introduced due to the requirement where the numerical calculations are performed on the threshold length scale smaller than the Kolmogorov length scale ($\Delta f < \eta$).

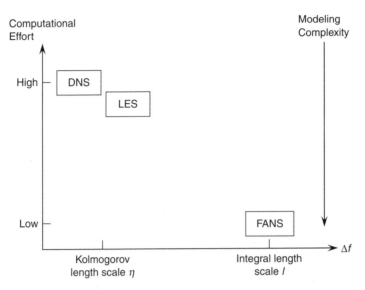

Figure 5.1 A representation of the trade-off between the computational effort and the modeling complexity of different approaches.

Numerical results obtained from a DNS or LES simulation, generally contain very detailed information about the flow. They have the capacity of attaining increasing realism (complexity and Reynolds numbers) by an accurate realization of the flow structure encapsulating the broad range of length and time scales that exists within. Because of the wealth of information, DNS and LES can provide a qualitative understanding of the flow physics, and construct a quantitative model thereby allowing other similar flows to be computed. They may also assist in some cases, to improve the performance of currently applied turbulence models in practice. The alternative approaches to handling turbulence via the DNS and LES are described in the next section.

5.2 Alternative Approach to Handling Turbulence

5.2.1 Direct Numerical Simulation (DNS)

DNS refers to computations where all relevant spatial and temporal scales are adequately resolved for the given application. In order to obtain a valid simulation, all the range of length scales including the smallest scales must be accommodated from which the viscosity is active. It is therefore imperative to capture all of the kinetic energy dissipation within the turbulent flow. DNS requires calculations to be carried out at length scales smaller than the Kolmogorov length scales (see Figure 5.1). From dimensional analysis, assuming dependence only upon viscosity ν and dissipation rate of kinetic energy ε, estimates for the so-called Kolmogorov micro-scale of length η as well as the Kolmogorov micro-scales of time τ and velocity υ can be obtained as

$$\eta = \left(\frac{\nu^3}{\varepsilon}\right)^{1/4} \tag{5.2.1}$$

$$\tau = \left(\frac{\nu}{\varepsilon}\right)^{1/2} \tag{5.2.2}$$

$$\upsilon = (\nu\varepsilon)^{1/4} \tag{5.2.3}$$

The Reynolds number (Re_l) at the micro-scale level ($\eta\nu/\nu$) is equivalent to unity, which indicates that the small-scale motion is rather viscous. On the basis of the integral length scale l and the characteristic root mean square value of the fluctuations u, the dissipation scales in the same way as production—that is, u^3/l. It can therefore be shown that the relations between the smallest and largest scales (Tennekes and Lumley, 1976) can be expressed according to

$$\frac{\eta}{l} = (ul/\nu)^{-3/4} = Re_l^{-3/4} \tag{5.2.4}$$

$$\frac{\tau}{t} = (ul/\nu)^{-1/2} = Re_l^{-1/2} \tag{5.2.5}$$

$$\frac{\upsilon}{u} = (ul/\nu)^{-1/4} = Re_l^{-1/4} \tag{5.2.6}$$

As the Reynolds number increases, the gap between the smallest and largest length scales widens. The expression in equation (5.2.4) also represents the ratio of the number of grid points in one dimension, such that the number of grid points in three-dimensional DNS scales according to $Re_l^{9/4}$. If the total computational time is assumed to be proportional to the total number of grid points (N) and the number of time steps, the computational cost scales proportional to Re_l^3.

The system of equations to be solved is exactly those formulated in Table 2.1, for the flow of a compressible Newtonian fluid in Cartesian coordinates under laminar conditions. DNS of turbulent flow takes this set of transport equations as a starting point and develops a transient solution on sufficiently fine mesh and small time steps to resolve the smallest turbulent eddies and the quickest fluctuations. In the area of combustion, DNS has provided significant fundamental insights into both non-premixed (Vervisch and Poinsot, 1998) and premixed flames (Bray and Cant, 1991). Luo (2005) has studied a fire-related phenomena concerning the dynamics of a free-standing buoyant diffusion flames from rectangular, square, and round fuel sources using a high-order DNS methodology based on the transport equation for variable-density and single step finite-rate Arrhenius chemistry. It is worthwhile to expound the many numerical issues in DNS that usually require special treatment. Some specific requirements in carrying out this type of computation are provided below.

Spatial and Temporal Resolutions

The main key issue that typically dictates any DNS simulations, is the mesh or spatial resolution within the flow domain. From the preceding, it is noted that DNS requires the requirement of $N \propto Re_l^{9/4}$ to resolve the largest geometrical structures at one end of the spectrum, and the finest turbulence scales at the other end. For a Reynolds number of 10000, the simulation would require in the order of 1000 grid points along each coordinate direction. Detailed grid refinement studies performed by Moin and Mahesh (1998) have shown that such requirement may be relaxed since most of the dissipation occurs at scales that are substantially larger than that of the Kolmogorov length scale, about 5η -15η. The number of grid points can be reduced so long as the bulk dissipation process is adequately represented; a reduction by a factor of 100 is possible without significant loss of accuracy. Treating the DNS studies of Luo (2005) as a guide to typical mesh systems that are to be sensibly utilized for fire-related problems, a mesh density of $192 \times 192 \times 288$ was used to resolve the computational domain for the cases having round and square fire source, while a finer mesh density of $256 \times 256 \times 384$ was employed for the case of a rectangular fire source. The flow Reynolds number of 1000 was recorded at the inlet, and increased up to 6185 downstream due to buoyancy acceleration. In order to aptly represent the bulk dissipation process, the simulation would require at least 150 grid points along each co-ordinate direction. It is evidently clear that the mesh densities adopted for the three cases adequately resolve all the relevant scales occurring within the turbulent reacting flow.

Another key issue is the need for accurate temporal resolution. In general, DNS requires an accurate time history. Since a wide range of time scales is experienced, the system of equations is inherently stiff and small time steps are inadvertently adopted in order to better accommodate all the relevant scales of turbulent motion and combustion. A special note is given on the strong influence of time step size on small-scale amplitude and phase error that has been addressed by Moin and Mahesh (1998).

Spatial and Temporal Discretisations

For DNS calculations adopting the finite difference approach, the importance of employing an energy conservative spatial differencing scheme is imperative. Unlike in FANS calculations where low-order upwind differencing schemes are commonly advocated, high-order *diffusion-free* central differencing schemes are used instead in many DNS simulations, to approximate the gradients in the governing equations. In Figure 5.2, the modified wave numbers of the second-, fourth-, and sixth-order central differences are presented as a function of the wave number for the first order derivative approximation. If the grid is sufficiently dense, only the coefficients of the small wave numbers are of

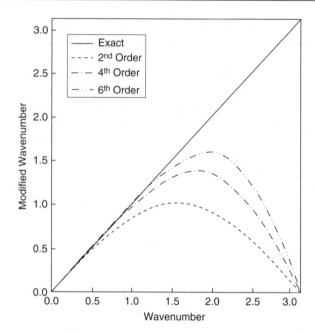

Figure 5.2 Plot of modified wave number against wave number for first order derivative approximations using second, fourth, and sixth order central differences with reference to the exact differentiation. (Adapted after Lele, 1992.)

significance; accurate results can thus be expected. However, the departure from the exact differentiation establishes at a higher wave number as the order of the schemes increases. Luo (2005) has used a sixth-order finite difference approximation for the spatial discretisation in his DNS calculations. Note that central differencing schemes are inherently unstable unless high spatial resolutions and small temporal resolutions are duly exercised. Rai and Moin (1991) have, however, employed the fifth-order upwind-biased scheme to their DNS calculations of a channel flow, and they have found that it was capable of yielding good agreement of the first- and second-order statistics of the flow characteristics. The high-order accurate upwind-biased scheme has a number of favorable features. It is stable and represents a good candidate for direct simulations of turbulent flows with complex geometries. Nevertheless, the presence of *false diffusion*, which is prevalent in all upwind differencing schemes, may still limit the extensive application of this scheme if the mesh resolution is not sufficiently fine.

In order to attain an accurate realization of the turbulent flow across a broad spectrum of time and length scales, DNS also requires the implementation of suitable time-marching methods. Because of the need for complete time resolution to aptly describe the energy dissipation process, *explicit* methods based on strict Courant-Friedrich-Levy (CFL) requirements are preferred in most simulations, instead of *implicit* time advancement and large time steps that are routinely used in FANS calculations. The most commonly used time-marching methods are the second order Adams-Bashforth and the third or fourth order Runge Kutta methods. The former method will be described in more detail in Section 5.4. Luo (1997) has employed the latter method to fully describe the transient flaming characteristics. More details on the numerical implementation of the third or fourth order Runge Kutta method can be found in Rai and Moin (1991). In practice, this method allows a larger time step to be adopted for the same order of accuracy to be achieved and thus marginally compensate for the increased amount of computations that are experienced during the numerical calculations.

Initial and Boundary Conditions

In comparison to FANS computations, DNS requires all details of the three-dimensional velocity field including the complete velocity field on a plane (or surface) for the *inflow* conditions of a turbulent flow at each time step. Owing to the low Froude numbers that are experienced in buoyant fires, it is rather common to prescribe laminar boundary conditions where the inlet velocity and temperature are not subjected to any external perturbations in order to eliminate the need to initiate arbitrarily specified inflow disturbances that could impair the downstream flow and thermal characteristics. At *outflow* boundaries, it is important that boundary conditions that prevent the pressure waves to be reflected off these boundaries and back into the interior of the domain are imposed. The so-called convective boundary condition can be derived from

$$\frac{\partial \phi}{\partial t} + \left(u_j \frac{\partial \phi}{\partial x_j} \right) \cdot n_j = 0 \qquad (5.2.7)$$

where n_j is the unit vector normal to the boundary. Concerning *open* boundaries, the traction-free boundary condition as described by Gresho (1991a, 1991b) is adopted:

$$\sigma_{ij} \cdot n_j = 0 \qquad (5.2.8)$$

where σ_{ij} is the stress tensor given by $\sigma_{ij} = -p\delta_{ij} + v (\partial u_i/\partial x_j + \partial u_j/\partial x_i)$ and v is the kinematic viscosity. This particular boundary condition allows the entrainment of the ambient fluid to be realized into the flow region of interest (Boersma et al., 1998). At solid walls, the boundary conditions follow the description as highlighted in Section 2.6. It is nevertheless noted that certain boundary conditions that are applicable in FANS are unsuitable for DNS such as the *symmetry* boundary condition.

Setting appropriate initial conditions are problematic in DNS, since the initial state is usually not known a priori for different geometry configurations and conditions in fire dynamics. It is customary to assume that the initial state takes upon an environment at ambient conditions with a quiescent velocity field being essentially zero. In spite of all attempts to prescribe the initial conditions to be as realistic as possible, a DNS simulation should be allowed to take its course for some lengths of time in order that the fluid flow and heat transfer develop with the correct characteristics of the fluid flow. Initially, the fluctuating quantity may reveal some systematic decreasing or increasing trends, but when the flame is fully developed, the value will exhibit sensible statistical fluctuations with time. At later stages of the fire simulations, statistical averaging over time can be performed on the transient results to obtain the mean and fluctuating characteristics of the velocity components, temperature, chemical species, and other variables of interest.

5.2.2 Large Eddy Simulation (LES)

The basic idea behind large eddy simulation (LES) is that the turbulent eddies that account for most of the mixing or large scale motion are large enough to be calculated with sufficient accuracy from the equations of fluid mechanics. The hope is that the small-scale eddy motion is approximated by some appropriate models, which must be ultimately justified by appeal to experiments. The establishment of the LES method has its roots in the prediction of atmospheric flows since the 1960s and recently in fire engineering, the development of the fire dynamics simulator (FDS) computer code by NIST, which is increasingly being adopted for practical engineering investigations of fires.

In LES, the governing equations are formally derived by applying a filtering operation, which proceeds according to

$$\bar{\phi}(x_i', t) = \int_\Delta \phi(x_i', t) G(|\, x_i - x_i' \,|) dx_i' \qquad (5.2.9)$$

where G is a filter function. The most common localized filter functions and their corresponding Fourier transform pairs, are represented in Table 5.1.

Table 5.1 Filters $G(|\, x_i - x_i' \,|)$ and their Fourier transform $G(k)$.

Filter	Filter Function $G(|\, x_i - x_i' \,|)$	Fourier Transform $G(k)$
Top Hat	$= \begin{cases} \dfrac{1}{\Delta} & \text{for } |\, x_i - x' \,| < \dfrac{\Delta}{2} \\ 0 & \text{otherwise} \end{cases}$	$\dfrac{\sin\left(\frac{\Delta k}{2}\right)}{\frac{\Delta k}{2}}$
Gaussian	$\sqrt{\dfrac{6}{\pi\Delta^2}} \exp\left(\dfrac{-6(x_i - x')^2}{\Delta^2}\right)$	$\exp\left(-\dfrac{(\Delta k)^2}{24}\right)$
Fourier Cut-Off	$\dfrac{\sin\left(k_c(x_i - x')\right)}{\pi(x_i - x')}, k_c = \dfrac{\pi}{\Delta}$	$= \begin{cases} 1 \text{ if } k < k_c \\ 0 \text{ otherwise} \end{cases}$

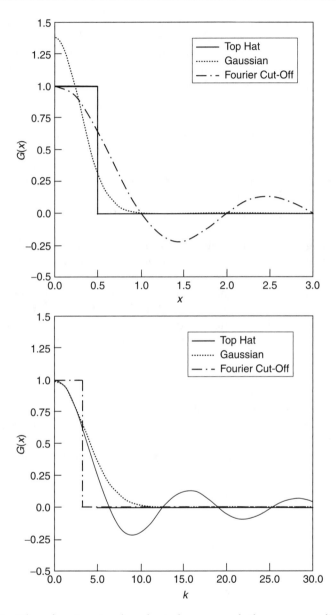

Figure 5.3 Filter function in the physical space and the corresponding Fourier transform presented in the wave number space.

Figure 5.3 represents the graphical representations of the filters in physical space and wave space. As observed, the top hat and Gaussian filters, which are straightforward to implement in the finite volume implementation and finite difference method, do not completely eliminate the component with the wave number greater than the cut-off wave number k_c. The components of k_c are also subsequently damped. On the other hand, the Fourier cut-off filter is ideal in the wave number spaces, but its operational length is infinitely long in the physical space. In spectral calculations (i.e., Fourier series to describe the flow variables), the finite number of modes automatically defines the cut-off filter and the method is attractive from the viewpoint of separating the large and small eddies. LES is usually conducted so that a finite computational mesh with the truncation error from the numerical discretisation of the flow equations is considered as the filtering operator. The advantage of this approach is that no explicit filtering operation is required. Nonetheless, the danger is that the truncation error at the smallest resolved scales, i.e. at the highest wave numbers, can be substantially large. Although explicit filters can be employed to remove these errors, *mesh refinement* has shown to significantly improve the numerical results at a much faster rate. In essence, a denser grid without imposing any explicit filtering produces better results albeit the smallest scales are influenced by the numerical error. This error is removed to the high wave numbers, whose contribution to the results is apparently small. Within the finite volume method, it is rather sensible to consider the filter width to be of the same order as the grid size. In three-dimensional computations with grid cells of different grid sizes along the Cartesian coordinate directions, the filter width is often taken to be the cube root of the grid volume

$$\Delta = \sqrt[3]{\Delta x \Delta y \Delta z} \qquad (5.2.11)$$

In a rough sense, the flow eddies larger than the filter width are considered to be *large eddies*, while eddies smaller than the filter width are *small eddies* requiring modeling.

The Favre-averaging is applied here in a similar way as in deriving the FANS equations, which is given by

$$\tilde{\phi}(x_i', t) = \frac{\overline{\rho \phi(x_i', t)}}{\bar{\rho}} \qquad (5.2.12)$$

The instantaneous property $\phi(x_i', t)$ may now be written according to

$$\phi(x_i', t) = \tilde{\phi}(x_i', t) + \phi''(x_i', t) \qquad (5.2.13)$$

where $\tilde{\phi}(x_i', t)$ represents the filtered or resolvable component (essentially a local average of the complete field) and $\phi''(x_i', t)$ is the subgrid scale component that accounts for unresolved spatial variations at a length smaller than the filter width Δ. When filtering is performed on the governing equations, the Filtered Favre-Averaged Navier-Stokes can be expressed in compact form, as represented in Table 5.2. The quantities τ_{ij}, q_{ij}, and s_{ij} as indicated, represent the unknown subgrid scale correlations, which require closure models.

Table 5.2 Filtered Favre-Averaged Navier-Stokes equations in Cartesian coordinates.

Favre-Averaged Mass

$$\frac{\partial \bar{\rho}}{\partial t} + \frac{\partial}{\partial x_j}(\bar{\rho}\tilde{u}_j) = 0 \quad j = 1, 2, 3$$

Favre-Averaged Momentum

$$\frac{\partial}{\partial t}(\bar{\rho}\tilde{u}_i) + \frac{\partial}{\partial x_j}(\bar{\rho}\tilde{u}_i\tilde{u}_j) = -\frac{\partial \bar{\sigma}_{ij}}{\partial x_j} - \frac{\partial}{\partial x_j}\underbrace{\left(\overline{\rho u_i u_j} - \bar{\rho}\tilde{u}_i\tilde{u}_j\right)}_{\tau_{ij}} + \bar{S}_{u_i}$$

where

$$\bar{\sigma}_{ij} = \bar{p}\delta_{ij} - \mu\left(\frac{\partial \tilde{u}_i}{\partial x_j} + \frac{\partial \tilde{u}_j}{\partial x_i}\right) + \frac{2}{3}\mu\frac{\partial \tilde{u}_i}{\partial x_j}\delta_{ij} \quad i, j = 1, 2, 3$$

Favre-Averaged Enthalpy

$$\frac{\partial}{\partial t}(\bar{\rho}\tilde{h}) + \frac{\partial}{\partial x_j}(\bar{\rho}\tilde{u}_j\tilde{h}) = \frac{\partial}{\partial x_j}\left[\frac{k}{C_p}\frac{\partial \tilde{h}}{\partial x_j}\right] - \frac{\partial}{\partial x_j}\underbrace{\left(\overline{\rho u_j h} - \bar{\rho}\tilde{u}_j\tilde{h}\right)}_{q_{ij}} + \bar{S}_h \quad j = 1, 2, 3$$

Favre-Averaged Scalar Property

$$\frac{\partial}{\partial t}(\bar{\rho}\tilde{\varphi}) + \frac{\partial}{\partial x_j}(\bar{\rho}\tilde{u}_j\tilde{\varphi}) = \frac{\partial}{\partial x_j}\left[\bar{\rho}D\frac{\partial \tilde{\varphi}}{\partial x_j}\right] - \frac{\partial}{\partial x_j}\underbrace{\left(\overline{\rho u_j \varphi} - \bar{\rho}\tilde{u}_j\tilde{\varphi}\right)}_{s_{ij}} + \bar{S}_\varphi \quad j = 1, 2, 3$$

Note: $(u_1,u_2,u_3) \equiv (u,v,w)$; $(x_1,x_2,x_3) \equiv (x,y,z)$

Basic SGS Models

In LES, the small dissipative scales are not solved accurately. The prime objective of the subgrid scale (SGS) models is to represent the kinetic energy losses due to the viscous forces and not attempt to produce the SGS stresses accurately but rather only account for their effect in a statistical sense. Most models are prescribed through the eddy-viscosity concept; it therefore shares many similarities to that used in FANS modeling. Smagrorinsky (1963) suggested that the Boussinesq hypothesis can be invoked to provide a good description of the unresolved eddies of the resolved flow, since the smallest turbulent eddies are almost isotropic. For the unresolved SGS turbulent stresses τ_{ij}, they are modeled accordingly as

$$\tau_{ij} = \overline{\rho \widetilde{u_i u_j}} - \overline{\rho} \tilde{u}_i \tilde{u}_j = -2\mu_T^{SGS} \tilde{S}_{ij} + \frac{1}{3}\tau_{kk}\delta_{ij} \qquad (5.2.14)$$

$$\tilde{S}_{ij} = \frac{1}{2}\left(\frac{\partial \tilde{u}_i}{\partial x_j} + \frac{\partial \tilde{u}_j}{\partial x_i}\right) - \frac{1}{3}\frac{\partial \tilde{u}_k}{\partial x_k}\delta_{ij}$$

where μ_T^{SGS} is the SGS eddy viscosity and $\tilde{S}_{ij}$ is the strain rate of the large-scale or resolved field. The Smagorinsky-Lilly model assumes that the SGS eddy viscosity can be described in terms of a length and a velocity scale. Taking the length scale to be the filter width Δ, the velocity scale can be expressed as the product of the length scale and the average strain rate of the resolved flow; the SGS eddy viscosity takes the following dependency:

$$\mu_T^{SGS} = \overline{\rho} C_1 \Delta^2 \mid \tilde{S} \mid \qquad (5.2.15)$$

where C_1 is an empirical constant and $\mid \tilde{S} \mid = \sqrt{2\tilde{S}_{ij}\tilde{S}_{ij}}$. The stress tensor τ_{kk} in equation (5.2.14) can be similarly modeled as

$$\tau_{kk} = 2\overline{\rho} C_k \Delta^2 \mid \tilde{S} \mid \qquad (5.2.16)$$

Erlebacher et al. (1992) have found that τ_{kk} may be ignored for practical calculations since $C_k \ll C_1$.

The Smagroinsky constant, $C_S = \sqrt{C_1}$, generally varies between 0.065 and 0.3, depending on the particular fluid flow problem. Difference in C_S is attributed to the effect of the mean flow strain or shear. This gives an indication whereby the behavior of the small eddies is not as universal as has been surmised in the beginning. On a theoretical analysis of the decay rates of isotropic turbulent eddies in the inertial subrange of the energy spectrum, Lilly (1966, 1967) has, for example, obtained values of C_S between 0.17 and 0.21. After reviewing other works, Rogallo and Moin (1984) suggested values between 0.19 and 0.24 across a range of grids and filter functions. Zhou et al. (2001) have indicated that a little larger C_s is more applicable for thermal flows of which they have employed a value of 0.23 in their LES study. For open buoyant fires with fully developed turbulence, the current authors have employed a value of C_S equivalent to 0.2 with much success. In most internal flow calculations, $C_S = 0.1 - 0.13$ is nonetheless commonly adopted in practice, as suggested by Piomelli et al. (1988) and Scotti et al. (1993). Note that there is a difference in the way the turbulent viscosity is evaluated between the LES and FANS approaches. From equation (5.2.15), LES determines the turbulent viscosity directly from the filtered velocity field. In FANS, by reference to equation (2.11.6) in Chapter 2, the turbulent viscosity is evaluated through the flow field containing two additionally derived variables, which are the turbulent kinetic energy k and its rate of dissipation ϵ values. In regions close to the solid surfaces, the turbulent viscosity can be damped by using a combination of mixing length minimum function and a viscosity damping function:

$$\mu_T^{SGS} = \bar{\rho}\min(\kappa y, f_\mu C_S \Delta)^2 \mid \tilde{S} \mid \qquad (5.2.17)$$

where κ is the model constant equivalent to 0.42 and y is the distance closest to the wall. The damping function f_μ can be set either according to van Driest (1956) wall damping function

$$f_\mu = 1 - \exp(-y^+/25) \qquad (5.2.18)$$

or formulated by Piomelli et al. (1987) as

$$f_\mu = \sqrt{1 - \exp[(-y^+/25)^3]} \qquad (5.2.19)$$

The use of wall functions in LES has shown to be a successful recipe for attached flow problems (Piomelli et al., 1989).

In addition to the Smagorinsky-Lilly model, other basic subgrid-viscosity models such as the Structure Function model by Métais and Lesieur (1992) and the Mixed Scale Model by Sagaut (1996), have also been proposed. In the Structure Function model, the subgrid eddy viscosity is alternatively evaluated according to

$$\mu_T^{SGS} = \bar{\rho} C_2 \Delta \sqrt{\overline{F_2}(\Delta)} \qquad (5.2.22)$$

where $\overline{F}_2$ is the second-order structure function constructed with the filtered velocity field

$$\overline{F}_2(\Delta) = \int_{|x'|=\Delta} [\tilde{u}(x) - \tilde{u}(x + x')]^2 d^3 x'$$
$$\approx \frac{1}{6}\sum_{i=1}^{3}\left\langle [\tilde{u}(x) - \tilde{u}(x + \Delta x_i)]^2 + [\tilde{u}(x) - \tilde{u}(x - \Delta x_i)]^2 \right\rangle \left(\frac{\Delta}{\Delta x_i}\right)^{2/3}$$
$$(5.2.23)$$

In equation (5.2.23), the structure function $\overline{F}_2$ has been approximated based on a local statistical average of the square (filtered) velocity differences, with the six immediately adjacent cells. A constant value of $C_2 = 0.063$ is prescribed as suggested by Métais and Lesieur (1992). The Mixed Scale Model (MSM), as proposed by Sagaut (1996), accounts for the contribution of the resolved field gradients, the kinetic energy of the highest resolved modes, and the cut-off length scale Δ. The viscosity is defined as

$$\mu_T^{SGS} = \bar{\rho} C_3 |\; \tilde{\omega}\; |^{1/2} \Delta^{3/2} (q_c^2)^{1/4} \tag{5.2.24}$$

where $\tilde{\omega}$ is the vorticity of the resolved scales defined by $\tilde{\omega} = \nabla \times \tilde{u} \equiv \mathrm{Curl}(\tilde{u})$ and the quantity $q_c^2 = \frac{1}{2} u_i' u_i'$ is the kinetic energy of the test field $u' = \tilde{u} - \hat{\tilde{u}}$, which is extracted from the resolved velocity field through the application of a test filter associated with the cut-off length scale $\hat{\Delta} > \Delta$, usually taken as $\hat{\Delta} = 2\Delta$. More discussions on the evaluation of the test filtered velocity $\hat{\tilde{u}}$ will be expounded below. The value of the constant C_3 according to Saguat (1996) is 0.1.

The SGS enthalpy flux and scalar flux correlations, q_{ij} and s_{ij}, are modeled in a manner similar to the SGS turbulence stresses by the *standard gradient diffusion hypothesis* as

$$q_{ij} = \overline{\rho u_j h} - \bar{\rho} \tilde{u}_j \tilde{h} = -\bar{\rho} C_{\alpha h} \Delta^2 |\; \tilde{S}\; | \frac{\partial \tilde{h}}{\partial x_j} \tag{5.2.20}$$

$$s_{ij} = \overline{\rho u_j \varphi} - \bar{\rho} \tilde{u}_j \tilde{\varphi} = -\bar{\rho} C_{\alpha \varphi} \Delta^2 |\; \tilde{S}\; | \frac{\partial \tilde{\varphi}}{\partial x_j} \tag{5.2.21}$$

Diffusive coefficients $C_{\alpha h}$ and $C_{\alpha \phi}$ are evaluated from $C_{\alpha h} = C_d / Pr_T^{SGS}$ and $C_{\alpha \varphi} = C_d / Sc_T^{SGS}$ where C_d denotes the constant C_1, C_2 or C_3 and Pr_T^{SGS} and Sc_T^{SGS} are the subgrid turbulent Prandtl and Schmidt numbers, respectively.

All the preceding models have been designed assuming that the simulated flow is turbulent, fully developed, and isotropic, and therefore do not incorporate any information related to an eventual departure of the simulated flow from these assumptions. In order to obtain an automatic adaptation of the models for inhomogeneous flows, simulations of engineering flows are more likely to be based on the dynamic formulations of these models.

Dynamic SGS Models

One possible approach to develop a self-adaptive SGS model, is the dynamic procedure proposed by Germano et al. (1991). This is based on the application of two different filters. In addition to the grid filter G, a test filter $\hat{G}$ is applied. The test filter width $\hat{\Delta}$ is usually taken to be larger than the grid filter width Δ. Defining the mass-weighted test filter operation by

$$\hat{\phi}(x_i', t) = \frac{\overline{\bar{\rho} \tilde{\phi}(x_i', t)}}{\hat{\bar{\rho}}} \tag{5.2.25}$$

and applying the grid filter and subsequently the test filter on the instantaneous Favre-averaged momentum, the following equation is attained:

$$\frac{\partial}{\partial t}(\hat{\bar{\rho}}\hat{\tilde{u}}_i) + \frac{\partial}{\partial x_j}(\hat{\bar{\rho}}\widehat{\tilde{u}_i\tilde{u}_j}) = -\frac{\partial}{\partial x_j}\left(\hat{\bar{p}}\delta_{ij} - \mu\left(\frac{\partial\hat{\tilde{u}}_i}{\partial x_j} + \frac{\partial\hat{\tilde{u}}_j}{\partial x_i}\right) + \frac{2}{3}\mu\frac{\partial\hat{\tilde{u}}_i}{\partial x_j}\delta_{ij}\right) - \frac{\partial T_{ij}}{\partial x_j} + \hat{\bar{S}}_{u_i}$$

$$(5.2.26)$$

where the subtest stresses are given by

$$T_{ij} = \hat{\bar{\rho}}\widehat{\widetilde{u_iu_j}} - \hat{\bar{\rho}}\widehat{\tilde{u}_i\tilde{u}_j}$$

$$(5.2.27)$$

If the test filter is now directly applied to the grid-filtered Favre-averaged momentum in Table 5.2, the equation becomes

$$\frac{\partial}{\partial t}(\hat{\bar{\rho}}\hat{\tilde{u}}_i) + \frac{\partial}{\partial x_j}(\hat{\bar{\rho}}\widehat{\tilde{u}_i\tilde{u}_j}) = -\frac{\partial}{\partial x_j}\left(\hat{\bar{p}}\delta_{ij} - \mu\left(\frac{\partial\hat{\tilde{u}}_i}{\partial x_j} + \frac{\partial\hat{\tilde{u}}_j}{\partial x_i}\right) + \frac{2}{3}\mu\frac{\partial\hat{\tilde{u}}_i}{\partial x_j}\delta_{ij}\right)$$
$$- \frac{\partial\hat{\tau}_{ij}}{\partial x_j} - \frac{\partial L_{ij}}{\partial x_j} + \hat{\bar{S}}_{u_i}$$

$$(5.2.28)$$

with L_{ij} and $\hat{\tau}_{ij}$ given as

$$L_{ij} = -(\hat{\bar{\rho}}\widehat{\tilde{u}_i\tilde{u}_j} - \widehat{\bar{\rho}\tilde{u}_i}\widehat{\bar{\rho}\tilde{u}_j}/\hat{\bar{\rho}})$$

$$(5.2.29)$$

$$\hat{\tau}_{ij} = \hat{\bar{\rho}}\widehat{\widetilde{u_iu_j}} - \widehat{\bar{\rho}\tilde{u}_i}\widehat{\bar{\rho}\tilde{u}_j}/\hat{\bar{\rho}}$$

$$(5.2.30)$$

On the basis of equations (5.2.27), (5.2.29), and (5.2.30), the Leonard term for the Favre-filtered case L_{ij} can be written as

$$L_{ij} = T_{ij} - \hat{\tau}_{ij}$$

$$(5.2.31)$$

In principal, the dynamic procedure can be applied to any of the basic SGS models. Table 5.3 highlights the different filtered and subtest kernels of the three models, as just described. Note that the test filtered strain rate tensor $\hat{\tilde{S}}_{ij}$ is given by

$$\hat{\tilde{S}}_{ij} = \frac{1}{2}\left(\frac{\partial\hat{\tilde{u}}_i}{\partial x_j} + \frac{\partial\hat{\tilde{u}}_j}{\partial x_i}\right) - \frac{1}{3}\frac{\partial\hat{\tilde{u}}_k}{\partial x_k}\delta_{ij}$$

and the quantity $|\hat{\tilde{S}}|$ is the contraction of the strain rate tensor at the test-level, defined as

$$|\hat{\tilde{S}}| = \sqrt{2(\widehat{\bar{\rho}\tilde{S}_{ij}/\hat{\bar{\rho}}})(\widehat{\bar{\rho}\tilde{S}_{ij}/\hat{\bar{\rho}}})}$$

On the basis of these, the filtered and subtest stresses can thus be represented as

$$\tau_{ij} - \frac{1}{3}\tau_{kk}\delta_{ij} = C_d\beta_{ij} \tag{5.2.32}$$

$$T_{ij} - \frac{1}{3}T_{kk}\delta_{ij} = C_d\alpha_{ij} \tag{5.2.33}$$

where C_d is a coefficient to be determined, which is associated with the respective model constants C_1, C_2, and C_3 in the previous section. Substituting the above stresses into equation (5.2.30) yields

$$L_{ij} - \frac{1}{3}L_{kk}\delta_{ij} \equiv L_{ij}^a = C_d\alpha_{ij} - \widehat{C_d\beta_{ij}} \tag{5.2.34}$$

For the enthalpy and scalar property, similar Leonard terms can also be derived according to

$$L_{ij}^q = -(\widehat{\bar{\rho}\tilde{u}_j\tilde{h}} - \widehat{\bar{\rho}\tilde{u}_j}\widehat{\bar{\rho}\tilde{h}}/\hat{\bar{\rho}}) = C_{\alpha h}\hat{\bar{\rho}}\hat{\Delta}^2 |\hat{\tilde{S}}| \frac{\partial(\widehat{\bar{\rho}\tilde{h}/\hat{\bar{\rho}}})}{\partial x_j} - \widehat{C_{\alpha h}\bar{\rho}\Delta^2 |\tilde{S}| \frac{\partial \tilde{h}}{\partial x_j}} \tag{5.2.35}$$

$$L_{ij}^s = -(\widehat{\bar{\rho}\tilde{u}_j\tilde{\varphi}} - \widehat{\bar{\rho}\tilde{u}_j}\widehat{\bar{\rho}\tilde{\varphi}}/\hat{\bar{\rho}}) = -C_{\alpha\varphi}\hat{\bar{\rho}}\hat{\Delta}^2 |\hat{\tilde{S}}| \frac{\partial(\widehat{\bar{\rho}\tilde{\varphi}/\hat{\bar{\rho}}})}{\partial x_j} + \widehat{C_{\alpha\varphi}\bar{\rho}\Delta^2 |\tilde{S}| \frac{\partial \tilde{\varphi}}{\partial x_j}} \tag{5.2.36}$$

From equation (5.2.33), Lilly (1992) suggested a least-squares approach to evaluate the local values of C_d. By assuming that C_d is the same for both filtering operations, the error

Table 5.3 Subgrid model kernels for the dynamics procedure.

Model	β_{ij}	α_{ij}
Smagorinsky	$-2\bar{\rho}\Delta^2 \| \tilde{S} \| \tilde{S}_{ij}$	$-2\hat{\bar{\rho}}\hat{\Delta}^2 \| \hat{\tilde{S}} \| \hat{\tilde{S}}_{ij}$
Structure Function	$-2\bar{\rho}\Delta\sqrt{\bar{F}_2(\Delta)}\tilde{S}_{ij}$	$-2\hat{\bar{\rho}}\hat{\Delta}\sqrt{\hat{\bar{F}}_2(\hat{\Delta})}\hat{\tilde{S}}_{ij}$
Mixed Scale	$-2\bar{\rho}\| \tilde{\omega} \|^{1/2}\Delta^{3/2}(q_c^2)^{1/4}\tilde{S}_{ij}$	$-2\hat{\bar{\rho}}\| \hat{\tilde{\omega}} \|^{1/2}\hat{\Delta}^{3/2}(\hat{q}_c^2)^{1/4}\hat{\tilde{S}}_{ij}$

$$e_{ij} = L^a_{ij} - C_d \alpha_{ij} + C_d \hat{\beta}_{ij} \qquad (5.2.37)$$

is minimized by requiring $\partial e_{ij} e_{ij} / \partial C_d = 0$, which in turn gives

$$C_d = \frac{L^a_{ij} M_{ij}}{M_{kl} M_{kl}} \qquad (5.2.38)$$

in which $M_{ij} = \alpha_{ij} - \hat{\beta}_{ij}$. Diffusive coefficients $C_{\alpha h}$ and $C_{\alpha \varphi}$ can also be similarly determined according to the procedure in equation (5.2.37). In order to speed up computations, these coefficients could be explicitly determined based on the available local values of C_d without the recourse of the dynamic procedure by the *a priori* specified subgrid turbulent Prandtl and Schmidt numbers. It is noted that the numerator in equation (5.2.37) can attain both positive and negative values. This indicates that the model allows the possibility of accounting the backscatter of the turbulent energy, which is the energy transferred from the *small eddies* to the *large eddies*. Such occurrences are prevalent in real flows, although the long time average energy transport is from the *large eddies* to the *small eddies*. Nevertheless, a negative viscosity has a tendency of causing severe numerical instability and the denominator may become zero, which would make the constant C_d indeterminate. It is rather common that averaging is performed to equation (5.2.35), in order to dampen large local fluctuations either by performing plane-averaging along a homogeneous direction or local-averaging over the test filter cell. In complex flows, an average over small time interval is used instead.

The apparently *ad hoc* averaging, which recovers the statistical notion of energy transfer from the resolved to the subgrid scales and removal of negative eddy velocity, effectively stabilizes the dynamic model. However, this fact still precludes the computation of a fully inhomogeneous flow. Ghosal et al. (1995) removed the mathematical inconsistency by generalizing the least square method into a constrained variational problem, consisting of the minimization of the integral of the error over the entire domain, with the additional constraint that C_1 be non-negative. This led to a rigorous problem of solving the Fredholm's integral equation of the second kind, which requires the integral to be iteratively solved using under-relaxation to improved convergence. The cost is comparable to the Poisson equation for the pressure, and can be rather expensive. Piomelli and Liu (1995), however, developed a simpler constrained model where equation (5.2.33) is recast in the form

$$L^a_{ij} = C_d \alpha_{ij} - \widehat{C^*_d \beta}_{ij} \qquad (5.2.39)$$

where an estimate of the coefficient denoted by C^*_d is assumed to be known. Equation (5.2.38) can be minimized locally by the following contraction:

$$C_1 = \frac{(L_{ij}^a - \widehat{C_d^* \beta_{ij}})\alpha_{ij}}{\alpha_{mn}\alpha_{mn}} \qquad (5.2.40)$$

It is noted that the denominator in the preceding expression is positive definite. Normally, the coefficient C_d^* can be obtained by either the *zeroth-order* approximation by taking the value at the previous time-step: $C_d^* = C_d^{n-1}$ or evaluated using a *first-order* approximation formulated in the form: $C_d^* = C_d^{n-1} + \dfrac{t_n - t_{n-1}}{t_{n-1} - t_{n-2}}(C_d^{n-1} - C_d^{n-2})$.

The Lagrangian Dynamic model proposed by Meneveau et al. (1996) that combines the features of statistical and local approaches, presents another model formulation capable of handling inhomogeneous flows in complex configurations. In a Lagrangian frame of reference, this model is derived by minimizing the error incurred by considering the Germano identity along fluid-particle trajectories. At a position x at time t, the trajectory of a fluid particle for times $t' < t$ is given by

$$z(t') = x - \int_{t'}^{t} \bar{u}[z(t''), t'']dt'' \qquad (5.2.41)$$

The error in equation (5.2.40), written in terms of the Lagrangian description to be minimized, becomes

$$e_{ij}(z, t') = L_{ij}^a(z, t') - C_1(z, t')M_{ij}(z, t') \qquad (5.2.42)$$

Note that the model coefficient C_1 (z, t') has been removed from the filter operation, which is equivalent to the linearization operation used in the Germano-Lilly procedure as above. This model coefficient to be used at time t and position x, is now determined by minimizing the error over the trajectory of the fluid particle. Defining the total error E, which is defined as the weighted integral along the trajectories of the error proposed by Lilly

$$E = \int_{-\infty}^{t} e_{ij}\left(z(t'), t'\right)e_{ij}\left(z(t'), t'\right)W(t - t')dt' \qquad (5.2.43)$$

where $W(t - t')$ is introduced to control the memory effect, the total error is minimized with respect to C_d by enforcing

$$\frac{\partial E}{\partial C_1} = \int_{-\infty}^{t} 2e_{ij}\left(z(t'), t'\right)\frac{\partial e_{ij}\left(z(t'), t'\right)}{\partial C_d}W(t - t')dt' = 0 \qquad (5.2.44)$$

Making use of equation (5.2.37), the model coefficient can be obtained as

$$C_d = \frac{J_{LM}}{J_{MM}} \tag{5.2.45}$$

where

$$J_{LM}(x, t') = \int_{-\infty}^{t} L_{ij}M_{ij}\Big(z(t'), t'\Big) W(t - t')dt' \tag{5.2.46}$$

$$J_{MM}(x, t') = \int_{-\infty}^{t} M_{ij}M_{ij}\Big(z(t'), t'\Big) W(t - t')dt' \tag{5.2.47}$$

According to Meneveau et al. (1996), an exponential weighting of the form $W(t - t') = T_{lag}^{-1}\exp[(t - t')/T_{lag}]$ in which T_{lag} is the Lagrangain correlation time, provides the distinct advantage whereby the integrals J_{LM} and J_{MM} are solutions to the following transport equations:

$$\frac{\partial J_{LM}}{\partial t} + \bar{u}_j \frac{\partial J_{LM}}{\partial x_j} = \frac{1}{T_{lag}}(L_{ij}M_{ij} - J_{LM}) \tag{5.2.48}$$

$$\frac{\partial J_{MM}}{\partial t} + \bar{u}_j \frac{\partial J_{MM}}{\partial x_j} = \frac{1}{T_{lag}}(M_{ij}M_{ij} - J_{MM}) \tag{5.2.49}$$

Solving the preceding transport equations directly would undoubtedly increase the computational expense in the context of LES. To alleviate the problem, a simpler formulation based on discretising the preceding equations in time should suffice, which results in

$$\frac{J_{LM}^{n+1}(x) - J_{LM}^{n}(x - \bar{u}^n\Delta t)}{\Delta t} = \frac{1}{T_{lag}}\Big([L_{ij}M_{ij}]^{n+1}(x) - J_{LM}^{n+1}(x)\Big) \tag{5.2.50}$$

$$\frac{J_{MM}^{n+1}(x) - J_{MM}^{n}(x - \bar{u}^n\Delta t)}{\Delta t} = \frac{1}{T_{lag}}\Big([M_{ij}M_{ij}]^{n+1}(x) - J_{MM}^{n+1}(x)\Big) \tag{5.2.51}$$

Positions x are coincident with the grid points of the simulation. The value of J_{LM} at the previous time step and at the upstream location $x - \bar{u}^n\Delta t$, can be obtained through a multi-linear interpolation procedure. Equations (5.2.49) and (5.2.50) can be re-arranged to yield

$$J_{LM}^{n+1}(x) = a[L_{ij}M_{ij}]^{n+1}(x) + (1 - a)J_{LM}^{n}(x - \bar{u}^n\Delta t) \tag{5.2.52}$$

$$J_{MM}^{n+1}(x) = a[M_{ij}M_{ij}]^{n+1}(x) + (1 - a)J_{MM}^{n}(x - \bar{u}^n\Delta t) \tag{5.2.53}$$

where

$$a = \frac{\Delta t / T_{lag}}{1 + \Delta t / T_{lag}} \qquad (5.2.54)$$

On the basis of isotropic homogeneous turbulence, the correlation time is T_{lag} can be estimated by

$$T_{lag} = 1.5\Delta (J_{LM}^n J_{MM}^n)^{-1/8} \qquad (5.2.55)$$

which is significantly reduced in the high-shear regions where J_{MM} is large, and those regions where the non-linear transfers are high—for example, for large J_{LM}.

   In order to improve the prediction of intermittent phenomena, another possibility for achieving self-adaptive SGS models is to combine the basic SGS models with a selection function. This selection function examines the structural properties of the test field $\tilde{u}'$ and turns off the SGS model when these properties do not correspond to those expected from a fully turbulent field. A turbulent velocity is expected to exist in practice, thereby requiring a SGS model when the local angular fluctuation of the instantaneous vorticity is higher than a given threshold θ_o. The selection criterion will therefore depend on the estimation of the angle θ between the resolved vorticity and local average vorticity, computed by applying a test filter associated with the cut-off length scale $\hat{\Delta} > \Delta$. Instead of the Boolean selection function proposed (Leisuer and Métais, 1996), which may pose serious problems in numerical calculations because of its discontinuous nature, the modified continuous function is adopted:

$$f_{\theta_o}(\theta) = \begin{cases} 1 & \text{if } \theta > \theta_o \\ r(\theta)^n & \text{otherwise} \end{cases} \qquad (5.2.56)$$

in which the function r is given by

$$r(\theta)^n = \frac{tan^2(\theta/2)}{tan^2(\theta_o/2)} \qquad (5.2.57)$$

and the exponent n in practice is taken to be equal to 2. According to Sagaut (2006), the quantity $tan^2 (\theta/2)$ can be estimated using the relation

$$tan^2(\theta/2) = \frac{2\hat{\tilde{\omega}}\tilde{\omega} - \hat{\tilde{\omega}}^2 - \tilde{\omega}^2 + \omega'^2}{2\hat{\tilde{\omega}}\tilde{\omega} + \hat{\tilde{\omega}}^2 + \tilde{\omega}^2 - \omega'^2} \qquad (5.2.58)$$

where $\omega' = \tilde{\omega} - \hat{\tilde{\omega}}$. The selection function is used as a multiplicative factor to the subgrid viscosity, leading to the definition of selective models:

$$\mu_T^{SGS} = -\mu_T^{SGS}(x,t)f_{\theta_o}\left(\theta(x)\right) \qquad (5.2.59)$$

On the right-hand side, the subgrid viscosity can be ascertained by any of the viscosity models aforementioned. In order to maintain the same average subgrid viscosity value over the entire fluid domain, a factor of 1.65 is multiplied to the model coefficients that appear in equations (5.2.15), (5.2.22), and (5.2.24). This factor has been evaluated on the basis of isotropic homogeneous turbulence simulations. A threshold angle of 20° is usually taken for most practical calculations.

One-Equation SGS Models

An alternative strategy to the basic SGS models, is the adoption of ideas already established in turbulence modeling within Chapter 2 to purposefully develop a one-equation model, which uses a transport equation for the SGS kinetic energy. On the basis of the Boussinesq hypothesis, the SGS viscosity can be expressed in terms of the SGS kinetic energy k_{SGS} as

$$\mu_T^{SGS} = \bar{\rho}C\Delta\sqrt{k_{SGS}} \qquad (5.2.60)$$

where the constant $C = 0.069$ represents a theoretically value (Saguat, 2004). Other values ranging between 0.04 and 1.0 have also been employed in a number of applications, as illustrated by Schmidt and Schumann (1989). A transport equation to determine the distribution of k_{SGS} accounting the effects of convection, diffusion, production, and destruction can be formulated as

$$\frac{\partial}{\partial t}(\bar{\rho}k_{SGS}) + \frac{\partial}{\partial x_j}(\bar{\rho}\tilde{u}_j k_{SGS}) = \frac{\partial}{\partial x_j}\left(\bar{\rho}C_{\alpha k}\Delta\sqrt{k_{SGS}}\frac{\partial k_{SGS}}{\partial x_j}\right)$$
$$+ P_{k_{SGS}} - D_{k_{SGS}} + B_{k_{SGS}} \qquad (5.2.61)$$

where $C_{\alpha k} = C/\sigma_k$, $P_{k_{SGS}}$ is the regular production term, $D_{k_{SGS}}$ is the destruction term, and $B_{k_{SGS}}$ is the production due to buoyancy. This is the LES equivalent of a one-equation RANS turbulence model, such as the one employed in the *two-layer k-ε model* for the viscous-dominated near-wall region. Similar to the *k*-equation, the production term $P_{k_{SGS}}$ is modeled according to

$$P_{k_{SGS}} = -\tau_{ij}\frac{\partial \tilde{u}_i}{\partial x_j} \qquad (5.2.62)$$

The destruction term $D_{k_{SGS}}$ is estimated based on the cut-off length scale Δ as

$$D_{k_{SGS}} = C_* \frac{\overline{\rho} k_{SGS}^{3/2}}{\Delta} \tag{5.2.63}$$

while the term $B_{k_{SGS}}$ represents the production due to buoyancy, which can be modeled according to the *standard gradient diffusion hypothesis* as

$$B_{k_{SGS}} = -\frac{C}{\sigma_\rho} \Delta \sqrt{k_{SGS}} \left(\frac{\partial \overline{\rho}}{\partial x_j} \cdot g_j \right) \tag{5.2.64}$$

where g_j is the gravity vector. For the other constants, values of unity for the coefficient C_* as well as the turbulent Prandtl numbers σ_k and σ_ρ are typically adopted.

In the consideration for a self-adaptive version of the one-equation model, the dynamic coefficient C can be ascertained by applying the Germano identity to the subgrid model. The grid-filtered and test-filtered stresses can be expressed as

$$\tau_{ij} - \frac{1}{3} \tau_{kk} \delta_{ij} = -2\overline{\rho} C \Delta \sqrt{k_{SGS}} \tilde{S}_{ij} \tag{5.2.65}$$

$$T_{ij} - \frac{1}{3} T_{kk} \delta_{ij} = -2\hat{\overline{\rho}} C \hat{\Delta} \sqrt{K} \hat{\tilde{S}}_{ij} \tag{5.2.66}$$

On the basis of the above stresses, the Leonard term is given by

$$L_{ij}^a = -2\hat{\overline{\rho}} C \hat{\Delta} \sqrt{K} \hat{\tilde{S}}_{ij} + \overline{2\overline{\rho} C \Delta \sqrt{k_{SGS}} \tilde{S}_{ij}} \tag{5.2.67}$$

The subgrid kinetic energy on the test level, K, can be evaluated either from the algebraic relation

$$K = k_{SGS} + \frac{1}{2} L_{ii} \tag{5.2.68}$$

or a transport equation given by

$$\frac{\partial}{\partial t} (\hat{\overline{\rho}} K) + \frac{\partial}{\partial x_j} (\hat{\overline{\rho}} \hat{\tilde{u}}_j K) = \frac{\partial}{\partial x_j} \left(\hat{\overline{\rho}} C_{\alpha k} \hat{\Delta} \sqrt{K} \frac{\partial K}{\partial x_j} \right) + P_{K_{TEST}} - \underbrace{C_* \frac{\hat{\overline{\rho}} K^{3/2}}{\hat{\Delta}}}_{D_{k_{TEST}}} + B_{K_{TEST}}$$

$$\tag{5.2.69}$$

Adopting the dynamic procedure based on the standard least-squares approach by Lilly (1992), assuming that C is the same for both filtering operations, the dynamic coefficient can be determined in a similar fashion as

$$C = -\frac{L_{ij}^a M_{ij}}{2M_{kl}M_{kl}}$$

(5.2.70)

where $M_{ij} = \overline{\rho}\hat{\Delta}\sqrt{\hat{K}}\hat{\tilde{S}}_{ij} - \overline{\rho}\Delta\sqrt{k_{SGS}}\tilde{S}_{ij}$. Assuming that the turbulent Prandtl numbers σ_k and σ_ρ are known *a priori*, Davidson (1997) proposed that the coefficient C_* can be estimated by equating the test-filtered production and destruction terms in equation (5.2.60) with the production and destruction terms in equation (5.2.68). In other words,

$$\hat{P}_{k_{SGS}} - \hat{D}_{k_{SGS}} = P_{K_{TEST}} - D_{K_{TEST}}$$

$$\hat{P}_{k_{SGS}} - \frac{1}{\Delta}\widehat{C_*\overline{\rho}k_{SGS}^{3/2}} = P_{K_{TEST}} - \frac{1}{\hat{\Delta}}C_*\hat{\overline{\rho}}K^{3/2}$$

$$C_*^{n+1} = \frac{\hat{\Delta}}{\hat{\overline{\rho}}K^{3/2}}\left(P_{K_{TEST}} - \hat{P}_{k_{SGS}} + \frac{1}{\Delta}\widehat{\overline{\rho}C_*^n k_{SGS}^{3/2}}\right)$$

(5.2.71)

Following Piomelli and Liu (1995), the dynamic coefficient under the filter is taken from the previous time level. To ensure numerical stability, a constant value of C in space is used in the momentum equations as well as in other transport equations throughout the whole computational domain. The constant $\langle C \rangle_{xyz}$ is determined by requiring that the global production remains the same throughout the entire domain—that is,

$$\left\langle 2C\Delta\sqrt{k_{SGS}}\tilde{S}_{ij}\tilde{S}_{ij}\right\rangle_{xyz} = 2\langle C\rangle_{xyz}\left\langle\Delta\sqrt{k_{SGS}}\tilde{S}_{ij}\tilde{S}_{ij}\right\rangle_{xyz}$$

(5.2.72)

The basic idea is to represent all the local dynamic information through the source terms of the transport equation for k_{SGS}. This allows the effect of the large fluctuations in the dynamic constant C to be effectively smoothed out, which reduces or removes the need to restrict or limit the dynamic coefficient. The spatial variation of C is included via the production term in the modeled k_{SGS} transport equation. Reverse energy transfer from the *small eddies* to *large eddies*, which is known as the *backscatter*, is therefore accounted for in an indirect way. Although the effect is not fed directly back to the resolved flow, it influences the resolved flow via the kinetic subgrid energy. A negative production reduces k_{SGS}, which subsequently influences the vicinity of the flow through the convection and diffusion of k_{SGS}.

Computational Issues

Various dynamic SGS models have been proposed, and it appears that the most efficient ones are based on the resolved modes of the highest frequency. This so-called *test field* is extracted from the resolved field through the application of a low-pass filter, commonly known as the *test filter*. Examples from the preceding are the Germano-Lilly dynamic procedure, based on the Germano relationship, which links without any approximation of the SGS tensors associated with different levels of filtering and some improved versions of the original structure function of Métais and Leisuer (1992), which involve the test field: the selective function model, which includes a test on the topology of the vorticity of the test field, and the filtered structure function model, which evaluates the model on the test field. The application of discrete test filters with compact stencils based on weighted averages, is attractive from the viewpoint of practical numerical calculations. Some common filters adopted in practice are described herein. Consider the one-dimensional finite volume representation as depicted in Figure 5.4. On the basis of applying the top hat *test filter*, the *test field* can be obtained from

$$\hat{\tilde{\phi}}_i = \frac{1}{\hat{\Delta}} \int_{\hat{\Delta}} \tilde{\phi}(x) dx \tag{5.2.73}$$

The integration of the resolved field can be approximated by applying respectively, the Trapezoidal rule and the Simpson rule. By taking $\hat{\Delta} = 2\Delta$, the two widely used three-point filters in the uniform one-dimensional case are

$$\hat{\tilde{\phi}}_i = \frac{1}{\underbrace{\hat{\Delta}}_{=2\Delta}} \left[\frac{\Delta}{2} (\tilde{\phi}_{i-1} + \tilde{\phi}_i) + \frac{\Delta}{2} (\tilde{\phi}_i + \tilde{\phi}_{i+1}) \right] = \frac{1}{4} (\tilde{\phi}_{i-1} + 2\tilde{\phi}_i + \tilde{\phi}_{i+1})$$

    Trapezoidal rule

or

$$\hat{\tilde{\phi}}_i = \frac{1}{\underbrace{\hat{\Delta}}_{=2\Delta}} \left[\frac{\Delta}{3} (\tilde{\phi}_{i-1} + 4\tilde{\phi}_i + +\tilde{\phi}_{i+1}) \right] = \frac{1}{6} (\tilde{\phi}_{i-1} + 4\tilde{\phi}_i + \tilde{\phi}_{i+1}) \text{ Simpson rule}$$

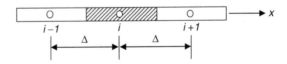

Figure 5.4 Illustration of three-point filters in one dimension.

These filters compute the average of the variable $\tilde{\phi}$ over the control volume cell surrounding the point i.

The filtered field at the ith grid point $\tilde{\phi}_i$ that has been obtained through a discrete filter can be formalized according to

$$\hat{\tilde{\phi}}_i = F\tilde{\phi}_i \equiv \sum_{l=-1}^{1} a_l \tilde{\phi}_{i+l} \qquad (5.2.74)$$

Preservation of the coefficients a_l is ensured under the condition

$$\sum_{l=-1}^{1} a_l = 1 \qquad (5.2.75)$$

For an extension to the three-dimensional case (see Figure 5.5), a multi-dimensional filter F^p (where p is the dimension of space) can be constructed from one-dimensional filter F by two methods: construction by linear combination and construction by product. The first method consists of filtering each direction in space independently, by the simultaneous use of each one-dimensional filter of which the multi-dimensional filter F^p can be written as

$$F^p = \frac{1}{p} \sum_{i=1}^{p} F^i \qquad (5.2.76)$$

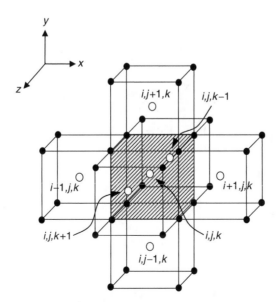

Figure 5.5 Illustration of three-point filters in three dimensions.

where F^p is the one-dimensional filter defined in the ith space direction. At the grid point (i, j, k), the *test field* for $\hat{\phi}$ in three dimensions by applying the Trapezoidal rule is evaluated as

$$\hat{\tilde{\phi}}_{i,j,k} = F^3 \tilde{\phi}_{i,j,k} \equiv \frac{1}{3} \sum_{l=-1}^{1} a_l \tilde{\phi}_{i+l}$$

$$= \frac{1}{3} \left[\frac{1}{4} (\tilde{\phi}_{i-1,j,k} + 2\tilde{\phi}_{i,j,k} + \tilde{\phi}_{i+1,j,k}) + \frac{1}{4} (\tilde{\phi}_{i,j-1,k} + 2\tilde{\phi}_{i,j,k} + \tilde{\phi}_{i,j+1,k}) \right.$$

$$\left. + \frac{1}{4} (\tilde{\phi}_{i,j,k-1} + 2\tilde{\phi}_{i,j,k} + \tilde{\phi}_{i,j,k+1}) \right]$$

$$= \frac{1}{12} (\tilde{\phi}_{i-1,j,k} + 2\tilde{\phi}_{i,j,k} + \tilde{\phi}_{i+1,j,k}) + \frac{1}{12} (\tilde{\phi}_{i-1,j,k} + 2\tilde{\phi}_{i,j,k} + \tilde{\phi}_{i+1,j,k})$$

$$+ \frac{1}{12} (\tilde{\phi}_{i-1,j,k} + 2\tilde{\phi}_{i,j,k} + \tilde{\phi}_{i+1,j,k})$$

(5.2.77)

The second method defines the multi-dimensional filter F^p as the composition of one-dimensional filters applied in each space direction, which is equivalent to a sequential application of the one-dimensional filter

$$F^p = \prod_{i=1}^{p} F^i \tag{5.2.78}$$

The *test field* for $\tilde{\phi}_{i,j,k}$, applying the Trapezoidal rule, can thus be achieved by

$$\hat{\tilde{\phi}}_{i,j,k} = F^3 \tilde{\phi}_{i,j,k} \equiv \sum_{l=-1}^{1} \sum_{m=-1}^{1} \sum_{n=-1}^{1} a_l a_m a_n \tilde{\phi}_{i+l,j+m.k+n}$$

$$= \frac{1}{4} (\tilde{\phi}_{i-1,j,k} + 2\tilde{\phi}_{i,j,k} + \tilde{\phi}_{i+1,j,k}) \times \frac{1}{4} (\tilde{\phi}_{i,j-1,k} + 2\tilde{\phi}_{i,j,k} + \tilde{\phi}_{i,j+1,k})$$

$$\times \frac{1}{4} (\tilde{\phi}_{i,j,k-1} + 2\tilde{\phi}_{i,j,k} + \tilde{\phi}_{i,j,k+1})$$

$$= \frac{1}{64} (\tilde{\phi}_{i-1,j,k} + 2\tilde{\phi}_{i,j,k} + \tilde{\phi}_{i+1,j,k})(\tilde{\phi}_{i-1,j,k} + 2\tilde{\phi}_{i,j,k} + \tilde{\phi}_{i+1,j,k})$$

$$(\tilde{\phi}_{i-1,j,k} + 2\tilde{\phi}_{i,j,k} + \tilde{\phi}_{i+1,j,k})$$

(5.2.79)

Similar discrete representations using the Simpson rule could also be appropriately formulated using the preceding two methods. On the basis of the elaborate analyses performed by Sagaut and Grohens (1999) on discrete filters for LES, other more sophisticated approaches to extend the discrete filters based on the equivalence class concept have been proposed, especially to better accommodate curvilinear structured and unstructured meshes. Alternatively, discrete approximations of convolution filters, which are best fitted to the continuous filter in a given sense, were also found to yield satisfactory results. Interested readers are encouraged to refer to this literature for further material in this subject area.

The issues concerning the spatial and temporal resolutions and discretisations as well as the initial and boundary conditions governing LES, are not different from those already discussed for the consideration of DNS. The requirements of a high mesh density, diffusion-free discretisation schemes, accurate temporal resolution, and efficient time-marching methods are just some of the important pre-requisites in carrying out any LES calculations. Subject to the availability of computational resources, there still remains the possibility of the lack of sufficient resolution in fully capturing all the important associated dynamics, especially at high wave numbers, which have not been filtered or modeled. Accumulated numerical truncation errors have a tendency to overwhelm the previous explicit SGS models, which in turn detracts from the original physics that is intended to simulate. The monotone-integrated-large-eddy-simulation (MILES) has been proposed to purposefully utilize the numerical truncation errors directly by the use of implicit higher-order schemes that belong to the family of TVD algorithms in Appendix A.2 in order to act as a SGS model instead of the widely developed explicit models. Interested readers are referred to Garnier et al. (1999), Sagaut (2004), and Hahn and Drikakis (2005) for a greater understanding on the background theory behind MILES and their applications in turbulent flow simulations.

5.3 Favre-Averaged Navier-Stokes versus Large Eddy Simulation

The inherent unsteady nature of LES suggests that the computational requirements are much larger than FANS. This is indeed the case when LES is compared to those of two-equation models such as k-ϵ and k-ω. The ability to better describe the fire dynamics via the consideration of appropriate sub-models describing the combustion, radiation, soot production, and pyrolysis, along with a classical turbulence model, has certainly provided the necessary means of widening the scope of investigations toward handling more complex fire processes such as flame spread, the interaction of water spray from sprinklers with fuel surfaces, and various other heat transfer mechanisms. Whereas the inclusion of these important features in most instances limits the spatial resolution of the computational grid in the past, sufficiently more powerful computers in the present are not only capable of performing such calculations with even a higher mesh density but also with increasing level of sophistications embedded within the sub-models. For most fire engineering purposes, details of the turbulent fluctuations are generally not required, and information emphasizing on the mean flow should suffice to quantify the characteristics of practical fires. In view of this, the FANS approach provides many advantages as a viable design tool, especially the rather quick turnaround of numerical results for numerous assessment and evaluation in fire-related problems with low computational costs.

It can nevertheless be argued that conventional k-ε models contain a number of dependent constants, and it is unclear what the effects of these various constants have on the numerical solution and what the solution actually represents for a variety of applications. Also, such models often include an empirical description, which relies heavily on the level of turbulence prescribed through the choice of turbulence models. In contrast, SGS models have fewer constants, for example, the Smagorinsky-Lilly model contains only one constant. Furthermore, the constants within these models may be dynamically determined and consequently adapted to the changing flow dynamics, based on the statistics of the resolved fluctuations. This represents a significant advantage over FANS, as the LES approach to field modeling fire phenomena seeks approximate solutions to the governing equations directly, and by considering convection, combustion, and thermal radiation in parallel, each is permitted to evolve separately in its own length scale and time scale. This approach, however, needs to be cultivated by the availability of ever-increasing power of computers, which emphasizes high spatial resolution and faster numerical algorithms. Impressive developments based on current-generation computers are already in progress for the former, while some efficient flow solving techniques developed throughout the years are readily applicable for the latter. Nowadays, simulations involving in excess of a million grid cells, are not overly difficult and fairly elaborate geometries can be considered without sacrificing the spatial resolution.

In order to suitably adopt LES in field modeling, the equations describing the transport of mass, momentum, and energy must be simplified so that they can be efficiently solved for the range of fire problems of actual interest. In order to better characterize fire-induced flows, the approximate form of the Navier-Stokes equations are solved to describe the low speed motion of a gas driven by chemical heat release and buoyancy. This approximation, which is essentially the *low Mach number* assumption, involves the filtering of acoustic waves while allowing for large variations in temperature and density. Adopting a characteristically explicit time-marching predictor-corrector projection method, and coupled with a fast *direct solver*, this rather efficient numerical strategy establishes the basis of many LES fire simulations, primarily exemplified by the fire dynamics simulator (FDS) computer code.

Similarly, the explicit predictor-corrector scheme based on the low-Mach-number variable-density formulation of Knio et al. (1999) adapted from Najm et al. (1998), has been demonstrated to be well suited for handling practical fires of which such a scheme can effectively handle the disparity of broadly ranging flow and chemical scales that may exist in such fires. Its construction is based on a two-stage predictor-corrector projection formulation;

other similar variant methodologies such as those proposed by McMurtry et al. (1986), Rutland et al. (1989), and Mahalingam et al. (1990) are also noted. Feasible LES simulations, especially on the treatment of turbulent buoyant fires, have been exemplified in our study carried out by ourselves and our co-workers in Cheung et al. (2007). In buoyant flows, the effects of density represent an integral part in buoyant fires. The most prevailing feature of this time-marching scheme is the strong coupling between the density and fluid flow equations through each stage of the predictor-corrector method, in order to appropriately account for the large density variation within various zones of the flame structure. To best handle non-premixed combustion, the combustion model is implemented in a manner consistent with the mixture fraction-based approach. Of particular difference against those adopted in FANS, is the formulation of a local filtered chemical heat release rate, which is inserted into the energy equation. Modeling of this term generally requires separate consideration from the explicit filtering of the flow field, since the chemical reactions for infinitely fast combustion take place within the unresolved small scales. Elaborate potential additions through the inclusion of appropriate models for radiation and soot, are also becoming more prevalent in recent LES fire simulations due to increasing computing power. More descriptions of the appropriate numerical algorithms and models in the context of LES are detailed in the next section.

5.4 Formulation of Numerical Algorithm

5.4.1 Explicit Predictor-Corrector Scheme

Consider a *weakly incompressible* or *thermal expandable* ideal gas driven by chemical heat release as well as radiation heat exchange for buoyant fires. The filtered Favre-averaged conservation equations of mass, momentum, and energy governing the motion of the fluid in a form suitable for low Mach number applications can be expressed as

$$\frac{\partial \overline{\rho}}{\partial t} + \frac{\partial (\overline{\rho}\tilde{u}_j)}{\partial x_j} = 0 \tag{5.4.1}$$

$$\frac{\partial}{\partial t}(\overline{\rho}\tilde{u}_i) + \frac{\partial}{\partial x_j}(\overline{\rho}\tilde{u}_i\tilde{u}_j) = -\frac{\partial \overline{\sigma}_{ij}}{\partial x_j} - \frac{\partial}{\partial x_j}(\overline{\rho\widetilde{u_i u_j}} - \overline{\rho}\tilde{u}_i\tilde{u}_j) + \overline{\rho}g \tag{5.4.2}$$

$$\overline{\rho}C_p\frac{\partial \tilde{T}}{\partial t} + \overline{\rho}C_p\tilde{u}_j\frac{\partial \tilde{T}}{\partial x_j} = \frac{\partial}{\partial x_j}\left(k\frac{\partial \tilde{T}}{\partial x_j}\right) - C_p\frac{\partial}{\partial x_j}(\overline{\rho\widetilde{u_j T}} - \overline{\rho}\tilde{u}_j\tilde{T}) + \overline{\omega}_T + S_{rad} \tag{5.4.3}$$

Buoyancy is accounted in the momentum equation (5.4.2) via $\bar{\rho}g$. Here again, g refers to the gravity vector. The term $\bar{\sigma}_{ij}$ represents the resolved stress tensor for compressible flow, as described in Table 5.2. In the low Mach number limit, the acoustic wave propagation is ignored. The pressure field can thus be decomposed into a spatially uniform component p_0 that varies only with time and a hydrodynamic component $\tilde{p}$ that changes both with time and space. This particular approximation is made to filter out the acoustic waves; these equations are thus considered as *weakly incompressible*. Considering the energy equation, the work due to pressure can be considered to be negligible because of the insignificant pressure gradient and the flow is assumed to be close to a divergence-free state. For the same reason, the volumic energy and enthalpy variations can be assumed to be equivalent, as they only differ through the addition of pressure. Hence, the low Mach number energy equation can be alternatively expressed in the form of the temperature equation as described in equation (5.4.3). The source term $\bar{\omega}_T$ represents the filtered heat release rate, while S_{rad} constitutes the radiation source contribution of the absorption/emission characteristics of combustion products, which will be further addressed at a later stage.

In the preceding equations, the thermal conductivity k may be evaluated according to the mass-weighted procedure described by equation (3.4.21) or *Wilke's Law* by equation (3.4.22) in Chapter 3. The corresponding mixture specific heat C_p is usually determined using equation (3.4.19) – $C_p = \sum \tilde{Y}_i C_{p,i}$, where $\tilde{Y}_i$ and $C_{p,i}$ are the corresponding mass fraction and specific heat of combustion gases of ith species at constant pressure. The perfect gas state equation can be expressed as $p_0 = \bar{\rho}R\tilde{T}$, where the zeroth-order pressure p_0 is taken as equivalent to the atmospheric pressure, while the gas constant R is determined through $R = R_u \sum \tilde{Y}_i / W_i$, where W_i represent the molecular weights for the ith species. For the purpose of numerical implementation, of which will become more apparent later in the development of the explicit predictor-corrector scheme, the time rate of change of density can be found by differentiating the equation of state as

$$\frac{\partial \bar{\rho}}{\partial t} = \bar{\rho}\left(-\frac{1}{\tilde{T}}\frac{\partial \tilde{T}}{\partial t} - \frac{1}{\sum Y_i/W_i}\sum \frac{1}{W_i}\frac{\partial \tilde{Y}_i}{\partial t} \right) \tag{5.4.6}$$

In the context of LES, the conserved scalar approach is often adopted in order to characterize the combustion of fires. The filtered forms of the Favre-averaged conservation equations of the resolved mixture fraction $\tilde{Z}$ and scalar variance of the mixture fraction $\widetilde{Z''^2}$ are given by

$$\frac{\partial(\bar{\rho}\tilde{Z})}{\partial t} + \frac{\partial(\bar{\rho}\tilde{u}_j\tilde{Z})}{\partial x_j} = \frac{\partial}{\partial x_j}\left(\bar{\rho}D_{\tilde{Z}}\frac{\partial \tilde{Z}}{\partial x_j} \right) - \frac{\partial}{\partial x_j}(\widetilde{\bar{\rho}u_jZ} - \bar{\rho}\tilde{u}_j\tilde{Z}) \tag{5.4.7}$$

$$\frac{\partial(\overline{\rho}\widetilde{Z''^2})}{\partial t} + \frac{\partial(\overline{\rho}\tilde{u}_j\widetilde{Z''^2})}{\partial x_j} = \frac{\partial}{\partial x_j}\left(\overline{\rho}D_{\widetilde{Z''^2}}\frac{\partial\widetilde{Z''^2}}{\partial x_j}\right) - \frac{\partial}{\partial x_j}(\overline{\rho u_j Z''} - \overline{\rho}\tilde{u}_j\widetilde{Z''^2}) + \overline{S}_{Z''^2}$$

$$(5.4.8)$$

The filtered source term $\overline{S}_{Z''^2}$ appearing in the scalar variance equation (5.4.8) will be further described in the next section. Similar to the consideration in FANS (see Chapter 3), the diffusion terms in the preceding equations may be modeled according to

$$\frac{\partial}{\partial x_j}\left(\overline{\rho}D_{\tilde{Z}}\frac{\partial\tilde{Z}}{\partial x_j}\right) \equiv \frac{\partial}{\partial x_j}\left(\frac{\mu}{Sc_Z}\frac{\partial\tilde{Z}}{\partial x_j}\right)$$

$$\frac{\partial}{\partial x_j}\left(\overline{\rho}D_{\widetilde{Z''^2}}\frac{\partial\widetilde{Z''^2}}{\partial x_j}\right) \equiv \frac{\partial}{\partial x_j}\left(\frac{\mu}{Sc_{Z''^2}}\frac{\partial\widetilde{Z''^2}}{\partial x_j}\right)$$

where Sc_Z and $Sc_{Z''^2}$ represent the laminar Schmidt numbers for the mixture fraction and scalar variance of the mixture fraction, respectively. The unresolved small-scale turbulence as represented by the unknown SGS correlations

$$\overline{\rho u_i u_j} - \overline{\rho}\tilde{u}_i\tilde{u}_j, \quad \overline{\rho u_j T} - \overline{\rho}\tilde{u}_j\tilde{T}, \quad \overline{\rho u_j Z} - \overline{\rho}\tilde{u}_j\tilde{Z}, \quad \overline{\rho u_j Z''} - \overline{\rho}\tilde{u}_j\widetilde{Z''^2}$$

require appropriate closure models. The subgrid momentum stress tensor can be modelled by invoking the Boussinesq hypothesis:

$$\overline{\rho u_i u_j} - \overline{\rho}\tilde{u}_i\tilde{u}_j = -2\mu_T^{SGS}\tilde{S}_{ij} + \frac{1}{3}\tau_{kk}\delta_{ij} \tag{5.4.9}$$

$$\tilde{S}_{ij} = \frac{1}{2}\left(\frac{\partial\tilde{u}_i}{\partial x_j} + \frac{\partial\tilde{u}_j}{\partial x_i}\right) - \frac{1}{3}\frac{\partial\tilde{u}_k}{\partial x_k}\delta_{ij}$$

Similar to the eddy viscosity assumption, the unknown SGS correlations for the temperature, mixture fraction, and scalar variance of the mixture fraction can be formulated according the *standard gradient diffusion hypothesis* as

$$\overline{\rho u_j T} - \overline{\rho}\tilde{u}_j\tilde{T} = -\frac{\mu_T^{SGS}}{Pr_T}\frac{\partial\tilde{T}}{\partial x_j} \tag{5.4.10}$$

$$\overline{\rho u_j Z} - \overline{\rho}\tilde{u}_j\tilde{Z} = -\frac{\mu_T^{SGS}}{SC_{T,Z}}\frac{\partial\tilde{Z}}{\partial x_j} \tag{5.4.11}$$

$$\widetilde{\overline{\rho u_j Z''^2}} - \overline{\rho} \widetilde{u}_j \widetilde{Z''^2} = -\frac{\mu_T^{SGS}}{SC_{T,Z''^2}} \frac{\partial \widetilde{Z''^2}}{\partial x_j} \tag{5.4.12}$$

where Pr_T is the subgrid turbulent Prandtl number for the temperature and $Sc_{T,Z}$ and Sc_{T,Z''^2} are the subgrid turbulent Schmidt numbers for the mixture fraction and scalar variance of the mixture fraction, respectively. Various SGS models as previously described in Section 5.2.2 may be adopted to evaluate the eddy viscosity μ_T^{SGS} in equations (5.2.14) through (5.2.17).

Numerical integration is now carried out on the governing equations (5.4.1), (5.4.2), (5.4.3), (5.4.7), and (5.4.8) using the predictor-corrector approach. The essence of the scheme is described as follows. In the predictor stage, a second-order Adams-Bashforth time integration scheme is employed to update the velocity, mixture fraction, and scalar variance of the mixture fraction, and incorporates a pressure correction step in order to satisfy the continuity equation. In the corrector stage, the momentum equation relies on a second-order quasi Crank-Nicolson integration, and also incorporates a pressure correction step. In both cases, the pressure correction step involves the inversion of a pressure correction Poisson equation, which can be solved using either a direct or iterative solvers. The *explicit predictor-corrector scheme* for the simulation of buoyant fires is implemented according to the following numerical steps:

Predictor
Step 1 The local time derivatives for the temperature and mass fraction of species at time level n, $\partial \widetilde{T}/\partial t \,|^n$ and $\partial \widetilde{Y}_i/\partial t \,|^n$ are first evaluated in order to determine the local time derivative of the density given in equation (5.4.6). Local species time derivatives of $\partial \widetilde{Y}_i/\partial t \,|^n$ can be obtained through

$$\left.\frac{\partial \widetilde{Y}_i}{\partial t}\right|^n = \left.\frac{\partial \widetilde{Z}}{\partial t}\right|^n \int_0^1 (dY_i/dZ)P(f)df \tag{5.4.14}$$

The time derivative $\partial \widetilde{Z}/\partial t \,|^n$ can be expressed in a similar form as exemplified in equation (5.4.7), while the instantaneous species gradient dY_i/dZ may be evaluated based upon the prescriptive state relationships in Section 3.4.2.4 or the laminar flamelet approach in Section 3.4.2.6.

Step 2 Intermediate values for the density, mixture fraction, and scalar variance of the mixture fraction are predicted by a second-order Adams-Bashforth time integration scheme at time levels n and $n-1$

$$\frac{\overline{\rho}^* - \overline{\rho}^n}{\Delta t} = \frac{3}{2}\left.\frac{\partial \overline{\rho}}{\partial t}\right|^n - \frac{1}{2}\left.\frac{\partial \overline{\rho}}{\partial t}\right|^{n-1} \tag{5.4.15}$$

$$\frac{\overline{\rho}^*\tilde{Z}^* - \overline{\rho}^n\tilde{Z}^{n-1}}{\Delta t} = \frac{3}{2}\frac{\partial(\overline{\rho}\tilde{Z})}{\partial t}\bigg|^n - \frac{1}{2}\frac{\partial(\overline{\rho}\tilde{Z})}{\partial t}\bigg|^{n-1} \tag{5.4.16}$$

$$\frac{\overline{\rho}^*\widetilde{Z^{n2}} - \overline{\rho}^n\widetilde{Z^{\prime\prime 2}}^{n-1}}{\Delta t} = \frac{3}{2}\frac{(\partial\overline{\rho}\widetilde{Z^{\prime\prime 2}})}{\partial t}\bigg|^n - \frac{1}{2}\frac{\partial(\overline{\rho}\widetilde{Z^{\prime\prime 2}})}{\partial t}\bigg|^{n-1} \tag{5.4.17}$$

The predicted intermediate temperature distribution is determined from the equation of state

$$\tilde{T}^* = p_0/(\overline{\rho}^* R^*) \tag{5.4.18}$$

Step 3 An intermediate velocity field is calculated by integrating the pressure-split momentum equations according to a second-order Adams-Bashforth time integration scheme

$$\frac{\overline{\rho}^*\tilde{u}'_i - \rho^n\tilde{u}^n_i}{\Delta t} = \frac{3}{2}R^n_i - \frac{1}{2}R^{n-1}_i - \frac{\partial\overline{p}^n}{\partial x_j} \tag{5.4.19}$$

$$\frac{\overline{\rho}^*\hat{\tilde{u}}_i - \overline{\rho}^*\tilde{u}'_i}{\Delta t} = \frac{\partial\overline{p}^n}{\partial x_j} \tag{5.4.20}$$

where

$$R_i = -\frac{\partial(\overline{\rho}\tilde{u}_i\tilde{u}_j)}{\partial x_j} + \frac{\partial}{\partial x_j}\left(\mu\left(\frac{\partial\tilde{u}_i}{\partial x_j} + \frac{\partial\tilde{u}_j}{\partial x_i}\right) - \frac{2}{3}\mu\frac{\partial\tilde{u}_k}{\partial x_k}\delta_{ij}\right)$$
$$+ \frac{\partial}{\partial x_j}\left(2\mu^{SGS}_T\tilde{S}_{ij} - \frac{1}{3}\tau_{kk}\delta_{ij}\right) + \overline{\rho}g \tag{5.4.21}$$

Step 4 The predicted velocity field in the predictor step is obtained using the projection step

$$\frac{\overline{\rho}^*\tilde{u}^*_j - \overline{\rho}^*\hat{\tilde{u}}_j}{\Delta t} = -\frac{\partial\overline{p}^*}{\partial x_j} \tag{5.4.22}$$

By defining the pressure correction $p' = \overline{p}^* - \overline{p}^n$, equations (5.4.20) and (5.4.22) can be combined to yield

$$\frac{\overline{\rho}^*\tilde{u}^*_j - \overline{\rho}^*\tilde{u}'_j}{\Delta t} = -\frac{\partial p'}{\partial x_j} \tag{5.4.23}$$

Instead of equation (5.4.22), the predicted velocity field is now obtained through equation (5.4.23). The local pressure correction p' is obtained from the solution of the Poisson equation

$$\frac{\partial^2 p'}{\partial x_j^2} = \frac{1}{\Delta t}\left[\frac{\partial(\bar{\rho}^* \tilde{u}'_j)}{\partial x_j} + \left.\frac{\partial\bar{\rho}}{\partial t}\right|^*\right] \qquad (5.4.24)$$

where $\partial\bar{\rho}/\partial t\,|^*$ may be given by a first-order approximation as

$$\left.\frac{\partial\bar{\rho}}{\partial t}\right|^* = \frac{1}{\Delta t}[\bar{\rho}^n - \bar{\rho}^{n-1}] \qquad (5.4.25)$$

It is noted that if equations (5.4.19) and (5.4.20) are combined, the original formulation of Knio et al. (1999) for the evaluation of the intermediate velocity field without the pressure gradient is recovered, of which represents the standard fractional-step technique employed by many (e.g., Chorin, 1968 and Yanenko, 1971). Establishing boundary conditions for these intermediate velocities is a common source of ambiguity. In time-splitting methods, only the boundary conditions for the velocity field are given at each complete time-step, and those of the intermediate velocity field are unknown. Kim and Moin (1985) have demonstrated that proper boundary conditions are required to be consistent with the governing equations or else the solution suffers from appreciable numerical errors. The introduction of an additional step of equation (5.4.20) removes the need to construct boundary conditions for the intermediate velocity field, since the pressure gradient is accounted in equation (5.4.19); boundary conditions that are employed for the velocity field can be immediately applied for the intermediate velocity field.

Corrector
Step 5 The temporal derivatives at the new time level $n + 1$ are estimated based on the predicted and corrected values obtained through a second-order quasi Crank-Nicolson integration. The density in this corrector step can be obtained as

$$\frac{\bar{\rho}^{n+1} - \bar{\rho}^n}{\Delta t} = \frac{1}{2}(C_\rho^n + D_\rho^n + G_\rho^n) + \frac{1}{2}(C_\rho^* + D_\rho^* + G_\rho^*) + R_\rho^* \qquad (5.4.26)$$

where

$$C_\rho = \frac{\bar{\rho}}{\tilde{T}}\tilde{u}_i\frac{\partial\tilde{T}}{\partial x_i}, \quad D_\rho = \frac{1}{C_p\tilde{T}}\left[\frac{\partial}{\partial x_j}\left(k\frac{\partial\tilde{T}}{\partial x_j}\right) + \frac{\partial}{\partial x_j}\left(\frac{\mu_T^{SGS}}{Pr_T}\frac{\partial\tilde{T}}{\partial x_j}\right)\right],$$

$$G_\rho = -\frac{\overline{\rho}}{\sum Y_i/W_i} \sum \frac{1}{W_i} \frac{\partial \tilde{Y}_i}{\partial t}, \quad R_\rho = \frac{1}{\overline{\rho} C_p} \overline{\omega}_T + \frac{1}{\overline{\rho} C_p} S_{rad}$$

Local species time derivatives of $\partial \tilde{Y}_i/\partial t \,|^*$ appearing in G_ρ^* are similarly evaluated according to equation (5.4.14)

$$\left.\frac{\partial \tilde{Y}_i}{\partial t}\right|^* = \left.\frac{\partial \tilde{Z}}{\partial t}\right|^* \int_0^1 (dY_i/dZ)P(f)df \qquad (5.4.27)$$

The mixture fraction and scalar variance are accordingly determined as

$$\frac{\overline{\rho}^{n+1}\tilde{Z}^{n+1} - \rho^n \tilde{Z}^n}{\Delta t} = \frac{1}{2}\left[\left(-\frac{\partial(\overline{\rho}\tilde{u}_j\tilde{Z})}{\partial x_j} + \frac{\partial}{\partial x_j}\left(\frac{\mu}{Sc_z}\frac{\partial \tilde{Z}}{\partial x_j}\right) + \frac{\partial}{\partial x_j}\left(\frac{\mu_T^{SGS}}{Sc_{T,Z}}\frac{\partial \tilde{Z}}{\partial x_j}\right)\right)^n \right.$$
$$\left. + \left(-\frac{\partial(\overline{\rho}\tilde{u}_j\tilde{Z})}{\partial x_j} + \frac{\partial}{\partial x_j}\left(\frac{\mu}{Sc_z}\frac{\partial \tilde{Z}}{\partial x_j}\right) + \frac{\partial}{\partial x_j}\left(\frac{\mu_T^{SGS}}{Sc_{T,Z}}\frac{\partial \tilde{Z}}{\partial x_j}\right)\right)^*\right]$$

$$(5.4.28)$$

$$\frac{\overline{\rho}^{n+1}\widetilde{Z''^2}^{n+1} - \rho^n \widetilde{Z''^2}^n}{\Delta t} = \frac{1}{2}\left[\left(-\frac{\partial(\overline{\rho}\tilde{u}_j\widetilde{Z''^2})}{\partial x_j} + \frac{\partial}{\partial x_j}\left(\left[\frac{\mu}{Sc_{Z''^2}} + \frac{\mu_T^{SGS}}{Sc_{T,Z''^2}}\right]\frac{\partial \widetilde{Z''^2}}{\partial x_j}\right)\right)^n \right.$$
$$\left. + \left(-\frac{\partial(\overline{\rho}\tilde{u}_j\widetilde{Z''^2})}{\partial x_j} + \frac{\partial}{\partial x_j}\left(\left[\frac{\mu}{Sc_{Z''^2}} + \frac{\mu_T^{SGS}}{Sc_{T,Z''^2}}\right]\frac{\partial \widetilde{Z''^2}}{\partial x_j}\right)\right)^*\right] + \overline{S}_{Z''^2}^*$$

$$(5.4.29)$$

while the new temperature distribution is found using the equation of state

$$\tilde{T}^{n+1} = p_0/(\overline{\rho}^{n+1}R^{n+1}) \qquad (5.4.30)$$

Step 6 A second intermediate velocity field is subsequently determined using the pressure-split momentum equations

$$\frac{\overline{\rho}^*\tilde{u}'_j - \rho^n\tilde{u}^n_j}{\Delta t} = \frac{3}{2}R_i^n - \frac{1}{2}R_i^{n-1} - \frac{\partial \overline{p}^n}{\partial x_j} \qquad (5.4.31)$$

Step 7 The pressure distribution at the new time level is obtained through the solution of the Poisson equation

$$\frac{\partial^2 p'}{\partial x_j^2} = \frac{1}{\Delta t}\left[\frac{\partial(\bar{\rho}^* \tilde{u}'_j)}{\partial x_j} + \frac{\partial \bar{\rho}}{\partial t}\Big|^{n+1}\right] \tag{5.4.32}$$

where $\partial \bar{\rho}/\partial t \,|^{n+1}$ is approximated by

$$\frac{\partial \bar{\rho}}{\partial t}\Big|^{n+1} = \frac{1}{\Delta t}[\bar{\rho}^{n+1} - \bar{\rho}^n] \tag{5.4.33}$$

Step 8 Finally, the predicted velocity field at the new time level is determined using

$$\frac{\bar{\rho}^{n+1}\tilde{u}_j^{n+1} - \bar{\rho}^* \tilde{u}'_j}{\Delta t} = -\frac{\partial p'}{\partial x_j} \tag{5.4.34}$$

5.4.2 Combustion Modeling

The role of a SGS reaction model for turbulent non-premixed combustion is designed to incorporate the effect of subgrid fluctuations in the thermo-chemical variables on the filtered chemical source term. On the basis of the mixture fraction-based approach, all the species mass fractions are taken to be only functions of the mixture fraction. Using this assumption, Bilger (1977) has derived the expression for the rate of reaction for the ith species, which can also be found in Kuo (1986) in the form of

$$\omega_i = -\frac{1}{2}\rho\chi\frac{d^2 Y_i}{df^2} \tag{5.4.35}$$

where χ is the instantaneous scalar dissipation given by $\chi = 2D \,(\partial Z/\partial x_j)^2$, which incidentally is an identical expression already defined in equation (3.4.89) under the laminar flamelet approach. In LES, the flame is typically not spatially resolved by the computational grid. It is therefore assumed that at the subgrid level there exists a statistical ensemble of laminar diffusion fla-melets, each satisfying universal state relationships. Under near-equilibrium conditions, the state relationships could be represented such as those of equilib-rium chemistry assumption or experimental state relationships established by Sivathanu and Faeth (1990). In order to predict highly non-equilibrium flame events such as lift-off or extinction, the state relationships are modified by the consideration of the scalar dissipation and to distinguish between burning and extinguished flamelets—the laminar flamelet approach. The heat release rate that is required for temperature equation (5.4.3) is determined for N species from

$$\omega_T = -\sum_{i=1}^{N} h_{fi}^o \omega_i \qquad (5.4.36)$$

where h_{fi}^o is the ith species standard heat of formation, and ω_i is given in equation (5.4.35).

In order to determine the filtered composition and subsequently the filtered heat release rate, models for subgrid fluctuations of the mixture fraction and its derivative are needed. Similar to the FANS approach, a subgrid Favre-filtered PDF for the mixture fraction $P(Z)$ could be applied to determine the filtered composition near-equilibrium conditions

$$\tilde{Y}_i = \int_0^1 Y_i(Z)P(Z)dZ \qquad (5.4.37)$$

The filtered heat release rate $\overline{\omega}_T$ is thus given by

$$\overline{\omega}_T = \frac{1}{2}\overline{\rho}\int_0^\infty \int_0^1 \left[\sum_{i=1}^{N} h_{fi}^o \frac{d^2 Y_i}{dZ^2}\right] \chi P(Z,\chi)dZd\chi \qquad (5.4.38)$$

where $P(Z, \chi)$ represents the joint Favre-filtered PDF for the mixture fraction and its scalar dissipation. Assuming statistical independence for the mixture fraction and the scalar dissipation, the filtered heat release rate in equation (5.4.38) becomes alternatively

$$\overline{\omega}_T = \frac{1}{2}\overline{\rho}\tilde{\chi}\sum_{i=1}^{N}\int_0^1 h_{fi}^o \frac{d^2 Y_i}{dZ^2} P(Z)dZ \qquad (5.4.39)$$

where $\tilde{\chi} = \int_0^\infty \chi P(\chi)d\chi$. To evaluate $P(Z)$ within the integration, the approach based on a presumed shape of the PDF can be adopted for computational simplicity. For a beta function PDF, its equivalent form is given by

$$P(Z) = \frac{Z^{\alpha-1}(1-Z)^{\beta-1}}{\int_0^1 Z^{\alpha-1}(1-Z)^{\beta-1}df} \qquad (5.4.40)$$

where a and b are the two parameters of the beta function: $a = \tilde{Z}\left(\tilde{Z}(1-\tilde{Z})/\widetilde{Z''^2} - 1\right)$ and $b = (1-\tilde{Z})\left(\tilde{Z}(1-\tilde{Z})/\widetilde{Z''^2} - 1\right)$, respectively.

The filtered scalar dissipation $\tilde{\chi}$ in equation (5.4.39) may be modeled according to the proposal of Jiménez et al. (2001). A simple and general model can be taken from FANS modeling of dissipation in terms of a characteristic mixing time, which is assumed to be proportional to the turbulent characteristics time. In LES, the SGS scalar mixing time can be defined as

$$\frac{1}{\tau_Z} = \frac{\tilde{\chi}}{\widetilde{Z''^2}} \qquad (5.4.41)$$

An equivalent SGS turbulent characteristic time $\bar{\tau}$ can be expressed on the basis of the ratio between the SGS kinetic energy $\bar{k} = \frac{1}{2}(\widetilde{u_i u_i} - \bar{u}_i \bar{u}_i)$ and the filtered kinetic energy dissipation rate $\bar{\varepsilon} = \overline{v(\partial u_i/\partial x_j \partial u_i/\partial x_j)}$. Assuming proportionality between both times, a model for $\tilde{\chi}$ can be derived as

$$\frac{\tilde{\chi}}{\widetilde{Z''^2}} = \frac{1}{\tau_Z} \sim \frac{C}{\bar{\tau}} = C\frac{\bar{\varepsilon}}{\bar{k}} \qquad (5.4.42)$$

The parameter C is assumed to be adequately represented by $C = 1/Sc$, where in our studies of turbulent buoyant fires, Sc is given by the turbulent Schmidt number for the scalar variance—that is, Sc_{T,Z''^2}. Unlike in FANS k-ϵ calculations, there are no transport equations to evaluate the quantities such as the SGS kinetic energy or its dissipation when it comes to a practical LES. Jiménez et al. (2001) proposed nonetheless in employing the approximations of $\bar{k}$ and $\bar{\varepsilon}$ derived from SGS turbulence models. By adopting the eddy viscosity model for the SGS stresses and the Yoshizawa model for the SGS kinetic energy given by

$$\bar{\varepsilon} = 2(\mu/\bar{\rho} + \mu_T^{SGS}/\bar{\rho})\tilde{S}_{ij}\tilde{S}_{ij}, \quad \bar{k} = 2C''\Delta^2\tilde{S}_{ij}\tilde{S}_{ij} \qquad (5.4.43)$$

the scalar dissipation $\tilde{\chi}$ is henceforth determined according to

$$\bar{\rho}\tilde{\chi} = \frac{(\mu + \mu_T^{SGS})}{Sc_{T,Z''^2}C''\Delta^2}\widetilde{Z''^2} \qquad (5.4.44)$$

which results in a well-conditioned expression, provided that the constants are not zero. Simulations performed by Jiménez et al. (2001) have indicated that C'' varied from a value of 0.09 to a value of 0.06. Their investigations have revealed that dissipation could be accurately predicted both locally and on average, when a constant intermediate value of 0.07 is adopted. Estimation of a locally varying parameter could also be achieved via the consideration of a dynamic procedure. The turbulent viscosity μ_T^{SGS} in equation (5.4.44) may be ascertained from the range of SGS models discussed in Section 5.2.2. Equation (5.4.44) allows the formulation of appropriate forcing terms to represent dissipation and effects of mixing in the scalar variance evolution, of which can be realized via the modeled transport equation as proposed in equation (5.4.8). Following Jiménez et al. (2001) proposal, the filtered source term $\bar{S}_{Z''^2}$ can be modeled according to

$$\bar{S}_{Z''^2} = 2\left(\frac{\mu}{Sc_{Z''^2}} + \frac{\mu_T^{SGS}}{Sc_{T,Z''^2}}\right)\frac{\partial \tilde{Z}}{\partial x_j}\frac{\partial \tilde{Z}}{\partial x_j} - \bar{\rho}\tilde{\chi} \tag{5.4.45}$$

Alternatively, the scalar variance may be determined using a scale similarity model instead of the additional computational requirement of solving a transport equation. The notion of scale similarity is based on the inference of the small-scale statistics of a scalar quantity from the statistics of the smallest resolved structures. On the basis of the fractal nature of turbulence, subgrid scale turbulent structures can be approximated by assuming that the unresolved scales are similar to the smallest scales, hence *scale similarity*. This method involves filtering the resolved mixture fraction field with a test filter $\hat{\Delta}$, in order to gather information about the fluctuation in mixture fraction on the smallest resolved scales. Invoking the scale similarity hypothesis, the subgrid scalar variance can be modeled as

$$\widetilde{Z''^2} \equiv \widetilde{(Z - \tilde{Z})^2} \approx K(\widehat{\tilde{Z}^2} - \hat{\tilde{Z}}^2) \tag{5.4.46}$$

where K is a model parameter and is determined from *a priori* analysis. Jiménez et al. (1997) have employed spectral reasoning to arrive at an appropriate expression for K, which is

$$K = (2^{\beta-1} - 1)^{-0.5} \tag{5.4.47}$$

Based on the spectral slope β equivalent to 5/3 in the Kolmogorov cascade, equation (5.4.47) yields a K value of about 1.3. In equation (5.4.46), $\widehat{\tilde{Z}^2}$ implies that the filtered mixture fraction is first squared and then the resulting field is test filtered, while $\hat{\tilde{Z}}^2$ simply means the squared of the test filtered mixture fraction field. In practice, the discrete three-dimensional test field for the mixture fraction $\hat{\tilde{Z}}$ at the cell location (i, j, k) may be evaluated by applying the Trapezoidal rule via the construction by linear combination as

$$\hat{\tilde{Z}}_{i,j,k} = \frac{1}{12}(\tilde{Z}_{i-1,j,k} + 2\tilde{Z}_{i,j,k} + \tilde{Z}_{i+1,j,k}) + \frac{1}{12}(\tilde{Z}_{i-1,j,k} + 2\tilde{Z}_{i,j,k} + \tilde{Z}_{i+1,j,k})$$
$$+ \frac{1}{12}(\tilde{Z}_{i-1,j,k} + 2\tilde{Z}_{i,j,k} + \tilde{Z}_{i+1,j,k})$$

$$\tag{5.4.48}$$

The value of $\widehat{\tilde{Z}^2}$ is simply evaluated by the square of the numerical representation of equation (5.4.48). In a similar manner, $\widehat{\tilde{Z}^2}$ is calculated accordingly by

$$\widehat{\tilde{Z}^2}_{i,j,k} = \frac{1}{12}(\tilde{Z}^2_{i-1,j,k} + 2\tilde{Z}^2_{i,j,k} + \tilde{Z}^2_{i+1,j,k}) + \frac{1}{12}(\tilde{Z}^2_{i-1,j,k} + 2\tilde{Z}^2_{i,j,k} + \tilde{Z}^2_{i+1,j,k})$$
$$+ \frac{1}{12}(\tilde{Z}^2_{i-1,j,k} + 2\tilde{Z}^2_{i,j,k} + \tilde{Z}^2_{i+1,j,k})$$

$$(5.4.49)$$

The local equilibrium assumption for the scalar variance allows the dissipation $\tilde{\chi}$ to be explicitly determined. By setting the filtered source term $\overline{S}_{Z''^2}$ in equation (5.4.45) equals to zero, $\tilde{\chi}$ can be evaluated as

$$\tilde{\chi} = 2\left(\frac{\mu}{\overline{\rho}Sc_{Z''^2}} + \frac{\mu_T^{SGS}}{\overline{\rho}Sc_{T,Z''^2}}\right)\frac{\partial \tilde{Z}}{\partial x_j}\frac{\partial \tilde{Z}}{\partial x_j}$$

$$(5.4.50)$$

which incidentally follows similar consideration as proposed by Girimaji and Zhou (1996) and de Bruyn Kops et al. (1998).

To incorporate more detail combustion chemistry, the approach based on subgrid modeling for turbulent reacting flows developed by Cook and Riley (1998) has been found to be useful for non-premixed turbulent flames. This method accounts for finite-rate chemistry by invoking the laminar flamelet approximation and applies the Large Eddy Probability Density Function (LEPDF) of a mixture fraction. By assuming that mixing and reaction occur in local thin regions of steady, one-dimensional, laminar counterflow flames, the instantaneous scalar dissipation χ can be determined analytically and is given by

$$\chi = \chi_o F(Z)$$

$$(5.4.51)$$

where the function F is the inverse of the Gaussian error function of the mixture fraction given by

$$F(Z) = \exp\left(-2[erf^{-1}(2Z-1)]^2\right)$$

$$(5.4.52)$$

In equation (5.4.49), χ_o refers to the local peak value of χ within the reaction layer. The modeling implies that χ_o is independent of Z; the filtered composition can thus be expressed as

$$\tilde{Y}_i = \int_0^1 \int_{\chi_o^{min}}^{\chi_o^{max}} Y_i(Z, \chi_o) P(\chi_o) P(Z) d\chi_o dZ \qquad (5.4.53)$$

where χ_o^{min} and χ_o^{min} are the minimum values of χ_o within the LES grid cell and $P(Z)$ and $P(\chi_o)$ are the respective subgrid scale probability density functions of Z and χ. Cook and Riley (1998) demonstrate that $Y_i(Z, \chi_o)$ is a slow function of χ and thus of χ_o (e.g., Mell et al., 1994). Assuming that the interval $\chi_o^{max} - \chi_o^{min}$ is not too large, $Y_i(Z, \chi_o)$ can be approximated via the Taylor series expansion about χ_o as

$$Y_i(Z, \chi_o) \approx Y_i(Z, \tilde{\chi}_o) + \left.\frac{\partial Y_i}{\partial \chi_o}\right|_{\tilde{\chi}_o} (\chi_o - \tilde{\chi}_o) \qquad (5.4.54)$$

Inserting equation (5.4.54) into equation (5.4.53) and integrating over χ_o yields

$$\tilde{Y}_i = \int_0^1 Y_i(Z, \tilde{\chi}_o) P(Z) dZ \qquad (5.4.55)$$

The integration in the preceding equation requires only the evaluation of $P(Z)$, to which a beta function PDF in the form of equation (5.4.39) may be applied. The filtered local peak value of χ within the layer χ_o ($\chi_o^{max} \leq \chi_o \leq \chi_o^{min}$) can be immediately obtained via equation (5.4.51) as

$$\tilde{\chi}_o = \frac{\tilde{\chi}}{\int_0^1 F(Z) P(Z) dZ} \qquad (5.4.56)$$

Prior to running LES, the strategy is to purposefully construct a flamelet library for $\tilde{Y}_i(\tilde{Z}, \widetilde{Z''^2}, \tilde{\chi}_o)$. Firstly, $\tilde{Z}$ and $\widetilde{Z''^2}$ are chosen and $P(Z)$ is determined from equation (5.4.39). Secondly, $\tilde{\chi}_o$ is chosen. By assuming that the local subgrid fluctuation on χ_o to be negligible, which gives $\tilde{\chi}_o = \chi_o$, equation (5.4.51) is used to replace χ in the steady, species equation of the laminar flamelet model. With reference to the laminar flamelet approach described in Section 3.4.2.6, equation (3.4.87) reduces to

$$\tilde{\chi}_o F(Z) \frac{\partial^2 Y_i}{\partial Z^2} = \tilde{\chi}_o \exp\left(-2[erf^{-1}(2Z - 1)]^2\right) \frac{\partial^2 Y_i}{\partial Z^2} = \omega_i \qquad (5.4.57)$$

With $P(Z)$ already known and solutions of $Y_i(Z, \tilde{\chi}_o)$ obtained from the preceding equation, the filtered mass species $\tilde{Y}_i$ are then computed from equation (5.4.55). In equation (5.4.56), the filtered scalar dissipation $\tilde{\chi}$ can be evaluated either using equation (5.4.44) or (5.4.50), depending on which approach is undertaken for the evaluation of the scalar variance during the LES computations. Local peak value of $\tilde{\chi}_o$ is henceforth determined and alongside with calculated values of $\tilde{Z}$ and $\widetilde{Z''^2}$, the filter composition $\tilde{Y}_i$ is subsequently ascertained from the generated library of $\tilde{Y}_i(\tilde{Z}, \widetilde{Z''^2}, \tilde{\chi}_o)$.

5.4.3 Inclusion of Other Physical Models

The consideration of radiation from absorbing/emitting combustion gaseous such as CO_2 and H_2O and soot can also be treated by solving the filtered radiative transfer equation for a non-scattering medium. If the radiation method is considered for the system of ith equations of transfer for the ith gray gases, as described in equation (3.11.2), the radiative transfer equation via spatial filtering can be expressed as

$$\frac{d\overline{I_i(\vec{r},\vec{s})}}{ds} = \overline{-k_i I_i(\vec{r},\vec{s}) + k_i a_{\varepsilon,i} I_{black}(\vec{r})} \qquad (5.4.58)$$

The net radiative heat loss (gain) of the filtered volumetric radiative source/sink term S_{rad} in equation (5.4.3), consists primarily of the energy absorbed from the incident radiation field minus the energy emitted to the surroundings

$$S_{rad} = \overline{Q}_a - \overline{Q}_e \qquad (5.4.59)$$

In equation (5.4.57), the energy emitted depends on the local absorption coefficient and temperature, while the energy absorbed depends on the distribution of these quantities over the entire flame.

For practical fire simulations, the effects of the SGS turbulence-radiation interaction in equations (5.4.58) and (5.4.59) are normally neglected in view of extensive computational costs associated with the evaluation of non-linear correlations. The filtered radiative transfer equation can thus be simplified to

$$\frac{\overline{dI_i(\vec{r},\vec{s})}}{ds} \approx -\overline{k}_i \overline{I_i(\vec{r},\vec{s})} + \overline{k}_i \overline{a}_{\varepsilon,i} \overline{I_{black}(\vec{r})} \qquad (5.4.60)$$

Note that Favre-averaged quantities are invariably used in the context of model formulation to evaluate the terms on the right-hand side of equation (5.4.60); implicit in this is obviously the neglect of the density-temperature correlation. For the filtered source/sink term S_{rad}, the energy emitted by an infinitesimally small volume in the flame over the solid angle $d\Omega$ is obtained by

$$\overline{Q}_e = \sum_i \int_\Omega \overline{k}_i \overline{a}_{\varepsilon,i} \frac{\sigma \tilde{T}^4}{\pi} d\Omega = \sum_i \int_0^{2\pi} \int_0^\pi \overline{k}_i \overline{a}_{\varepsilon,i} \frac{\sigma \tilde{T}^4}{\pi} \sin\theta d\theta d\phi = 4 \sum_i \overline{k}_i \overline{a}_{\varepsilon,i} \sigma \tilde{T}^4$$

(5.4.61)

The filtered absorbed energy is taken as the sum of absorbed portion of the ith gray gas intensities incident on the volume from all directions

$$\overline{Q}_a = \sum_i \int_\Omega \overline{k}_i \overline{I}_i d\Omega = \sum_i \int_0^{2\pi} \int_0^\pi \overline{k}_i \overline{I}_i \sin\theta d\theta d\phi$$

(5.4.62)

Based on the formulation of equations (5.4.60), (5.4.61), and (5.4.62), standard solution methods such as the Monte Carlo, P-1 radiation model, discrete transfer radiative model, discrete ordinates model, and finite volume method in Chapter 3 can be immediately employed for LES.

To account for the radiation contribution due to soot particles, the filtered transport equation, depending on which model is employed, can be written in the general form as

$$\frac{\partial(\overline{\rho}\tilde{\phi})}{\partial t} + \frac{\partial(\overline{\rho}\tilde{u}_j\tilde{\phi})}{\partial x_j} + 0.55 \frac{\partial}{\partial x_j}\left[\tilde{\phi}\frac{\mu}{\tilde{T}}\frac{\partial\tilde{T}}{\partial x_j}\right] = -\frac{\partial}{\partial x_j}(\overline{\rho u_j \phi} - \overline{\rho}\tilde{u}_j\tilde{\phi}) + \overline{S}\tilde{\phi}$$

(5.4.63)

where the scalar variable ϕ denotes the soot mass fraction, soot volume fraction, soot number density, or soot moments. The subgrid correlation of the scalar variable with velocity in equation (5.4.63) can be closed by using the Smagorinsky turbulence model. In order to avoid the unduly complicated evaluation of additional non-linear correlation terms, Desjardin and Frankel (1999) suggested that the filtered source term $\overline{S}_{\tilde{\phi}}$ is closed in terms of only the filtered composition and filtered temperature, thus neglecting some effects of the SGS fluctuations on this term. From a practical perspective in field modeling, soot production is predominantly determined from large-scale turbulent advection, diffusion of soot caused by thermophoresis, SGS turbulent diffusion, and finite-rate soot chemistry effects.

5.5   Worked Examples on Large Eddy Simulation Applications

5.5.1   A Freestanding Buoyant Fire

Buoyant fires are typically characterized by a very low initial momentum and are strongly affected by buoyancy effects. Experiments by Cox and Chitty (1980) and McCaffrey (1983) have clearly observed and distinguished three distinct regions for buoyant fires: a persistent flame, an intermittent flame, and a buoyant plume such as depicted in Figure 5.6. It is well recognized that this buoyant flame exhibits an oscillatory behavior. The occurrence of this oscillation, also known as the "puffing" effect, stems from the presence of coherent structures above a fire plume, generally as a consequence of the developing buoyancy driven instabilities, which in turn leads to vortex shedding, especially through the formation of large flaming vortices that rise up until they burn out at the top of the flame. The pulsating characteristics of such fires are strongly governed by the rate of air entrainment into the flame, flame height, combustion efficiency, and radiation heat output of flames.

An in-house developed large eddy simulation fire model, which involves direct numerical simulation of the large-scale turbulence and subgrid scale modeling of the mixture fraction-based combustion model, radiation heat transfer via the discrete ordinates method and finite-rate soot chemistry model of Moss et al. (1988) and Syed et al. (1990), is demonstrated in this worked example to capture and identify several key features or physical aspects of the buoyant fire as aforementioned—three-zone flame structure. The pulsating instability via the frequency of pulsation is also examined. Numerical results

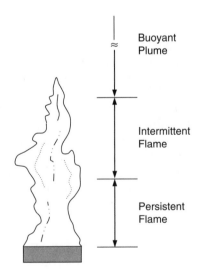

Figure 5.6   Schematic drawing illustrating the three distinct regimes for a buoyant fire.

from the model as reported in Cheung et al. (2007a) and Cheung et al. (2007) are validated against the predictions of the macroscopic observables against correlation obtained from Baum and McCaffrey (1989) and McCaffrey (1983) experimental data and verified against the results obtained from the FDS computer code.

The fire experiment by McCaffrey (1983) used for the comparison exercise centers on a buoyant fire in an open environment. The burner using natural gas (methane) at various controlled rates was constructed of a porous refractory material of 0.3 m square. Different heat release rates of the buoyant fire were investigated; the case of 45 kW is adopted in the present investigation. Conditions at the centerline of the burner attracted particular attention, and detailed measurements of the time-averaged vertical velocity and temperature have been obtained using the bi-directional pressure probe and thermocouple. Visual observation of the flame behavior was also recorded using a collection of photographic images to ascertain the three regimes of the flame structure exhibited by the buoyant fire.

Numerical features: The low-Mach-number Favre-filtered mass, momentum, energy, and species (mixture fraction, scalar variance of the mixture fraction, soot particulate number density, and soot volume fraction) conservation equations, which have been described in previous sections, are solved. In these equations, the molecular viscosity is assumed to be a function of the temperature such that $\mu = \mu_{ref}(\tilde{T}/T_{ref})^{0.7}$. The molecular Prandtl number is set to a value of 0.7, while the molecular Schmidt numbers for the mixture fraction, the mixture fraction variance, soot particulate number density, and soot volume fraction are prescribed at values of 0.7, 0.7, 700, and 700, respectively. Schmidt numbers for the soot have been attained from Sivathanu and Gore (1994).

On the basis of the application of the standard Smagorinsky-Lilly model, the Smagorinsky constant C_s is prescribed at 0.2, while the turbulent Prandtl and all the scalar turbulent Schmidt numbers of 0.3 are imposed. Zhou et al. (2001) have indicated that a little larger C_s is generally used for many thermal flows; C_s for cold jets is usually taken to lie within a range of 0.1–0.13. They have employed a value of 0.23 for their large eddy simulation study. The scalar turbulent Schmidt numbers correspond to the combustion studies on turbulent diffusion flame recently performed by Yaga et al. (2002), and confirmed through direct numerical simulation data in Jiménez et al. (2001).

In order to realize a true predicative capability of the fire model, it is imperative to understand the range of length scales that are required to be resolved for a large eddy simulation For a fire plume, the characteristic length scale can be related to the total heat release rate $\dot{Q}$ (W) by the following relationship as suggested by McGrattan et al. (1998):

$$L^* = \left(\frac{\dot{Q}}{\rho_{ref} T_{ref} C_p \sqrt{g}} \right)^{2/5} \tag{5.5.1}$$

In general, the large-scale structure that is controlled by the inviscid terms can be completely described when this characteristic length L^* is adequately resolved. For the heat release considered in this present investigation, the characteristic length L^* is approximately in the order of 0.3 m. McGrattan et al. (1998) have ascertained that the large-scale structure can be completely described when L^* is spanned by roughly ten computational cells. This implies that adequate resolution of the fire plume, particularly above the porous square burner, can be achieved with a spatial resolution of about 0.03 m. With the spatial value of 0.03 m used as reference, two non-uniform mesh distributions of $96 \times 96 \times 96$ and $116 \times 116 \times 116$ cells have been tested within the computational domain with finer grid cells centered above the burner to better capture all the necessary macroscopic large-scale features of the flaming fire. No significant difference of the predicted results is observed when simulations are performed on the two grid resolutions. For the best trade-off between numerical accuracy and cost, the mesh of $96 \times 96 \times 96$ cells is therefore employed. A transient analysis is preformed via the two-stage predictor-corrector approach for low Mach number compressible flows to account for the strong coupling between the density and fluid flow equations. The computational is set to 35 seconds to ensure that it reaches the stable and converged status. The time step is determined by employing a CFL number of 0.35 to achieve a time-accurate solution via

$$dt = \frac{0.35}{\max\left(\left|\dfrac{\tilde{u}}{\Delta x}\right| + \left|\dfrac{\tilde{v}}{\Delta y}\right| + \left|\dfrac{\tilde{w}}{\Delta z}\right|\right)} \qquad (5.5.2)$$

Numerical results: A closer examination of the dynamic behavior of the buoyant fire can be analyzed by tracing the instantaneous velocity and temperature quantities at the centerline axial locations $y/D = 1, 2, 3, 4$, and 5, above the porous burner in Figures 5.7 and 5.8, respectively. It is noted that the parameter D refers to the burner width that is 0.3 m. It can be seen the instantaneous velocity and temperature quantities at axial locations of $y/D = 1, 2$, and 3 display a highly repeatable flickering behavior at the same frequency but with different amplitudes. These results clearly show the puffing mechanism being actively maintained within the persistent and intermittent flame regions. At axial locations of $y/D = 4$ and 5, the oscillatory behaviors of the instantaneous velocity and temperature quantities are observed to be more random though some repeatable flickering behavior of the flame is still sustained. The increasing randomness of these results demonstrates the transition of the flow into the buoyant plume region.

The frequency spectra obtained from the time histories of the instantaneous velocity and temperature quantities at the centreline axial locations $y/D = 2$ by Fast Fourier Transform (FFT), are respectively shown in Figure 5.9. The phase period over one flickering cycle is obtained about 2 Hz. Regardless of whether

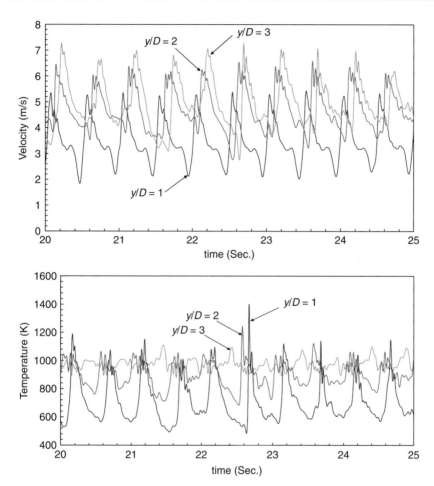

Figure 5.7   Time-tracing of the centerline axial velocity and temperature at $y/D = 1, 2,$ and 3.

FFT is performed on the instantaneous velocity or temperature quantities, the dominating frequency was nearly the same at each of the axial locations. Several sub-harmonics could also be observed aside from the dominating frequency. The same frequency being obtained from either the instantaneous velocity or temperature quantities, clearly demonstrates the strong coupling that existed within the predictor-corrector method where the fluid motion was driven predominantly by the buoyancy effects (varying density).

Figure 5.10 illustrates a series of frames capturing a flickering period, by comparing the macroscopic predictions using the large eddy simulation fire model against the photographic images obtained from McCaffrey (1983). The naturally flickering behavior demonstrates the flow and undergoes a phase shift along the axial direction. Numerically predicted images in Figure 5.10

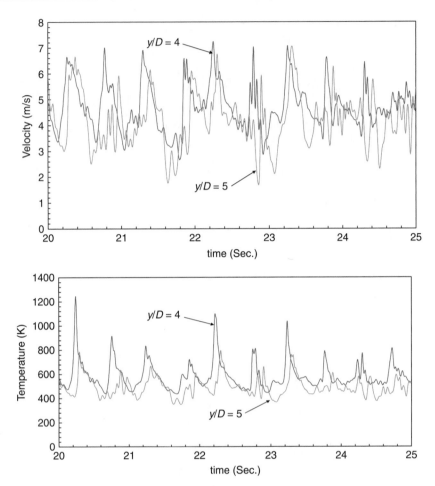

Figure 5.8   Time-tracing of the centerline axial velocity and temperature at $y/D = 4$ and 5.

correspond to three iso-surfaces of temperatures at 800 K (visible flame), 450 K, and 310 K (cold smoke). Two distinct regions of the flame that comprise of the upper flame separating from the lower persistent flame as depicted in frames 1 and 8 by the present model, show a remarkable resemblance to the associated photographic images. The upward surging flame structures in frames 6 and 7 before flame separation in frame 8 are also adequately captured by the model when compared against the corresponding images of the developing fire observed during the experiment.

A snapshot of the instantaneous temperature contours of the simulated flame from the in-house large eddy simulation fire model and FDS computer code is illustrated in Figure 5.11. Three distinct regions of the flame could be identified in both model predictions. The two models clearly exemplify a persistent flame

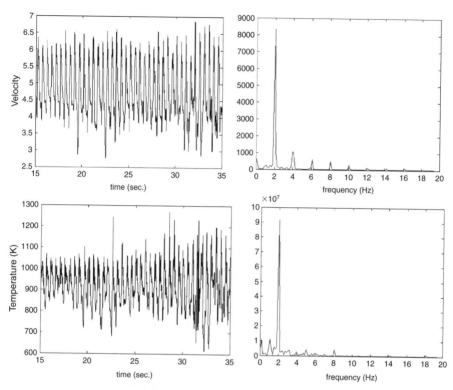

Figure 5.9 Time history of the centerline axial velocity and temperature at $y/D = 2$ accompanied by the respective frequency spectra.

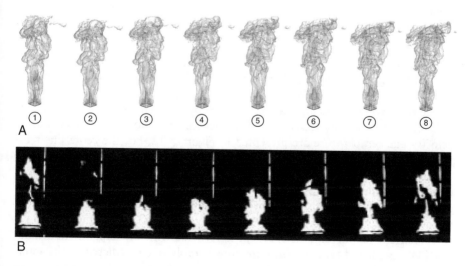

Figure 5.10 Demonstration of the puffing effect during one flickering cycle: (a) Predicted temperature iso-surfaces and (b) Experimentally visualized photographic images McCaffrey (1983).

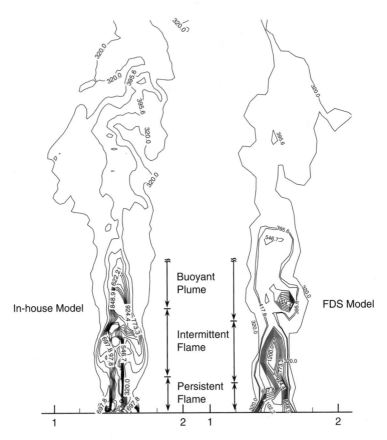

Figure 5.11 A snapshot of the instantaneous temperature contour predicted by the in-house model (left) and the FDS computer code (right).

region prevailing just above the porous burner, an intermittent flame region encapsulating the flame "puffing" effect and beyond that a buoyant plume region.

The time-averaged quantities are obtained by averaging over 20 puffing periods after a statistically steady state is achieved. The centerline predicted axial velocity and temperature are presented in Figure 5.12. For the centerline temperature in Figure 5.12 (top), it is observed that temperatures are slightly under-predicted at the persistent flame regime by the in-house model. This discrepancy could be due to the neglect of the oxidation of soot particles. Soot quantities may therefore be over-predicted; temperatures at the persistent regime are subsequently under-predicted. Nonetheless, temperatures above the persistent regime exhibit excellent agreement with the experimental data. For FDS, predicted temperatures are considerably over-predicted at the intermittent flame regime. A possible source of error could be the prescription of the empirical soot production constant. In FDS, a fix fraction of fuel with

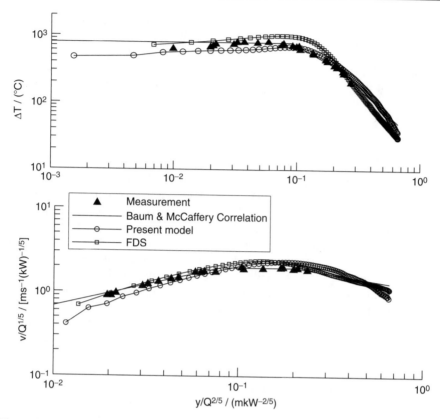

Figure 5.12 Comparison between the predicted time average centerline temperatures and velocities and Baum & McCaffrey correlation.

the value of 1% was assumed by default to be converted as soot particles. Such constant fraction of soot production most probably under-estimates the amount of soot produced within the persistent and intermittent flame regimes. The centerline velocity, as illustrated in Figure 5.12 (bottom) by the two models, gave nevertheless similar results and agreed well with the measured value and correlation.

Conclusion: The fire model based on large eddy simulation has shown to be capable of reproducing the temporal and spatial evolution of the self-excited large toroidal flaming vortex structures, that correspond to the experimentally observed vortex shedding phenomenon of a typical buoyant diffusion fire. The three-zone flame structure of a persistent flame, an intermittent flame, and a buoyant plume is well captured. Numerous pulsation frequency values have been reported; pertinent data can be obtained from experimental works by McCaffrey (1983), Protscht (1975), Byram and Nelson (1970), and Zukoski et al. (1984). For a fire bed of 0.3 m, experimental investigation by Byram and Nelson (1970) gave a frequency of about 2.5 Hz, while Protscht (1975) showed a characteristic frequency of 2 Hz. Zukoski et al. (1984), however, revealed a range of frequencies between 2 Hz and 3 Hz. Incidentally, based on the visual photographic images, McCaffrey (1983) provided a pulsation frequency of about 3 Hz corresponding to the higher end of the frequency range. The current in-house model prediction corresponds to the lower end of the frequency range. In general, the agreement can be considered to be rather reasonable to the expected measured frequency for this particular fire bed size.

5.5.2   Fire in a Single-room Compartment

In this worked example, the large eddy simulation fire model is further applied to examine the apparently confined rigid boundaries surrounding a steady fire in an enclosure. The full-scale single-room compartment fire experiment performed by Steckler et al. (1984) is employed as the test case. In addition to the detailed measurements of temperature using aspirated thermocouples and velocity by bi-directional probes at the doorway in validating the model predictions, physical insights are provided by the large eddy simulation fire model for the temporal fire behavior and smoke layer development within the compartment with a heat release rate $\dot{Q}$ of 62.9 kW was also adopted in previous worked examples illustrating the field model based on Favre-averaging. Numerical simulations are performed through the FDS computer code.

Numerical features: FDS solve numerically the Navier-Stokes equations, which are based on the assumption of low speed, thermally-driven flow with a special emphasis on smoke, and heat transport from fires. The partial derivatives of the conservation equations of mass, momentum, and energy are approximated by the finite difference method. An explicit predictor-corrector scheme, second order in time and space, is adopted, which embodies the core algorithm of the computer code. Turbulence is treated by means of the Smagorinsky (1963) form of large eddy simulation. FDS employs the conserved scalar approach (mixture fraction based model) to handle the combustion of fires. The model thus assumes that combustion is mixing-controlled and that the reaction of fuel and oxygen is infinitely fast. Mass fractions of all the major reactants and products can be derived from the mixture fraction by means of state relationships, empirical expressions arrived at by a combination of simplified analysis and measurement. Radiative heat transfer is solved via the solution of the radiation transport equation for a non-scattering gray gas. The radiation transfer is effectively handled through the Finite Volume Method. In retrospect, the finite volume solver requires about 15 percent of the total CPU time of a calculation using approximately 100 discrete angles (12 angles for one octant of a sphere), which represents a modest cost given the complexity of the radiation heat transfer. More details can be referred to in the user guide manual of FDS/Smokeview version 4.0 released in July 2004.

The schematic drawing of the computational geometry is shown in Figure 5.13, which has been made possible by Smokeview, a post-processor graphical-user-interface application. In contrast with the field model based

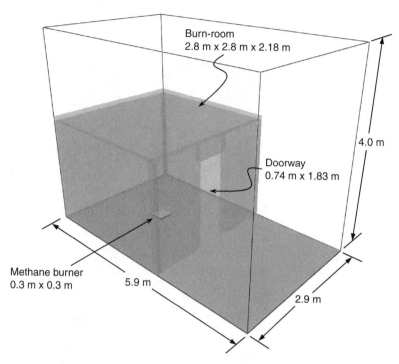

Figure 5.13 Schematic drawing of the computational domain for the single-room compartment fire.

on Favre-averaging, which allowed the use of a symmetry boundary condition to simplify the single-room compartment fire, large eddy simulation calculations require the consideration of realistic solutions to be obtained in a full three-dimensional geometry. On the basis of the mesh criterion in equation (5.5.1) stipulated in section 5.5.1, a mesh resolution of approximately 0.03 m is constructed for the entire burn room as well as the extended region resulting in a total mesh of 159 (along the length) × 89 (along the width) × 128 (along the height), which also corresponds to the similar resolution adopted in the numerical simulation carried out by McGratten et al. (1998) on this particular fire problem. The solution is marched in time on a three-dimensional rectilinear grid, which is dynamically adjusted by the instantaneous velocities $\left(dt < \min\left(\frac{\Delta x}{\bar{u}}, \frac{\Delta y}{\bar{v}}, \frac{\Delta w}{\bar{w}}\right)\right)$. Instantaneous results are monitored regularly and the fire is assumed to have reached quasi-steady state when the instantaneous values appear periodically. The ambient temperature is taken to be at 20°C.

Numerical results: The quasi-steady state behavior of the methane flame driven by buoyancy and the smoke-filling in the single-room configuration, are illustrated in Figure 5.14 by the series of snapshots representing the development of instantaneous temperature contours in time via Smokeview on a

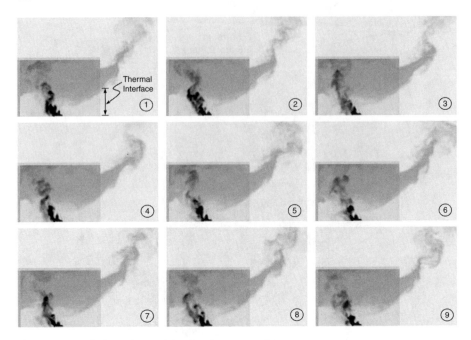

Figure 5.14 Series of transient development of instantaneous temperature contours separated by 0.4 seconds time interval. The dark colored contours indicate temperatures above 320°C.

vertical plane cutting through the middle of the burner and doorway. As shown from these results, the commonly observed characteristic pertaining to a compartment fire is depicted by a distinct thermal interface (see definition in Chapter 2), separating the well-established upper hot layer of combustion products from the lower cold layer consisting of entraining ambient air through the doorway. The prevalence of a leaning fire toward the back wall of the compartment corresponds to another important feature, which has also been confirmed in numerical studies of previous worked examples in Chapters 2 and 3 and through experimental observations for a compartment fire with a heat release rate of 62.9 kW.

More importantly, the large eddy simulation results have provided the capacity of attaining increasing realism of the turbulent buoyant flow by an accurate realization of the flow structure. Firstly, the occurrence of the oscillatory or puffing effect, which stems from the presence of coherent structures above a fire plume as a consequence of the developing buoyancy driven instabilities, is clearly evident in the results illustrated in Figure 5.14. Vortex shedding, especially through the formation of large flaming vortices that rise up until they burn out at the top of the flame, is observed at the top left-hand corner of the compartment. Secondly, the pulsating characteristic of such a fire, which is also governed by the incoming ambient air into the flame, clearly contributes to a persistent leaning flame structure experienced throughout the burning process. Thirdly, the vortical motion of a buoyant plume of hot gases spilling out from the burn room just below the soffit, reflects a smoke movement that is typically manifested in a real fire situation during the pre-flashover stage of a compartment fire. This wealth of information by means of additional insights into the transient behaviors of the flame emanating from porous bed gas burner and flow physics surpasses the capability of the field model based on the Favre-averaging approach.

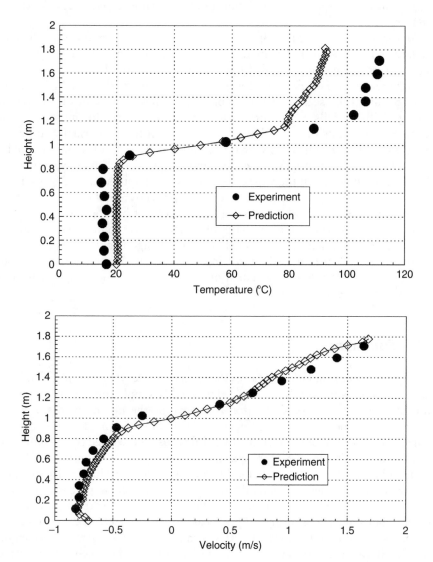

Figure 5.15 Comparison of centerline time-averaged temperature and velocity profiles at the doorway.

The predicted time-averaged centerline temperature and velocity profiles at the doorway, plotted alongside with Steckler et al. (1984) experimental data, are given in Figure 5.15. In spite of the temperatures being marginally under-predicted at the upper part of the doorway, the relatively good agreement between the measured and calculated results is an indicator of the predictive capability of the present approach. This worked example demonstrates that not only can the smoke movement be described with reasonably high mesh resolution but also the importance of including the combustion and radiation models into the large eddy simulation model for simulating practical fires of interest.

Conclusion: Even for the simple enclosure fire problem demonstrated herein, the large eddy simulation fire model in the FDS computer code has shown the ability of providing very detailed information about the flow physics and an accurate realization of the turbulent buoyant flow encapsulating a broad range of length and time scales. The question arises whether it is an imperative to resolve all the details of the turbulent fluctuations, where for most practical purposes, the effects of the turbulence on the mean flow are usually sufficient to quantify the turbulent flow characteristics. This type of solution can be attained rather efficiently via the field model based on Favre-averaging, as already demonstrated in the worked examples of Chapters 2 and 3. Nevertheless, it should be emphasized that even a perfectly resolved computation employing the k-ε or similar turbulent models, can at best produce a solution of the *model* equations. On the basis of increasing computational resources, the large eddy simulation fire model, which has significantly less adjustable parameters, can systematically and progressively capture the dynamic range contained in the Navier-Stokes equations as the spatial and temporal resolutions are improved.

5.6 Summary

Direct numerical simulation requires the spatial and temporal scales of the turbulent structures to be fully resolved. Large eddy simulation entails the direct simulation of the large-scale motion that governs the mixing of gases, while the small-scale motion is approximated via a suitable SGS models. The macroscopic consideration of the turbulent fluid flow using the large eddy simulation coupled with suitable SGS representation of combustion, radiation, and soot chemistry interactions, are particularly emphasized in this chapter. Description of appropriate numerical methodologies and inclusion of practical worked examples are intended to demonstrate the feasibility of carrying out three-dimensional simulations of fires of practical interest with a high degree of accuracy. Present computational capabilities permit simulations to be performed on sufficiently fine mesh resolution, in the order of one million cells of capturing in particular the temporal and spatial evolution of self-excited oscillatory fluid flow typically attributed to turbulent buoyant fires—namely the puffing or flickering behavior. Numerical results for the freestanding and compartment fires obtained through the large eddy simulation fire model, compare favorably with experimental data as have been exemplified in the worked examples. The good agreement indicates the usefulness of the model as an emergent predictive tool in fire safety engineering investigations.

Review Questions

5.1. In CFD, what are other possible approaches to turbulent simulation besides adopting the Favre-Averaged Navier-Stokes approach?

5.2. Define the Kolmogorov micro-scales of length, time, and velocity. How can they be applied in the numerical computation?

5.3. What are the key issues that govern the spatial and temporal resolutions and discretisations in direct numerical simulation?

5.4. Describe the basic idea behind the approach based on large eddy simulation. What is the main difference between direct numerical simulation and large eddy simulation?

5.5. Do the key issues that govern the spatial and temporal resolutions and discretisations in direct numerical simulation apply to large eddy simulation? If so, why?

5.6. What are the typical filters that can be applied in large eddy simulation?

5.7. In large eddy simulation, the small dissipative scales are, in general, not solved accurately. How are they modeled?

5.8. What is the dynamic procedure? Why are such models useful?

5.9. An alternative strategy to characterize the turbulence is through the use of the one-equation subgrid scale model. What does it entail?

5.10. What are the advantages and disadvantages between the Favre-Averaged Navier-Stokes and large eddy simulation approaches?

5.11. When employing large eddy simulation, what assumption is commonly imposed in characterizing the fire induced flows? Based on the assumption, how are the governing equations treated and solved numerically?

5.12. How is combustion modeled in large eddy simulation? Compare against the Favre-Averaged Navier-Stokes approach in Chapter 3.

6 Other Challenges in Fire Safety Engineering

Abstract

The required performance-based methodologies for the fire safety engineering approach are described in this chapter. Central to this fire safety engineering approach is the use of CFD-based fire models in determining the necessary performance requirements for fire safety. For a complete fire safety assessment and evaluation of the whole building, deterministic calculations including the use of artificial neural network, evacuation modeling, and probabilistic methods necessitate the total evaluation of the risk of life and property in the building and fire protection systems. An overview of the emerging technique of artificial neural network, suitable evacuation models, and probabilistic analyzes is provided.

6.1 Fire Safety Evaluation and Assessment

6.1.1 Deviation from Prescriptive-Based Statutory Requirements

Prescriptive codes are still being used in most countries, because they enable a straightforward evaluation of determining whether the requirements for the fire protection systems in buildings have been met or have not been met. There are a number of benefits in retaining the prescriptive codes. Firstly, they are easier to implement by the Fire Authority. Secondly, fire officers have been well trained over the years to enforce these codes, and engineers and professionals are familiar with their requirements. Thirdly, they have been developed for many years and have evolved to accommodate for newer requirements, which may supersede existing ones. Nevertheless, prescriptive codes have become rather complex and are often becoming exceedingly difficult to employ for new technologies and for changing engineering practices. Some drawbacks of the prescriptive codes are:

- Design requirements are specified without any statement of objectives
- Cost-effective designs are not promoted
- Very little or no flexibility that can be exercised for innovative solutions and unusual situations
- There is only one design of providing the level of safety, which in itself is not stated

Copyright © 2009 by Academic Press. Inc. All rights of reproduction in any form reserved.

- They cannot be directly used for most of today's large structural buildings of complex architecture and construction; traditional regulations are no longer applicable to new building designs such as for high-rise construction, large compartmental volume, and excessive traveling distance

In recent years, building codes, regulations, and standards in a number of countries such as Japan, Canada, UK, Scandinavia, Australia, New Zealand, USA, and Hong Kong, China have been undergoing a steady transition from prescriptive-based to performance-based. The increasing tendency to embrace the performance approach is driven by the expected advantages that performance-based fire safety design can offer over the prescriptive-based design. These advantages are:

- The fire safety goals are clearly defined at the beginning, and the means of achieving the goals are left to the designer
- Innovative design solutions that can meet the established performance requirements are promoted
- International harmonization of regulation systems are allowed
- The premise to use new knowledge as it becomes available is permitted
- Cost-effectiveness and flexibility in designs are encouraged
- Prompt introduction of new technologies to the marketplace can be realized
- Complexity associated with existing prescriptive regulations can be eliminated

In relation to performance-based codes, the *fire safety engineering approach* is described in the next section. This particular approach is tantamount to the use of fire safety engineering design tools or performance-based methodologies to better assess the performance of any number of design alternatives against established safety levels and provide improved design alternatives in compliance with performance-based fire safety regulations.

6.1.2 Adopting Performance-Based Methodologies

By definition, fire safety engineering involves "the application of scientific and engineering principles, rules, and expert judgement, based on an understanding of the phenomena and effects of fire and of the reaction and behaviour of people to fire, to protect people, property, and the environment from the destructive effects of fire." The overall objectives of fire safety engineering are *life safety*, such as to ensure that the occupants leave the building without any exposure to hazardous or untenable conditions, as well as firefighters are able to effectively carry out rescue and prevent the extensive spread of fire and *property protection*. A fire safety engineering design should therefore provide a framework in demonstrating whether the performance requirements of legislation can be met, even though the design solutions adopted fall outside the prescriptive recommendations. Essentially, the fire safety engineering approach can be described by the flowchart shown in Figure 6.1.

For minor non-compliance with the prescriptive requirements, the like-to-like substitution or equivalence is adopted. Most prescriptive codes allow the provision of alternative designs as long as the safety levels they provide are

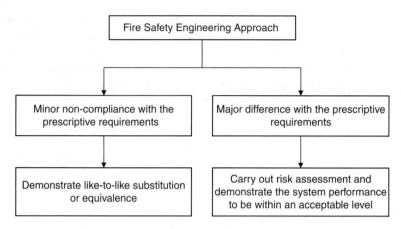

Figure 6.1 The fire safety engineering approach.

equivalent to or even better than that have been intended. Because of this equivalency consideration, a number of buildings are currently designed based on engineering calculations rather than following the prescriptive requirements. Consider the example of a semi-enclosed atrium, as depicted in Figure 6.2, to purposefully illustrate the like-to-like substitution approach. For this particular configuration, the placement of a steel roof is intended to provide a covering for the open atrium area in order to increase the utilization of the area at the bottom of the atrium irrespective of weather conditions. According to the prescriptive codes, the fire safety design of the building complies with the prescriptive requirement if the steel roof is omitted. Evaluations and assessments of fire safety are investigated for the fire and smoke spread with and without the steel roof. Deterministic calculations using the CFD-based fire model via the large eddy simulation FDS computer code, are performed on the domain of an approximate volume of $28,000 \text{ m}^3$ represented

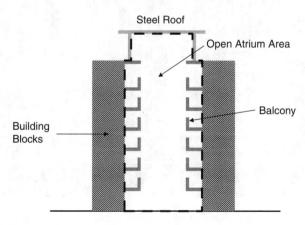

Figure 6.2 Schematic drawing of the semi-enclosed atrium with a steel roof.

by the dotted lines, as shown in Figure 6.2. The predicted numerical results of a sectional view of the time-averaged velocity and temperature distributions of centrally located fire are shown in Figures 6.3 and 6.4, respectively. On the basis of the field model predictions, the results of the calculations demonstrate that the smoke exhausting at the top of the open atrium area is not significantly impeded by the presence of the steel roof. Also, the temperature contours at the occupant's space are almost the same in both conditions. This

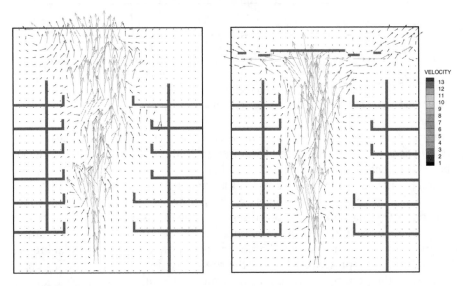

Figure 6.3 Time-averaged distribution of velocity vectors with and without a steel roof in the atrium.

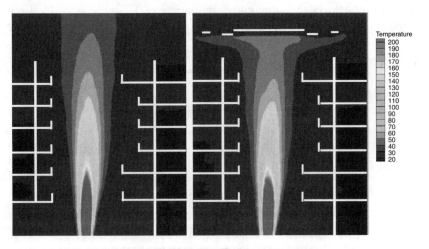

Figure 6.4 Time-averaged distribution of temperature contours with and without a steel roof in the atrium.

can be concluded that the steel roof does not significantly affect the spread of smoke and consequently the evacuation of occupants in the event of a fire within the open atrium area.

For major or significant differences with the prescriptive requirements, it is recommended that a more complete fire safety engineering approach be adopted. The use of performance-based methodologies based upon deterministic computations and stochastic or probabilistic analysis, is usually required to obtain a thorough fire safety evaluation and assessment of a building. In order to determine whether the fire safety level can be achieved, the performance criteria should ensure that the designs are verifiable by meeting the necessary performance-based requirements and enforceable by the code authority. Table 6.1 presents a summary of lower and upper limits for the deterministic criteria for pre-flashover fires (based on the data in Hadjisophocleous and Bénichou, 2000), which can be used for design considerations and in fire safety engineering design tools. Note that these limits should not be construed as exact values, but rather be regarded as industry-consensus agreed limits and should only serve as a guide to establish practical ways of quantifying the overall safety level in a building.

The quantification process of satisfying the performance requirements of the different components of the fire safety design can be achieved by:

- The use of fire safety engineering design tools to investigate a specific fire scenario in isolation, in order to analyze a particular aspect of the design
- Hazard analysis of multiple fire scenarios
- Fire risk analysis where the probabilities of the scenarios are added up proportionally to determine the most safe and cost-effective design

On the basis of the current state of available computing power, field modeling based on CFD techniques clearly represents the dominant method in tackling the first requirement of the quantification process. CFD-based fire models allow a realistic description for the interaction among the components of the fire safety system, which includes the fire outbreak, fire growth, and fire and smoke spread and the means of predicting the level of life safety for building design. Predicted outputs such as the concentrations of products, radiant heat fluxes, and hot gases temperatures determined through these models can be assessed against the limits in Table 6.1, to determine whether compliance with the performance criteria is met.

Considerable strides have nonetheless been achieved in the development of a fire engineering model through the concept of artificial neural network, which demonstrates the prospect of significantly shortening the design cycle process in arriving at optimal building design solutions for fire safety. It is recognized that field models, in general, still require extensive computational resources to provide useful engineering information, especially for the second requirement of the quantification process. The use of artificial neural network models along with field models presents enormous potential of carrying out a wider consideration of hazard analysis of multiple fire scenarios, in contrast to only a selected few through field modeling alone.

Table 6.1 Lower and upper limits of deterministic criteria for pre-flashover fires.

Stage	Suggested Deterministic Criteria	Lower Limit	Upper Limit
Pre-flashover (ignition and fire growth)	Radiant heat flux for ignition (kW m^{-2})	12	27
	• Pilot	–	28
	• Spontaneous		
	Surface temperature for ignition (°C)		
	• Pilot	270	350
	• Spontaneous	–	600
	Heat flux for ignitability (kW m^{-2})	10	40
	Maximum heat release rate (kW m^{-2})	250	600
Pre-flashover (life safety)	Convection heat (°C)	65	190
	Radiant heat (kW m^{-2})	2.5	2.5
	Oxygen (%)	10	15
	Carbon monoxide (ppm)	1400	1700
	Dioxide monoxide (%)	5	6
	Hydrogen cyanide (ppm)	–	80
	Upper gas layer temperature (°C)	183	200
	Visibility (m)		
	• Primary fire compartments	2	3
	• Other rooms	10	–
	Critical time to reach untenable limits (minutes)		
	• Unprotected zones	2	6
	• Partially protected zones	5	10
	• Protected zones	30	60

The basic fire safety design principle is that occupants are able to leave the building before the fire reaches a stage such that it is impossible to remain in it (i.e., untenable condition). Additional components of the fire safety system, which include the occupant response to fire and the fire service response to fire, require the consideration of evacuation modeling. In essence, the evaluation of the fire safety level of a design should depend upon the comparison between the *time available* for people to reach a place of safety—the Available Safe Egress Time (ASET)—and the *time required* by the people to reach the safe place—the Required Safe Egress Time (RSET). ASET represents the time required for the fire to develop from ignition to a condition causing the environment to be untenable. It is thus predicted that occupants inside or entering an enclosure are likely to save themselves due to the effects of exposure to smoke and heat. The value of ASET can be defined as the shortest time reaching either one of the performance criteria in Table 6.1. ASET is usually accomplished through the use of deterministic models. RSET is the time required for the occupants to travel to a place of safety. The value of RSET comprises of two major components, namely the total response time and travel time. The latter is determined through the application of suitable evacuation models in simulating the evacuation flow pattern of the occupants in the building. The conceptual framework of comparing ASET against RSET can be summarized as shown in Figure 6.5.

If ASET is significantly larger than RSET, occupants in the buildings are likely to leave or reach a safe place before the onset of untenable condition. The illustration of the ASET/RSET timeline is shown in Figure 6.6. ASET, as ascertained through deterministic models, is governed by the ignition and fire growth and the spread of fire and smoke. These greatly depend upon the fire load, the reaction to fire properties of the combustible lining materials and contents, the height and ventilation of the compartment, and the fire protection systems. In Figure 6.6, RSET can generally be evaluated based on the sum of four constituent times

$$\text{RSET} = \underbrace{t_d + t_a + t_p}_{\text{Total Response Time}} + \underbrace{t_m}_{\text{Travel Time}} \qquad (6.1.1)$$

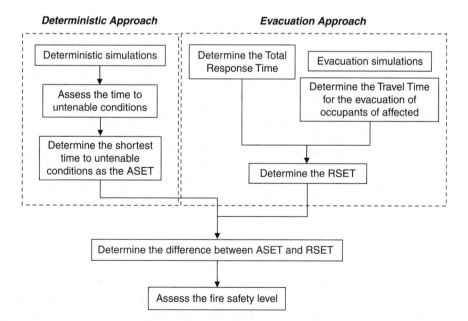

Figure 6.5 The conceptual framework comparing ASET and RSET.

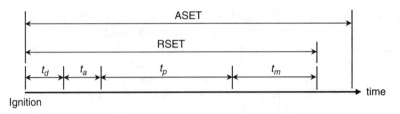

Figure 6.6   ASET/RSET timeline.

where the total response time consists mainly of the detection time (t_d), the alarm time (t_a), the occupant pre-movement time (t_p), and the occupant movement time (t_m), which is essentially the travel time. The detection time refers to the time between ignition of fire and detection, while the alarm time denotes the time between detection to warning occupants and evacuate. The occupant pre-movement time defines the time after an alarm or cue is evident, but before the occupants of a building begin to move toward the exits. In order to model a complete situation of the building, total evacuation of the whole building needs to be simulated. The travel time is the required egress time for all the occupants entering into the fire protected areas of the buildings or reaching an ultimate safe place under simultaneous evacuation strategy. It is also defined as the time when the last person entering into the fire protected area of the building or reaching an ultimate safe place. Evacuation models are applied to determine the travel time (t_m).

   Although the use of deterministic and evacuation calculations provide physical insights of what the conditions may be at a given time in compartment fires, it has limited ability in considering the entire building with its fire protection systems, functions, and occupants as a global system. A comparison of alternative designs using deterministic and evacuation approaches is limited only to specific elements. In contrast, probabilistic methods may be applied to provide the quantification of the whole fire safety level for the building (not element by element evaluation). Here, fire risk levels are estimated by using the likelihood of a fire incident occurring and the possible consequences (injury, death, and so forth). These calculated risk levels that are obtained through probabilistic risk assessment methods can be compared to the risk criteria (which can be established through statistical data) to determine whether the proposed designs meet the performance-based fire safety regulations. It is expected that the probabilistic approach, which is the third requirement of the quantification process, will be increasingly considered in performance-based design, as it not only quantifies the risk levels but also allows identification of designs that will have acceptable risk levels at minimum costs.

6.2   Overview of Emerging Technique in Field Modeling

The concept of artificial neural network can be best explained by none other than to mimic the operation of the human brain—the most complex biological machine. In the human brain, a brain cell is called a *neuron*. Our brain composes of a huge number of neurons being interconnected to each other to form a rather sophisticated network. Electrical signals are transmitted throughout the neural network to perform tasks associated with thinking, emotion, cognition, perception, and many other useful functions. In hindsight, mathematical models that propose to simulate the human brain system are called the *artificial neural network*. Specifically, artificial neural network models are designed to capture the system's behavior by learning its historical information of the system. They are therefore very reliant on the quality of the training data. Upon the completion of network training, they can be applied for carrying out tasks of predictions or classifications.

Better known by its acronym ANN, application of ANN to fire research and engineering is still relatively new. Pioneering works by Milke and McAvoy (1995) and Okayama (1991) have demonstrated the feasibility of applying ANN to the modeling of fire detector responses. Their models were based on the procedures of feed-forward multi-layer perceptron (MLP), recurrent networks for time series prediction, and self-organizing map (SOM) to predict the actuation time of sprinklers in compartment fires. On the basis of these encouraging studies, more sophisticated ANN models relevant to fire dynamics have been developed and extensively investigated by Lee et al. (2000, 2001, 2002, 2004a). The MLP and Fuzzy ARTMAP (FAM) were applied to predict the actuation time of sprinklers in compartment fires (Lee et al., 2000), while occurrence of flashover in compartment fires was successfully predicted by the Fuzzy ARTMAP (Lee et al., 2001). PEMap developed by Lee et al. (2002) was also used to predict the occurrence of flashover in compartment fires. The performance of the PEMap was shown to be comparable to that of the Fuzzy ARTMAP but with a simpler network structure and mechanism. A comprehensive review on the application of these ANN models can be found in the dissertation of Lee (2003). Another model, denoted as GRNNFA (Lee et al., 2004a), represents the very first ANN model that has been specifically designed for fire studies. The most important characteristic of the GRNNFA model is the removal of the noise embedded in the training samples. This particular feature is extremely suitable, since collected fire experiment data are normally noisy. The model has been critically validated by a series of benchmarking problems and proven to be superior to other well-known models that are designed for working in noisy environments (Lee et al., 2004b).

In essence, the technique based on the general regression neural network (GRNN) (Specht, 1991) requires no predefinition of the number of kernels of the network, since the network recruits all training samples during the course of training. Owing to the simple network structure, rapid network training, powerful in regression and ease of implementation, the GRNN model is widely established in a variety of fields that include image processing (Rzempoluck, 1997), nonlinear adaptive control (Schaffner and Schroder, 1993), and financial prediction (Leung et al., 2000). Specht (1991) applied a clustering technique to reduce the number of kernels to hold and process the information of all the training samples in the operation of the GRNN. As the preprocessor to the GRNN model, classification models such as Fuzzy k-mean clustering (Bezdek, 1980) and Kohonen Self-organization Map (Kohonen, 1990) have been applied. However, these models using Euclidean distance, as shown in Moore (1989), to measure the similarities between the input sample and the kernels for clustering have been perceived to yield unstable boundaries between kernels for the particular cluster region during network training. Nevertheless, development of Fuzzy ART (FA) by Carpenter et al. (1991), based on the adaptive resonance theory (ART) (Grossberg, 1976), has been demonstrated to be a more stable clustering tool in network training and more importantly, has the feature of growing autonomously with the network structure.

Detailed model formulations of the GRNN and FA networks have been highlighted elsewhere in Specht (1991), Tomandl and Schober (2001), and Carpenter et al. (1991) and will not be repeated here. In this section, we describe the formulation of the new and unique hybrid GRNNFA model developed for fire predictions. The enhancement introduced to this new model employs FA by clustering the training sample data to fewer numbers of prototypes, which are then converted to representative kernels for the nonlinear regression to be performed by the GRNN model. The architecture of the GRNNFA is detailed in Figure 6.7.

The GRNNFA model comprises of two modules: (i) FA is used for network training to create the prototypes according to the distribution of training data

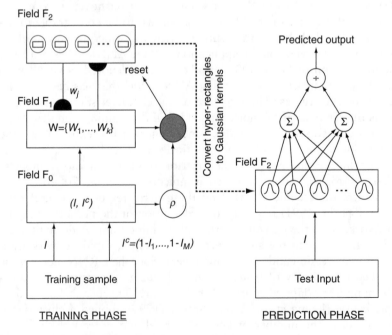

Figure 6.7 Architecture of the GRNNFA model.

in the input domain, and (ii) prediction is carried out via the GRNN. It is well perceived that the prototypes created by the FA cannot be employed directly as the kernels for the GRNN, since the prototypes are only represented by the vertices of hyper-rectangles. The compression scheme by which the prototypes are converted to a set of representative Gaussian kernels for the GRNN is developed.

Details of this compression scheme are subsequently described. Let $X_i \in \mathfrak{R}^m$ be a subspace of the input domain covering the prototype i created by the FA to which the samples $\{x_{i1}, x_{i2}, \ldots, x_{in}\} \in X_i$ are clustered. In the meantime, let $y = f(x)$ and $\tilde{y}_{ij}$ be respectively the underlying scalar function and the noise corrupted output corresponding to the input x_{ij} ($j = 1, 2, \ldots, n$). The corrupted output can be separated into clean and noisy components as

$$\tilde{y}_{ji} = y(x_{ij}) + \varepsilon(x_{ij}) \quad \forall x_{ij} \in X_i \tag{6.2.1}$$

where the variable ε is the symmetrically distributed noise with zero mean. By integrating equation (6.2.1) over the subspace X_i, the noise is removed giving

$$\int_{X_i} \tilde{y} \, dx = \int_{X_i} y \, dx \tag{6.2.2}$$

The preceding equation can be discretised and formulated according to

$$\sum_{j=1}^{n} \tilde{y}_{ij}' = \sum_{j=1}^{n} y_{ij} \tag{6.2.3}$$

From the preceding n refers to the number of samples clustered into prototype X_i. The centroid of the clean output μ_{yi} can be obtained by dividing the terms in equation (6.2.3) by n resulting in

$$\mu_{yi} = \frac{\sum_{j=1}^{n} y_{ij}}{n} = \frac{\sum_{j=1}^{n} \tilde{y}_{ij}}{n} \tag{6.2.4}$$

This implies that the centroid of the clean outputs over X_i can be obtained by taking the centroid of the noisy data points in X_i to the output domain. The value of μ_{yi} is taken as the representative output of domain X_i. Similarly, the representative input μ_i of X_i could be attained by taking the centroid of the input vectors of the noisy data points in X_i, henceforth

$$\mu_i = \frac{\sum_{j=1}^{n} x_{ij}}{n} \tag{6.2.5}$$

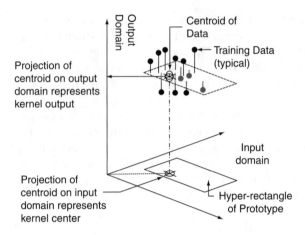

Figure 6.8   Compression scheme for noise removal.

Equations (6.2.4) and (6.2.5) compress all data points in $\mathbf{X}_i$ to $\boldsymbol{\mu}_i$ and μ_{yi}, which are later employed as kernel center and output, respectively in the GRNNFA model for prediction. Figure 6.8 illustrates the concept of the compression scheme. Each of these kernel centers within the physical domain represents only localized behavior based on the compression scheme performed through equations (6.2.4) and (6.2.5) on a window cluster of sample points within the whole physically meaningful domain. Information lost through the compression scheme can be recovered through the determination of the kernel widths for each of these window clusters. A straightforward representation can be attained by returning to the window clusters containing the compressed kernel centers originally evaluated. However, such simple inverse evaluation does not properly account for the entire physical domain response. It is therefore essential that the kernel widths be varied to match the actual responsive behavior of the training data. A novel approach is proposed by evaluating the kernel widths on a particular window clusters based on the influence from other window cluster to better capture or reflect the essence of the original representative behaviors of the sample data points within the physical domain. In this study, we apply the K-nearest-neighbor (K–NN) (Lim and Harrison, 1997) to assist the GRNN model to construct the regression surface by tuning the kernel widths, which is further described following.

In the traditional GRNN model, all the kernels are taken to be identical and hyper-spherical in shape. The error surface with respect to the global kernel width is simpler than that of multiple kernels with different radii. Since gradient descent, conjugate gradient, or similar iterative methods may easily be trapped in the local minima, it is proposed to determine all kernel widths indirectly. The proposal in this current model formulation is to adopt the K–NN approach as used in Lim and Harrison (1997). The width of a kernel j, σ_j, is taken as half of the average Euclidean distance over K nearest neighbors to kernel j yielding

$$\sigma_j = \frac{1}{2K}\sum_{k=1}^{K} \|x_j - x_k\| \quad j \neq k, \quad 1 \leq K \leq N - 1 \tag{6.2.6}$$

where x_j and x_k are position vectors of kernels j and k, respectively, and N is the total number of kernels. Instead of optimizing all kernel widths, only the value of K is determined. A heuristic procedure, based on gradient descent, is proposed. The predictive model is defined as

$$P = \Psi(T_{in} \mid K, \Omega) \tag{6.2.7}$$

where P, Ψ, T_{in}, and Ω are the predicted output row vector, prediction model, input vector, and network structure, respectively, while the prediction error is $e = (P - T_{out})(P - T_{out})^{\mathrm{T}}$, where T_{out} is the corresponding target output vector of T_{in}. Since FA is a stable network, the network structure Ω becomes increasingly stable during the course of network training. Equation (6.2.7) can therefore be approximated as

$$P = \Psi(T_{in} \mid K) \tag{6.2.8}$$

where e_i is the prediction error of the ith sample. For stable convergence, small values of α are usually employed. Since K is an integer, the smallest change of K is unity. Hence,

$$K_{i+1} = K_i - \Lambda\left(\frac{e_i - e_{i-1}}{K_i - K_{i-1}}\right) \quad \text{where} \quad \Lambda(a) = \begin{cases} 1 & \text{if } a > 0 \\ 0 & \text{if } a = 0 \\ -1 & \text{if } a < 0 \end{cases} \tag{6.2.11}$$

is proposed. The value of K is clipped during the course of network training by

$$K = \begin{cases} 1 & \text{if } K < 1 \\ K & \text{if } 1 \leq K \leq N - 1 \\ N - 1 & \text{if } K > N - 1 \end{cases} \tag{6.2.12}$$

where N is the total number of prototypes created by the FA. The overall training mechanism of the GRNNFA is shown in Figure 6.9.

The aforementioned formulation, however, only allows for a single scalar prediction of the fire scenario. For multi-dimensional predictions such as the

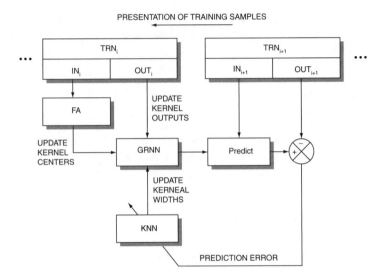

Figure 6.9 Overall training mechanism of the GRNNFA model.

three-dimensional flow field, the GRNNFA model can be modified according
to

$$
\tilde{y}(x) = \frac{\sum\limits_{j=1}^{s} \frac{A_j}{\sigma_j^m} exp - \left[-\frac{1}{2} \sum\limits_{p=1}^{m} \left(\frac{x_p - \mu_{jp}}{\sigma_j} \right)^2 \right]}{\sum\limits_{j=1}^{s} \frac{B_j}{\sigma_j^m} exp - \left[-\frac{1}{2} \sum\limits_{p=1}^{m} \left(\frac{x_p - \mu_{jp}}{\sigma_j} \right)^2 \right]} \tag{6.2.13}
$$

The values of A_j and B_j are incrementally updated from step t to step $t + 1$ by
the scheme

$$
\begin{aligned}
A_j^{(t+1)} &= A_j^{(t)} + b_j^{(t)} \\
B_j^{(t+1)} &= B_j^{(t)} + 1
\end{aligned} \tag{6.2.14}
$$

The format is very similar to the original GRNNFA model except that the
sample outputs $\{b_j\}_{j=1}^{s}$ and the predicted output $\hat{y}$ are vectors. Similar to the
original GRNNFA model, the value of $K \in I^+$ of the K-nearest-neighbor must
be updated adaptively. The stochastic updating formula of the value of K is

$$
K^{(t+1)} = K^{(t)} - \Lambda \left(\frac{\hat{e}^{(t)} - \hat{e}^{(t-1)}}{K^{(t)} - K^{(t-1)}} \right) \quad \text{where} \quad \Lambda(a) = \begin{cases} 1 & \text{if } a > 0 \\ 0 & \text{if } a = 0 \\ -1 & \text{if } a < 0 \end{cases}
$$

$$\tag{6.2.15}$$

Equation (6.1.15) is the same form as formulated for the original GRNNFA model except for the error term $\hat{e}$, which is defined as the sum-or-square of the normalized difference between the target value and the predicted output

$$\hat{e}^{(t)} = \sum_{q=1}^{n} \left(\frac{\tilde{y}_q^{(t)} - v_q^{(t)}}{max(v_q) - min(v_q)} \right)^2 \qquad (6.2.16)$$

6.3 Overview of Evacuation Modeling

Evacuation or egress models are specifically developed to predict the traveling time for occupants of a structure to evacuate. Focusing mainly on evacuation from buildings in this book, these models are increasingly becoming a part of performance-based analyses to assess the level of life safety provided in such structures. Fire safety design on the basis of performance criteria can lead not only to a more cost-effective fire protection design but also to a more rational and practical design. Engineers have been looking into the use of computer models for viable predictions on the escape pattern of the occupants in a building in order to better assist in the evaluation of the adequacy of the escape system. An evacuation model can thus be regarded as the *skeleton* for fire safety design in buildings.

It should be emphasized that the actual evacuation process of occupants in buildings is extremely complicated, since human behaviors are highly involved (e.g., wayfinding, psychological response, panic, and so forth). In modern buildings of complex architecture, it can be difficult to determine the escape pattern by the mere back-of-the-envelope (hand) calculations, of which equations given in the Emergency Movement Chapter of the Society of Fire Protection Engineers (SPFE) Handbook (2002) are adopted to calculate mass flow evacuation from any height of building. With the advent of digital computers, many computer-based evacuation models, each with unique characteristics and specialties, have been developed especially for the consideration of the evacuation of large group of occupants through unique geometries and varying fire scenarios.

Watts (1987) have performed an early review on evacuation analysis, which concentrated on the introduction of early network, queuing, and simulation models. The first survey of computer models for fire and smoke, which included egress models, was conducted by Friedman (1992). Gwynne et al. (1999a, 1999b) have by far provided the most substantial review to date on the methodologies used in evacuation modeling. In addition, Kuligowski and Peacock (2005) have provided an updated, unbiased, and more detailed review of the many evacuation models that are currently available. This report supplies useful information on newly developed evacuation models, a more detailed explanation of model features, the inner workings of each model, and each model's validation methods and limitations. They have managed to systematically categorize the many building evacuation models by their availability in practice. Additional information is further provided on the particular features of each model according to the *modeling method, purpose, model structure and perspective, methods for simulating movement and behavior, model output, use of fire data, use of visualization,* and *use of CAD drawings.* The authors strongly encourage the reader to use this comprehensive review as a practical guide toward the selection of appropriate model or models for his or her evacuation calculations.

A total of 30 computer models that focus on providing evacuation data from buildings have been reviewed by Kuligowski and Peacock (2005). In hindsight, models such as CRISP (Boyce et al., 1998, Fraser-Mitchell, 1996,), EvacSim (Poon, 1995), and EXITT (Levin, 1988a, 1988b) possess the features of simulating the movement of occupants by determining the walking speed from crowdedness and the walking direction from a pre-defined set of rules. Other models such as Magnetic Model (Okasaki and Matsushita, 1994) and Fluid Model (Takahashi et al., 1988) apply functional analogy behavioral approaches to simulate the behavior of occupants in buildings. These approaches are, however, untenable because the actual evacuation processes consist of the neglect of the conservation of momentum and the occupants are taken to stop and start at will such as highlighted by Still (2000). Furthermore, such models are incapable of simulating the herding behavior such as experienced in the actual evacuation process (Pan et al., 2005). Gwynne et al. (2001) incorporated a socio-psychological feature of evacuation into their computer model "buildingEXODUS" in order that the decision-making process of an occupant may be influenced by the actions of the surrounding population. The cellular automata evacuation model represents another computer model, which is able to qualitatively reproduce the characteristics of the cooperative and non-cooperative behaviors of the occupants during evacuation. Different cellular automata evacuation models such as PedGo (Klüpfel and Meyer-König, 2003), EGRESS (Ketchell et al., 1994, 1995), and SGEM (Lo and Fang, 2000, Lo et al., 2004) have been developed to simulate the movement of individuals in buildings. The behavior is governed by a set of pre-defined rules, which ignores the individual degree of panic of occupants.

In addition to the aforementioned evacuation models, it is worth mentioning the stochastic social-forces model proposed by Helbing et al. (2000), which described the escape panic of a large group of people through a model of pedestrian behavior to investigate the mechanisms of jamming by uncoordinated motion in crowds. Their computer simulations for the crowd dynamics of pedestrians have been based on a generalized force model developed in Helbing et al. (2000). By assuming a mixture of socio-psychological and physical forces influencing the behavior in a crowd, the collective phenomenon of escape panic was determined in the framework of self-driven many-particle systems. Some promising prospects of this model for evacuation calculations in buildings have been demonstrated. In order to employ the model for fire studies, the essential element of describing the coordinative behaviors of the occupants during evacuation needs to be incorporated into the model.

6.4 Overview of Probabilistic Approach

In ISO/CD (13388), the probabilistic approach has been adopted in the proposed international standard for fire safety engineering as a procedure for identifying fire scenarios for design purposes. For compartment fires, fuel characteristics and building environment are generally considered as the dominant factors affecting the fire spread and smoke movement. The purpose of a probabilistic approach is to evaluate the many associated fire risks within. For any modern building construction, the operational reliability of the fire protection system is an important and significant consideration. In carrying out any fire risk analysis, it is inadequate to assume that the system will always function without failure, as more often than not some components of the system may invariably malfunction in the event of a fire.

In assessing the operational reliability of the fire protection system, a fault tree that provides a logic graphical description of the possible probable fire scenarios can be used to determine the probability/frequency of fire occurrence. The fire consequence can be determined via an event tree. An event tree is a diagrammatic representation of different occurrences leading to fire via different events (e.g., failure of fire detection, failure of fire extinguishing system, failure of fire compartmentation system, and so forth). Depending on the initial conditions of the event, different outcomes can be realized. It is often necessary to examine a large number of scenarios with different chains of events. Each final event, outcome, or sub-scenario can be assigned a probability of occurrence as a consequence of the uncertainty in which event will actually occur. Franztich (1998) has demonstrated a simple example of an event tree for an installation of an automatic fire alarm system that will either operate or fail. The three important features that account for the operational reliability of the system are: *smoke alarm*, *sprinkler*, and *emergency door blocked*. This particular simple fire risk analysis resulted in a total of eight final events or outcomes denoted as sub-scenarios. In an extended consideration by Chu et al. (2007), the important features that account for the effect of operational reliability of the fire protection system on smoke movement are *sprinkler*, *automatic detention*, *manual detection*, and *mechanical smoke exhaust fan*. The more complicated event tree structure proposed by Chu et al. (2007), designed specifically for the evacuation of occupants in a supermarket, is illustrated in Figure 6.4. It has been assumed according to Chu et al. (2007) that if the sprinkler or smoke detector functions properly, the fire will be extinguished automatically. If neither the sprinkler nor the smoke detector sets off the alarm, it may be started manually. As seen in Figure 6.10, the event tree shows an aggregation of 10 sub-scenarios after the occurrence of fire.

The event tree structures the scenarios according to three questions:

- What can happen?
- What is the probability of each sub-scenario?
- What are the consequences of each sub-scenario?

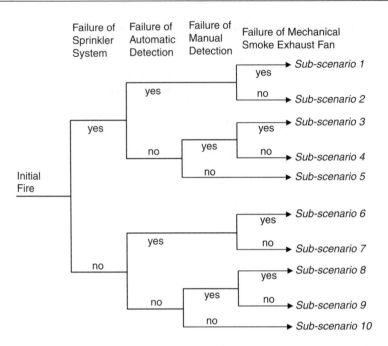

Figure 6.10 A sample event tree for the protection fire system.

Each sub-scenario is defined by its probability and its consequence. At each branching point, the possible outcome probabilities for a two-way branch can be described as $p_{failure}$ and $p_{success} = 1 - p_{failure}$. The probability of the final sub-scenario for each branch is merely the product of the branch probabilities leading to that sub-scenario. Note that the probability of the initial event (i.e., probability of fire occurrence) should also be included into the probability of the final sub-scenario. Usually, both the outcome probability of the sub-scenario and the description of the consequences are subject to some uncertainty. The expected value of the consequences of the sub-scenarios represents the consequence of the fire occurrence. This consequence, together with the frequency of fire occurrence obtained from the fault tree, determines the fire risk level of the study. Information concerning the state of knowledge of the variables must be included in both the probability sub-scenario and the consequence. The state of knowledge in the probability of each sub-scenario can be expressed by assuming a single value, or follows a probability density function such as suggested in Franztich (1998).

The failure of an event is an indispensable parameter in determining the branch probabilities. The fault tree represents an effective method in feasibly attaining the reliability of probability of failure of the fire protection system. Figure 6.11 presents an example of a fault tree for the sprinkler system. It begins with the top fault or event, which in this case is the sprinklers failing

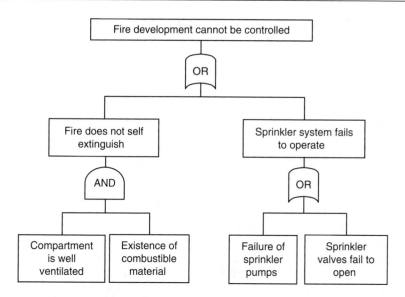

Figure 6.11 A sample fault tree for the control of fire development.

to extinguish the fire. They may never get activated, or they are activated but fail to operate. For the latter, it is connected to the top fault with an "OR" gate, because it only takes one or the other to cause the fault. A branch is introduced tracing the possible causes, which is either water does not reach the sprinklers and if the valves fail to open. For the former, the branch where automatic sensors fail and no manual activation is linked to an "AND" gate, because it undertakes both of them to cause the fault. The fault tree in Figure 6.5 should only be seen as means of illustrating the method, and is not complete. Other considerations such as water pump and piping could also be incorporated into the fault tree to ascertain the failure probability of the system. On the basis of the event tree in Figure 6.4 and the operational reliability of fire protection system determined either through statistical results or estimation, the occurrence of probability of every fire scenario can thus be calculated.

6.5 Case Studies

A selection of case studies is presented to demonstrate the combined application of CFD-based fire model with other fire modeling approaches in fire engineering. It is important to note that the results from these test cases should not be construed as to provide any endorsement or the acceptance of the computer model for this particular purpose. Likewise, the absence of a test case from any particular model does not provide a statement on the unsuitability of the model for this particular application.

6.5.1   The Predictive Capability of Artificial Neural Network Fire Model in a Single-Room Compartment Fire

The GRNNFA model is applied to predict the single scalar prediction of the location of thermal interface and the multi-dimensional prediction of the temperature and velocity profiles at the center of the doorway of a single compartment fire.

Numerical features: Experimental data from Steckler et al. (1984), which consists of a total of 55 experiments, are employed for training and evaluation of the GRNNFA. The controlled parameters and measured results are presented in Table 6.2. The dataset includes six controlled parameters: width and height of the sill of the opening, parallel and perpendicular distances from the center of the fire bed to the vertical centerline of the opening, fire strength, and ambient temperature, which are employed as the sample input data for the network. Only mean values of the measured heights of the thermal interface are selected to be the target values for network training, and the errors are only used to evaluate the performance of GRNNFA.

The well-known conventional approach, "Leave-One-Out Cross-Validation," which is frequently used for ANN model performance evaluation, is applied. This entails that out of the total number of 55 trials, for every trial, 54 trials are presented to the network for training to predict the "taken out" sample data. A total number of 20 sets of training and prediction are performed; each order of the training samples is randomly shuffled. Bootstrapping with 5000 re-samplings is applied to the 20 sets of prediction results to obtain the

Table 6.2 Subgrid model kernels for the dynamics procedure.

Controlled Variables	• Width of opening
	• Height of the sill of the opening
	• Fire Strength
	• Distance from the vertical centerline of the opening to the center of the fire bed (parallel to the opening)
	• Distance from the vertical centerline of the opening to the center of the fire load (perpendicular to the opening)
	• Ambient temperature
Measured Results	• Air mass flow rate
	• Neutral plane location
	• Thermal interface height
	• Average temperature of the upper gas layer
	• Average temperature of the lower air layer
	• Maximum mixing rate
	• Air velocity profile at opening
	• Temperature profile at opening

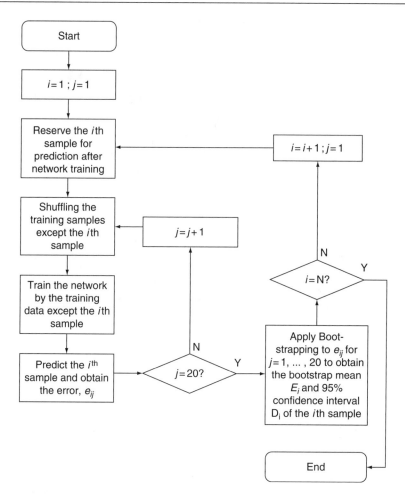

Figure 6.12 A sample fault tree for the sprinkler system.

bootstrap mean and the 95% confidence limits of that trial. Figure 6.12 shows the procedure of the leave-one-out validation with bootstrapping techniques.

The bootstrap means of the predicted outputs are plotted against the target values in Figure 6.13. Good agreement is achieved between the predicted outputs and targeted values. The correlation coefficient between the experimental and the predicted results by the GRNNFA yields a value of 0.929. The 95% bootstrap confidence intervals and the ranges of the target values are illustrated in Figure 6.14, where the prediction results have been arranged in ascending order with the target values. It is succinctly observed that, except the 3 samples indicated in Figure 6.14, the remaining 52 samples fall within the range envelope (i.e., mean ± error) of the target output values. We can conclude that the statistical percentage of correct prediction lies at 94.5%. The majority of fire

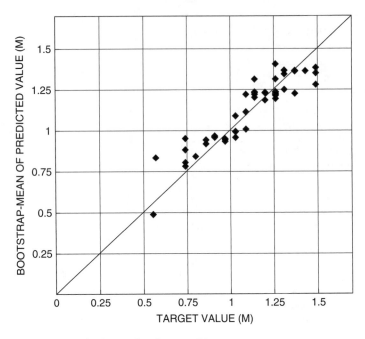

Figure 6.13   A sample fault tree for the sprinkler system.

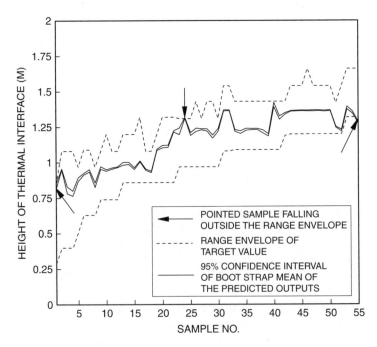

Figure 6.14   A sample fault tree for the sprinkler system.

measurements, including the experimental data of Steckler et al. (1984), have implicitly various degrees of embedded noise due to the fluctuations present in the fluid flow and heat transfer processes. The proposed GRNNFA model for fire predictions, being enhanced with the embedded effective noise removal feature, has responded positively toward the accurate predictions of the thermal interface. This is confirmed by the high confidence levels attained through the GRNNFA predictions despite the limited sample input data available for training purposes.

Numerical results: The performance of the GRNNFA model is evaluated against five different test cases, which are depicted in Figure 6.15. Predicted results of the GRNNFA model are compared against the results obtained from the CFD-based fire model, FDS. Table 6.3 presents the predicted results attained via the GRNNFA and FDS models. FDS has shown to consistently under-predict the experimental results of the thermal interface height. In order to compensate for the under-predicted values, an absolute difference, averaged to be approximately 0.187 m, is added as a correction to the FDS simulation

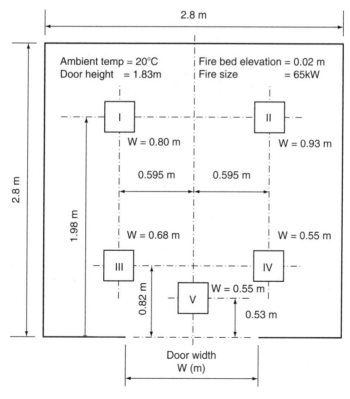

Figure 6.15 Configuration of cases of which the parameters are unseen from experiments.

Table 6.3 Prediction results by GRNNFA and FDS on the five unseen cases.

Test Case	GRNNFA (m)	FDS (m)	FDS (corrected) (m)	Absolute Difference (m)
I	1.207 (1.198,1.214)	0.917	1.104	0.103 < 0.17
II	1.225 (1.214,1.236)	1.028	1.215	0.010 < 0.17
III	0.954 (0.949,0.962)	0.850	1.037	0.083 < 0.17
IV	0.897 (0.886,0.912)	0.917	1.104	0.207 > 0.17
V	0.913 (0.904,0.923)	0.593	0.780	0.133 < 0.17

Note: The bracketed figures are the upper and lower limits of the 95% confidence interval of bootstrapping (Lee et al., 2004).

results. Except for case IV, the results have been shown to be well within the minimum error range of the experimental data (i.e., 0.17m). The significant difference between the results in case IV could be attributed to a number of factors.

ANN is recognized as a knowledge-driven prediction model. If the knowledge recruited from a domain is less than the others (i.e., less dense data distribution), the performance of this model in this domain is expected to be poorer than others. This is particularly applied in the knowledge distribution of the Steckler et al. (1984) experiment. The original data distribution is shown in Figure 6.16. The bracketed values shown in the figure indicate the number of samples in each fire location. It is seen that the majority of the samples are located at A, B, and C. Comparing to the five test cases, case I and II benefit from these samples. Although most of the samples are remote from this location, there are only two samples (i.e., locations F and G) for case V contributing to the knowledge of this fire location. For cases III and IV, they receive less benefit from the remote samples. However, there is a sample at location H that contributes to case III. Although this sample also contributes to case IV (mirror at the center of opening), the door width of this sample (i.e., 0.74 m) is closer to case III (i.e., 0.68 m) than case IV (i.e., 0.55 m). This explains the significant difference between the predictions by the GRNNFA and FDS in case IV. It is also revealed that the performance of the GRNNFA model can be further improved by introducing additional samples (obtained either by experiment or simulation) into the domain with less dense data distribution. Tests carried out against the five cases that are unseen during the experiments, demonstrate the excellent performance of the GRNNFA model. It has been shown to be capable of capturing the system behavior from the training of a limited number of noisy samples.

The multi-dimensional GRNNFA predictions of the temperature and velocity profiles at the center of the doorway of a single compartment fire are illustrated in Figures 6.17 and 6.18, respectively. Results predicted by the GRNNFA model are initially compared against the results simulated by the CFD model, FLOW3D (the earlier version of CFX) by Kerrison et al. (1994b) using the volumetric heat source approach. The profiles of the case of a centrally located

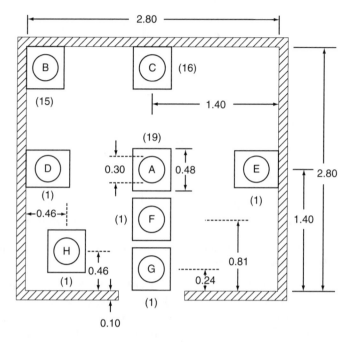

Figure 6.16 Location of fire samples in the Steckler et al. (1984) experiment with number of samples (bracketed) indicated.

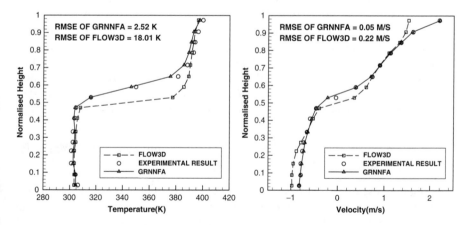

Figure 6.17 Location of fire samples in the Steckler et al. (1984) experiment with number of samples (bracketed) indicated.

fire with a door width of 0.74 m are shown in Figure 6.17. The Root Mean Square Errors (RMSEs) of the velocity and temperature profiles predicted by the GRNNFA model are shown to be much less than that of FLOW3D. Yeoh et al. (2003b) also simulated the velocity and temperature profiles of the same case, with the consideration of combustion and radiation models.

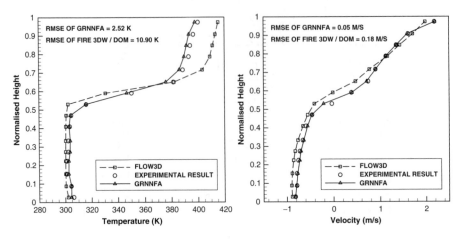

Figure 6.18   Location of fire samples in the Steckler et al. (1984) experiment with number of samples (bracketed) indicated.

The profiles are shown in Figure 6.18. RMSEs of the GRNNFA have been determined to be smaller than those of FIRE3D. The GRNNFA performs better because of additional cases with the same door width and same location of the fire-bed that have been used in aptly training the model, in order to provide sufficient knowledge of the system behavior to the GRNNFA model.

More extensive results on the predictive capability of the GRNNFA model for a range of other fire scenarios can be found in Lee et al. (2004a) and Yuen et al. (2006). It should be noted that the performance of the GRNNFA model may deteriorate in the local region at which the knowledge is sparsely distributed. This means that the training samples in that local region may not be sufficient to describe the general behavior of the system. As the *sufficiency* of the training samples is critical to the success of the model application, a novel network training scheme that employs the knowledge of human experts to supplement that of the limited training samples needs to be appropriately developed.

6.5.2   The Application of CFD-Based Fire Model and Evacuation Model for Fire Safety Evaluation and Assessment

The possible use of a CFD-based fire model, coupled with an evacuation model to assess the progress of smoke and the hazard it may represent to the occupants in a multi-story building, is demonstrated in this case study.

Numerical features: For the field modeling calculations, the FDS computer code is adopted to predict the transient development of the extent of spread of the smoke layer and combustion products. Relevant numerical models for the current study are identical to those already described in the worked example of section 5.5.2. The schematic representations of the computational geometry of a generic multi-storey building in isometric and plan views are shown in Figures 6.19 and 6.20, respectively. On the basis of the mesh criterion in equation (5.5.1) predetermined

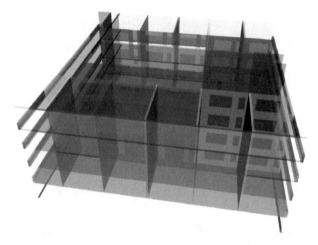

Figure 6.19 Isometric view of the multi-story building.

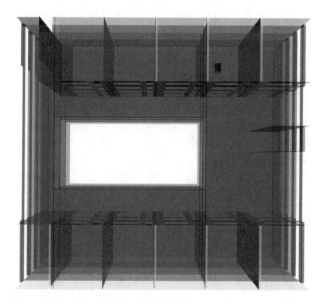

Figure 6.20 Plan view of the multi-story building.

in section 5.5.1, a total mesh of $143 \times 139 \times 67$ is constructed. The solution is marched in time on a three-dimensional rectilinear grid until it reaches 300 seconds in order to allow sufficient time for the fire to achieve untenable conditions.

In general, the possible fire sizes that can be applied in practice are related to the usage of the zone and fire bed location. A summary of possible combustibles obtained from the SFPE handbook (1996) is provided in Table 6.4. In the current study, the design fire has been assumed to have a maximum

Table 6.4 Heat release rates for different materials extract.

Possible combustibles	Maximum Possible Heat Release Rates
Mail bags (1.5 m² surface area)	0.9 MW
Trash bags	0.3 MW
Christmas tree (7.4 kg)	0.5 MW
Wardrobe with 1.93 kg of clothing and paper	3.5 MW
Idle pallet fires (Six stacked of mixed wooden pallets of about 167 kg up to 1325 mm high were used for the test.)	2.4 MW

possible heat release rate $\dot{Q}$ of 3.5 MW, which represents a wardrobe with 1.93 kg of clothing and paper. A fast t-square growth of the fire is considered:

$$\dot{Q} = 4.69 \times 10^{-5} \, t^2 \tag{6.5.1}$$

From the preceding equation, the energy release rate from the fire bed initially reaches the heat release rate of 3.5 MW at about 273 seconds after ignition. As a conservative consideration, the decay of the design fire is not considered. The numerical predictions are assessed against the deterministic criteria described in Table 6.5. It should be emphasized that the Available Safe Egress Time (ASET) for the occupants to reach a safe place has been chosen in this case study to correspond conservatively to the time when the descending smoke layer enters the staircase.

For the evacuation calculations, the spatial-grid evacuation model (SGEM) developed by Lo and his co-workers is employed. The basic theory of the model is described in Lo and Fang (2000), while enhancements of the model in making use of CAD plans for building up evacuation networks is presented in Lo et al. (2004). In brief, the model resolves the setting of a building into a

Table 6.5 Time to untenable conditions.

Untenable conditions	Time to untenable conditions (s)
Descending of smoke layer entering the staircase	166
Exposure to radiant heat intensity of 2.5 kW/m² at head level (i.e., 2.0 m above finished floor level)	>300
Exposure to convective heat of 115°C at head level (i.e., 2.0 m above finished floor level)	>300
Available Safety Egress Time (ASET)	166

network with nodes representing zones that may represent rooms, corridors, or halls. These nodes are connected to their neighboring nodes by way of openings such as doors, exits, staircases, and so forth. The possible escape direction of each zone can be found by analyzing the function of each zone, and the geometrical location as well as way-finding tendency of the evacuees. Movement of the evacuees is solved by a series of difference equations within a finite grid of cells that is generated within a zone. The movement trajectory of each individual is recorded, and the evacuation patterns of the evacuees upon reaching the final exit points can be determined by various environmental stimuli, such as the distance to the exit, the presence of exit signs, visual accessibility, and personal characteristics such as familiarity of routes, and so forth. More details on the model's specific features can be found in the aforementioned articles. As will be illustrated in the predicted results following, the movement of the people is modeled by considering the density and the position of the evacuees and the position of other people around them.

Numerical results: The numerical calculations are performed in two different stages. Firstly, the FDS computer code is solved to initially provide the time-dependant results of the fire growth and the migration of smoke within the confined space. The ASET requirement is subsequently evaluated. From the CFD simulations, predicted results on the development of the hot smoke layer, temperature at the level of 2.0 m from above finished floor level, and radiant heat flux at also the same level of 2.0 m from above finished floor level are extracted, which are then employed as inputs for the SGEM, the second stage, to determine the shortest and nearest evacuation path of the occupants, according to the particular architectural layout and the tenability limit of each different zone (i.e., occupants will not evacuate across any region where untenable conditions prevail).

Figure 6.21 illustrates a series of snapshots representing the time-dependant development of the instantaneous surface contours of the hot gas layer of 60 °C within the confined space at the same floor level of which the fire source is located. Note that it has been ascertained in Yuen et al. (1999) that this particular criterion could be feasibly employed to identify the onset of hazardous conditions due to smoke filling. As seen from the CFD predicted results, the hot gas layer leaves the burn room and eventually fills the adjacent corridor and the right exit at about 210 seconds. The left exit, however, remains smoke free, even for the entire duration of 300 seconds.

For the evacuation calculations, the SGEM is applied to determine the traveling time of the occupants. In order to determine Required Safe Egress Time (RSET), which is equivalent to the total response time and traveling time, it shall be assumed that the total response time is quantified solely by the detection time. On the basis of the following requirements: (i) height measured from the fire to ceiling – 2.1 m, (ii) maximum distance between the fire bed and sprinkler – 2.5 m, (iii) sprinkler actuation temperature – 68°C, (iv) sprinkler response time index (RTI) – 135 $(ms)^{0.5}$, (v) fast growth t-square fire, and (vi) ambient temperature – 25°C, the detection time according to Vettori and

Madrzykowski (2000) is analytically determined to be 139 seconds. Since the spread of fire and smoke is taken as not to be affected by the evacuation of the occupants, the pairing between the CFD-based fire and evacuation models is effectively a one-way coupling process. For the purpose of illustration, it shall be assumed that the occupants and the location of the fire are at the same floor level with the two possible exit points.

Figure 6.22 depicts the migration patterns of the occupants at different traveling times, from the moment they are called upon or alerted to evacuate. At the traveling time of 30 seconds, which corresponds to a RSET of 169 seconds, the occupants have reached the required exit points and are steadily moving down the staircases of the left and right exits. Nevertheless, the predicted hot gas layer at 300 seconds as shown in Figure 6.21 covers a substantial area below the ceiling adjacent to the right exit. This therefore poses a

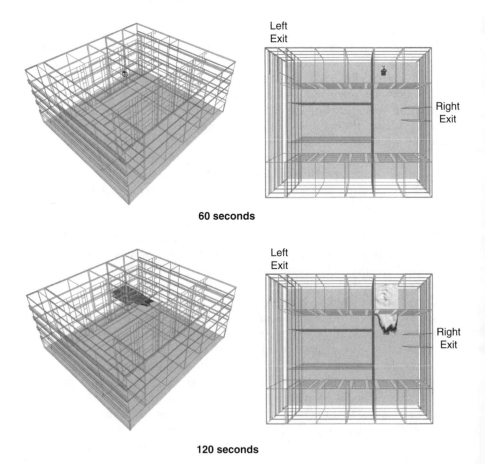

Figure 6.21 Surface contour plots representing the transient development of the hot gas layer of 60 °C seen from the isometric (left) and plan (right) views.

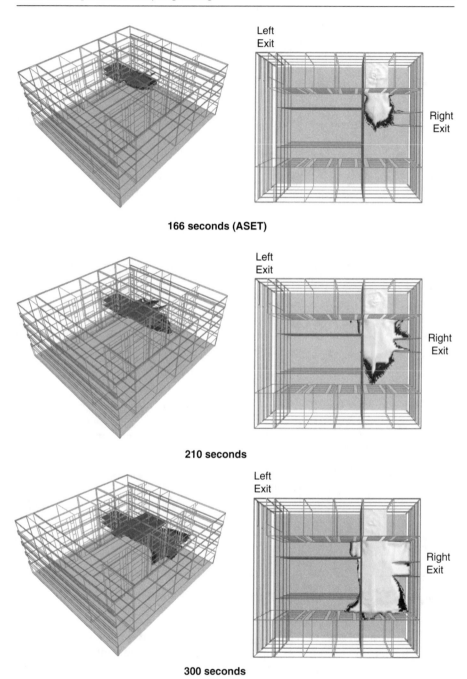

Figure 6.21 Cont'd.

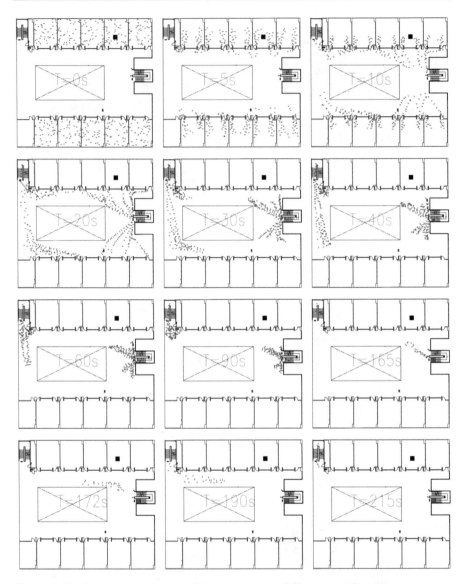

Figure 6.22 Evacuation patterns of the evacuees at different traveling times.

significant threat for the occupants, who will continue to persist to travel down the particular escape route as the smoke begins to descend into the staircase. When the traveling time reaches 165 seconds and beyond, which corresponds to a RSET greater than 304 seconds, the SGEM appropriately demonstrates the likelihood of occupants, sensing the possible prevailing untenable condition, seeking an alternate route of escape by traveling across the floor level to the left exit.

6.6 Future Developments in Fire Predictive and Assessment Models

With the increasing complexity of modern buildings, the need to adopt the high precision and growing validation of the deterministic CFD-based fire models is growing, especially for performance-based analyses of assessing numerous fire safety aspects related to intricate structural designs. General-purpose CFD commercial software packages that can be used for fire modeling applications and specific field modeling computer codes that are intended only for modeling fires have the propensity nowadays of simulating rather complex fire scenarios. In addition to the basic transport equations for mass, momentum, and energy, along with appropriate turbulent models that can be readily applied to resolve the turbulent fluid flow, physical characteristics such as combustion, non-luminous and luminous radiation, soot production, and even solid pyrolysis are progressively considered as essential requirements to be included in many field modeling investigations of compartment fires.

Nevertheless, advancements to the models are still required, especially to better resolve complicated flaming conditions in practical fire scenarios. Current combustion models that have been applied to solve a whole range of fire problems, need to be further improved beyond the fast chemistry assumption. Development of combustion models able to accommodate a wide range of chemical and turbulent time scales, will certainly assist in the predictive capability of the field model in simulating the growth and spread of fires of not only in a well-ventilated environment but also in an under-ventilated environment such as the depletion of oxygen supply in a room with the door shut. The latter aspect has serious implication toward the possible dire consequence of a backdraft fire incident, usually persisting only few seconds before exhausting its fuel supply. Most commonly applied soot models in field modeling can still be regarded as very empirical in nature. The spread of smoke in a confined enclosure poses enormous threat to occupants who may be exposed to the hot sooty gases during evacuation. Complex and detailed attempts, especially through the population balance approach to characterize the soot process via detailed models that seek to solve the rate of equations for elementary reactions, leading to soot and to predict the evolution of the size distribution of the soot particles generated by chemical reaction and/or undergoing chemical and physical processes, could provide better prediction of the soot concentration levels and thus a more realistic representation of the smoke barrier within the confined area. The development of soot pyrolysis models to cater for a wider range of condensed fuels, in addition to the model for simple types of solid fuel such as cellulosic fuels and wood products already considered in this book, is necessary to predict the fire spread in a more fundamental way in order to characterize actual flaming behaviors in real fires.

For the consideration of evacuation, the coupling between the CFD-based fire models and evacuation models remains a very challenging prospect. Both models, in general, still require extensive computational time and resources. For example, typical turnaround times for large eddy simulations of fires in complex geometries can still amount to hundreds of hours. For a practical fire engineering approach, it is recommended that ANN models should be included as a part of the whole design process, by speeding up the procedure in evaluating the appropriate fire safety design of a building. Consider the structure of an integrated platform of an optimization procedure for evaluating the range of feasible design parameters, as depicted in Figure 6.23. It is observed that ANN models, which can be trained to capture the time-dependent smoke development behavior, have the capacity of evaluating the performance of each individual fire safety design, based on the smoke propagation speed with minimal computational cost. An automatic optimization procedure can be later applied to identify targeted designs having the most likelihood to be the optimal solution. These targeted designs can be assessed via the CFD-based fire models, of which the corresponding ASET and RSET values can be determined. Design options with the largest fire safety margin are ascertained and returned to the designers as the final optimal design. As illustrated in Figure 6.24, the smoke barrier predicted by the CFD-based fire models will allow the smoke propagation history to be accounted in a more direct fashion into the evacuation calculations, such as the response of occupants adopting an alternative route under the presence of smoke. This effective implementation will not only provide a more complete representation of the behavioral reactions of occupants confronted with a smoke barrier but will also give more creditability to the evacuation simulations with a more realistic travel time of evacuation.

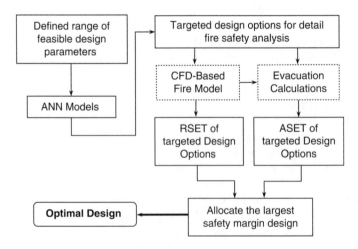

Figure 6.23 Integrated framework of a fire safety design optimization procedure.

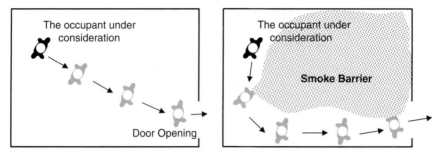

Figure 6.24 Occupant's tendency toward adhering close to the walls in order to obtain guidance during evacuation when the room is engulfed by smoke.

As aforementioned, more complex models of combustion, soot, and solid pyrolysis in field modeling will eventually provide the means of better predicting not only the growth and spread of fires in the pre-flashover stages, but also fully developed fires in the post-flashover stage. The latter fires are generally relevant to structural behavior, which could lead to the development of more sophisticated fire models. The effective coupling between the fire and its impact on structural components is still very much in its infancy stages. Thus far, the fluid-structure interaction has been achieved by either replacing the gas phase simulation by a gas temperature history by the standard temperature-time curve and studying the structural response in detail, or using the CFD-based fire models for the gas phase, calculating heat transfer into the structure but making simplified estimates of the structural response. The feasible coupling between the CFD-based fire models and structural analysis models under these conditions, needs to be thoroughly assessed and validated against appropriate physical experiments. Before a seamless coupling can be realized, significant challenges await toward the effective coupling between the two models through suitable interface methodologies that are still yet to be developed.

6.7 Summary

In this chapter, the required performance-based methodologies that contribute to the framework of the fire safety engineering approach, can be summarized according to the illustration in Figure 6.25. It is to be expected that CFD-based fire models will feature more dominantly in ascertaining the necessary performance requirements of the fire safety aspect in building designs, as faster and cheaper computers become more prevalent in the not too distant future. Also, the high precision and growing validation of the CFD-based fire models have the affinity of providing useful information for performance assessment against

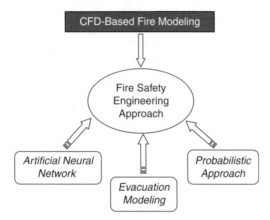

Figure 6.25 Performance-based methodologies for the fire safety engineering approach.

accepted performance criteria, and the increasing realism of predicting the fire growth and smoke spread. Nevertheless, such deterministic models are generally restricted to only carrying out performance analyses on specific elements of the fire safety designs. The use of ANN models will allow the scope of analyzing numerous fire scenarios to be achieved, thereby shortening the design evaluation process for the specific elements. To account for the whole fire safety level in the building and the emergence of more novel and innovative designs, it is imperative that both evacuation modeling and probabilistic methods are supplemented into the framework of the fire safety engineering approach in order to better evaluate the risk of life and property in the building and the fire protection systems.

Review Questions

6.1. What are some drawbacks of the use of prescriptive codes in evaluating the fire protection systems?

6.2. What is the performance-based approach? How is it different from the fire safety engineering approach?

6.3. Define and describe the fire safety engineering approach.

6.4. In a complete fire safety engineering approach, what are the methodologies available to satisfy the performance-based requirements?

6.5. The quantification process of satisfying the performance requirements of the fire safety design consists of three components. What are they?

6.6. What is the basic fire safety design principle?

6.7. Define the Available Safe Egress Time (ASET) and Required Safe Egress Time (RSET). What is the difference between these times?

6.8. How can ASET and RSET be accomplished within the conceptual framework of the fire safety engineering approach?

6.9. What is artificial neural network (ANN)? How can ANN be appropriately applied in fire engineering?

6.10. What are evacuation models? Describe some features of the models in determining the escape pattern of occupants.

6.11. In the probabilistic approach, what are event and fault trees? How are they applied in the probability analyses?

6.12. How can CFD-based fire modeling contribute to the framework of fire safety engineering approach?

6.13. How can ANN, evacuation modeling, and the probabilistic approach be integrated in the framework of the fire safety engineering approach?

Appendix A   Higher-Order Differencing Schemes and Time-Marching Methods

A.1   Higher-Order Differencing Schemes

The formulation of the second order upwind and third order QUICK schemes is described following. Improvements to the first order upwind scheme can be enhanced via the calculations of the interface properties at cell faces of w and e by the consideration of additional field variables located at the neighboring grid nodal points indicated by the properties at points WW and EE, as shown Figure A.1. Evaluations of interface values along coordinate directions of y and z may also be carried out in a similar manner.

For the second order upwind scheme (see illustration in Figure A.2), assuming uniform distribution of the grid nodal points, additional information of the fluid flow is introduced into the approximation by considering an extra upstream variable point. In other words,

$$\phi_w = \frac{3}{2}\phi_W - \frac{1}{2}\phi_{WW}$$
$$\phi_e = \frac{3}{2}\phi_P - \frac{1}{2}\phi_W \qquad \text{if} \quad u_w > 0 \quad \text{and} \quad u_e > 0 \qquad \text{(A.1.1)}$$

$$\phi_w = \frac{3}{2}\phi_P - \frac{1}{2}\phi_E$$
$$\phi_e = \frac{3}{2}\phi_E - \frac{1}{2}\phi_{EE} \qquad \text{if} \quad uw < 0 \quad \text{and} \quad ue < 0 \qquad \text{(A.1.2)}$$

For the third order QUICK scheme (see illustration in Figure A.3), a quadratic approximation is introduced across two variable points at the upstream and one at the downstream, depending on the flow direction. The unequal weighting influence of this particular scheme still hinges on the knowledge biased toward the upstream flow information. The interface values ϕ_w and ϕ_e based on a uniform grid nodal point distribution can be determined as

Computational Fluid Dynamics in Fire Engineering
Copyright © 2009 by Academic Press. Inc. All rights of reproduction in any form reserved.

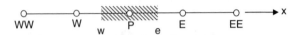

Figure A.1 A schematic representation of a control volume around a node P with surrounding grid nodal points of WW, W, E, and EE along the x direction.

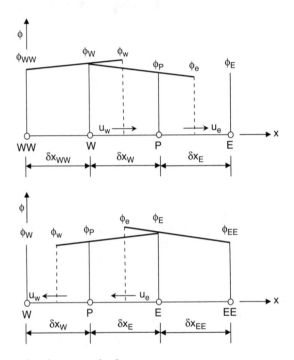

Figure A.2 Second order upwind scheme.

$$\phi_w = -\frac{1}{8}\phi_{WW} + \frac{6}{8}\phi_W + \frac{3}{8}\phi_P$$

$$\phi_e = -\frac{1}{8}\phi_W - \frac{6}{8}\phi_P + \frac{3}{8}\phi_E$$

if $u_w > 0$ and $u_e > 0$ (A.1.3)

$$\phi_w = -\frac{1}{8}\phi_E + \frac{6}{8}\phi_P + \frac{3}{8}\phi_W$$

$$\phi_e = -\frac{1}{8}\phi_{EE} + \frac{6}{8}\phi_E + \frac{3}{8}\phi_P$$

if $uw < 0$ and $ue < 0$ (A.1.4)

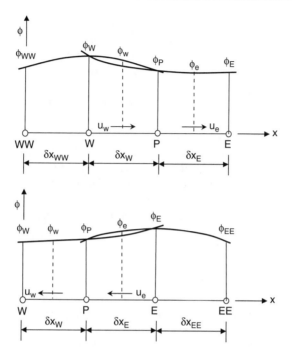

Figure A.3 Third order QUICK scheme.

A.2 Total Variable Diminishing (TVD) Schemes

The difficulties associated with the development of reliable higher-order schemes stem from the conflicting requirements to satisfy the properties of *accuracy, stability,* and *boundedness*. Solutions that are predicted through the higher-order schemes as described in the previous section are generally more accurate than the first order upwind scheme and more stable than the second order central differencing scheme. Nonetheless, they have a tendency to provoke unwanted oscillations when the local Peclet number is high and the gradients of the flow properties are steep. In order to suppress these oscillations, the composite flux limiter approach has proven to be rather effective without significantly affecting the accuracy of the predicted solutions. Through this approach, the numerical flux at the cell interface is now modified by employing a flux limiter that enforces the *boundedness* criterion. The schemes based on TVD flux limiters are typical examples.

In this section, the normalized variable and space formulation (NVSF) methodology proposed by Darwish and Moukalled (1994), which is an extension of the normalized variable formulation (NVF) by Leonard (1988), is adopted as a framework for the development of the high-resolution schemes. This particular methodology allows the composite high-resolution schemes to be applied for flow problems involving non-uniform, distorted, or non-Cartesian grids. For

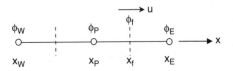

Figure A.4 Interpolation grid nodal points used in evaluating the interface property ϕ_f.

the sake of simplicity, consider the requirement to evaluate the property value at the control volume face f, as shown in Figure A.4 along the coordinate x direction. The upstream, central, interface, and downstream dependent variables of ϕ_W, ϕ_P, ϕ_f, and ϕ_E are taken to be located at distances x_W, x_P, x_f, and x_E from the origin, respectively. Since a normalized variable and space formulation is sought, the following normalized variables can be defined as

$$\tilde{\phi} = \frac{\phi - \phi_W}{\phi_E - \phi_W} \quad \tilde{x} = \frac{x - x_W}{x_E - x_W} \tag{A.2.1}$$

From the preceding definitions, the normalized parameters simplify the functional representation of the simple and composite high-resolution schemes and greatly assist in defining the *stability* and *boundedness* conditions. Upon normalizing, the value for interface $\tilde{\phi}_f$ is simply a function of

$$\tilde{\phi}_f = f(\tilde{\phi}_P, \tilde{x}_P, \tilde{x}_f) \tag{A.2.2}$$

In order to satisfy the *boundedness* property, the convection *boundedness* criterion formulated by Gaskell and Lau (1988) is applied herein. This criterion states that its functional relationship $f(\tilde{\phi}_P)$ should (i) be continuous and bounded from below by $\tilde{\phi}_f = \tilde{\phi}_P$ and from above by unity, (ii) pass through the points (0,0) and (1,1) in the monotonic range $0 < \tilde{\phi}_P < 1$, and (iii) equal $\tilde{\phi}_P$ for $\tilde{\phi}_P < 0$ and $\tilde{\phi}_P > 1$. These conditions can be expressed mathematically as

$$\begin{cases} f(\tilde{\phi}_P) & \text{is continuous} \\ f(\tilde{\phi}_P) = 0 & \text{for } \tilde{\phi}_P = 0 \\ f(\tilde{\phi}_P) = 1 & \text{for } \tilde{\phi}_P = 1 \\ f(\tilde{\phi}_P) < 1 \text{ and } f(\tilde{\phi}_P) > \tilde{\phi}_P & \text{for } 0 < \tilde{\phi}_P < 1 \\ f(\tilde{\phi}_P) = \tilde{\phi}_P & \text{for } \tilde{\phi}_P < 0 \text{ and } \tilde{\phi}_P > 1 \end{cases} \tag{A.2.3}$$

The preceding conditions may be described graphically on a normalized variable diagram (NVD), as shown in Figure A.5. From this figure, it is clearly demonstrated that the only scheme satisfying the *boundedness* criterion is the first order upwind scheme—that is, $\tilde{\phi}_f = \tilde{\phi}_P$. Other schemes such as the second order upwind, central difference, and QUICK schemes may in general yield

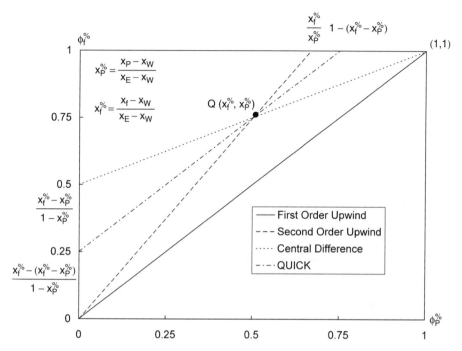

Figure A.5 NVD for the first order upwind, second order upwind, central difference, and QUICK schemes using the NVSF.

physically unrealistic results. It should be noted that an NVD relationship nearer to the first order upwind NVD tends to be highly diffusive, while schemes whose NVD relationships are closer to the first order downwind NVD as represented by the line $\tilde{\phi}_f = 1$ are highly compressive.

Concerning the property of *accuracy*, Leonard (1988) stipulated that the necessary and sufficient condition for a scheme to be second order is for its functional relationship to pass through the point Q $(0.5, 0.75)$. In addition, if it passes through Q with a slope of 0.75, the scheme is third order. For the case of NVSF, these conditions can be derived by noting that all second and third order schemes may be represented according to the functional relationship

$$\tilde{\phi}_f = \tilde{x}_f + M(\tilde{\phi}_P - \tilde{x}_P) \tag{A.2.4}$$

where M is the slope of the function. From Figure A.5,

Second order upwind scheme: $M = \dfrac{\tilde{x}_f}{\tilde{x}_P}$

Central difference scheme: $M = \dfrac{\tilde{x}_f - 1}{\tilde{x}_P - 1}$

$$QUICK \ scheme: \ M = \frac{\tilde{x}_f(\tilde{x}_f - 1)}{\tilde{x}_P(\tilde{x}_P - 1)}$$

Following the formulation carried out for the preceding common schemes, the NVSF of composite high-resolution schemes on a non-uniform grid may also be similarly derived. Some popular TVD schemes, such as the MUSCL scheme of Van Leer (1977b), the MINMOD scheme of Harten (1983), the OSHER scheme of Chakravarthy and Osher (1983), and the SMART scheme of Gaskell and Lau (1988), are described following. Only the final forms of their functional relationships are provided. More details on their derivations can be found in the article by Darwish and Moukalled (1994).

MUSCL:

$$\tilde{\phi}_f = \frac{2\tilde{x}_f - \tilde{x}_P}{\tilde{x}_P}\tilde{\phi}_P \quad 0 < \tilde{\phi}_P < \frac{\tilde{x}_P}{2\tilde{x}_f}$$

$$\tilde{\phi}_f = \tilde{x}_f - \tilde{x}_P + \tilde{\phi}_P \quad \frac{\tilde{x}_P}{2\tilde{x}_f} < \tilde{\phi}_P < 1 + \tilde{x}_P - \tilde{x}_f$$

$$\tilde{\phi}_f = 1 \quad\quad\quad\quad 1 + \tilde{x}_P - \tilde{x}_f < \tilde{\phi}_P < 1$$

$$\tilde{\phi}_f = \tilde{\phi}_P \quad\quad\quad \text{elsewhere}$$

MINMOD:

$$\tilde{\phi}_f = \frac{\tilde{x}_f}{\tilde{x}_P}\tilde{\phi}_P \quad\quad\quad\quad\quad 0 < \tilde{\phi}_P < \tilde{x}_P$$

$$\tilde{\phi}_f = \frac{\tilde{x}_f - \tilde{x}_P}{1 - \tilde{x}_P} + \frac{(\tilde{x}_f - 1)}{(\tilde{x}_P - 1)}\tilde{\phi}_P \quad \tilde{x}_P < \tilde{\phi}_P < 1$$

$$\tilde{\phi}_f = \tilde{\phi}_P \quad\quad\quad\quad\quad\quad\quad \text{elsewhere}$$

OSHER:

$$\tilde{\phi}_f = \frac{\tilde{x}_f}{\tilde{x}_P}\tilde{\phi}_P \quad 0 < \tilde{\phi}_P < \frac{\tilde{x}_P}{\tilde{x}_f}$$

$$\tilde{\phi}_f = 1 \quad\quad \frac{\tilde{x}_P}{\tilde{x}_f} < \tilde{\phi}_P < 1$$

$$\tilde{\phi}_f = \tilde{\phi}_P \quad\quad \text{elsewhere}$$

SMART:

$$\tilde{\phi}_f = \frac{\tilde{x}_f(1 - 3\tilde{x}_P + 2\tilde{x}_f)}{\tilde{x}_P(\tilde{x}_P - 1)} \tilde{\phi}_P \qquad 0 < \tilde{\phi}_P < \frac{\tilde{x}_P}{3}$$

$$\tilde{\phi}_f = \frac{\tilde{x}_f(\tilde{x}_f - \tilde{x}_P)}{1 - \tilde{x}_P} + \frac{\tilde{x}_f(\tilde{x}_f - 1)}{\tilde{x}_P(\tilde{x}_P - 1)} \tilde{\phi}_P \qquad \frac{\tilde{x}_P}{3} < \tilde{\phi}_P < \frac{\tilde{x}_P}{3}(1 + \tilde{x}_f - \tilde{x}_P)$$

$$\tilde{\phi}_f = 1 \qquad\qquad\qquad 1 + \tilde{x}_P - \tilde{x}_f < \tilde{\phi}_P < 1$$

$$\tilde{\phi}_f = \tilde{\phi}_P \qquad\qquad\qquad \text{elsewhere}$$

A.3 Higher-Order Time-Marching Methods

The formulation of the second order explicit Adams-Bashford, semi-implicit Crank-Nicolson methods, and second order fully implicit methods are described following.

As an extension to the first order explicit method, the second order explicit Adams-Bashford requires the values at time level n as well as at time level $n - 1$. The unsteady transport equation of property ϕ can be formulated according to

$$
\left(\frac{(\rho\phi)^{n+1} - (\rho\phi)^n}{\Delta t} \right) \Delta V = \left(\frac{3}{2} \frac{\partial(\rho\phi)}{\partial t} \bigg|^n - \frac{1}{2} \frac{\partial(\rho\phi)}{\partial t} \bigg|^{n-1} \right) \Delta V
$$

$$
= -\frac{3}{2} \left(\sum_{i=1}^{N}(\rho u\phi)_i A_i^x + \sum_{j=1}^{N}(\rho v\phi)_j A_j^y + \sum_{k=1}^{N}(\rho w\phi)_k A_k^z \right)^n
$$

$$
+ \frac{1}{2} \left(\sum_{i=1}^{N}(\rho u\phi)_i A_i^x + \sum_{j=1}^{N}(\rho v\phi)_j A_j^y + \sum_{k=1}^{N}(\rho w\phi)_k A_k^z \right)^{n-1}
$$

$$
+ \frac{3}{2} \left(\sum_{i=1}^{N} \left(\Gamma\frac{\partial\phi}{\partial x} \right)_i A_i^x + \sum_{j=1}^{N} \left(\Gamma\frac{\partial\phi}{\partial y} \right)_j A_j^y + \sum_{k=1}^{N} \left(\Gamma\frac{\partial\phi}{\partial z} \right)_k A_k^z \right)^n
$$

$$
- \frac{1}{2} \left(\sum_{i=1}^{N} \left(\Gamma\frac{\partial\phi}{\partial x} \right)_i A_i^x + \sum_{j=1}^{N} \left(\Gamma\frac{\partial\phi}{\partial y} \right)_j A_j^y + \sum_{k=1}^{N} \left(\Gamma\frac{\partial\phi}{\partial z} \right)_k A_k^z \right)^{n-1}
$$

$$
+ \left(\frac{3}{2}S_\phi^n - \frac{1}{2}S_\phi^{n+1} \right) \Delta V
$$

$$(A.3.1)$$

For the second order Crank-Nicolson method, this special type of differencing in time requires the solution of ϕ_P^{n+1} to be attained by averaging the properties between time levels n and $n + 1$. The weighting parameter in equation (2.7.23) is prescribed midway between time levels n and $n + 1$—that is, $\theta = \frac{1}{2}$. The unsteady transport equation becomes

$$
\left(\frac{(\rho\phi)^{n+1} - (\rho\phi)^n}{\Delta t} \right) \Delta V
$$

$$
= -\frac{1}{2} \left(\sum_{i=1}^{N} (\rho u\phi)_i \, A_i^x + \sum_{j=1}^{N} (\rho v\phi)_j \, A_j^y + \sum_{k=1}^{N} (\rho w\phi)_k \, A_k^z \right)^n
$$

$$
-\frac{1}{2} \left(\sum_{i=1}^{N} (\rho u\phi)_i \, A_i^x + \sum_{j=1}^{N} (\rho v\phi)_j \, A_j^y + \sum_{k=1}^{N} (\rho w\phi)_k \, A_k^z \right)^{n+1}
$$

$$
+\frac{1}{2} \left(\sum_{i=1}^{N} \left(\Gamma \frac{\partial\phi}{\partial x} \right)_i A_i^x + \sum_{j=1}^{N} \left(\Gamma \frac{\partial\phi}{\partial y} \right)_j A_j^y + \sum_{k=1}^{N} \left(\Gamma \frac{\partial\phi}{\partial z} \right)_k A_k^z \right)^n
$$

$$
+\frac{1}{2} \left(\sum_{i=1}^{N} \left(\Gamma \frac{\partial\phi}{\partial x} \right)_i A_i^x + \sum_{j=1}^{N} \left(\Gamma \frac{\partial\phi}{\partial y} \right)_j A_j^y + \sum_{k=1}^{N} \left(\Gamma \frac{\partial\phi}{\partial z} \right)_k A_k^z \right)^{n+1}
$$

$$
+\frac{1}{2} (S_\phi^n + S_\phi^{n+1}) \Delta V
$$

$$
\text{(A.3.2)}
$$

In contrast to the first order implicit method, the second order implicit method involves the time derivative of the transport equation of property ϕ to be approximated according to

$$
\int_{t}^{t+\Delta t} \frac{\partial(\rho\phi)}{\partial t} \, dt = \frac{3(\rho\phi)^{n+1} - 4(\rho\phi)^n - (\rho\phi)^{n-1}}{2\Delta t} \tag{A.3.3}
$$

Using the preceding equation (A.3.3) in place of the first order approximation and setting the weighting parameter $\theta = 1$, equation (2.7.23) reduces to

$$
\left(\frac{3(\rho\phi)^{n+1} - 4(\rho\phi)^{n} - (\rho\phi)^{n-1}}{2\Delta t} \right) \Delta V
$$

$$
= -\left(\sum_{i=1}^{N} (\rho u \phi)_i \, A_i^x + \sum_{j=1}^{N} (\rho v \phi)_j \, A_j^y + \sum_{k=1}^{N} (\rho w \phi)_k \, A_k^z \right)^{n+1}
$$

$$
+ \left(\sum_{i=1}^{N} \left(\Gamma \frac{\partial \phi}{\partial x} \right)_i A_i^x + \sum_{j=1}^{N} \left(\Gamma \frac{\partial \phi}{\partial y} \right)_j A_j^y + \sum_{k=1}^{N} \left(\Gamma \frac{\partial \phi}{\partial z} \right)_k A_k^z \right)^{n+1}
$$

$$
+ S_\phi^{n+1} \Delta V
$$

$$
\text{(A.3.4)}
$$

Appendix B Algebraic Equation System and CFD-Based Fire Model

B.1 Conversion of Governing Equation to Algebraic Equation System Using the Finite Volume Method

On the basis of equation (2.7.23), the first order fully implicit, unsteady transport equation of property ϕ can be expressed as

$$
\left(\frac{(\rho\phi)^{n+1} - (\rho\phi)^n}{\Delta t} \right) \Delta V
$$

$$
+ \left(\sum_{i=1}^{N} (\rho u\phi)_i A_i^x + \sum_{j=1}^{N} (\rho v\phi)_j A_j^y + \sum_{k=1}^{N} (\rho w\phi)_k A_k^z \right)^{n+1}
$$

$$
- \left(\sum_{i=1}^{N} \left(\Gamma \frac{\partial\phi}{\partial x} \right)_i A_i^x + \sum_{j=1}^{N} \left(\Gamma \frac{\partial\phi}{\partial y} \right)_j A_j^y + \sum_{k=1}^{N} \left(\Gamma \frac{\partial\phi}{\partial z} \right)_k A_k^z \right)^{n+1}
$$

$$
= S_\phi^{n+1} \Delta V \tag{B.1.1}
$$

For a structured grid arrangement as shown in Figure 2.9, the projected areas A_i^x along the x direction are given by $A_1^x = -A_w$ and $A_2^x = A_e$. Similarly, the projected areas A_j^y and A_k^z along the y and z directions are $A_1^y = -A_s$ and $A_2^y = A_n$ and $A_1^z = -A_b$ and $A_2^z = A_t$, respectively. Assuming piecewise-linear gradient profiles spanning the nodal points between the central point P and neighboring points W, E, S, N, B, and T, the first order derivatives at the control volume faces of the diffusive fluxes in equation (B.1.1) can be approximated by

Computational Fluid Dynamics in Fire Engineering
Copyright © 2009 by Academic Press. Inc. All rights of reproduction in any form reserved.

$$
-\Gamma_e A_e \left(\frac{\phi_E - \phi_P}{\delta x_E}\right)^{n+1} + \Gamma_w A_w \left(\frac{\phi_P - \phi_W}{\delta x_W}\right)^{n+1} - \Gamma_n A_n \left(\frac{\phi_N - \phi_P}{\delta y_N}\right)^{n+1}
$$

$$
+ \Gamma_s A_s \left(\frac{\phi_P - \phi_S}{\delta y_S}\right)^{n+1} - \Gamma_t A_t \left(\frac{\phi_T - \phi_P}{\delta z_T}\right)^{n+1} + \Gamma_b A_b \left(\frac{\phi_P - \phi_B}{\delta z_B}\right)^{n+1}
$$

$$
= -D_e(\phi_E - \phi_P)^{n+1} + D_w(\phi_P - \phi_W) - D_n(\phi_N - \phi_P)^{n+1}
$$

$$
+ D_s(\phi_P - \phi_S)^{n+1} - D_n(\phi_T - \phi_P)^{n+1} + D_s(\phi_P - \phi_B)^{n+1}
$$

$$
\tag{B.1.2}
$$

Subsequently, the convective flux in equation (B.1.1) is given as

$$
(\rho u \phi)_e^{n+1} A_e - (\rho u \phi)_w^{n+1} A_w + (\rho v \phi)_n^{n+1} A_n - (\rho v \phi)_s^{n+1} A_s + (\rho w \phi)_t^{n+1} A_t
$$

$$
- (\rho w \phi)_b^{n+1} A_b
$$

$$
= F_e \phi_e^{n+1} - F_w \phi_w^{n+1} + F_n \phi_n^{n+1} - F_s \phi_s^{n+1} + F_t \phi_t^{n+1} - F_b \phi_b^{n+1}
$$

$$
\tag{B.1.3}
$$

In the preceding equation, F signifies the mass flux across each control volume face.

By approximating the interface values of ϕ at the control volume faces by central difference, equation (B.1.3) results in the realization between the central and neighboring points according to

$$
\frac{F_e}{2}\phi_P^{n+1} + \frac{F_e}{2}\phi_E^{n+1} - \frac{F_w}{2}\phi_W^{n+1} - \frac{F_w}{2}\phi_P^{n+1} + \frac{F_n}{2}\phi_P^{n+1} + \frac{F_n}{2}\phi_N^{n+1} - \frac{F_s}{2}\phi_S^{n+1}
$$

$$
- \frac{F_s}{2}\phi_P^{n+1} + \frac{F_t}{2}\phi_P^{n+1} + \frac{F_t}{2}\phi_T^{n+1} - \frac{F_b}{2}\phi_B^{n+1} - \frac{F_b}{2}\phi_P^{n+1}
$$

$$
\tag{B.1.4}
$$

By substituting equation (B.1.2) and (B.1.4) into equation (B.1.1), the discrete form of the transport equation after some algebraic manipulation can be written as

$$
A_P \phi_P^{n+1} = A_E \phi_E^{n+1} + A_W \phi_W^{n+1} + A_N \phi_N^{n+1} + A_S \phi_S^{n+1} + A_T \phi_T^{n+1} + A_B \phi_B^{n+1} + b
$$

$$
\tag{B.1.5}
$$

where the matrix coefficients and source term is given by

$$A_E = D_e - \frac{F_e}{2}; \quad A_W = D_w + \frac{F_w}{2};$$

$$A_N = D_n - \frac{F_n}{2}; \quad A_S = D_s + \frac{F_s}{2};$$

$$A_T = D_t - \frac{F_t}{2}; \quad A_B = D_b + \frac{F_b}{2};$$

$$A_P = D_e + \frac{F_e}{2} + D_w - \frac{F_w}{2} + D_n + \frac{F_n}{2} + D_s - \frac{F_s}{2} + D_t + \frac{F_t}{2} + D_b - \frac{F_b}{2}$$

$$+ \frac{\rho_p^{n+1} \Delta V}{\Delta t}$$

$$= A_E + A_W + A_N + A_S + A_T + A_B + (F_e - F_w) + (F_n - F_s) + (F_t - F_b)$$

$$+ \frac{\rho_p^{n+1} \Delta V}{\Delta t};$$

$$b = S_\phi^{n+1} \Delta V + \frac{(\rho \phi)_P^n \Delta V}{\Delta t} \tag{B.1.6}$$

It should be noted that whereas D always remains positive, F can either take positive or negative values, depending on the direction of the fluid flow. Hence, there is a distinctive possibility where the coefficients of matrix A in equation (2.7.26) become negative and ill behaved, causing the solution of equation (B.1.5) to possibly diverge.

Alternatively, the coefficients of the transport equation by applying the first order upwind difference can be formulated as

$$A_E = D_e + max(-F_e, 0.0); A_W = D_w + max(F_w, 0.0);$$

$$A_N = D_n + max(-F_n, 0.0); A_S = D_s + max(F_s, 0.0);$$

$$A_T = D_t + max(-F_t, 0.0); A_B = D_b + max(F_b, 0.0);$$

$$A_P = D_e + max(F_e, 0.0) + D_w + max(-F_w, 0.0) + D_n + max(F_n, 0.0)$$

$$+ D_s + max(-F_s, 0.0) + D_t + max(F_t, 0.0) + D_b + max(-F_b, 0.0)$$

$$+ \frac{\rho_p^{n+1} \Delta V}{\Delta t}$$

$$= A_E + A_W + A_N + A_S + A_T + A_B + (F_e - F_w) + (F_n - F_s) + (F_t - F_b)$$

$$+ \frac{\rho_p^{n+1} \Delta V}{\Delta t}$$

$$b = S_\phi^{n+1}\Delta V + \frac{(\rho\phi)_P^n \Delta V}{\Delta t} \tag{B.1.7}$$

It is evident from equation (B.1.7) that there are no negative coefficients. Therefore, the matrix A in equation (2.7.26) will invariably be well behaved, and the solution is always physically realistic.

One difficulty in implementing a higher-order scheme is that the coefficients such as A_B, A_S, A_W, A_P, A_E, A_N, and A_T in matrix A of equation (2.7.26) may lose their diagonal dominance under conditions of highly convective flows, as a result of applying the finite volume method. The first order upwind difference as just demonstrated is seen to be always stable because of its positive coefficients and results in a diagonally dominant matrix. Taking into account the favorable properties, Khosla and Rubin (1974) proposed a procedure that entails using the standard first order upwind scheme to obtain the central and neighboring coefficients of matrix A, with an extra deferred correction of the higher-order scheme being treated explicitly as a source term. The coefficients A_E, A_W, A_N, A_S, A_T, A_B, and A_P are thus identical to those represented in equation (B.1.7), except that the source term b contains now the additional correction term b^{dc} given by

$$b = S_\phi^{n+1}\Delta V + \frac{(\rho\phi)_P^n \Delta V}{\Delta t} + b^{dc} \tag{B.1.8}$$

According to Khosla and Rubin (1974), the correction term b^{dc} takes the form

$$\begin{aligned}
b^{dc} = & -max(F_e, 0.0)(\phi_e^{higher\text{-}order} - \phi_P^{n+1}) + max(-F_e, 0.0)(\phi_e^{higher\text{-}order} - \phi_E^{n+1}) \\
& -max(-F_w, 0.0)(\phi_w^{higher\text{-}order} - \phi_P^{n+1}) + max(F_w, 0.0)(\phi_w^{higher\text{-}order} - \phi_W^{n+1}) \\
& -max(F_n, 0.0)(\phi_n^{higher\text{-}order} - \phi_P^{n+1}) + max(-F_n, 0.0)(\phi_n^{higher\text{-}order} - \phi_N^{n+1}) \\
& -max(-F_s, 0.0)(\phi_s^{higher\text{-}order} - \phi_P^{n+1}) + max(F_s, 0.0)(\phi_s^{higher\text{-}order} - \phi_S^{n+1}) \\
& -max(F_t, 0.0)(\phi_t^{higher\text{-}order} - \phi_P^{n+1}) + max(-F_t, 0.0)(\phi_n^{higher\text{-}order} - \phi_T^{n+1}) \\
& -max(-F_b, 0.0)(\phi_b^{higher\text{-}order} - \phi_P^{n+1}) + max(F_b, 0.0)(\phi_b^{higher\text{-}order} - \phi_B^{n+1})
\end{aligned} \tag{B.1.9}$$

The source term in equation (B.1.9) may be viewed as the means of reducing the excessive false diffusion caused by the first order upwind difference. Appropriate values of property ϕ at the respective faces of the control volume in the preceding equation are determined via suitable choice of higher-order differencing schemes such as exemplified in Appendix A. The present implementation offers enormous flexibility and ease of incorporating various schemes to improve the accuracy of the numerical solution.

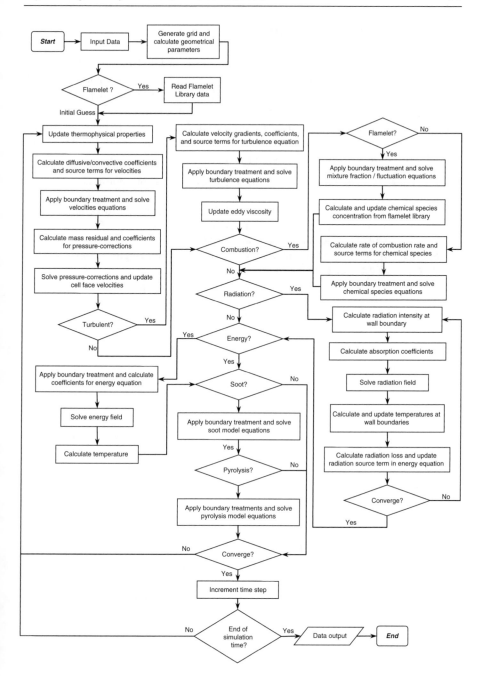

Figure B.1 A CFD-based fire model for solving the turbulent reacting flow with radiation, soot, and pyrolysis in a compartment fire situation.

B.2   CFD-Based Fire Model

A flow chart illustrating the specific operation of a CFD-based fire model to resolve the turbulent reacting flow with radiation, soot, and pyrolysis considerations in compartment fire situations is illustrated in Figure B.1. The methodology, based on an implicit procedure, is solved iteratively via efficient numerical solvers. An in-house computer code (FIRE3D) has been developed based on the FORTRAN computer language. Basically, the CFD-based fire model can be divided into six distinct modules:

*   Gas phase laminar flow module with the consideration of appropriate turbulent models
*   Combustion module incorporating eddy-dissipation and presumed PDF models
*   Radiation module using discrete ordinates method
*   Energy module for calculating the heat transfer
*   Soot module incorporating single-step empirical and semi-empirical soot models
*   Pyrolysis module for charring materials

Appendix C Advanced Combustion Modeling

C.1 Probability Density Function Method

Probability density function (PDF) represents a general statistical description of the turbulent reacting flow. This PDF can be considered to be proportional to the fraction of time the fluid spends at each chemical species, temperature, and pressure state. From the PDF, any thermo-chemical moment such as the mean and RMS chemical species or temperature can be determined. The transport equation for the joint PDF of velocities and reactive scalars can be derived from the Navier-Stokes equations and the convection-diffusion equation for the reactive scalar as

$$
\frac{\partial(\bar{\rho}\tilde{P})}{\partial t} + \frac{\partial(\bar{\rho}\tilde{u}_j\tilde{P})}{\partial x_j} + \frac{\partial(w_k\tilde{P})}{\partial \psi_k} = -\frac{\partial}{\partial x_j}\{\bar{\rho}\langle u_j'' \mid \psi\rangle\tilde{P}\}
$$
$$
+ \frac{\partial}{\partial \psi_k}\left\{\bar{\rho}\left\langle\frac{1}{\bar{\rho}}\frac{\partial J_{j,k}}{\partial x_j}\bigg|\psi\right\rangle\tilde{P}\right\} - \frac{\partial(\bar{\rho}w_{rad}\tilde{P})}{\partial \psi_k}
$$

$$(C.1.1)$$

where $\tilde{P}$ is the Favred joint PDF, ψ is the composition space vector, u_j'' is the fluid velocity fluctuation vector, and $J_{j,k}$ is the molecular diffusion flux vector. Note that $\langle A|B\rangle$ is the conditional probability of event A, given that event B occurs. In equation (C.1.1), summation is implied over repeated indices of j and k within the terms, with j and k representing the physical and composition space, respectively.

In equation (C.1.1), the first and second terms on the left-hand side represents the unsteady rate of change and convection of the PDF in physical space. The third terms on both sides represent the sources that depend on the local composition (e.g., chemical reaction rate and radiation emission). Since all the terms on the left-hand side are readily solved including the reaction rate term, the principal strength of the PDF approach in comparison to the Reynolds-averaged or Favred-averaged approach is that the highly non-linear reaction rate term is completely closed and therefore requires no modeling. However, the two terms on the right-hand side are not readily solved and require modeling, which they are represented by the scalar convection by

Computational Fluid Dynamics in Fire Engineering
Copyright © 2009 by Academic Press. Inc. All rights of reproduction in any form reserved.

turbulent and molecular mixing/diffusion, respectively. Appropriate models are generally provided for these terms. It should be emphasized that the model for the molecular mixing/diffusion is the weakest link in the PDF transport approach. Modeling the effect of molecular mixing/diffusion on the composition poses significant challenge as exemplified by the development of many models such as the IEM (Interaction by Exchange with the Mean), LMSE (Linear Mean-Square Estimation), MC (Modified Curl), and EMST (Euclidian Minimum Spanning Tree) (Peters, 2000, Pope, 2000). This is considered to be a critical component of the PDF transport approach, because of the combustion occurring at the smallest scales when the reactants and heat are diffused together. None of the aforementioned models have thus far been satisfactory, and considerable development of a suitable model is still required, especially to yield physically realistic representation of the scalar dissipation rate to capture the local flame extinction and re-ignition for non-premixed combustion (Bilger, 2000).

C.2 Conditional Moment Closure

According to the conditional moment closure (CMC) approach, the instantaneous mass fraction Y_i can be decomposed into a conditional mean and fluctuation around the conditional mean as

$$Y_i(x,t) = Q_i(\eta, x, t) + Y_i''(x,t) \qquad (C.2.1)$$

where $Q_i(\eta, x, t)$ can be defined as $Q_i(\eta, x, t) \equiv \langle Y_i(x,t)|Z(x,t) = \eta \rangle$ at location x, time t and η indicates the space variable for the mixture fraction. The governing equation for the ith species $Q_\alpha \equiv \langle \rho Y_i|\eta \rangle / \langle \rho|\eta \rangle$ can be derived via conditional averaging as

$$\frac{\partial Q_\alpha}{\partial t} + \langle u_j \mid \eta \rangle \frac{\partial Q_\alpha}{\partial x_j} = \langle \chi \mid \eta \rangle \frac{\partial^2 Q_\alpha}{\partial \eta^2} + \langle W_\alpha \mid \eta \rangle + e_Q + e_Y \qquad (C.2.2)$$

with

$$e_Q \equiv \left\langle \frac{1}{\rho} \frac{\partial}{\partial x_j} \left(\rho D \frac{\partial Q_\alpha}{\partial x_j} \right) + D \frac{\partial Z}{\partial x_j} \left(\frac{\partial}{\partial x_j} \frac{\partial Q_\alpha}{\partial \eta} \right) \bigg| \eta \right\rangle \qquad (C.2.3)$$

$$e_Y \equiv -\left\langle \frac{\partial Y_i''}{\partial t} + u_j \frac{\partial Y_i''}{\partial x_j} - \frac{1}{\rho} \frac{\partial}{\partial x_j} \left(\rho D \frac{\partial Y_i''}{\partial x_j} \right) \bigg| \eta \right\rangle \qquad (C.2.4)$$

Similarly, the CMC transport equation for the enthalpy $Q_h \equiv \langle \rho h|\eta \rangle / \langle \rho|\eta \rangle$ can be derived as

$$\frac{\partial Q_h}{\partial t} + \langle u_j \mid \eta \rangle \frac{\partial Q_h}{\partial x_j} = \langle \chi \mid \eta \rangle \frac{\partial^2 Q_h}{\partial \eta^2} - \langle W_{rad} \mid \eta \rangle + e_Q + e_Y \qquad (C.2.5)$$

where W_{rad} is the heat loss due to radiation. According to Bilger (1993) and Li and Bilger (1993), the unclosed e_Q can be assumed to be negligible for the high Reynolds number, since it is of the order Re^{-1} and there is no differential diffusion. On the basis of the proposal by Kim (2004), the closure hypotheses for the unclosed term e_Y through the decomposition approach is given by

$$e_Y = -\frac{1}{\bar{\rho}P(\eta)} \frac{\partial}{\partial x_j} \left(\langle u_j'' Y_\alpha'' \mid \eta \rangle \bar{\rho} P(\eta) \right) \qquad (C.2.6)$$

where $P(\eta)$ is the probability density function (PDF), which involves the use of a prescribed beta function. In equations (C.2.2) and (C.2.5), the first and second terms on the right-hand side correspond to the diffusion in the mixture fraction space and are quantified by the scalar dissipation rate χ and the conditional expectation on the chemical reaction or heat loss due to radiation, respectively. The last term on the right-hand side and the left-hand side in both equations represents the spatial diffusion and convection. These terms that are absent in the laminar flamelet model constitute the main difference between the two methods.

In order to close the system of equations, the remaining terms such as the conditional velocity $\langle u_j|\eta \rangle$, conditional scalar dissipation rate χ, and conditional fluxes $\langle u_i'' \phi'' \mid \eta \rangle$ are usually solved according to the appropriate models. Firstly, two different models, the *Linear model* by Kilmenko and Bilger (1999) and Swaminathan and Bilger (2001) and the *PDF gradient model* by Colucci et al. (1998) and Bilger (2000), have been considered for the conditional velocity. In the inhomogeneous case, the Linear model has shown to perform better than the PDF gradient model. Secondly, the Grimanji *model* (Grimanji, 1992) assuming local homogeneity is commonly used for the conditional scalar dissipation rate. Alternatively, the *Amplitude Mapping Closure (AMC) model* (O'Brien and Jiang, 1991), which may be applied in conjunction with beta PDF, has shown to yield more favorable results in the inhomogeneous case. Thirdly, the conditional fluxes are normally modeled by the *standard gradient diffusion hypothesis*.

For the reaction rate, the *first order CMC closure* can be written as

$$\langle W_\alpha \mid \eta \rangle \approx W_\alpha(Q_\alpha, Q_T, \bar{p}) \qquad (C.2.7)$$

where $Q_T \equiv \langle \rho T|\eta \rangle / \langle \rho|\eta \rangle$. This approach is valid only if the conditional fluctuations of the reactive scalars are small enough for the higher-order terms to be negligible. *Second order CMC closure* employs conditional variances and co-variances to improve the estimate of $\langle W_\alpha|\eta\rangle$. In the assumed PDF method, the condition expectation of the reaction rate is given by

$$\langle W_\alpha \mid \eta \rangle \approx \int W_\alpha(\zeta_1,\ldots,\zeta_n,\eta)P(\zeta_1,\ldots,\zeta_n \mid \eta)d\zeta_1,\ldots,d\zeta_n \qquad \text{(C.2.8)}$$

where n is the number of reactions, ζ_i is the sample variable for the reaction progress variable, and P is the conditional joint PDF. There are also other models with CMC, which could also be similarly employed for more sophisticated considerations. Examples are the *multi-conditional moment closure model* by Kilmenko and Bilger (1999) and *conditional source-term estimation model* by Steiner and Bushe (1999).

Appendix D   Relevant Tables for Combustion and Radiation Modeling

Table D.1 Heats of combustion of selected gases, liquids, and solids at 25°C (298.15K)[a].

		$-\Delta H_c \times 10^6$ $(J\ kg^{-1})$	$-\Delta H_{c,air} \times 10^6$ $(J\ kg\ (air)^{-1})$	$-\Delta H_{c,ox} \times 10^6$ $(J\ kg\ (O_2^{-1})$
Hydrogen	H_2	120.97	3.41	15.12
Carbon monoxide	CO	10.11	4.11	17.69
Methane	CH_4	50.01	2.90	12.50
Ethane	C_2H_6	47.79	2.97	12.80
Propane	C_3H_8	46.36	2.96	12.75
n-Butane	n-C_4H_{10}	45.75	2.96	12.76
i-Butane	i-C_4H_{10}	45.61	2.95	12.72
n-Pentane	n-C_5H_{12}	45.36	2.96	12.76
n-Hexane	n-C_6H_{14}	44.75	2.94	12.66
n-Octane	n-C_8H_{18}	44.77	2.97	12.76
i-Octane	i-C_8H_{18}	44.30	2.93	12.63
Acetylene	C_2H_2	48.24	3.64	15.68
Ethylene	C_2H_4	47.20	3.20	13.77
Propylene	C_3H_6	45.80	3.10	13.36
Benzene	C_6H_6	40.17	3.03	13.06
Toluene	C_7H_8	40.59	3.01	12.97
Formaldehyde	CH_2O	17.31	3.77	16.23
Methanol	CH_4O	19.94	3.09	13.29
Ethanol	C_2H_6O	28.87	3.21	13.83
Cellulose	$(C_2H_4O)_n$	16.09	3.15	13.59
Polyethylene	$(C_2H_4)_n$	43.28	2.93	12.65

Continued

Computational Fluid Dynamics in Fire Engineering
Copyright © 2009 by Academic Press. Inc. All rights of reproduction in any form reserved.

Table D.1   Heats of combustion of selected gases, liquids, and solids at 25°C (298.15K)[a].—Cont'd

		$-\Delta H_c \times 10^6$ (J kg^{-1})	$-\Delta H_{c,air} \times 10^6$ (J kg (air)$^{-1}$)	$-\Delta H_{c,ox} \times 10^6$ (J kg (O$_2$)$^{-1}$)
Polypropylene	$(C_3H_6)_n$	43.31	2.94	12.66
Polystyrene	$(C_8H_8)_n$	39.85	3.01	12.97
Polyvinylchloride	$(C_2H_3Cl)_n$	16.43	2.98	12.84
Polymethylmethacrylate	$(C_5H_8O_2)_n$	24.89	3.01	12.98

[a]All products are taken to be in their gaseous states. Values of heats of combustion in air and in oxygen can be obtained based on the knowledge of the completer reaction in air and from the heats of combustion determined in a bomb calorimeter. For example, $\Delta H_{c,air}$ of methane according to the complete combustion reaction $CH_4 + O_2 + 2 \times 3.76 N_2 \rightarrow CO_2 + H_2O + 2 \times 3.76 N_2$ can be obtained from ΔH_c by

$$\Delta H_{c,air} = \frac{\Delta H_c \times \text{molecular weight of fuel}}{\text{number of moles of air} \times \text{molecular weight of air}} = \frac{50.01 \times 16.0}{9.52 \times 28.95} = 2.90MJ \text{ kg(air)}^{-1}.$$

Similarly, $\Delta H_{c,ox}$ of methane according to the preceding complete combustion reaction can be obtained from ΔH_c by.

$$\Delta H_{c,ox} = \frac{\Delta H_c \times \text{molecular weight of fuel}}{\text{number of moles of oxygen} \times \text{molecular weight of oxygen}} = \frac{50.01 \times 16.0}{2.0 \times 32.0} = 12.5MJ \text{ kg(O}_2)^{-1}.$$

Note that values of heats of combustion for cellulose, polyethylene, polypropylene, polystyrene, polyvinylchloride, and polymethylmethacrylate have been obtained from Drysdale (1999).

TABLE D.2   Standard heats of formation of selected gases at 25°C (298.15K).

		$\Delta H_f^o \times 10^6 \text{(J kg}^{-1})$
Carbon monoxide	CO	−3.95
Carbon dioxide	CO_2	−8.95
Hydrogen	H_2	0.0
Water vapor	H_2O	−13.44
Nitrogen	N_2	0.0
Oxygen	O_2	0.0
Methane	CH_4	−4.68
Ethane	C_2H_6	−2.82
Propane	C_3H_8	−2.36
n-Butane	$n\text{-}C_4H_{10}$	−2.15
i-Butane	$i\text{-}C_4H_{10}$	−2.27
n-Pentane	$n\text{-}C_5H_{12}$	−2.03
n-Hexane	$n\text{-}C_6H_{14}$	−1.94
n-Octane	$n\text{-}C_8H_{18}$	−1.82
i-Octane	$i\text{-}C_8H_{18}$	−1.96

Continued

TABLE D.2 Standard heats of formation of selected gases at 25°C (298.15K).—Cont'd

		$\Delta H_f^o \times 10^6 (\text{J kg}^{-1})$
Acetylene	C_2H_2	+8.72
Ethylene	C_2H_4	+1.87
Propylene	C_3H_6	+0.48
Benzene	C_6H_6	+1.06
Toluene	C_7H_8	+0.54
Formaldehyde	CH_2O	−3.86
Methanol	CH_4O	−6.28
Ethanol	C_2H_6O	−5.11

Table D.3 Adiabatic flame temperatures of selected fuels at 25°C (298.15K)[b].

		$T_{adiabatic}$ (K)
Hydrogen	H_2	2430.7
Methane	CH_4	2326.5
Ethane	C_2H_6	2381.3
Propane	C_3H_8	2393.4
n-Butane	$n\text{-}C_4H_{10}$	2393.4
i-Butane	$i\text{-}C_4H_{10}$	2392.3
n-Pentane	$n\text{-}C_5H_{12}$	2402.7
n-Hexane	$n\text{-}C_6H_{14}$	2405.5
n-Octane	$n\text{-}C_8H_{18}$	2409.1
i-Octane	$i\text{-}C_8H_{18}$	2409.1
Acetylene	C_2H_2	2907.5
Ethylene	C_2H_4	2565.8
Propylene	C_3H_6	2505.7
Benzene	C_6H_6	2528.2
Toluene	C_7H_8	2503.3
Formaldehyde	CH_2O	2574.3
Methanol	CH_4O	2330.9
Ethanol	C_2H_6O	2355.0

[b]The adiabatic flame temperatures have been determined via the routine GASEQ – Chemical Equilibrium Program for Windows.

Table D.4 Leonard-Jones parameters of selected gases[c].

		$\sigma_{o,l}$	ε / k
Air		3.711	78.6
Oxygen (atomic)	O	3.050	106.7
Oxygen (molecule)	O_2	3.467	106.7
Hydroxyl	OH	3.147	79.8
Carbon monoxide	CO	3.690	91.7
Carbon dioxide	CO_2	3.941	195.2
Hydrogen (atomic)	H	2.708	37.0
Hydrogen (molecule)	H_2	2.827	59.7
Water vapor	H_2O	2.641	809.1
Nitrogen	N_2	3.798	71.4
Methane	CH_4	3.758	148.6
Ethane	C_2H_6	4.418	230.0
Propane	C_3H_8	5.061	254.0
i-Butane	i-C_4H_{10}	5.341	313.0
n-Pentane	n-C_5H_{12}	6.769	345.0
n-Hexane	n-C_6H_{14}	5.909	413.0
n-Octane	n-C_8H_{18}	7.451	320.0
Acetylene	C_2H_2	4.033	231.8
Ethylene	C_2H_4	4.163	224.7
Propylene	C_3H_6	5.061	254.0
Benzene	C_6H_6	5.270	440.0

[c]According to Lautenberger (2002), the diffusion collision integral $\Omega_{D_{l,inert}}$ can be approximated empirically as $\Omega_{D_{l,inert}} A = 1.080794A^{-0.16033} + 0.605009\ exp(-0.88524A) + 2.115672\ exp(-2.98308A)$, where A = $T/(\varepsilon/k)_{l,inert}$ and $(\varepsilon/k)_{l,inert}$ is evaluated according to $(\varepsilon/k)_{l,inert} = \sqrt{(\varepsilon/k)_l (\varepsilon/k)_{inert}}$. The variables $\sigma_{o,l}$ and ε/k are known as the Lenoard-Jones parameters, which can be obtained from Table D.4. Also according to Lautenberger (2002), the diffusion integral Ω_μ can be approximated as $\Omega_\mu B = 1.16145B^{-0.14874} + 0.52487$ exp $(-0.7732B) + 2.16178$ exp$(-2.43787B)$ where $B = T/(\varepsilon/k)$.

Table D.5 Low-order polynomials of specific heats at constant pressure of selected gases (constructed from Sandler, 1989)[d].

		a	$b \times 10^1$	$c \times 10^{-2}$	$d \times 10^{-6}$	Temperature Range (K)
Oxygen	O_2	880.60	0.0197	−0.0023	—	273–3800
Carbon monoxide	CO	968.73	0.0234	−0.0036	—	273–3800
Carbon dioxide	CO_2	See below				
Hydrogen	H_2	13444.95	0.2175	−0.0016	—	273–3800
Water vapor	H_2O	1620.85	0.0806	−0.0112	—	273–3800
Nitrogen	N_2	976.05	0.0222	−0.0034	—	273–3800
Methane	CH_4	1242.69	0.3139	0.0793	−0.6881	273–1500
Ethane	C_2H_6	229.88	0.5754	−0.2135	0.2428	273–1500
Propane	C_3H_8	−91.33	0.7353	−0.3572	0.7211	273–1500
n-Butane	$n\text{-}C_4H_{10}$	68.2	0.6404	−0.3161	0.6033	273–1500
i-Butane	$i\text{-}C_4H_{10}$	−136.40	0.7171	−0.3966	0.8603	273–1500
n-Pentane	$n\text{-}C_5H_{12}$	95.81	0.6308	−0.3119	0.5872	273–1500
n-Hexane	$n\text{-}C_6H_{14}$	80.65	0.6420	−0.3331	0.6707	273–1500
Acetylene	C_2H_2	838.78	0.3543	−0.2510	0.7002	273–1500
Ethylene	C_2H_4	141.12	0.5045	−0.2979	0.6309	273–1500
Propylene	C_3H_6	75.05	0.5672	−0.2900	0.5860	273–1500
Benzene	C_6H_6	−464.20	0.6755	−0.4046	0.9949	273–1500
Toluene	C_7H_8	−373.68	0.6077	−0.3745	0.8736	273–1500
Formaldehyde	CH_2O	760.01	0.1359	0.0238	−0.2899	273–1500
Methanol	CH_4O	595.18	0.2860	−0.0381	−0.2512	273–1500
Ethanol	C_2H_6O	432.24	0.4555	−0.2256	0.4359	273–1500

[d]Constants in the table are for the equation $C_p = a + bT + cT^2 + dT^3$, where T is in Kelvin and C_p is in J (kg K) − 1. The equation for CO_2 in the temperature range 273-3800 K is $C_p = 1715.82 - 4.2562 \times 10^{-3}\,T - 15.107 \times 10^3/\sqrt{T}$ J(kgK)$^{-1}$.

Table D.6 Generalized state relation functions for hydrocarbon fuels of the form C_nH_m (after Sivathanu and Faeth, 1990).

Species	$\psi(Y_i)$	$Y_i(\psi)$
N_2	$\left(\dfrac{Y_{N_2}}{Y_{N_2,st}}\right)\left(\dfrac{Y_{N_2,A} - Y_{N_2,st}}{Y_{N_2,A} - Y_{N_2}}\right)$	$\dfrac{\psi Y_{N_2,A} Y_{N_2,st}}{Y_{N_2,A} - Y_{N_2,st} + \psi Y_{N_2,st}}$
O_2	$\left(\dfrac{Y_{O_2}}{Y_{O_2,A}}\right)\left(\dfrac{32n + 8m + M_{fu}Y_{O_2,A}}{32n + 8m + M_{fu}Y_{O_2}}\right)$	$\dfrac{\psi Y_{O_2,A}(32n + 8m)}{(32n + 8m) - \psi Y_{O_2,A}M_{fu}}$
CO_2	$\left(\dfrac{Y_{CO_2}}{Y_{CO_2,st}}\right)\left(\dfrac{44n - M_{fu}Y_{CO_2,st}}{44n - M_{fu}Y_{CO_2}}\right)$	$\dfrac{\psi Y_{CO_2,st}44n}{44n + \psi Y_{CO_2,st}M_{fu} - Y_{CO_2,st}M_{fu}}$
H_2O	$\left(\dfrac{Y_{H_2O}}{Y_{H_2O,st}}\right)\left(\dfrac{9m - M_{fu}Y_{H_2O,st}}{9m - M_{fu}Y_{H_2O}}\right)$	$\dfrac{\psi Y_{H_2O,st}9m}{9m + \psi Y_{H_2O,st}M_{fu} - Y_{H_2O,st}M_{fu}}$
CO	$\left(\dfrac{Y_{CO}}{Y_{CO_2,st}}\right)\left(\dfrac{44n - M_{fu}Y_{CO_2,st}}{44n - M_{fu}Y_{CO}}\right)$	$\dfrac{\psi Y_{CO_2,st}44n}{44n + \psi Y_{CO_2,st}M_{fu} - Y_{CO_2,st}M_{fu}}$
H_2	$\left(\dfrac{Y_{H_2}}{Y_{H_2O,st}}\right)\left(\dfrac{9m - M_{fu}Y_{H_2O,st}}{9m - M_{fu}Y_{H_2}}\right)$	$\dfrac{\psi Y_{H_2O,st}9m}{9m + \psi Y_{H_2O,st}M_{fu} - Y_{H_2O,st}M_{fu}}$
Fuel	Y_{fu}	ψ

Table D.7 Values for parameter ψ in terms of the equivalence ratio (after Sivathanu and Faeth, 1990).

			$\psi(Y_i,\phi)$				
ϕ	N_2	O_2	CO_2	H_2O	CO	H_2	Fuel
0.0	100.0	1.0	0.0	0.0	0.0	0.0	0.0
0.01	100.0	0.99	0.01	0.01	0.0	0.0	0.0
0.02	50.0	0.98	0.02	0.02	0.0	0.0	0.0
0.05	20.0	0.95	0.05	0.05	0.0	0.0	0.0
0.1	10.0	0.9	0.1	0.1	0.0	0.0	0.0
0.2	5.0	0.8	0.2	0.2	0.0	0.0	0.0
0.5	2.0	0.51	0.48	0.5	0.015	0.0	0.0
0.8	1.25	0.25	0.7	0.78	0.03	0.004	0.0
1.0	1.0	0.11	0.8	0.96	0.115	0.008	0.0
1.5	0.667	0.065	0.82	0.98	0.25	0.018	0.0
2.0	0.5	0.051	0.8	0.97	0.3	0.022	0.028
5.0	0.2	0.41	0.58	0.86	0.26	0.022	0.185
10.0	0.1	0.035	0.4	0.7	0.18	0.02	0.33
20.0	0.05	0.025	0.27	0.49	0.125	0.017	0.55
50.0	0.02	0.018	0.14	0.23	0.07	0.012	0.75
100.0	0.01	0.008	0.06	0.13	0.04	0.0094	0.87
∞	0.0	0.0	0.0	0	0.0	0.0	1.0

Table D.8 Coefficients for emissivity according to Smith et al. (1982).

J	i	$P_{H_2O}/P_{CO_2} = 2$				
		$b_{\varepsilon,i.1}$	$b_{\varepsilon,i.2}$	$b_{\varepsilon,i.3}$	$b_{\varepsilon,i.4}$	k_i
3	1	6.508	-5.551	3.029	-5.353	0.4201
	2	-0.2504	6.112	-3.882	6.528	6.516
	3	2.718	-3.118	1.221	-1.612	131.9

J	i	$P_{H_2O}/P_{CO_2} = 1$				
		$b_{\varepsilon,i.1}$	$b_{\varepsilon,i.2}$	$b_{\varepsilon,i.3}$	$b_{\varepsilon,i.4}$	k_i
3	1	5.150	-2.303	0.9779	-1.494	0.4303
	2	0.7749	3.399	-2.297	3.730	7.055
	3	1.907	-1.824	0.5608	-0.5122	178.1

Table D.9 Coefficients for emissivity according to Beer, Foster, and Siddall (1971).

J	i	$P_{H_2O}/P_{CO_2} = 2$				$P_{H_2O}/P_{CO_2} = 2$			
		$b_{\varepsilon,i.1}$	$b_{\varepsilon,i.2}$	k_i	k_{HC_i}	$b_{\varepsilon,i.1}$	$b_{\varepsilon,i.2}$	k_i	k_{HC_i}
3	1	0.437	7.13	0.0	3.85	0.486	8.97	0.0	3.41
	2	0.390	−0.52	1.88	0.0	0.381	−3.96	2.5	0.0
	3	1.173	−6.61	68.83	0.0	0.133	−5.01	109.0	0.0
4	1	0.364	4.74	0.0	3.85	0.4092	7.53	0.0	3.41
	2	0.266	7.19	0.69	0.0	0.284	2.58	0.91	0.0
	3	0.252	−7.41	7.4	0.0	0.211	−6.54	9.4	0.0
	4	0.118	−4.52	80.0	0.0	0.0958	−3.57	130.0	0.0

Table D.10 S_2, S_4, S_6, and S_8 quadratures in one octant for rectangular enclosures (after Jamaluddin and Smith, 1988).

Designation		ξ_n	μ_n	η_n	w_n
S_2		0.57735	0.57735	0.57735	1.57080
S_4	$S_{4,1}$	0.29588	0.90825	0.29588	0.52360
	$S_{4,2}$	0.90825	0.29588	0.29588	0.52360
	$S_{4,3}$	0.29588	0.29588	0.90825	0.52360
S_6	$S_{6,1}$	0.18387	0.96560	0.18387	0.16095
	$S_{6,2}$	0.69505	0.69505	0.18387	0.36265
	$S_{6,3}$	0.96560	0.18387	0.18387	0.16095
	$S_{6,4}$	0.18387	0.69505	0.69505	0.36265
	$S_{6,5}$	0.69505	0.18387	0.69505	0.36265
	$S_{6,6}$	0.96560	0.18387	0.96560	0.16095
S_8	$S_{8,1}$	0.14426	0.97955	0.14426	0.17124
	$S_{8,2}$	0.57735	0.80401	0.14426	0.09923
	$S_{8,3}$	0.80401	0.57735	0.14426	0.09923
	$S_{8,4}$	0.97955	0.14426	0.14426	0.17124
	$S_{8,5}$	0.14426	0.80401	0.57735	0.09923
	$S_{8,6}$	0.57735	0.57735	0.57735	0.46172
	$S_{8,7}$	0.80401	0.14426	0.57735	0.09923
	$S_{8,8}$	0.14426	0.57735	0.80401	0.09923
	$S_{8,9}$	0.57735	0.14426	0.80401	0.09923
	$S_{8,10}$	0.14426	0.14426	0.97955	0.17124

References

1. Abou-Ellail, M. M. M. and Salem, H. (1990). A Skewed PDF Combustion Model for Jet Diffusion Flame, *ASME J. Heat Transfer*, Vol. 112, pp. 1002-1007.
2. Abramowitz, M. and Stegun, I. A. (1964). Handbook of Mathematical Functions, *NBS Publication AMS 55*, Washington, p. 260.
3. Alves, S. S. (1988). *Modelacao da Pirolise de Madeira e outros Materiais Linhoce-lulosicos*, Ph.D. Thesis, Instituto Superior Tecnico, Lisboa.
4. Alves, S. S. and Figueiredo, J. L. (1989). A Model for Pyrolysis of Wet Wood, *Chem. Eng. Sci.* Vol. 44, pp. 2861-2869.
5. Anderson, W., Thomas, J. L. and Van Leer, B. (1986). Comparison of Finite Volume Flux Vector Splittings for the Euler Equations, *AIAA J.*, Vol. 24, pp. 1453-1460.
6. Annele, K. K., Virtanen, A. K. K., Ristimäki, J. M., Vaaraslahti, K. M. and Keskinen, J. (2004). Effect of Engine Load on Diesel Soot Particles, *Environ. Sci. Tech.*, Vol. 38, pp. 2551-2556.
7. Appel, J., Bockhorn, H. and Frenklach, M. (2000). Kinetic Modeling of Soot Formation with Detailed Chemistry and Physics: Laminar Premixed Flames of C_2 Hydrocarbons, *Combust. Flame*, Vol. 121, pp. 122-136.
8. Archaya, S. and Moukalled, F. (1989). Improvements to Incompressible Flow Calculation on a Non-Staggered Grid, *Numer. Heat Transfer, Part B*, Vol. 15, pp. 131-152.
9. Arking, A. and Grossman, K. (1972). The Influence of Line Shape and Band Structure on Temperatures in Planetary Atmospheres, *J. Atmos. Sci.*, Vol. 29, pp. 937-949.
10. Artelt, C., Schmid, H. J. and Peukert, W. (2003). On the Relevance of Accounting for the Evolution of the Fractal Dimension in Aerosol Process Simulations, *J. Aerosol Sci.*, Vol. 34, pp. 511-534.
11. Atreya, A. (1983). Pyrolysis, Ignition and Fire Spread on Horizontal Surfaces of Wood, Ph.D. Thesis, Harvard University.
12. ASTM (1993). Standard Test Method for Heat and Visible Smoke Release Rates for Materials and Products Using an Oxygen Consumption Calorimeter, E1354-92, Annual Books of ASTM Standards, 04/07, 1044-1060.
13. Bamford, C. H., Crank, J. and Malan, D. H. (1946). The Combustion of Wood: Part I, *Proc. Camb. Phil. Soc.*, Vol. 42, pp. 166-182.
14. Babrauskas, V. (1979). COMPF2- A Program for Calculating Post-Flashover Fire Temperatures, National Bureau of Standards (US), NIST Technical Note 991, Gaithersburg, MD.
15. Barrett, J. C. and Webb, N. A. (1998). A Comparison of Some Approximate Methods for Solving the Aerosol General Dynamics Equation, *J Aerosol Sci.*, Vol. 29, pp. 31-39.

16. Baum, H. R., Ezekoye, O. A., Mcgrattan, K. B. and Reum, R. G. (1994). Mathematical-Modeling and Computer-Simulation of Fire Phenomena, *Theo. & Comp. Fluid Dynamics*, Vol. 6, pp. 125-139.

17. Baum, H. R. and McCaffrey, B. J. (1989). Fire Induced Flow Field-Theory and Experiment, Fire Safety Science – *Proceedings of Second International Symposium*, Hemisphere, New York, Vol. 2, pp. 129-148.

18. Beard, A. N. (1997). Fire Models and Design, *Fire Safety J.*, Vol. 28, pp. 117-138.

19. Beer, J. M., Foster, P. J. and Siddall, R. G. (1971). Calculation Methods of Radiative Heat Transfer, HFTS Design Report No. 22, AEA Technology.

20. Bezdek, J. C. (1980). A Convergence Theorem for the Fuzzy ISODATA Clustering Algorithms, *IEEE Trans. Pattern Anal. Machine Intelligence*, PAMI-2, pp. 1-8.

21. Bilger, R. W. and Kent, J. H. (1972). Measurements in Turbulent Diffusion Flames, *Report F41*, Dept. of Mechanical Engineering, University of Sydney.

22. Bilger, R. W. (1975). A Note on Favre Averaging in Variable Density Flows, *Combust. Sci. Tech.*, Vol. 11, pp. 215-217.

23. Bilger, R. W. (1977). Reaction rates in diffusion flames, *Combust. Flame*, Vol. 30, pp. 277-284.

24. Bilger, R. W. (1980). Turbulent Flows with Non-premixed Reactions, *Turbulent Reacting Flows*, Libby, P. A. and Williams, F. A. (eds.), Academic Press, New York, pp. 65-114.

25. Bilger, R. W. (1988). The Structure of Turbulent Non-Premixed Flames, *Twenty-Second Symposium (International) on Combustion*, The Combustion Institute, University of Washington, Seattle, pp. 475-488.

26. Bilger, R. W. (1993). Conditional Moment Closure for Turbulent Reacting Flow, *Phys. Fluids*, Vol. 5, pp. 436-444.

27. Bilger, R. W. (2000). Future Progress in Turbulent Combustion Research, *Prog. Energy Combust. Sci.*, Vol. 26, pp. 367-380.

28. Boersma, B. J., Brethouwer, G. and Nieuwstadt, F. T. M. (1998). A Numerical Investigation on the Effect of the Inflow Conditions on the Self-Similar Region of a Round Jet, *Phys. Fluids*, Vol. 10, pp. 899-909.

29. Bonnefoy, F., Gilot, P. and Prado, G. (1993). A Three-Dimensional Model for the Determination of Kinetic Data from the Pyrolysis of Beech Wood, *J. Analy. Appl. Pyrolysis*, Vol. 25, pp. 387-394.

30. Borghi, R. (1973). Etude Théorique de L'evolution Residulle de Produits Pollutants dans les Jets de Turboreacteurs, AGARD Meeting CP 125.

31. Boussinesq, J. (1877). Théorie de l'Ecoulement Tourbillant, *Mem. Présentés par Divers Savants Acad. Sci. Inst. Fr.*, Vol. 23, pp. 46-50.

32. Bove, S., Solberg, T. and Hjertager, B. H. (2005). A Novel Algorithm for Solving Population Balance Equations: The Parallel Parent and Daughter Classes. Derivation, Analysis and Testing, *Chem. Eng. Sci*, Vol. 60, pp. 1449-1464.

33. Boyce, K., Fraser-Mitchell, J. N. and Shields, J. (1998). Survey Analysis and Modeling of Office Evacuation Using the CRISP Model, *Human Behavior in Fire – Proceedings of First International Symposium*, pp. 691-702.

34. Bradbury, A. G. W., Sakai, Y. and Shafizadeh, F. (1979). A Kinetic Model for Pyrolysis of Cellulose, *J. Appl. Polymer Sci*, Vol. 23, pp. 3271-3280.

35. Bray, K. N. C. (1979). The Interaction between Turbulence and Combustion, *Seventeenth Symposium (International) on Combustion*, The Combustion Institute, Pittsburgh PA, pp. 223-233.

36. Bray, K. N. C. and Cant, R. S. (1991). Some Applications of Kolmogorov's Turbulence Research in the Field of Combustion, *Proc. Roy. Soc. Lond.*, Vol. A 434, pp. 217-240.

37. Broido, A. and Nelson, M. A. (1975). Char Yield on Pyrolysis of Cellulose, *Combust. Flame*, Vol. 24, pp. 263-268.

38. Brookes, S. J. and Moss, J. B. (1999). Predictions of Soot and Thermal Radiation Properties in Confined Turbulent Jet Diffusion Flames, *Combust. Flame*, Vol. 116, pp. 486-503.

39. Bryan, G. M. and Nelson, R. M. (1970). The Modelling of Pulsating Fires, *Fire. Tech.*, Vol. 6, pp. 102-110.

40. Burke, S. P. and Schumann, T. E. W. (1928). Diffusion Flames, *J. Ind. Eng. Chem.*, Vol. 20, pp. 998-1004.

41. Carpenter, G. A., Grossberg, S. and David, B. R. (1991). Fuzzy ART: Fast Stable Learning and Categorization of Analog Patterns by an Adaptive Resonance System, *Neural Network*, Vol. 4, pp. 759-771.

42. Case, K. M. and Zweifel, P. F. (1967). *Linear Transport Theory*, Addison-Wesley, Reading, MA.

43. Chai, J. C., Patankar, S. V. and Lee, H. S. (1994). Evaluation of Spatial Differencing Practices for the Discrete Ordinates Method, *J. Thermophys. Heat Transfer*, Vol. 8, pp. 140-144.

44. Chakravarthy, S. R. and Osher, S. (1983). High Resolution Applications of the OSHER Upwind Scheme for the Euler Equations, AIAA Paper 83-1943.

45. Chan, W. R., Kelbon, M. and Krieger, B. B. (1985). Modeling and Experimental Verification of Physical and Chemical Processes during Pyrolysis of a Large Biomass Particle, *Fuel*, Vol. 64, pp. 1505-1513.

46. Chandrasekhar, S. (1960). *Radiative Transfer*, Dover, New York.

47. Cheung, S. C. P., Yuen, R. K. K., Yeoh, G. H. and Cheng, G. W. Y. (2004). Contribution of Soot Particles on Global Radiative Heat Transfer in a Two-Compartment Fire, *Fire Safety J.*, Vol. 39, pp. 412-428.

48. Cheung, S. C. P., Lo, S. M., Yeoh, G. H. and Yuen, R. K. K. (2006). The Influence of Gaps of Fire-Resisting Doors on the Smoke Spread in a Building Fire, *Fire Safety J.*, Vol. 41, pp. 539-546.

49. Cheung, S. C. P., Yeoh, G. H., Cheung, A. L. K. and Yuen, R. K. K. (2007a). Flickering Behavior of Turbulent Buoyant Fires Using Large-Eddy Simulation, *Numer. Heat Transfer, Part A*, Vol. 52, pp. 679-712.

50. Cheung, S. C. P., Yeoh, G. H. and Tu, J. Y. (2007b). On the Numerical Study of Isothermal Vertical Bubbly Flow Using Two Population Balance Approaches, *Chem. Eng. Sci.*, Vol. 62, pp. 4659-4674.

51. Cheung, A. L. K., Lee, E. W. M., Yuen, R. K. K., Yeoh, G. H. and Cheung, S. C. P. (2007). Capturing the Pulsation Frequency of a Buoyant Pool Fire Using the Large Eddy Simulation Approach, *Numer. Heat Transfer, Part A*, Vol. 53, pp. 561-576.

52. Chien, K.-Y. (1980). Predictions of Channel and Boundary Layer Flows with a Low-Reynolds-Number Two-equation Model of Turbulence, AIAA paper 80-0134.

53. Chitty, R. and Cox, G. (1979). A Method of Measuring Combustion Intermittency in Fires, *Fire Materials*, Vol. 3, pp. 238-242.

54. Coppalle, A. and Vervisch, P. (1983). The Total Emissivities of High-Temperature Flames, *Combust. Flame*, Vol. 49, pp. 101-108.

55. Chorin, A. J. (1968). Numerical Solution of Navier-Stokes Equations, *Math. Comp.*, Vol. 22, pp. 745-762.

56. Chow, W. K. and Wong, W. K. (1991). A Study of the Fire Aspect of Atrium Buildings in Hong Kong, *Proceedings of the Third International Symposium on Fire Safety Science*, University of Edinburgh, Scotland, pp. 335-344.

57. Chow, W. K. and Zou, G. W. (2005). Correlation Equations on Fire-induced Air Flow Rates through Doorways Derived by Large Eddy Simulation, *Build. Environ.*, Vol. 40, pp. 897-906.

58. Chu, G. Q., Chen, T., Sun, Z. H. and Sun, J. H. (2007). Probabilistic Risk Assessment for Evacuees in Building Fires, *Build. Environ.*, Vol. 42, pp. 1283-1290.

59. Colket, M. B. and Hall, R. J. (1994). Success and Uncertainties in Modeling Soot Formation in Laminar, Premixed Flames, *Soot Formation in Combustion Mechanisms and Models*, Bockhom, H. (ed.), Springer-Verlag, Berlin, pp. 442-470.

60. Colucci, P. J., Jaberi, F. A., Givi, P. and Pope, S. B. (1998). Filtered Density Function for Large Eddy Simulation of Turbulent Reacting Flows, *Phys. Fluids*, Vol. 10, pp. 499-515.

61. Cook, A. W. and Riley, J. J. (1998). Subgrid-Scale Modeling for Turbulent Reacting Flows, *Combust. Flame*, Vol. 112, pp. 593-606.

62. Concus, P., Golub, G. and O'Leary, D. (1976). A Generalized Conjugate Gradient Method for the Numerical Solution of Elliptic Partial Differential Equations, *Sparse Matrix Computations*, Bunch, J. and Rose, D. (eds.), Academic Press, New York, pp. 309-332.

63. Cooper, L. Y., Harkleroad, M., Quintiere, J. and Rinkinen, W. (1982). An Experimental Study of Upper Hot Layer Stratification in Full-scale Multi-room Fire Scenarios, *ASME J. Heat Transfer*, Vol. 104, pp. 741-749.

64. Cox, G. (1995). Compartment Fire Modeling, *Combustion Fundamentals of Fire*, Cox, G. (ed.), Academic Press, London, pp. 329-404.

65. Cox, G. and Chitty, R. (1980). A Study of the Deterministic Properties of Unbounded Fire Plumes, *Combust. Flame*, Vol. 39, pp. 191-209.

66. Cox, G., Chitty, R. and Kumar, S. (1989). Fire Modeling and the King's Cross Fire Investigation, *Fire Safety J.*, Vol. 15, pp. 103-106.

67. Daru, V. and Tenaud, C. (2004). High Order One-Step Monotonicity-Preserving Schemes for Unsteady Compressible Flow Calculations, *J. Comp. Phys.*, Vol. 193, pp. 563-594.

68. Darwish, M. S. and Moukalled, F. H. (1994). Normalized Variable and Space Formulation Methodology for High Resolution Schemes, *Numer. Heat Transfer, Part B*, Vol. 26, pp. 79-96.

69. Davidson, L. (1997). Large Eddy Simulation: A Dynamic One-Equation Subgrid Model For Three-Dimensional Recirculation Flow, *11th Symposium on Turbulent Shear Flows*, Grenoble, France, pp. 26-1-26-6.

70. Davis, W. D. and Cooper, L. Y. (1991). Computer model for estimating the response of sprinkler links to compartment fires with draft curtains and fusible link-actuated ceiling vents, *Fire Technology*, Vol. 27, pp. 113-127.

71. Davis, R. J. (1978). The Effect of Finite Geometry on the Three-Dimensional Transfer of Solar Irradiance in Clouds, *J. Atmos. Sci.*, Vol. 35, pp. 1712-1725.

72. De Bruyn Kops, S. M., Riley, J. J., Kosály, G. and Cook, A. W. (1998). Investigation of Modeling for Non-Premixed Turbulent Combustion, *Flow. Turb. Comb.*, Vol. 60, pp. 105-122.

73. Delichatsios, M. A. (1993). Comments on "A comparison of numerical and analytical solution of the creeping flame spread over a thermally thin material" by S. Bhattacharjee, *Combust. Flame*, Vol. 93, pp. 336-339.

74. Di Blasi, C. (1994a). On the Influence of Physical Processes on the Transient Pyrolysis of Cellulosic Samples, Fire safety Science - *Proceedings of Fourth International Symposium*, International Association for Fire Safety Science, pp. 229-240.

75. Di Blasi, C. (1994b). Processes of Flames Spreading Over the Surface of Charring Fuels: Effects of the Solid Thickness, *Combust. Flame*, Vol. 97, pp. 225-239.

76. Di Blasi, C. (1996). Flammability Characteristics of Cellulosic Materials Under Slow Concurrent Flows, *Proceedings of the Seventh Fire Safety and Engineering Conference - Interflam, 96*, pp. 113-122.

77. Di Blasi, C., Branca, C., Santoro, A. and Hernandez, E. R. (2001). Pyrolytic Behavior and Products of Some Wood Varieties, *Combust. Flame*, Vol. 124, pp. 165-177.

78. Di Blasi, C. and Branca, C. (2002). Kinetics of Primary Product Formation from Wood Pyrolysis, *Ind. Eng. Chem. Res.*, Vol. 41, pp. 4201-4208.

79. Dobbins, R. A. and Mulholland, G. W. (1984). Interpretation of Optical measurements of Flame Generated Particles, *Combust. Sci. Tech.*, Vol. 40, pp. 175-191.

80. Docherty, P. and Fairweather, M. (1988). Predictions of Radiative Transfer from Nonhomogeneous Combustion Products Using the Discrete Transfer Method, *Combust. Flame*, Vol. 71, pp. 79-87.

81. Dopanzo, C. (1994). Recent Developments in PDF Methods, *Turbulent Reacting Flows*, Libby, P. A. and Williams, F. A. (eds.), Academic Press, New York.

82. Drysdale, D. (1986). *An Introduction to Fire Dynamics*, John Wiley & Sons, England.

83. Edelmen, R. B. and Fortune, O. F. (1969). A Quasi-Global Chemical Kinetic Model for the Finite Rate Combustion of Hydrocarbon Fuels with Application to Turbulent Burning and Mixing in Hypersonic Engines and Nozzles, AIAA paper 69-66.

84. Eddington, A. S. (1988). *The Internal Constitution of the Stars*, Cambridge University Press, UK.

85. Edwards, D. K. (1976). Molecular Gas Band Radiation, *Advances in Heat Transfer*, Irvine, T. F. and Hartnett, J. P. (eds.), Academic Press, New York, Vol. 12, pp. 115-193.

86. Edwards, D. K. and Balakrishnan, M. (1973). Thermal Radiation by Combustion Gases, *Int. J. Heat Mass Transfer*, Vol. 16, pp. 25-40.

87. El Ghobashi, E. E. (1974). Characteristics of Gaseous Turbulent Diffusion Flames in Cylindrical Chambers, A Theoretical and Experimental Investigation, Ph.D. Thesis, London University.

88. Erlebacher, G., Hussaini, M. Y., Speziale, V. G. and Zang, T. A. (1992). Towards the Large-Eddy Simulations of Compressible Turbulent Flows, *J. Fluid Mech.*, Vol. 238, pp. 155-185.

89. Fairweather, M., Jones, W. P. and Lindstedt, R. P. (1992). Predictions of Radiative Transfer from a Turbulent Reacting Jet in Cross-Wind, *Combust. Flame*, Vol. 89, pp. 45-63.

90. Favre, A. (1965). Equation des gaz Turbulents Compressibles, *J. Mechanique*, Vol. 4, pp. 361-392.

91. Favre, A. (1969). Statistical Equations of Turbulent Gases, *Problems of Hydrodynamics Continuum Mechanics*, SIAM, Philadelphia, p. 231.

92. FDS – Fire Dynamics Simulator (version 4) (2004). Technical Reference and User's Guide and User's Guide for Smokeview, National Institute Standards and Technology (US), Gaithersburg, MD.

93. Feinmore, C. P. and Jones, G. W. (1967). Oxidation of Soot by Hydroxyl Radicals, *J. Phys. Chem.*, Vol. 71, pp. 593-597.

94. Felske, J. D. and Tien, C. L. (1974). A Theoretical Closed Form Expression for the Total Band Radiating Gases, *Int. J. Heat Mass Transfer*, Vol. 17, pp. 155-158.

95. Firedman, R. (1991). *Survey of Computer Models for Fire and Smoke*, Second Edition, Factory Mutual Research Corporation, Norwood, MA.

96. Fleischmann, C. M., Dod, R. L., Brown, N. J., Novakov, T., Mowrer, F. W. and Williamson, R. B. (1990). The Use of Medium Scale Experiment to Determine Smoke Characteristic, *Characterization and Toxicity of Smoke*, ASTM STP 1082, Hasegawa, H. K. (ed.), America Society for Testing and Materials, pp. 147-164.

97. Fleischmann, C. M., Pagni, P. J. and Williamson, R. B. (1994). Quantitative Backdraft Experiments, Fire Safety Science – *Proceedings of Fourth International Symposium*, International Association for Fire Safety Science, 337-348.

98. Fletcher, D. F., Kent, J. H., Apte, V. B. and Green, A. R. (1994). Numerical Simulations of Smoke Movement from A Pool Fire in a vertical Tunnel, *Fire Safety J.*, Vol. 23, pp. 305-325.

99. Franztich, H. (1998). Uncertainty and Risk Analysis in Fire Safety, Ph.D. Thesis Lund University, Sweden.

100. Fraser-Mitchell, J. N. (1996). The Lessons Learnt During the Development of CRISPII: A Monte Carlo Simulation for Fire Risk Assessment, *Proceedings of Seventh International Fire Science and Engineering Conference, Interflam, 96*, pp. 631-639.

101. Fredlund, B. (1988). A Model for Heat and Mass Transfer in Timber Structures during Fire, A Theoretical, Numerical and Experimental Study, Ph.D. Thesis, *Report* LUTVDG/(TVBB-1003), Lund University, Sweden.

102. Fredlund, B. (1993). Modeling of Heat and Mass Transfers in Wood Structures during Fire, *Fire Safety J.*, Vol. 20, pp. 39-69.

103. Frenklach, M. and Wang, H. (1990). Detailed Modeling of Soot Particle Nucleation and Growth, *Twenty-third Symposium (International) on Combustion*, The Combustion Institute, University of Orlans, France, pp. 1559-1566.

104. Frenklach, M. (2002). Method of Moments with Interpolative closure, *Chem. Eng. Sci.*, Vol. 57, pp. 2229-2239.

105. Friedman, R. (1992). An International Survey of Computer Models for Fire and Smoke, *J. Fire Protection Eng.*, Vol. 4, pp. 81-92.

106. Fuchs, N. A. (1964). *The Mechanics of Aerosols*, Pergamon Press, Oxford.

107. Fusegi, T. and Farouk, B. (1989). Laminar and Turbulent Natural Convection Interaction in a Square Enclosure Filled with a Non Gray Gas, *Numer. Heat Transfer*, Vol. 15, pp. 303-322.

108. Gao, P. Z., Liu, S. L., Chow, W. K. and Fong, N. K. (2004). Large Eddy Simulations for Studying Tunnel Smoke Ventilation, *Tunneling Underground Space Tech.*, Vol. 19, pp. 557-586.

109. Garo, A., Prado, G. and Lahaye, J. (1990). Chemical Aspects of Soot Particles Oxidation in a Laminar Methane-Air Diffusion Flame, *Combust. Flame*, Vol. 79, pp. 226-233.

110. Garnier, E., Mossi, M., Sagaut, P., Comte, P. and Deville, M. (1999). On the Use of Shock-Capturing Schemes for Large-Eddy Simulation, *J. Comp. Phys.*, Vol. 153, pp. 273-311.

111. Garside, J. E., Ha., A. R. and Townend, D. T. A. (1943). Flow States in Emergent Gas Streams, *Nature*, Vol. 152, pp. 7-48.

112. Gaskell, P. H. and Lau, A. K. C. (1988). Curvature Compensated Convective Transport: SMART, A New Boundness Preserving Transport Algorithm, *Int. J. Numer. Meth. Eng.*, Vol. 8, pp. 617-641.

113. Germano, A., Piomelli, U., Moin, P. and Cabot, W. H. (1991). A Dynamic Subgrid-Scale Eddy Viscosity Model, *Phys. Fluids A*, Vol. 3, pp. 1760-1765.

114. Ghosal, S., Lund, T. S., Moin, P. and Akselvoll, K. (1995). A Dynamic Localization Model for Large-Eddy Simulation of Turbulent Flows, *J. Fluid Mech.*, Vol. 286, pp. 229-255.

115. Girimaji, S. S. (1992). On the Modeling of Scalar Diffusion in Isotropic Turbulence, *Phys. Fluids A*, Vol. 4, pp. 2529-2537.

116. Girimaji, S. S. and Zhou, Y. (1996). Analysis and Modeling Subgrid Scalar Mixing Using Numerical Data, *Phys. Fluids A*, Vol. 8, pp. 1224-1236.

117. Godunov, S. K. (1959). A Finite Difference Method for the Numerical Computation of Discontinuous Solutions of the Equations of Fluid Dynamics, *Mat. Sb.*, Vol. 47, p. 357.

118. Goody, R. M. (1952). A Statistical Model for Water Vapor Absorption, *Quart. J. Roy. Meteor. Soc.*, Vol. 78, pp. 165-169.

119. Goody, R. M. (1964). *Atmospheric Radiation*, Clarendon Press, Oxford.

120. Goody, R. M. and Yung, Y. L. (1989). *Atmospheric Radiation*, Second Edition, Oxford University Press, New York.

121. Goody, R. M., West, R., Chen, L. and Crisp, D. (1989). The Correlated-k Method for Radiation Calculations in Nonhomogeneous Atmospheres, *J. Quant. Spectrosc. Radiat. Transfer*, Vol. 42, pp. 539-550.

122. Gordon, S. and McBride, B. J. (1994). *Computer Program for Calculation of Complex Chemical Equilibrium Compositions and Applications*, NASA Reference Publication 1311, NASA, USA.

123. Gran, I. R. and Magnussen, B. F. (1996). A Numerical Study of a Bluff-Body Stabilized Diffusion Flame, Part 2, Influence of Combustion Modeling and Finite-Rate Chemistry, *Combust. Sci. Tech*, Vol. 119, pp. 191-217.

124. Grossberg, S. (1976). Adaptive Pattern Recognition and Universal Recoding II: Feedback, Expectation, *Olfaction and Illusions, Bio. Cyber.*, Vol. 23, pp. 187-202.

125. Gordon, R. G. (1968). Error Bounds in Equilibrium Statistical Mechanics, *J. Math. Phys*, Vol. 9, pp. 655-672.

126. Goring, D. A. I. and Timell, T. E. (1962). Molecular Weight of Native Cellulose, *TAPPI*, Vol. 45, pp. 454-459.

127. Gresho, P. M. (1991a). Some Current CFD Issues Relevant to the Incompressible Navier-Stokes Equations, *Comp. Methd. Appl. Mech. Eng.*, Vol. 87, pp. 201-223.

128. Gresho, P. M. (1991b). Incompressible Fluid Dynamics: Some Fundamental Formulation Issues, *Ann. Rev. Fluid Mech.*, Vol. 23, pp. 182-188.

129. Grosshandler, W. L. (1993). RADCAL: A Narrow-Band Model for Radiation Calculations in a Combustion Environment, National Institute and Standards of Technology (US), NIST *Technical Note 1402*, Gaithersburg, MD.

130. Gwynne, S., Galea, E. R., Owen, M., Lawrence, P. J. and Filippidis, L. (1999a). A Review of the Methodologies Used in Evacuation Modeling, *Fire Mater.*, Vol. 23, pp. 383-388.

131. Gwynne, S., Galea, E. R., Lawernce, P. J., Owen, M. and Filippidis, L. (1999b). A Review of the Methodologies Used in the Computer Simulation of Evacuation from the Built Environment, *Build Environ*, Vol. 34, pp. 741-749.

132. Hadjisophocleous, G. V. and Bénichou, N. (2000). Development of Perfomance-Based Codes, Performance Criteria and Fire Safety Engineering Methods, *Int. J. Eng. Performance-Based Fire Codes*, Vol. 2, pp. 127-142.

133. Hahn, M. and Drikakis, D. (2005). Large Eddy simulation of Compressible Turbulence Using High-Resolution Method, *Int. J. Numer. Meth. Fluids*, Vol. 47, pp. 971-977.

134. Hanjalic, K. and Jakirlic, S. (1993). A Model of Stress Dissipation in Second Moment Closures, *Appl. Sci. Res.*, Vol. 51, pp. 513-528.

135. Harris, S. J. and Maricq, M. M. (2002). The Role of Fragmentation in Defining the Signature Size Distribution of Diesel Soot, *J. Aerosol Sci.*, Vol. 33, pp. 935-942.

136. Harris, S. J. and Weiner, A. M. (1983a). Surface Growth of Soot Particles in Premixed Ethylene/Air Flames, *Combust. Sci. Tech.*, Vol. 31, pp. 155-167.

137. Harris, S. J. and Weiner, A. M. (1983b). Determination of the Rate Constant for Soot Surface Growth, *Combust. Sci. Tech.*, Vol. 32, pp. 267-275.

138. HArten, A. (1983). High Resolution Schemes for Hyperbolic Conservation Laws, *J. Comp. Phys.*, Vol. 67, pp. 355-366.

139. Haynes, B. S. and Wagner, H. G. (1981). Soot Formation, *Prog. Energy Combust. Sci.*, Vol. 7, pp. 229-273.

140. He, Y. P. and Beck, V. (1997). Smoke Spread Experiment in a Multi-Storey Building and Computer Modelling, *Fire Safety J.*, Vol. 28, pp. 139-164.

141. Helbing, D., Farkas, I. and Vicsek, T. (2000). Simulating Dynamical Features of Escape Panic, *Nature*, Vol. 407, pp. 487-490.

142. Heselden, A. J. M. (1971). Fire Problem of Pedestrian Precincts Part 1, The Smoke Production of Various Materials, Fire Research Note 856, Joint Fire Research Organization, Fire Research Station.

143. Hinds, W. C. (1999). *Aerosol Technology: Properties, Behavior, and Measurement of Airborne Particles*, Second Edition, Wiley–Interscience, New York.

144. Hong, S., Wooldridge, M. S., Im, H. G., Assanis, D. N. and Pitsch, H. (2005). Development and application of a comprehensive soot model for 3D CFD reacting flow studies in a diesel engine, *Combust. Flame*, Vol. 143, pp. 11-26.

145. Hottel, H. C. (1954). Radiant Heat Transmission, *Heat Transmission*, McAdams, W. H. (ed.), McGraw-Hill, New York.

146. Hottel, H. C. and Sarofim, A. F. (1967). *Radiative Transfer*, McGraw-Hill, New York.

147. Hu, B. and Koylu, U. O. (2004). Size and Morphology of Soot Particulates Sampled from a Turbulent Nonpremixed Acetylene Flame, *Aerosol Sci Tech.*, Vol. 38, pp. 1009-1018.

148. Hubbard, G. L. and Tien, C. L. (1978). Infrared Mean Absorption Coefficients of Luminous Flames and Smoke, *ASME J. Heat Transfer*, Vol. 100, pp. 235-239.

149. Hulburt, H. M. and Katz, S. (1964). Some Problems in Particle Technology: A Statistical Mechanical Formulation, *Chem. Eng. Sci.*, Vol. 19, pp. 55-574.

150. Hutchinson, P., Khalil, E. E. and Whitelaw, J. H. (1977). Measurement and Calculation of Furnace Flow Properties, *J. Energy*, Vol. 1, pp. 212-219.

151. Issa, R. I. (1986). Solution of the Implicitly Discretised Fluid Flow Equations by Operator-Splitting, *J. Comp. Phys.*, Vol. 62, pp. 40-65.

152. ISO/CD 13388 (1997). Fire Safety Engineering. Design Fire Scenarios and Design Fire, ISO/TC92/SC4, International Standards Organization (ISO).

153. ISO/DIS 9705 (1990). Fire Tests - Full Scale Room Test for Surface Products, International Standards Organization (ISO).

154. Jamaluddin, S. and Smith, P. J. (1988). Predicting Radiative Transfer in Rectangular Enclosures Using the Discrete Ordinates Method, *Combust. Sci. Tech.*, Vol. 59, pp. 321-340.

155. Jayatilleke, C. L. V. (1969). The Influence of Prandtl Number and Surface Roughness on the Resistance of the Laminar Sublayer to Momentum and Heat Transfer, *Prog. Heat Mass Transfer*, Vol. 1, pp. 193-321.

156. Jia, F., Galea, E. R. and Patel, M. K. (1999). The Numerical Simulation of the Noncharring Pyrolysis Process and Fire Development within a Compartment, *Appl. Math. Modeling*, Vol. 23, pp. 587-607.

157. Jiang, G.-S. and Shu, C.-W. (1996). Efficient Implementation of Weighted ENO Schemes, *J. Comp. Phys.*, Vol. 126, pp. 202-228.

158. Jiménez, J., Liñán, A., Rogers, M. and Higuera, F. (1997). A Priori Testing of Subgrid Models for Chemically Reacting Non-Premixed Turbulent Shear Flows, *J. Fluid Mech.*, Vol. 349, pp. 149-171.

159. Jiménez, C., Ducros, F., Cuenot, B. and Bédat, B. (2001). Subgrid Scale Variance and Dissipation of a Scalar Field in Large Eddy Simulations, *Phys. Fluids*, Vol. 13, pp. 1748-1754.

160. Jeans, J. H. (1917). The Equations of Radiative Transfer of Energy, *Mon. Not. Roy. Astr. Soc.*, Vol. 78, pp. 28-36.

161. Jones, W. P. and Launder, B. E. (1972). The Prediction of Laminarization with a Two-Equation Model of Turbulence, *Int. J. Heat Mass Transfer*, Vol. 15, pp. 301-314.

162. Jones, W. P. (1980). Models for Turbulent Flows with Variable Density, VK-I Lecture Series 1979-2, *Prediction Methods for Turbulent Flows*, Kollmann, W. (ed.), Hemisphere Publishing, New York.

163. Jones, W. P. and Linstedt, R. P. (1988). Global Reaction Schemes for Hydrocarbon Combustion, *Combust. Flame*, Vol. 73, pp. 233-249.

164. Kader, B. (1993). Temperature and Concentration Profiles in Fully Turbulent Boundary Layers, *Int. J. Heat Mass Transfer*, Vol. 24, pp. 1541-1544.

165. Kang, Y. and Wen, J. X. (2004). Large Eddy Simulation of a Small Pool Fire, *Combust. Sci. Tech.*, Vol. 176, pp. 2193-2223.

166. Kanury, A. M. and Blackshear, P. L., Jr. (1970a). Some Considerations Pertaining to the Problem of Wood-Burning, *Combust. Sci. Tech.*, Vol. 1, pp. 339-355.

167. Kanury, A. M. and Blackshear, P. L. (1970b). On the Combustion of Wood, II: The Influence of Internal Convection on the Transient Pyrolysis of Cellulose, *Combust. Sci. Tech.*, Vol. 2, pp. 5-9.

168. Kanury, A. M. (1972a). Thermal Decomposition Kinetics of Wood Pyrolysis, *Combust. Flame*, Vol. 18, pp. 75-83.

169. Kanury, A. M. (1972b). Rate of Burning of Wood, *Combust. Sci. Tech.*, Vol. 5, pp. 135-146.

170. Kennedy, I. M. and Kent, J. H. (1978). Measurements of a Conserved Scalar in Turbulent Jet Diffusion Flames, *Seventeenth Symposium (International) on Combustion*, The Combustion Institute, Pittsburgh, PA, pp. 279-287.

171. Kennedy, I. M. (1997). Models of Soot Formation and Oxidation, *Prog. Energy Combust. Sci.*, Vol. 23, pp. 95-132.

172. Kent, J. H. and Bilger, R. W. (1977). The Prediction of Turbulent Diffusion Flame Fields and Nitric Oxide Formation, *Sixteenth Symposium (International) on Combustion*, The Combustion Institute, Pittsburgh, PA, pp. 1643-1656.

173. Kent, J. H. and Honnery, D. R. (1990). A Soot Formation Rate Map for a Laminar Ethylene Diffusion Flame, *Combust. Flame*, Vol. 79, pp. 287-299.

174. Kerrison, L., Mawhinney, N., Galea, E. R., Hoffmann, N. and Patel, M. K. (1994a). A Comparison of Two Fire Field Models with Experimental Room Fire Data, *Fire Safety Science – Proceedings of Fourth International Symposium*, International Association for Fire Safety Science, pp. 161-172.

175. Kerrison, L., Galea, E. R., Hoffmann, N. and Patel, M. K. (1994b). A Comparison of a FLOW3D Based Fire Field Model with Experimental Room Fire Data, *Fire Safety J.*, Vol. 23, pp. 387-411.

176. Kershaw, D. S. (1978). The Incomplete Cholesky-Conjugate Gradient Method for the Iterative Solution of Systems of Linear Equation, *J. Comp. Phys.*, Vol. 26, pp. 43-65.

177. Ketchell, N., Cole, S. S. and Webber, D. M. (1994). The EGRESS Code for Human Movement and Behavior in Emergency Evacuation, *Engineering for Crowd Safety*, Smith, R. A. and Dickie, J. F. (eds.), Elsevier, London, pp. 361-370.

178. Ketchell, N., Bamford, G. J. and Kandola, B. (1995). Evacuation Modeling: A New Approach, *Proceedings of First International Conference on Fire Science and Engineering*, ASIAFLAM, 95, pp. 449-505.

179. Khalil, E. E. (1977). *Flow and Combustion in Axisymmetric Furnaces*, Ph.D. Thesis, London University.

180. Khan, M. and Greeves, G. A. (1974). A Method for Calculating the Formation and Combustion of Soot in Diesel Engines, *Heat Transfer in Flame*, Afgan, N. H. and Beer, J. M. (eds.), Scripta Book, Washington, pp. 391-402.

181. Kim, I. S. (2004). *Non-Premixed Turbulent Combustion*, Ph. D. Thesis, University of Cambridge.

182. Kim, S. E. and Choudbury, D. (1995). A Near-Wall Treatment Using Wall Functions Sensitized to Pressure Gradient, *ASME FED*, 217, *Separated and Complex Flows*, ASME.

183. Kim, S. H. and Huh, K. Y. (2002). Use of Conditional Moment Closure to Predict NO Formation in a Turbulent CH_4/H_2 Flame over a Bluff-Body, *Combust. Flame*, Vol. 130, pp. 94-111.

184. Kim, J. and Moin, P. (1985). Application of a Fractional-Step Method to Incompressible Navier-Stokes Equations, *J. Comp. Phys.*, Vol. 59, pp. 308-323.

185. Kim, T. K. and Lee, H. S. (1988). Effect of Anisotropic Scattering on Radiative Heat Transfer in Two-Dimensional Rectangular Enclosures, *Int. J. Heat Mass Transfer*, Vol. 31, pp. 1711-1721.

186. Kirkpatrick, S. W., Bocchieri, R. T., Sadek, F., MAcNeill, R. A., Holmes, R., Peterson, B. D., Cilke, R. W. and Navarro, C. (2005). Federal Building and Fire Safety Investigation of the World Trade Center Disaster: Computer Simulation of the Fires in the World Trade Center, National Institute of Standards and Technology (US), *NCSTAR 1-2B*, Gaithersburg, MD.

187. Klimenko, A. Y. and Bilger, R. W. (1999). Conditional Moment Closure for Turbulent Combustion, *Prog. Energy Combust. Sci*, Vol. 25, pp. 595-687.

188. Klüpfel, H. and Meyer-König, T. (2003). Characteristics of the PEdGo Software for Crowd Movement and Egress Simulation, *Second Conference in Pedestrian and Evacuation Dynamics*, University of Greenwich, UK.

189. Knio, O. M., Najm, H. B. and Wyckoff, P. S. (1999). A Semi-Implicit Numerical Scheme for Reacting Flow: II Stiff, Operator-Split Formulation, *J. Comp. Phys.*, Vol. 154, pp. 428-467.

190. Kohonen, T. (1990). The Self-Organizing Map, *Proceedings of the IEEE*, Vol. 78, pp. 464-1480.

191. Krambeck, F. J., Katz, S. and Shinnar, R. (1972). The Effects of Perturbations in Flow-Rate on a Stirred Combustor, *Combust. Sci. Tech.*, Vol. 4, pp. 221-225.

192. Kulogowski, E. D. and Peacock, R. D. (2005). A Review of Building Evacuation Model, National Institute of Standards and Technology (US), *NIST Technical Note 1471*, Gaithersburg, MD.

193. Kumar, S., Gupta, A. K. and Cox, G. (1991). Effects of Thermal Radiation on the Fluid Dynamics of Compartment Fires, Fire Safety Science – *Proceedings of Third International Symposium*, International Association for Fire Safety Science, pp. 345-354.

194. Kung, H. C. (1972). A Mathematical Model of Wood Pyrolysis, *Combust. Flame*, Vol. 18, pp. 185-195.

195. Kung, H. C. (1974). The Burning of Vertical Wooden Slabs, *Fifteenth Symposium (International) on Combustion*, The Combustion Institute, Pittsburg, PA, pp. 243-252.

196. Kung, H. C. and Kalelkar, A. S. (1973). On the Heat of Reaction in Wood Pyrolysis, *Combust. Flame*, Vol. 20, pp. 91-103.

197. Kuo, K. K. (1986). *Principles of Combustion*, John Wiley & Sons, New York.

198. Lacis, A. A. and Oinas, V. (1991). A Description of the Correlated-k Distribution Method for Modeling Nongray Gaseous Absorption, Thermal Emission, and Multiple Scattering in Vertically Inhomogeneous Atmospheres, *J. Geophys. Res.*, Vol. 96, pp. 9027-9063.

199. Lakshminarasimhan, K., Clemens, N. T. and Ezekoye, O. A. (2006). Characteristics of Strongly-Forced Turbulent Jets and Non-Premixed Jet Flames, *Exps. Fluids*, Vol. 42, pp. 523-542.

200. Lam, G. C. K. and Bremhorst, K. (1981). A Modified Form of the k-ε Model for Predicting Wall Turbulence, *J. Fluids Eng.*, Vol. 103, pp. 456-460.

201. Lathrop, K. D. (1976). THREETRAN – A Program to Solve the Multigroup Discrete Ordinates Transport Equation in (x,y,z) Geometry, Report *LA-4848-MS*, Los Alamos Scientific Laboratory.

202. Launder, B. E. and Spalding, D. B. (1974). The Numerical Computation of Turbulent Flows, *Comp. Meth. Appl. Mech. Eng.*, Vol. 3, pp. 269-289.

203. Launder, B. E. (1989). Second-moment Closures: Present and Future, *Int. J. Heat Fluid Flow*, Vol. 10, pp. 282-300.

204. Lautenberger, C. W. (2002). CFD Simulation of Soot Formation and Flame Radiation, M.Sc. Thesis Worcester Polytechnic Institute, Worcester.

205. Lautenberger, C. W., de Ris, J. L., Dembsey, N. A., Barnett, J. R. and Baum, H. R. (2005). A Simplified Model for Soot Formation and Oxidation in CFD Simulation of Non-Premixed Hydrocarbon Flames, *Fire Safety J.*, Vol. 40, pp. 141-176.

206. Leisuer, M. and Métais, O. (1996). New Trends in Large-Eddy Simulations of Turbulence, *Ann. Rev. Fluid Mech.*, Vol. 28, pp. 45-82.

207. Lee, C. K., Chaiken, R. F. and Singer, J. M. (1976). Charring Pyrolysis of Wood in Fires by Laser Simulation, *Sixteenth Symposium (International) on Combustion*, The Combustion Institute, Pittsburgh, P. A., pp. 1459-1470.

208. Lee, C. K. and Diehl, J. H. (1981). Combustion of Irradiated Dry and Wet Oak, *Combust. Flame*, Vol. 42, pp. 123-138.

209. Lee, K. B., Thring, M. W. and Beer, J. M. (1962). On the Rate of Combustion of Soot in a Laminar Soot Flame, *Combust. Flame*, Vol. 6, pp. 137-145.

210. Leonard, B. P. (1979). A Stable and Accurate Convective Modeling Procedure Based on Quadratic Upstream Interpolation, *Comp. Meth. Appl. Mech. Eng.*, Vol. 19, pp. 59-98.

211. Leonard, B. P. (1988). Simple High-Accuracy Resolution Program for Convective Modeling of Discontinuities, *Int. J. Numer. Meth. Eng.*, Vol. 8, pp. 1291-1318.

212. Leung, K. M., Lindstedt, R. P. and Jones, W. P. (1991). A Simplified Reaction Mechanism for Soot Formation in Nonpremixed Flame, *Combust. Flame*, Vol. 87, pp. 289-305.

213. Lewellen, P. C., Peters, W. A. and Howard, J. B. (1976). Cellulose Pyrolysis Kinetics and Char Formation Mechanism, *Sixteenth Symposium (International) on Combustion*, The Combustion Institute, Pittsburgh, PA, pp. 1471-1480.

214. Lewis, E. E. and Miller, W. F., Jr. (1984). *Computational Methods of Neutron Transport*, John Wiley & Sons, New York.

215. Lewis, M. J., Moss, M. B. and Rubini, P. A. (1997). CFD Modeling of Combustion and Heat Transfer in Compartment Fire, Fire Safety Science - *Proceedings of Fifth International Symposium*, Melbourne, pp. 463-474.

216. Liew, S. K., Bray, K. N. C. and Mss, J. B. (1981). A Flamelet Model of Turbulent Non-Premixed Combustion, *Combust. Sci. Tech.*, Vol. 27, pp. 69-73.

217. Lilly, D. K. (1966). *On the Application of the Eddy Viscosity Concept in the Inertial Sub-Range of Turbulence*, NCAR Report No. 123.

218. Lilly, D. K. (1967). The Representation of Small-Scale Turbulence in Numerical Simulation Experiments, *Proceedings of the IBM Scientific Computing Symposium and Enviromental Science*, pp. 195-210.

219. Lipska, A. E. and Parker, W. J. (1966). Kinetics of the Pyrolysis of Cellulose in the Temperature Range 250–300°C, *J. Appl. Polymer Sci.*, Vol. 10, pp. 1439-1453.

220. Liu, Y., Lau, K. S., Chan, C. K., Guo, Y. C. and Lin, W. Y. (2003). Structures of Scalar Transport in 2D Transitional Jet Diffusion Flames by LES, *Int. J. Heat Mass Transfer*, Vol. 46, pp. 3841-3851.

221. Liu, F. and Wen, J. (2002). The Effect of Turbulence Modeling on the CFD Simulation of Buoyant Diffusion Flames, *Fire Safety J.*, Vol. 37, pp. 125-150.

222. Lee, K. W. (1983). Change of Particle Size Distribution during Browning Coagulation, *J. Colloid Interface Sci.*, Vol. 92, pp. 315-325.

223. Lee, W. M., Yuen, R. K. K., Lo, S. M. and Lam, K. C. (2000). Prediction of Sprinkler Actuation Time Using the Artificial Neural Networks, *J. Build. Surv.*, Vol. 2, pp. 10-13.

224. Lee, W. M., Yuen, R. K. K., Lo, S. M. and Lam, K. C. (2001). Application of Fuzzy ARTMAP for Prediction of Flashover in Compartmental Fire, *Proceedings of International Conference on Construction*, Hong Kong, pp. 301-311.

225. Lee, W. M., Yuen, R. K. K., Lo, S. M. and Lam, K. C. (2002). Probabilistic Inference with Maximum Entropy for Prediction of Flashover in Single Compartment Fire, *Adv. Eng. Inform.*, Vol. 16, pp. 179-191.

226. Lee, W. M., Yuen, R. K. K., Lo, S. M., Lam, K. C. and Yeoh, G. H. (2004a). A Novel Artificial Neural Network Fire Model for Prediction of Thermal Interface Location in Single Compartment Fire, *Fire Safety, J.*, Vol. 39, pp. 67-87.

227. Lee, W. M., Lim, C. P., Yuen, R. K. K. and Lo, S. M. (2004b). A Hybrid Artificial Neural Network Model for Online Data Regression, *IEEE Trans. Syst. Man. Cybernet, Part B*, Vol. 34, pp. 951-960.

228. Lele, S. K. (1992). Compact Finite Difference Schemes with Spectral-Like Resolution, *J. Comp. Phys.*, Vol. 103, pp. 16-42.

229. Levin, B. M. (1988a). EXITT: A Simulation Model of Occupant Decisions and Actions in Residential Fires, National Institute of Standards and Technology (US), *Rep. No. NBSIR 88-3753*, Washington.

230. Levin, B. M. (1988b). EXITT: A Simulation Model of Occupant Decisions and Actions in Residential Fires, Fire Safety Science - *Proceedings of Second International Symposium*, International Association for Fire Safety Science, pp. 561-570.

231. Li, J. D. and Bilger, R. W. (1993). Measurement and Prediction of the Conditional Variance in a Turbulent Reactive-Scalar Mixing Layer, *Phys. Fluids A*, Vol. 5, pp. 3255-3264.

232. Lim, C. P. and Harrison, R. F. (1997). An Incremental Adaptive Network for On-Line Supervised Learning and Probability Estimation, *Neural Network*, Vol. 10, pp. 925-939.

233. Lo, S. M. and Fang, Z. (2000). A Spatial-grid Evacuation Model for Buildings, *J. Fire Sci.*, Vol. 18, pp. 376-394.

234. Lo, S. M., Fang, Z., Lin, P. and Zhi, G. S. (2004). An Evacuation Model: The SGEM package, *Fire Safety J.*, Vol. 39, pp. 169-190.

235. Lockwood, F. C. and Naguib, A. S. (1975). The Prediction of Fluctuations in the Properties of Free, Round Jet Turbulent Diffusion Flames, *Combust. Flame*, Vol. 38, pp. 1-15.

236. Lockwood, F. C. and Shah, N. C. (1981). A New Radiation Solution Method for Incorporation in General Combustion Prediction Procedures, *Eighteenth Symposium (International) on Combustion*, The Combustion Institute, Pittsburgh, PA, pp. 1405-1414.

237. Lockwood, F. C. and Malalasekera, W. M. G. (1988) Fire Computation: The 'Flashover' Phenomenon, *Twenty-second Symposium (International) on Combustion*, The Combustion Institute, University of Washington, Seattle, pp. 1319-1328.

238. Lidwug, C. B., Malkmus, W., Reardon, J. E. and Thompson, J. A. (1973). *Handbook of Infrared Radiation from Combustion Gases*, NASA SP-3080.

239. Luo, K. H. (2005). Axis Switching in Turbulent Buoyant Diffusion Flames, *Prog. Combust. Inst.*, Vol. 30, pp. 603-610.

240. Luo, M. and Beck, V. (1994). Fire Environment in a Multi-room Building, *Fire Safety J.*, Vol. 23, pp. 413-38.

241. Luo, M. and Beck, V. (1996). A Study of Non-flashover and Flashover Fires in a Full-Scale Multi-room Building, *Fire Safety J.*, Vol. 26, pp. 191-219.

242. Luo, M., He, Y. and Beck, V. (1997). Application of Field Model and Two-zone Model to Flashover Fires in a Full-scale Multi-room Single Level Building, *Fire Safety J.*, Vol. 29, pp. 1-25.

243. Lutz, A., Kee, R. J., Grcar, J. F. and Rupley, F. M. (1997), *OPPDIF: A FORTRAN Program for Computing Opposed-Flow Diffusion Flames*, Sandia Report SAND96-8243, Livermore.

244. Magnussen, B. F. and Hjertager, B. H. (1976). On Mathematical Modeling of Turbulent Combustion with Special Emphasis on Soot Formation and Combustion, *Sixteenth Symposium (International) on Combustion*, The Combustion Institute, Pittsburgh, PA, pp. 719-729.

245. Magnussen, B. F., Hjertager, B. H., Olsen, J. G. and Bhaduri, D. (1979). Effect of Turbulent Structure and Local Concentrations on Soot Formation and Combustion in Acetylene Diffusion Flames, *Seventeenth Symposium (International) on Combustion*, The Combustion Institute, Pittsburgh, PA, pp. 1383-1393.

246. Magnussen, B. F. (1981). On the Structure of Turbulence and a Generalized Eddy Dissipation Concept for Chemical Reaction in Turbulent Flow, *Nineteenth AIAA Aerospace Science Meeting*, St Louis, Missouri.

247. Malkmus, W. (1963a). Infrared Emissivity of Carbon Dioxide (2.7 μm band), General Dynamics/Astronautics AE63-0047.

248. Malkmus, W. (1963b). Infrared Emissivity of Carbon Dioxide (4.3 μm band), *J. Opt. Soc. Am.*, Vol. 53, pp. 951-961.

249. Marchisio, D. L., Vigil, D. R. and Fox, R. O. (2003a). Quadrature Method of Moments for Aggregation-Breakage Processes, *J. Colloid Interface Sci.*, Vol. 258, pp. 322-324.

250. Marchisio, D. L., Pikturna, J. T., Fox, R. O. and Vigil, R. D. (2003b). Quadrature Method of Moments for Population-Balance Equations, *AIChE J.*, Vol. 49, pp. 1266-1276.

251. Marchisio, D. L. and Fox, R. O. (2005). Solution of Population Balance Equations Using the Direct Quadrature Method of Moments, *J. Aerosol Sci.*, Vol. 36, pp. 43-73.

252. Mahalingam, S., Cantwell, B. J. and Ferziger, J. H. (1990). Numerical Simulations of Coflowing, Axisymmetric Jet Diffusion Flames, *Phys. Fluids A*, Vol. 2, pp. 720-728.

253. Markatos, N. C., Malin, M. R. and Cox, G. (1982). Mathematical Modeling of Buoyancy-Induced Smoke Flow in Enclosures, *Int. J. Heat Mass Transfer*, Vol. 25, pp. 63-75.

254. Markestein, G. H. (1986). Correlation of Heat and Smoke Points and Radiant Emission of Laminar Hydrocarbon Diffusion Flames, *Twenty-Second Symposium (International) on Combustion*, The Combustion Institute, University of Washington, Seattle, pp. 363-370.

255. Markatou, P., Wang, H. and Frenklach, M. (1993). A Computational Study of Sooting Limits in Laminar Premixed Flames of Ethane, Ethylene and Acetylene, *Combust. Flame*, Vol. 93, pp. 467-482.

256. Martin, S. (1965). Diffusion-Controlled Ignition of Cellulosic Materials by Intense Radiant Energy, *Tenth Symposium (International) on Combustion*, The Combustion Institute, Pittsburgh, P. A., pp. 877-896.

257. Mass, U. and Pope, S. B. (1992). Simplifying Chemical Kinetics: Intrinsic Low-Dimensional Manifolds in Composition Space, *Combust. Flame*, Vol. 88, pp. 239-264.

258. McCaffery, B. J. (1979). Purely Buoyant Diffusion Flames: Some Experimental Results, National Bureau of Standards (US), *NBSIR 79-1910*, Washington.

259. McCaffery, B. J. (1983). Momentum Implications for Buoyant Diffusion Flames, *Combust. Flame*, Vol. 52, pp. 149-167.

260. McCaffery, B. J. (1995). Flame Height, *SFPE Handbook of Fire Protection Engineering*, Second Edition, Boston, Society of Fire protection Engineers, pp. 2.1-2.8.

261. McGrattan, K. B., Rehm, R. G. and Baum, H. R. (1994). Fire-Driven in Enclosures, *J. Comp. Phys.*, Vol. 110, pp. 285-291.

262. McGrattan, K. B., Rehm, R. G. and Baum, H. R. (1996). Numerical Simulation of Smoke Plumes from Large Oil Fires, *Atmos. Environ.*, Vol. 30, pp. 4125-4136.

263. McGrattan, K. B., Rehm, R. G. and Baum, H. R. (1998). Large Eddy Simulation of Smoke Movement, *Fire Safety J.*, Vol. 30, pp. 161-178.

264. McGrattan, K. B., Bouldin, C. and Forney, G. P. (2005). Federal Building and Fire Safety Investigation of the World Trade Center Disaster: Computer Simulation of the Fires in the World Trade Center, National Institute of Standards and Technology (US), *NCSTAR 1–5F*, Gaithersburg, MD.

265. McGraw, R. (1997). Description of Aerosol Dynamics by the Quadrature Method of Moments, *Aerosol Sci, Tech*, Vol. 27, pp. 255-265.

266. McGraw, E. and Wright, D. L. (2003). Chemically Resolved Aerosol Dynamics for Internal Mixtures by the Quadrature Method of Moments, *J. Aerosol Sci*, Vol. 34, pp. 189-209.

267. McMurtry, P. A., Jou, W. H., Riley, J. J. and Metcalfe, R. W. (1986). Direct Numerical Simulations of a Reacting Mixing Layer with Chemical Heat Release, *AIAA J.*, Vol. 24, pp. 962-970.

268. McKee, T. B. and Cox, S. K. (1974). Scattering of Visible Radiation by Finite Clouds, *J. Atmos. Sci.*, Vol. 21, pp. 1885-1892.

269. Mell, W. E., Nilsem, V., Kosály, G. and Riley, J. J. (1994). Investigation of Closure Models for Nonpremixed Turbulent Reacting Flows, *Phys. Fluids A*, Vol. 6, pp. 1331-1356.

270. Meneveau, C., Lund, T. S. and Cabot, W. H. (1996). A Lagrangian Dynamic Subgrid-Scale Model of Turbulence, *J. Fluid Mech*, Vol. 319, pp. 353-385.

271. Meylan, B. A. and Butterfield, B. G. (1972). *Three-Dimensional Structure of Wood: A Scanning Electron Microscope Study*, Chapman and Hall, London.

272. Menter, F. R. (1993). Zonal Two Equation k-ω Turbulence Models for Aerodynamics Flows, AIAA paper 93-2906.

273. Menter, F. R. (1996). A Comparison of Some Recent Eddy-Viscosity Turbulence Models, *J. Fluids Eng.*, Vol. 118, pp. 514-519.

274. Mengüç, M. P. and Viskanta, R. (1986). Radiative Transfer in Axisymmetric Finite Cylindrical Enclosures, *ASME J. Heat Transfer*, Vol. 108, pp. 271-276.

275. Métais, O. and Lesieur, M. (1992). Spectral Large_eddy Simulation of Isotropic and Stably Stratified Turbulence, *J. Fluid Mech.*, Vol. 256, pp. 475-503.

276. Milke, J. A. and Mcavoy, T. J. (1995). Analysis of Signature Patterns for Discriminating Fire Detection with Multiple Sensors, *Fire Tech.*, Vol. 31, pp. 120-136.

277. Mitler, H. E. and Emmons, H. W. (1981). Documentation for CFC-V, the Fifth Harvard Computer Code, National Bureau of Standards (US), *NBS-GCR-81-344*, Washington.

278. Mitler, H. E. and Rockett, J. (1987) A User's Guide for First, a Comprehensive Single Room Fire Model, National Bureau of Standards (US), *NBS-GCR-87-3595*, Washington.

279. Modak, A. T. (1979). Radiation from Products of Combustion, *Fire Res.*, Vol. 1, pp. 339-361.

280. Modest, M. F. (1991). The Weighted-Sum-of-Gray Gases Model for Arbitrary Solution Methods in Radiative Transfer, *ASME J. Heat Transfer*, Vol. 113, pp. 650-656.

281. Moin, P. and Mahesh, K. (1998). Direct Numerical SimulationL A Tool in Turbulence Research, *Ann. Rev. Fluid Mech.*, Vol. 30, pp. 539-578.

282. Moore, W. J. (1972). *Physical Chemistry*, Fifth Edition, Longman, London.

283. Morvan, D., Porterie, B., Loraud, J. C. and Larini, M. (2000). Numerical Simulation of a Methane/Air Radiating Turbulent Diffusion Flame, *Int. J. Numer. Meth. Heat Fluid Flow*, Vol. 10, pp. 196-227.

284. Moore, B. (1989). ART 1 and Pattern Clustering, *Proceedings of the 1988 Connectionist Models Summer School*. San Mateo, Touretzky, D., Hinton, G. and Sejnowski, T. (eds.), Morgan Kaufmann Publishers, pp. 174-185.

285. Moss, J. B., Stewart, C. D. and Syed, K. J. (1988). Flowfield Modeling of Soot Formation at Elevated Pressure, *Twenty-Second Symposium (International) on Combustion*, The Combustion Institute, University of Washington, Seattle, pp. 413-423.

286. Moss, J. B. and Stewart, C. D. (1998). Flamelet-Based Smoke Properties for the Field Modeling, *Fire Safety J.*, Vol. 30, pp. 229-250.

287. Nagle, J. and Strickland-Constable, R. F. (1962). Oxidation of Carbon between 1000-2000°C, *Proceedings of Fifth Carbon Conference*, Vol. 1, pp. 154-164.

288. Najm, H. H., Wyckoff, P. S. and Knio, O. M. (1998). A Semi-implicit Numerical Scheme for Reacting Flow: I. Stiff Chemistry, *J. Comp. Phys*, Vol. 143, pp. 381-402.

289. Nakaya, I., Tanaka, T., Yoshida, M. and Steckler, K. (1985). Doorway Flow Induced by Propane Fire, *Fire Safety J.*, Vol. 10, pp. 185-195.

290. Neoh, K. G. Howard, J. B. Sarofim, A. F. (1981). *Soot Oxidation in Flames, Particulate Carbon: Formation during Combustion*, Siegla, D. C. and Smith, G. W. (eds.), Plenum, New York, pp. 261-282.

291. Nielsen, C. and Fleischmann, C. (2000). An Analysis of Pre-Flashover Fire Experiments with Field Modeling Comparison, Fire Engineering Research Report, ISSN 1173-5996, University of Canterbury, NZ.

292. Nelson, H. E. (1986). FIREFORM - A Computerized Collection of Convenient Fire Safety Computations, National Bureau of Standards (US), *NBSIR-86-3308*, Gaithersburg, MD.

293. Nelson, H. E. (1990). FPETOOL: Fire Protection Engineering Tools for Hazard Estimation, National Institute of Standards and Technology (US), *NISTIR 4380*, Gaithersburg, MD.

294. Norton, T. S., Smyth, K. C., Miller, J. H. and Smooke, M. D. (1993). Comparison of Experimental and Computed Species Concentration and Temperature Profiles in Laminar, Two-Dimensional Methane/Air Diffusion Flames, *Combust. Sci. Tech.*, Vol. 90, pp. 1-34.

295. Novozhilov, V., Moghtaderi, B., Fletcher, D. F. and Kent, J. H. (1996). Computational Fluid Dynamics Modeling of Wood Combustion, *Fire Safety J.*, Vol. 27, pp. 69-84.

296. Novozhilov, V. (2001). Computational Fluid Dynamics Modeling of Compartment Fire, *Prog. Energy Combust. Sci.*, Vol. 27, pp. 611-666.

297. O'Brien, E. E. and Jiang, T. (1991). The Conditional Dissipation Rate of an Initially Binary Scalar in Homogeneous Turbulence, *Phys. Fluids A*, Vol. 3, pp. 3121-3123.

298. Ohlemiller, T. J., Kashiwagi, T. and Werner, K. (1985). Products of Wood Gasification, National Bureau of Standards (US), *NBSIR 85-3127*, Gaithersburg, MD.

299. Okasaki, S. and Matsushita, S. (1994). A Study of Simulation Model for Pedestrian Movement with Evacuation and Queuing, *Engineering for Crowd Safety*, Smith, R. A. and Dickie, J. F. (eds.), Elsevier, London, pp. 271-280.

300. Okayama, Y. (1991). A Primitive Study of a Fire Detection Method Controlled by Artificial Neural Net, *Fire Safety J.*, Vol. 17, pp. 535-553.

301. Ozisik, M. N. (1973). *Radiative Transfer and Interactions with Conduction and Convection*, John Wiley & Sons, New York,.

302. Pan, X., Han, C. S., Dauber, K. and Law, K. H. (2005). Human and Social Behavior in Computational Modeling and Analysis of Egress, *Auto. Construct.*, Vol. 15, pp. 448-461.

303. Pape, R., Waterman, T. E. and Eichler, T. V. (1981). Development of a Fire in a Room from Ignition to Full Room Involvement – RFIRES, National Bureau of Standards (US), *NBS-GCR-81-301*, Washington.

304. Patankar, S. V. and Spalding, D. B. (1972). A Calculation Procedure for Heat, Mass and Momentum Transfer in Three-Dimensional Parabolic Flows, *Int. J. Heat Mass Transfer*, Vol. 15, pp. 1787-1806.

305. Piomelli, U., Ferziger, J. H. and Moin, P. (1987). Models for Large Eddy Simulation of Turbulent Channel Flows Including Transpiration, Technical Report, Report TF-32, Dept. Mech. Eng., Stanford University.

306. Piomelli, U., Moin, P. and Ferziger, J. H. (1988). Model Consistency in Large Eddy Simulation of Turbulent Channel Flow, *Phys. Fluids*, Vol. 31, pp. 1884-1891.

307. Piomelli, U., Ferziger, J. H., Moin, P. and Kim, J. (1989). New Approximate Boundary Conditions for Large Eddy Simulations of Wall-Bounded Flow, *Phys. Fluids A*, Vol. 1, pp. 1061-1068.

308. Piomelli, U. and Liu, J. (1995). Large-Eddy Simulation of Rotating Channel Flows Using a Localized Dynamics Model, *Phys. Fluids*, Vol. 7, pp. 839-848.

309. Peaceman, D. W. and Rachford, H. H. Jr. (1955). The Numerical Solution of Parabolic and Elliptic Differential Equations, *J. Soc. Ind.. Appl. Math.*, Vol. 3, No. 1, pp. 28-41.

310. Peacock, R. D., Forney, G. P., Reneke, P., Portier, R. and Jones, W. W. (1993). CFAST, the Consolidated Model of Fire Growth and Smoke Transport, National Institute of Standards and Technology (US), *NIST Technical Note 1299*, Gaithersburg, MD.

311. Penner, S. S. (1955). *Introduction to the Study of Chemical Reactions in Flow Systems*, Agardograph No. 7, Butterworth Scientific Publications, London.

312. Perry, R. H. and Chilton, C. H. (1997). *Chemical Engineers Handbook*, Seventh Edition, McGraw-Hill, New York:.

313. Peters, N. (1984). Laminar Diffsion Flamelet Models in Non-Premixed Turbulent Combustion, *Prog. Energy Combust. Sci.*, Vol. 10, pp. 319-339.

314. Peters, N. (1986). Laminar Flamelet Concepts in Turbulent Combustion, *Prog. Combust. Inst.*, Vol. 21, pp. 1231-1250.

315. Peters, N. (2000). *Turbulent Combustion*, Cambridge University Press, UK.

316. Pitts, W. M. (1994). Application of Thermodynamics and Detailed Chemical Kinetic Modeling to Understanding Combustion Product Generation in Enclosure Fires, *Fire Safety J.*, Vol. 23, pp. 273-303.

317. Portscht, R. (1975). Studies on Characteristic Fluctuations of the Flame Radiation Emitted by Fires, *Combust. Sci. Tech.*, Vol. 10, pp. 73-84.

318. Poon, L. S. (1995). Numerical Modeling of Human Behavior during Egress in Multi-Storey Office Building Fires using EvacSim - Some Validation Studies, *Proceedings of nternational Conference on Fire Science and Engineering*, ASIAFLAM, Vol. 95, pp. 163-74.

319. Pope, S. B. (1985). PDF Methods for Turbulent Reacting Flows, *Prog. Energy Combust. Sci.*, Vol. 11, pp. 119-192.

320. Pope, S. B. (2000). *Turbulent Flows*, Cambridge University Press, UK.

321. Pratsinis, S. E. (1988). Simultaneous Nucleation, Condensation, and Coagulation in Aerosol Reactors, *J. Colloid Interface Sci.*, Vol. 124, pp. 416-427.

322. Puri, R. and Santoro, R. J. (1991). The Role of Soot Particle Formation on the Production of Carbon Monoxides in Fires, *Proceedings of the Third International Symposium on Fire Safety Science*, pp. 595-604.

323. Puri, R., Santoro, R. J. and Smyth, K. C. (1994). The Oxidation of Soot and Carbon Monoxide in Hydrocarbon Diffusioin Flames, *Combust. Flame*, Vol. 97, pp. 125-144.

324. Qin, T. X., Guo, Y. C., Chan, C. K., Lau, K. S. and Lin, W. Y. (2005). Numerical Simulation of Fire-induced Flow through a Stairwell, *Build. Environ.*, Vol. 40, pp. 183-194.

325. Quintiere, J. (1977). Growth of Fires in Building Compartments, *ASTM STP 614*, American Society for Testing and Materials, Philadelphia, PA.

326. Quintiere, J., Rinkinen, W. J. and And Jones, W. W. (1981). The Effect of Room Openings on Fire Plume Entrainment, *Combust. Sci. Tech.*, Vol. 26, pp. 193-201.

327. Rai, M. M. and Moin, P. (1991). Direct Simulation of Turbulent Flow Using Finite-Difference Schemes, *J. Comp. Phys.*, Vol. 96, pp. 15-53.

328. Ramkrishna, D. and Mahoney, A. W. (2002). Population balance modeling. Promise for the future, *Chem. Eng. Sci.*, Vol. 57, pp. 595-606.

329. Rasbash, D. J. and Drysdale, D. D. (1982). Fundamentals of Smoke Production, *Fire Safety J.*, Vol. 5, pp. 77-86.

330. Rhie, C. M. and Chow, W. L. (1983). A Numerical Study of the Turbulent Flow Past An Isolated Airfoil with Trailing Edge Separation, *AIAA J.*, Vol. 21, pp. 1525-1532.

331. Roberts, A. F. (1970a). The Kinetic Behavior of Intermediate Compounds during the Pyrolysis of Cellulose, *J. Appl. Polymer Sci.*, Vol. 14, pp. 244-247.

332. Roberts, A. F. (1970b). A Review of Kinetics Data for the Pyrolysis of Wood and Related Substances, *Combust. Flame*, Vol. 14, pp. 261-272.

333. Rodi, W. (1993). *Turbulence Models and Their Application in Hydraulics*, Balkema, Rotterdam.

334. Rogallo, R. S. and Moin, P. (1984). Numerical Simulation of Turbulent Flows, *Ann Rev. Fluid Mech*, Vol. 16, pp. 99-137.

335. Rogers, G. F. C. and Mayhew, Y. R. (1980). Thermodynamic and Transport Properties of Fluids, Basil Blackwell, Oxford.

336. Rogg, B. (1993) RUN-1DL: The Cambridge Universal Laminar Flamelet Code, *Reduced Kinetic Mechanisms for Applications in Combustion Systems*, Peters, N. and Rogg, B. (eds.), Springer-Verlag, Berlin.

337. Rogg, B. and Wang, W. (1997). *The Laminar Flame and Flamelet Computer Code, User Manual*, Lehrsthul Strömungsmechanik, Institut für Thermo-und Fluiddynamik, Ruhr-Universität Bochum, Bochum.

338. Rosner, D. E. and Yu, S. (2001). MC Simulation of Aerosol Aggregation and Simultaneous Spheroidization, *AIChE J.*, Vol. 47, pp. 545-561.

339. Rotta, J. C. (1951). Statistische Theorie Nichthomogener Turbulenz, 1, *Z. Phys.*, Vol. 129, pp. 547-572.

340. Rutland, C., Ferziger, J. H. and Cantwell, B. J. (1989). Report TF-44, Thermosciences Div., Mech. Eng., Stanford Uni., Stanford.

341. Rzempoluck, E. J. (1997). Neural Network Classification of EEG during Camouflaged Object Identification, *In. J. Med. Inform.*, Vol. 44, pp. 169-175.

342. Said, R., Garo, A. and Borghi, R. (1997). Soot Formation Modeling for Turbulent Flames, *Combust. Flame*, Vol. 108, pp. 71-86.
343. Sagaut, P. (1996). Numerical Simulations of Separated Flows with Subgrid Models, *Rech. Aéro*, Vol. 1, pp. 51-63.
344. Sagaut, P. and Grohens, R. (1999). Discrete Filters for Large_eddy Simulation, *Intt. J. Numer. Meth. Fluids*, Vol. 31, pp. 1195-1220.
345. Sagaut, P. (2004). *Large Eddy Simulation for Incompressible Flows*, First Edition, Springer-Verlag, Berlin.
346. Sagaut, P. (2006). *Large Eddy Simulation for Incompressible Flows*, Second Edition, Springer-Verlag, Berlin.
347. Saka, S. (1993). Structure and Chemical Composition of Wood as a Natural Composite Material, *Recent Research on Wood and Wood-based Materials*, Current Japanese Materials Research, The Society of Materials Science, Japan, N. Shiraishi, H. Kajita and M. Norimoto (eds.), Elsevier Applied Science, Vol. 11, pp. 1-20.
348. Sandler, S. I. (1989). *Chemical and Engineering Thermodynamics*, John Wiley & Sons, USA.
349. Schaffner, C. and Schroder, D. (1993). An Application of General Regression Neural Network to Nonlinear Adaptive Control, *Fifth European Conference on Power Electronics and Applications*, Vol. 4, pp. 219-224.
350. Scheider, G. E. and Zedan, M. (1981). A Modified Strongly Implicit Procedure for the Numerical Solution Field Problems, *Numer. Heat Transfer, Part A*, Vol. 4, pp. 1-19.
351. Schmidt, H. and Schumann, U. (1989). Coherent Structure of the Convective Boundary Layer Derived from Large-Eddy Simulations, *J. Fluid Mech.*, Vol. 200, pp. 511-562.
352. Scotti, A., Meneveau, C. and Lilly, D. K. (1993). Generalized Smagorinsky Model for Anisotropic Grids, *Phys. Fluids A*, Vol. 5, pp. 1229-1248.
353. Seshadri, K. and Peters, N. (1988). Asymptotic Structure and Extinction of Methane-Air Diffusion Flames, *Combust. Flame*, Vol. 73, pp. 23-44.
354. Shafizadeh, F. (1968). Pyrolysis and Combustion of Cellulosic Materials, *Advances in Carbohydrate Chemistry*, Wolfrom, M. L. and Tipson, R. S. (eds.), Academic Press, Vol. 23, pp. 419-474.
355. Shafizadeh, F. and Lai, Y. Z. (1972). Thermal Degradation of 1–6-anhydro-β-D-glucopyranose, *J. Org. Chem.*, Vol. 37, pp. 278-284.
356. Shafizadeh, F. and Chin, P. P. S. (1977). Thermal Deterioration of Wood, *Wood Technology: Chemical Aspects*, Goldstein, I. S. (ed.), ACS Symposium Series, Vol. 23, ASC Washington, pp. 57-81.
357. Shafizadeh, F., Furneaux, R. H., Cochran, T. G., Scholl, J. P. and Sakai, Y. (1979). Production of Levoglucosan and Glucose from Pyrolysis of Cellulosic Materials, *J. Appl. Polymer Sci.*, Vol. 23, pp. 3525-3539.
358. Shah, N. G. (1979). *New Method of Computation of Radiation Heat Transfer in Combustion Chambers*, Ph. D. Thesis, London University.
359. Shields, T. J., Silcock, W. H. and Murray, J. J. (1993). The Effects of Geometry and Ignition Mode on Ignition Times Obtained using a Cone Calorimeter and ISO Ignitability Apparatus, *Fire Mater.*, Vol. 17, pp. 25-32.
360. Shih, T.-H., Liou, W. W., Shabbir, A., Yang, Z. and Zhu, J. (1995). A New k-ε Eddy Viscosity Model for High Reynolds Number Turbulent Flows, *Comp. Fluids*, Vol. 24, pp. 227-238.

361. Shu, C.-W. and Osher, S. (1988). Efficient Implementation of Essentially Non-Oscillatory Schemes. I., *J. Comp. Phys.*, Vol. 77, pp. 439-471.

362. Shu, C.-W. and Osher, S. (1989). Efficient Implementation of Essentially Non-Oscillatory Schemes. II., *J. Comp. Phys.*, Vol. 83, pp. 32-78.

363. Shvab, V. A. (1948). Relation between the Temperature and Velocity Fields of the Flame of a Gas Burner, *Gos. Energ. Izd.*, Moscow-Leningrad.

364. Siau, J. F. (1984). *Transport Processes in Wood*, Springer-Verlag, Berlin.

365. Simcox, S., Wilkes, N. S. and Jones, I. P. (1992). Computer Simulation of the Flows of Hot Gases from the Fire at King's Cross Underground Station, *Fire Safety J.*, Vol. 18, pp. 49-73.

366. Sivathanu, Y. R. and Faeth, G. M. (1990). Generalized State Relationships for Scalar Properties in Nonpremixed Hydrocarbon/Air Flames, *Combust. Flame*, Vol. 82, pp. 211-230.

367. Sjostrom, E. (1981). *Wood Chemistry: Fundamentals and Applications*, Academic Press, New York.

368. Skaar, C. (1988). Wood-Water Relations, Springer-Verlag, Berlin, pp. 73-104.

369. Smagorinsky, J. (1963). General Circulation Experiment with the Primitive Equations: Part I. The Basic Experiment, *Mon. Weather Rev.*, Vol. 91, pp. 99-164.

370. Smith, T. F., Shen, X. F. and Friedman, J. N. (1982). Evaluation of Coefficients for the Weighted Sum of Gray Gases, *ASME J. Heat Transfer*, Vol. 104, pp. 602-608.

371. Society of Fire Protection Engineers and National Fire Protection Association. (1996). *The SFPE Handbook of Fire Protection Engineering* (2002), Second Edition, Bethesda, MD.

372. Society of Fire Protection Engineers and National Fire Protection Association. (2002). *The SFPE Handbook of Fire Protection Engineering* (2002), Third Edition, Bethesda, MD.

373. Soufiani, A., Hartmann, J. M. and Taine, J. (1985). Validity of Band-Model Calculations for CO_2 and H_2O Applied to Radiative Properties and Conductive-Radiative Transfer, *J. Quant. Spectrosc. Radiat. Transfer*, Vol. 33, pp. 243-257.

374. Spalding, D. B. (1971) Mixing and Chemical Reaction in Steady Confined Turbulent Flames, *Thirteenth Symposium (International) on Combustion*, The Combustion Institute, Pittsburgh, PA, pp. 649-657.

375. Spalding, D. B. (1972). A Novel Finite-Difference Formulation for Differential Expressions Involving Both First and Second Derivatives, *Int. J. Num. Meth. Eng.*, Vol. 4, pp. 551-559.

376. Spalding, D. B. (1976). Development of the Eddy Break-Up Model of Turbulent Combustion, *Sixteenth Symposium (International) on Combustion*, The Combustion Institute, Pittsburgh, PA, pp. 1657-1663.

377. Spalding, D. B. (1980). Numerical Computation of Multi-Phase Fluid Flow and Heat Transfer, *Recent Advances in Numerical Methods in Fluid*, Taylor, C. and Morgan, K. (eds.), Vol. 1, pp. 139-167.

378. Specht, D. F. (1991). A General Regression Neural Network, *IEEE Trans. Neural Networks*, Vol. 2, pp. 568-576.

379. Speziale, C. G., Sarkar, S. and Gatski, T. B. (1991). Modeling the Pressure-Strain Correlation of Turbulence: An Invariant Dynamical Systems Approach, *J. Fluid Mech.*, Vol. 227, pp. 245-272.

380. Steckler, K. D., Quintiere, J. G. and Rinkinen, W. J. (1984). Flow Induced by Fire in a Compartment, National Bureau of Standards (US), *NBSIR 82-2520*, Washington.

381. Steiner, H. and Bushe, W. K. (1999). Large Eddy Simulation of a Turbulent Reacting Jet with Conditional Source-Term Estimation, *Phys. Fluids*, Vol. 11, pp. 1896-1906.

382. Still, G. (2000). *Crowd Dynamics*, Ph.D. Thesis, University of Warwick.

383. Strehlow, R. A. (1984). *Combustion Fundamentals*, McGraw-Hill, New York.

384. Stone, H. L. (1968). Iterative Solution of Implicit Approximations of Multi-dimensional Partial Differential Equations, *SIAM. J. Numer. Anal.*, Vol. 5, pp. 530-558.

385. Swaminathan, N. and Bilger, R. W. (2001). Analyses of Conditional Moment Closure for Turbulent Premixed Fames, *Combust. Theory Model.*, Vol. 5, pp. 241-260.

386. Sweby, P. K. (1984). High Resolution Schemes using Flux Limiters for Hyperbolic Conservation Laws, *SIAM J. Num. Anal.*, Vol. 21, pp. 995-1011.

387. Syed, K. J., Stewart, C. D. and Moss, J. B. (1990). Modeling Soot Formation and Thermal Radiation in Buoyant Turbulent Diffusion Flames, *Twenty-Third Symposium (International) on Combustion*, The Combustion Institute, University of Orlans, France, pp. 1533-1541.

388. Symth, K. C., Miller, J. H., Dorfman, R. C., Mallard, W. G. and Santoro, R. J. (1985). Soot Inception in a Methane Air Diffusion Flame as Characterized by Detail Species Profiles, *Combust. Flame*, Vol. 62, pp. 157-181.

389. Takahashi, K., Tanaka, T. and Kose, S. (1989). An Evacuation Model for Use in Fire Safety Design of Building, *Proceedings of the Second International Symposium on Fire Safety Science*, Hemisphere, New York, pp. 551-560.

390. Taniguchi, H., Kudo, K., Otaka, M. and Sumarsono, M. (1992). Development of a Monte Carlo Method for Numerical Analysis on Radiative Energy Transfer Through Non-Grey-Gas Layer, *Int. J. Num. Meth. Eng.*, Vol. 35, pp. 883-891.

391. Taylor, P. B. and Foster, P. J. (1974). Some Gray Weighting Coefficients for CO_2-H_2O-Soot Mixtures, *Int. J. Heat Mass Transfer*, Vol. 18, pp. 1331-1332.

392. Tinney, E. R. (1965). The Combustion of Wooden Dowels in Heated Air, *Tenth Symposium (International) on Combustion*, The Combustion Institute, Pittsburg, PA, pp. 925-930.

393. Tennekes, H. and Lumley, J. L. (1976). *A First Course in Turbulence*, MIT Press, Cambridge MA.

394. Tesner, P. A., Snegiriova, T. D. and Knorre, V. G. (1971a). Kinetics of Dispersed Carbon Formation, *Combust. Flame*, Vol. 17, pp. 253-260.

395. Tesner, P. A., Tsygankova, E. I., Guilazetdinov, E. I., Zuyev, V. P. and Loshakova, G. V. (1971b). The Formation of Soot from Aromatic Hydrocarbons ina Diffusion Flames of Hydrocarbon-Hydrogen Mixtures, *Combust. Flame*, Vol. 17, pp. 279-285.

396. Thomas, J. L., Diskin, B. and Brandt, A. (2003). Textbook Multigrid Efficiency for Fluid Simulations, *Ann. Rev. Fluid Mech.*, Vol. 35, pp. 317-340.

397. Timmermann, G. (2000). A Cascadic Multigrid Algorithm for Semilinear Elliptic Problems, *Numerische Mathematik*, Vol. 86, pp. 717-731.

398. Tomandl, D. and Schober, A. (2001). A Modified General Regression Neural Network (MGRNN) with New, Efficient Training Algorithms as a Robust 'Black Box' – Tool for Data Analysis, *Neural Networks*, Vol. 14, pp. 1023-1034.

399. Toro, E. F. (1997). *Riemann Solvers and Numerical Methods for Fluid Dynamics: A Practical Introduction*, Springer-Verlag, Berlin.

400. Tran, H. C. and Janssens, M. L. (1989). Room Fire Test for Fire Growth Modeling – A Sensitivity Study, *J. Fire Sci.*, Vol. 7, pp. 217-236.

401. Tran, H. C. and White, R. H. (1992). Burning Rate of Solid Wood Measured in a Heat Release Rate Calorimeter, *Fire Mater.*, Vol. 16, pp. 197-206.

402. Tsuji, H. and Yamaoka, I. (1967). The Counterflow Diffusion Flame in the Forward Stagnation Region of a Porous Cylinder, *Eleventh Symposium (International) on Combustion*, The Combustion Institute, Pittsburgh, PA, pp. 979-984.

403. Tzeng, L. S. and Atreya, A. (1991). Theoretical Investigation of Pilot Ignition of Wood National Institute of Standards and Technology (US), *NIST Report No: NIST-GCR-91-595*, Gaithersburg, MD.

404. Vanni, M. (2000). Approximate Population Balance Equations for Aggregation-Breakage Processes, *J. Colloid Interface Sci.*, Vol. 221, pp. 143-160.

405. Van Doormal, J. P. and Raithby, G. D. (1984). Enhancements of the SIMPLE Method for Predicting Incompressible Fluid Flows, *Numer. Heat Transfer*, Vol. 7, pp. 147-163.

406. Van Doormal, J. P. and Raithby, G. D. (1985). An Evaluation of the Segregated Approach for Predicting Incompressible Fluid Flows, ASME Paper 85-HT-9, *National Heat Transfer Conference, Denver, Colorado*.

407. Van Driest, E. R. (1956). On Turbulent Flow Near a Wall, *J. Aero. Sci.*, Vol. 23, pp. 1007-1011.

408. Van Leer, B. (1974). Towards the Ultimate Conservative Difference Scheme. II. Monotonicity and Conservation Combined in a Second-Order Scheme, *J. Comp. Phys.*, Vol. 14, pp. 361-370.

409. Van Leer, B. (1977a). Towards the Ultimate Conservative Difference Scheme. III. Upstream-Centered Finite Difference Schemes for Ideal Compressible Flow, *J. Comp. Phys.*, Vol. 23, pp. 263-275.

410. Van Leer, B. (1977b). Towards the Ultimate Conservative Difference Scheme. IV. A Second Order Sequel to Godunov's Method, *J. Comp. Phys.*, Vol. 23, pp. 276-299.

411. Van Leer, B. (1979). Towards the Ultimate Conservative Difference Scheme. V. A New Approach to Numerical Convection, *J. Comp. Phys.*, Vol. 32, pp. 101-136.

412. Venkatesh, S., Ito, A. and Saito, K.(1996). Flame Base Structure of Small-Scale Pool Fires, *Twenth-Sixth Symposium (International) on Combustion*, The Combustion Institute, Napoli, Italy, pp. 1437-1443.

413. Vervisch, L. and Poinsot, T. (1998). Direct Numerical Simulation of Non-Premixed Turbulent Combustion, *Ann. Rev. Fluid. Mech.*, Vol. 30, pp. 655-692.

414. Vettori, R. L. and Madrzykowski, D. (2000). Comparison of FPETool: FIRE SIMULATOR with Data from Full Scale Experiments, National Institute of Standards and Technology (US), *NISTIR 6470*, Gaithersburg, MD.

415. Viskanta, R. and Mengüc, M. P. (1987). Radiation Heat Transfer in Combustion Systems, *Prog. Energy Combust. Sci.*, Vol. 13, pp. 97-160.

416. Walton, D. W. (1985). ASET-B A Room Fire Program for Personal Computers, National Bureau of Standards (US), *NBSIR 85-3144*, Washington.

417. Walton, D. W. (1995). Zone Computer Fire Models for Enclosure, *SFPE Handbook of Fire Protection Engineering*, National Fire Protection Association, pp. 3-148-3-151.

418. Wang, H. Y. and Joulain, P. (1996). Three-dimensional Modelling for Prediction of Wall Fires with Buoyancy-induced Flow along a Vertical Rectangular Channel, *Combust. Flame*, Vol. 105, pp. 391-406.

419. Wang, H. Y., Coutin, M. and Most, J. M. (2002). Large Eddy Simulation of Buoyancy-driven Fire Propagation behind a Pyrolysis Zone along a Vertical Wall, *Fire Safety J.*, Vol. 37, pp. 259-285.

420. Warantz, J. (1984). Rate Coefficients in the C/H/O System Combustion Chemistry, *Combustion Chemistry*, Gardiner, W. C., Jr. (ed.), Springer-Verlag, New York.
421. Watts, J. M. (1987). Computer Models for Evacuation Analysis, *Fire Safety J.*, Vol. 12, pp. 237-245.
422. Wen, J. X. and Huang, L. Y. (2000). CFD Modeling of Confined Jet Fires Under Ventilation-Controlled Conditions, *Fire Safety J.*, Vol. 34, pp. 1-24.
423. Wen, J. X., Huang, K. Y. and Roberts, J. (2001). The Effect of Microscopic and Global Radiative Heat Exchange on the Field Predictions of Compartment Fires, *Fire Safety J.*, Vol. 36, pp. 205-223.
424. Westbrook, C. K. and Dryer, F. L. (1981). Simplified Reaction Mechanisms for the Oxidants of Hydrocarbon Fuels in Flames, *Combust. Sci. Tech.*, Vol. 27, pp. 31-43.
425. Wesseling, P. (1995). Introduction to Multi – Grid Methods, CR – 195045 ICASE 95-11, NASA.
426. Wieringa, J. A., Elich, J. J. Ph. And Hoogendoorn, C. J. (1991) Spectral Gas Effects in Gas-Fired Furnaces, *Heat Transfer in Radiating and Combusting Systems*, Carvalho, M. G. and Lockwood, J. T. (eds.), Springer-Verlag, Berlin.
427. Wilcox, D. C. (1998). *Turbulence Modelling for CFD*, DCW Industries, Inc.
428. Williams, F. A. (1965). *Combustion Theory*, Addison-Wesley, Reading, MA.
429. Williams, F. A. (1975). *Turbulent Mixing in Nonreactive and Reactive Flows*, Plenum Press, New York.
430. Woodburn, P. J. and Britter, R. E. (1996a). CFD Simulation of a Tunnel Fire – Part I, *Fire Safety J.*, Vol. 26, pp. 35-62.
431. Woodburn, P. J. and Britter, R. E. (1996b). CFD Simulation of a Tunnel Fire – Part II, *Fire Safety J.*, Vol. 26, pp. 63-90.
432. Woodburn, P. J. and Drysdale, D. D. (1998). Fires in Inclined Trenches: The Dependence of the Critical Angle on the Trench and Burner Geometry, *Fire Safety J.*, Vol. 31, pp. 143-164.
433. Xin, Y., Gore, J. P., Mcgrattan, K. B., Rehm, R. G. and Baum, H. R. (2002). Large Eddy Simulation of Buoyant Turbulent Pool Fires, *Twenth-Ninth Symposium (International) on Combustion*, The Combustion Institute, Sapporo, Japan, pp. 259-266.
434. Xin, Y., Gore, J. P., Mcgrattan, K. B., Rehm, R. G. and Baum, H. R. (2005). Fire Dynamic Simulation of a Turbulent Buoyant Flame using a Mixture-Fraction-Based Combustion Model, *Combust. Flame*, Vol. 141, pp. 329-335.
435. Xue, H., Ho, J. C. and Cheng, Y. M. (2001). Comparison of Different Combustion models in Enclosure Fire Simulation, *Fire Safety J.*, Vol. 36, pp. 37-54.
436. Yaga, M., Endo, H., Yamamoto, T., Aoki, H. and Miura, T. (2002). Modeling of Eddy Characteristic Time in LES for Calculating Turbulent Diffusion Flame, *Int. J. Heat Mass Transfer*, Vol. 45, pp. 2343-2349.
437. Yakhot, V., Prszag, S. A., Tangham, S., Gatski, T. B. and Speciale, C. G. (1992). Development of Turbulence Models for Shear Flows by a Double Expansion Technique, *Phys Fluids A: Fluid Dynamics*, Vol. 4, pp. 1510-1520.
438. Yakhot, V. and Orszag, S. A. (1986). Renormalization Group Analysis of Turbulence. I. Basic Theory, *J. Sci. Comp.*, Vol. 1, pp. 1-15.
439. Yan, Z. Y. and Holmstedt, G. (1996). CFD and Experimental Studies of Room Fire Growth on Wall Lining Materials, *Fire Safety J.*, Vol. 27, pp. 201-238.
440. Yanenko, N. N. (1971). *The Method of Fractional Steps*, Springer-Verlag, Berlin.
441. Yang, K. T. and Chang, J. C. (1977). UNDSAFE-I. A Computer Code for Buoyant Flow in an Enclosure, University of Notre Dame Technical Report TR 79002-77-1.

442. Yeoh, G. H., Yuen, R. K. K., Chen, D. H. and Kwok, W. K. (2002a). Combustion and Heat Transfer in Compartment Fires, *Numer. Heat Transfer, Part A*, Vol. 42, pp. 153-172.

443. Yeoh, G. H., Lee, E. W. M., Yuen, R. K. K. and Kwok, W. K. (2002b). Fire and Smoke Distribution in a Two-Room Compartment Structure, *Int. J. Num. Meth. Heat Fluid Flow*, Vol. 12, pp. 178-194.

444. Yeoh, G. H., Yuen, R. K. K., Lo, S. M. and Chen, D. H. (2003a). On Numerical Comparison of Enclosure Fire in a Multi-Compartment Building, *Fire Safety J.*, Vol. 38, pp. 85-94.

445. Yeoh, G. H., Yuen, R. K. K., Cheung, S. C. P. and Kwok, W. K. (2003b). On Modeling Combustion, Radiation and Soot Processes in Compartment Fires, *Build. Environ.*, Vol. 38, pp. 771-785.

446. Yeoh, G. H. and Tu, J. Y. (2004). Population Balance Modeling of Bubbly Flows with Heat and Mass Transfer, *Chem. Eng. Sci.*, Vol. 59, pp. 3125-3139.

447. Yeoh, G. H. and Tu, J. Y. (2005). Thermal-Hydrodynamics Modeling of Bubbly Flows with Heat and Mass Transfer, *AIChE J.*, Vol. 51, pp. 8-27.

448. Yeoh, G. H. and Tu, J. Y. (2006). Numerical Modelling of Bubbly Flows with and without Heat and Mass Transfer, *Appl. Math. Modeling*, Vol. 30, pp. 1067-1095.

449. You, H. Z. and Faeth, G. M. (1979). Ceiling Heat Transfer during Fire Plume and Fire Impingement, *Fire Materials*, Vol. 3, pp. 140-147.

450. Young, S. J. (1977). Nonisothermal Band Model Theory, *J. Quant. Spectrosc. Radiat. Transfer*, Vol. 18, pp. 1-28.

451. Yuen, R., Casey, R., de Vahl Davis, G., Leonardi, E., Yeoh, G. H., Chandrasekaran, V. and Grubits, S. J. (1995). Three-Dimensional Numerical Prediction of Burning of Wood in a Cone Calorimeter, *Proceedings of the International Conference on Fire Research and Engineering*, Orlando, USA, pp. 172-177.

452. Yuen, R., Casey, R., de Vahl Davis, G., Leonardi, E., Yeoh, G. H., Chandrasekaran, V. and Grubits, S. J. (1997). A Three-dimensional Mathematical Model for the Pyrolysis of Wet Wood, *Fire Safety Science - Proceedings of Fifth International Symposium*, International Association for Fire Safety Science, pp. 189-200.

453. Yuen, R., Casey, R., de Vahl Davis, G., Leonardi, E., Yeoh, G. H., Chandrasekaran, V. and Grubits, S. J. (1998). Thermophysical Properties for a Three-dimensional Pyrolysis Model for Wood, *Heat and Mass Transfer - Australasia*, Leonardi, E. and Madhusudana, C. V. (eds.), Begell House, New York, pp. 257-268.

454. Yuen, R. K. K., Lo, S. M. and Cheung, T. C. (1999). Calculation of Smoke Filling Time in a Fire Room, *J. Build. Surv.*, Vol. 1, pp. 33-37.

455. Yuen, R. K. K., de Vahl Davis, G., Leonardi, E. and Yeoh, G. H. (2000). The Influence of Moisture on the Combustion of Wood, *Numer. Heat Transfer, Part A*, Vol. 8, pp. 257-280.

456. Yuen, R. K. K., Lee, E. W. M., Lo, G. M. and Yeoh, G. H. (2006). Prediction of Temperature and Velocity Profiles in a Single Compartment Fire by an Improved Neural Network Analysis, *Fire Safety J.*, Vol. 41, pp. 478-485.

457. Zarghamee, M. S., Kitane, Y., Erbay, O. O., McAllister, T. P. and P. Gross, J. L. (2005a). Federal Building and Fire Safety Investigation of the World Trade Center Disaster: Global Structural Analysis of the Response of the World Trade Center Towers to Impact Damage and Fire, National Institute of Standards and Technology (US), *NCSTAR 1-6D*, Gaithersburg, MD.

458. Zarghamee, M. S., Bolourchi, S., Eggers, D. W., Kan, F. W., Kitane, Y., Liepins, A. A., Mudlock, M., Naguib, W. I., Ojdrovic, R. P., Sarawit, A. T., Barrett,

P. R., Gross, J. P. and McAllister, T. L. (2005b). Federal Building and Fire Safety Investigation of the World Trade Center Disaster: Component, Connection and Sunsystem Structural Analysis, National Institute of Standards and Technology (US), *NCSTAR 1-6C*, Gaithersburg, MD.

459. Zeldovich, Y. B. (1949). On the Theory on Combustion of Initially Unmixed Gases, *Zhur. Tekhn. Fiz.*, 19, p. 1199. See also in *NACA Tech. Memo No. 1296* (1951).

460. Zhou, X., Luo, K. H. and Williams, J. J. R. (2001). Numerical Studies on Vortex Structures in the Near-Field of Oscillating Diffusion Flames, *Heat Mass Transfer*, Vol. 137, pp. 101-110.

461. Zucca, A., Marchisio, D. L., Barresi, A. A. and Fox, R. O. (2006). Implementation of the Population Balance Equation in CFD Codes for Modeling Soot Formation in Turbulent Flames, *Chem. Eng. Sci.*, Vol. 61, pp. 87-95.

462. Zukoski, E. E., Cetegen, B. M. and Kubota, K. (1984). Visible Structures of Buoyant Diffusion Flames, *Twenty Symposium (International) on Combustion*, The Combustion Institute, Pittsburgh, PA, Vol. 20, pp. 361-366.

Further Suggested Reading

1. Anderson, J. D., Jr. (1995). *Computational Fluid Dynamics – The Basics with Applications*, McGraw-Hill, New York.

2. Chung, T. J. (2002). *Computational Fluid Dynamics*, Cambridge University Press, UK.

3. Drysdale, D. (1999) *An Introduction to Fire Dynamics*, Second Edition, John Wiley & Sons, England.

4. Ferziger, J. H. and Perić, M. (1999). *Computational Methods for Fluid Dynamics*, Springer-Verlag, Berlin.

5. Fletcher, C. A. J. (1991). *Computational Techniques for Fluid Dynamics*, Volumes I and II, Springer-Verlag, Berlin.

6. Kuo, K. K. (2005). *Principles of Combustion*, Second Edition, John Wiley & Sons, New York.

7. Modest, M. F. (2003). *Radiative Heat Transfer*, Academic Press, London.

8. Patankar, S. V. (1980). *Numerical Heat Transfer and Fluid Flow*, Hemisphere Publishing Corporation, Taylor & Francis Group, New York.

9. Quintiere, J. G. (2006). *Fundamentals of Fire Phenomena*, John Wiley & Sons, New York.

10. Siegel, R. and Howell, J. R. (2002). *Thermal Radiation Heat Transfer*, Fourth Edition, Taylor & Francis, New York.

11. Tu, J. Y., Yeoh, G. H. and Liu, G. Q. (2008). *Computational Fluid Dynamics – A Practical Approach*, Butterworth-Heinemann, Elsevier Science and Technology, USA.

12. Versteeg, H. K. and Malalasekera, W. (2007) *An Introduction to Computational Fluid Dynamics – The Finite Volume Method*, Second Edition, Prentice Hall, Pearson Education Ltd., England.

Index